T0290907

Atmospheric Thermodynamics

Atmospheric Thermodynamics

Second Edition

Craig F. Bohren
Department of Meteorology and Atmospheric Science, Pennsylvania State University

Bruce A. Albrecht
Department of Atmospheric Sciences, University of Miami

OXFORD
UNIVERSITY PRESS

OXFORD
UNIVERSITY PRESS

Great Clarendon Street, Oxford, OX2 6DP,
United Kingdom

Oxford University Press is a department of the University of Oxford.
It furthers the University's objective of excellence in research, scholarship,
and education by publishing worldwide. Oxford is a registered trade mark of
Oxford University Press in the UK and in certain other countries

First Edition published in 1998

Second Edition published in 2023

Published in the United States of America by Oxford University Press
198 Madison Avenue, New York, NY 10016, United States of America

British Library Cataloguing in Publication Data

Data available

Library of Congress Control Number: 2022946525

ISBN 978–0–19–887270–2
ISBN 978–0–19–887271–9 (pbk.)

DOI: 10.1093/oso/9780198872702.001.0001

Printed and bound by
CPI Group (UK) Ltd, Croydon, CR0 4YY

To our families Nanette, Gail, Birch, and Spritz; Nancy, Michael, and Adam

Preface to Second Edition

The first edition of our book is a well-regarded treatise on atmospheric thermodynamics, as evidenced by the many times it has been cited. Although intended as a textbook in atmospheric science, it could be used as a reference or textbook for classical thermodynamics taught in physics or engineering departments. It brings real-life problems and applications to a subject that otherwise can be somewhat dry. Earth's atmosphere is a freely accessible thermodynamics laboratory in which we all live. Many of our experiments were done in kitchens and on porches using inexpensive instruments.

The impetus for a second edition was to lower the price to what students can afford. It also allowed us to correct errors and pass on what we have learned in the 24 years since the first edition. We didn't stop thinking after it was published. We used it as a template but rethought every sentence. We removed a subsection that was just plain wrong, although about a side issue (phew!). A few subsections were removed, a few added. Minor errors were corrected.

We took more pains to state all assumptions, idealizations, and approximations notably absent from textbooks. Energy is conserved, entropy increases (probably), but everything else is contingent. We rearranged Chapters 3 and 4. Chapter 7 was expanded to include more about the interaction of humans with their environment. We discuss more about what could be called engineering heat transfer—a subject unknown to many physicists and meteorologists. Whenever possible we did simple experiments and made observations to support our assertions. We put considerable effort into making explanations impossible to misunderstand, an unattainable goal but worth aiming at. We doubled the number of problems to more than 400. They arose from questions we asked ourselves or were asked by students, colleagues, and correspondents. We profited from answering them. We included more suggestions for experiments and observations. Some of these experiments were done by our eminent predecessors more than 200 years ago. In the first edition, we criticized textbook history. After many years of reading and reflecting, we are even more convinced that there are few historically correct statements in textbooks. Ask any historian of science. Textbooks are not much influenced by original papers, ones criticizing textbook mythology, or papers by historians of science. We read many books and papers by, for example, Gay-Lussac, Fourier, Joseph Black, Joule, Robert Boyle, and others. Not only did this allow us to give credit where it is due, but also we acquired physical understanding and intuition from the thoughts of long-dead authors. We give more references than is the norm in textbooks to support what we write, rather than uncritically pass on errors. Our references are not window dressing. We read the papers we cite, sometimes entire books or at least relevant chapters, and took notes on them. Failure to cite references, even in textbooks, contributes to error propagation. We added about 150 references to bring our book up to date and because we happened upon many books and papers worth reading.

Because of our interest in the history of science, we call attention to a fairly recent journal, *History of Meteorology*, which at present is freely available.

There are many long-lived demonstrably incorrect explanations in meteorology. They originate in elementary and middle school, continue uncorrected into high school, university, and beyond. We do more debunking of such explanations. As H. L. Mencken wrote, "Nine times out of ten, in the arts as in life, there is actually no truth to be discovered; there is only error to be exposed."

We emphasize even more than in the first edition physical reasoning and interpretation, which are more likely to stick in the minds of students long after they have forgotten equations. As the book review editor of *American Journal of Physics* for about eight years, the senior author has seen many books that are only one equation after another. No interpretation, no experiments, no observations, no connection to reality.

We are heartened that heretical ideas in the first edition have subsequently been made by others. Robert Romer, as editor of *American Journal of Physics*, wrote a scathing editorial "Heat is not a noun" in which he cites us. Fierce and prolonged criticism of the notion that "entropy is just disorder" by the late chemist Frank Lambert in the last 20 years of his long life goaded chemists into removing this mantra from chemistry textbooks. It also has been attacked by Harvey Leff in physics journals. Many physical chemists no longer refer to "heats" of vaporization or combustion, but instead to enthalpies (as in our first edition), another sign of progress. Merely because we can find otherwise respectable teachers and authors who share our views does not mean that we are right—only that we are not lone crackpots. We corrected errors only if we found them in several textbooks or popularizations, especially if they have been made over decades. For many years, the *Journal of Chemical Education* had a regular column, "Textbook Errors," the purpose of which was to halt the spread of demonstrable errors. For an error to be exposed it had to occur in "at least two independent standard texts," which were not cited.

Sections and subsections marked with an asterisk (*) can be omitted on a first reading of the text.

We updated our discussion of instruments for measuring temperature, dew point, humidity, and pressure. In the past two decades instruments have gotten better and cheaper. Infrared thermometers are inexpensive. Students can (and should) do their own experiments and make observations so that they own them. "What you inherit from your father must first be earned before it is yours" (Goethe).

Following the text of each chapter is a list of annotated references. Each entry contains tidbits that could have been in the text. Read them to get the maximum benefit from this book even if you do not read the references cited. Work as many of the problems as possible, especially those requiring simple experiments.

Underlying all laws of physics (to be distinguished from laws of nature) are a great many assumptions and idealizations, often unstated. Critics of the space we "waste" on belaboring what "everyone knows" should read the first ever guest editorial in *European Journal of Physics*, Vol. 40 (2019) 030201. It begins "A major challenge [to] ... imparting what a teacher knows to students who do not know is overcoming the curse of knowledge."

The late Clifford Truesdell, a prolific historian of thermodynamics and contributor to rational thermodynamics, wrote in 1969 that "from the first thermodynamics was a dismal swamp of obscurity" and today "in common instruction it [still] is." This book is our contribution to draining the swamp.

Acknowledgments

First and foremost, we are indebted to Bert de With for being the only reader to write to us about a blatant error, thereby saving us possible further embarrassment. Dan Schroeder criticized our treatment of the temperature dependence of the enthalpy of vaporization. We still got the textbook result (Kirchhoff's equation), but now we understand it better, why it is a good approximation (within limits), and why it is *not* so good for the enthalpy of fusion. Roger Davies suggested that the decrease of downward infrared radiation with increasing altitude could contribute to sensations of cold at high altitudes. Peter Pilewskie encouraged Bruce Kindel to do radiation calculations to support this (Chapter 7). Raymond Shaw kept us from making blunders about turbulence and homogeneous nucleation, and also passed on some of Alex Kostinski's musings about the usual (superficial) physical explanation of why $C_p > C_v$. Shermila Singham straightened us out on the Arrhenius equation. Gregory McRae encouraged us to convert almost all our units to SI and shared with us his fascination with indicator diagrams. Sharan Majumdar spotted many typographical errors and asked for clarification. John Kenny encouraged us to note the huge difference in the reaction times of combustion of hydrogen in oxygen, a factor of about 10^{31} with and without a catalyst. From time to time, we asked Bob Hummer questions about wet chemistry and tried out ideas on him. Conversations with James Fleming continue to pique our interest in the history of meteorology. Despite our bruises from wrestling with Glenn Shaw over the coefficient of performance of a refrigerator, we emerged with a better understanding of what is admittedly a confusing concept. We also thank Shane Mayor for correspondence about various aspects of thermodynamics, Lee Grenci for many discussions about misconceptions he heroically tries to dispel, and John Wyngaard for preventing us from passing on the widespread but faulty physical interpretation of the Grashof number. Finally, we are indebted to Sonke Adlung for enthusiastic editorial support.

Craig F. Bohren
Tŷ'n y Coed
Oak Hall, Pennsylvania

Bruce A. Albrecht
St. Petersburg, Florida

March 2022

Preface to the First Edition

In *The End of Education* (p. 115), Neil Postman asserts that "We can improve the quality of teaching and learning overnight by getting rid of all textbooks. Most textbooks are badly written and, therefore, give the impression that they are boring. Most textbooks are also impersonally written. They ... reveal no human personality." This will not endear Postman to the authors of textbooks, but, sad to say, he tells the truth, unpalatable though it may be. When confronted with searing criticism like this, how will budding textbook authors react? Which path will they take: denial, or introspection and reform?

We wrote this book inspired by the highly radical notion that textbooks ought to be rollicking good literature given that their intended readers are mostly young people, those through whom the juices of life flow strongest, and hence, those most repelled by dry and lifeless textbook fodder. This notion is radical because the prevailing view of textbooks is that they should be as boring as possible. Written in the passive voice, humorless, without a whiff of controversy, textbooks are painstakingly purged of any human touch. They seem to be written with the aim of making science as uninteresting as possible so that their authors can then wring their hands over the lack of interest shown by young people in science.

We tried to make our writing as lively as possible by using active constructions and by not hiding the fact that we are humans first, scientists second. We do not shrink from criticizing fallacies, no matter how hallowed by frequent repetition. We use humor, irony, and sarcasm, all the techniques of the writing trade, to keep our readers interested. After all, they are mostly young people who are bombarded continually by striking visual images and vivid entertainment in many forms. Today's audience for textbooks is the most distracted and the least attentive in the history of literate humanity.

One of the most depressing features of textbooks, aside from their blandness, is their monotonous sameness. An original textbook is almost a contradiction in terms. Thermodynamics provides some prime examples. Every thermodynamics textbook finds it necessary to include a ponderous section belaboring exact and inexact differentials, which students do not understand and which, moreover, are unnecessary. The late, eminent mathematician Karl Menger almost half a century ago gently pointed out that differentials of any kind are unnecessary in teaching thermodynamics. To our knowledge, no subsequent author of a textbook on thermodynamics has paid any heed to Menger. Clifford Truesdell, an astute student of the foundations and history of thermodynamics, dips his pen in the most corrosive acid when writing about differentials in thermodynamics. Even mathematicians view them with disdain. Yet, differentials continue to appear in thermodynamics with depressing regularity. The mathematics of thermodynamics is really quite simple, child's play compared with that of, say, electromagnetic theory and fluid mechanics. To compensate for this, the mathematical notation in thermodynamics seems to have been specially designed to be confusing, to create the impression that it is the work of Martians.

No book, especially one about so hoary a topic as thermodynamics, should be written except by those who immodestly think that they can do better than their predecessors and who are willing to make a fresh start instead of merely rearranging what has been done previously. This goal is likely to be reached only if authors are aware of the ailments of existing books and can propose cures for these ailments.

To avoid committing the sins of our predecessors, we were guided by the accumulated wisdom in articles on thermodynamics and related topics published in *American Journal of Physics* from its inception about 60 years ago to the present. This journal is devoted to expository articles on physics written mostly by college and university teachers. It is a pedagogical gold mine, yet not used to its fullest extent by writers of textbooks. Before we were content with a way of discussing a concept, we carefully pondered the recommendations of those who have thought long and hard about how thermodynamics should be taught. Other sources of fruitful ideas are *Journal of Chemical Education, The Physics Teacher, European Journal of Physics, Physics Education, American Mathematical Monthly,* and *Contemporary Physics.* Given the treasure trove of thoughtful and critical expository articles, most of them highly readable, in these journals, the rubbish that continues to be propagated in textbooks is inexcusable. We only wish that we had had more time to take advantage of this huge but largely untapped reservoir of ideas.

What distinguishes a textbook from a technical monograph is problems in the former but not in the latter. And what distinguishes thermodynamics textbooks is problems even more boring than the text that precedes them. A typical problem is something like the following: "What is the final temperature of two kilograms of nitrogen initially at room temperature when its pressure is doubled adiabatically?" This is the kind of drill question rarely (if ever) asked outside the pages of textbooks, a question to which no one wants to know the answer, what students refer to contemptuously as "plug and chug." Yet, questions in textbooks should be genuine, by which we mean ones that inquisitive people ask in the course of their attempts to understand the world around them. An example of a genuine question is one we were asked by a writer for a motorcycle magazine. He called to ask about wind chill. Our response was duly published in the magazine. A question about thermodynamics asked by tattooed and bearded men wearing nose rings and riding chrome-plated Harley-Davidson motorcycles is by our reckoning a good question.

The more than 200 problems in this book are not optional, tacked onto the end of chapters as afterthoughts, but mandatory. Yet we look upon them as rewards, not punishments. The payoff for struggling with ideas is the satisfaction acquired from using them. Problems provide an opportunity for readers to test what they have learned, to go beyond the text, to generate new ideas. Indeed, some problems are a vehicle for presenting topics that could have been included in the text. We hope that most of these problems are interesting to students curious about how the world works. Many problems were inspired by questions asked by colleagues, students, and friends, Thermodynamic knowledge can be applied to countless genuine problems. There is no need for contrived ones.

At the end of each chapter, problems are given in random order, not arranged according to section. This was a compromise between putting all problems at the end of the book and grouping them according to section. The disadvantage of pairing problems with sections is that it encourages students to search frantically for a magic formula in a section that will enable them to solve a corresponding problem. Students should learn that knowledge is cumulative and that solving physical problems requires drawing upon resources from different directions.

Although our intended readers are undergraduate students, we expect this book to be suitable for graduate courses in atmospheric thermodynamics. This is especially true given that graduate students in atmospheric science often are drawn from physics, mathematics, engineering, and other technical fields. Thus, even at the graduate level, especially in departments without undergraduate programs, students are taught the same material as in undergraduate courses but at a faster pace and with higher expectations. Sections and subsections marked with an asterisk (*) can be omitted on a first reading of the text. All we expect of our readers in the way of prerequisites is that they be firmly grounded in the elements of calculus, at least know what a differential equation is, and are familiar with vectors. No prior knowledge of thermodynamics is assumed. Indeed, we would prefer readers to be innocent of thermodynamics so that they have less to unlearn.

The boundaries between scientific disciplines are now as tightly drawn and fiercely defended as the boundary between North and South Korea, yet we still can hope that teachers of physics and engineering might find our book a useful supplement in their courses.

It is sometimes said that thermodynamics is model independent, that if tomorrow we were to decide that molecules do not exist, the principles of thermodynamics would not need to be altered in the slightest. This is true but largely irrelevant. All college students today have at least heard of molecules and cannot be restored to their previous state of blissful ignorance. But what these students have heard about molecules is often bilge. For example, we have yet to encounter a student who did not fervently believe that the absolute zero of temperature is the temperature at which all molecules come grinding to a halt. This is not true and has been known not to be true for many decades. Meteorology was spawned from physics (or what is now physics but used to be natural philosophy). The parent and child have gone their separate ways, as a consequence of which what meteorologists are taught is sometimes ludicrous. As it happens, temperatures in the atmosphere never even approach absolute zero; so meteorologists can entertain wildly incorrect ideas about absolute zero with no practical consequences—except that they make themselves out to be fools in the eyes of people who know better.

Thermodynamics, we believe, tastes best when spiced with kinetic theory, and so we wove threads of the kinetic theory of gases into the fabric of classical thermodynamics. Although students cannot be prevented from invoking molecules to explain natural phenomena, they can be taught to make molecular arguments correctly. Moreover, invoking molecules does seem to give us a better grasp of physical phenomena. Despite all the current blather about holism, most people naturally gravitate toward reductionism. Although sometimes carried to extremes, which has given it a bad name, when restrained by common sense and prudence, reductionism is a highly effective way of acquiring understanding.

We have learned through experience that our students often look upon derivations of equations as ends in themselves. We view derivations as means, the ends of which are physical interpretations. We forget the details of derivations, which is not surprising given that they often are tedious. The product of a derivation is an equation, which must be interpreted physically, and its meaning expressed in words: we think with words, not with equations. Whenever possible, we state a mathematical result in words chosen to stick in the minds of readers. They will forget the mathematical details, but we can hope that they will retain their physical essence.

We took pains to derive equations carefully, exposing to the clear light of day all assumptions and approximations (we abhor the irritating textbook dodge, "it can be shown"). This approach is not likely to find favor with those who just want formulas to be chanted as mantras when faced with problems. Although it would be economical of space to pull equations out of thin air, ignorance of their antecedents is not merely useless; it is dangerous. If you don't know what underlies an equation, you can't use it correctly, nor can you fix it when it breaks. And virtually every equation in atmospheric thermodynamics is an approximation with a limited range of validity. We strongly adhere to the view that students should be encouraged to make as many observations of atmospheric phenomena as possible and even to do simple experiments. One of the great appeals of atmospheric science is that we always are in its laboratory. Accordingly, we suggest simple experiments and try not to miss opportunities to point out how life can be breathed into inert symbols on the pages of this book. A good example is provided by mixing clouds (Section 6.8). To understand these clouds requires conservation of enthalpy (in an adiabatic constant-pressure process) and the Clausius–Clapeyron equation, which itself embodies considerable thermodynamic reasoning. It is highly effective to link these perhaps inherently uninteresting relations with readily observable phenomena. Mixing clouds are everywhere, and hence, can be and should be used to teach thermodynamics.

Almost as a corollary to the sentiments expressed in the preceding paragraph, we include as much as possible of what we sometimes refer to facetiously as "real data." By this we mean data obtained by means other than computer simulations. For example, one of the most effective ways of driving home the message that pressure and temperature are not inextricably linked in the atmosphere is to show records of daily pressure and temperature variations. Our students, almost without exception, firmly believe that pressure invariably increases as temperature increases. Our efforts to convince them otherwise by way of the ideal gas law are often fruitless. A plot of actual temperatures and pressures (see Section 2.1) is more convincing.

Strictly speaking, thermodynamics deals with equilibrium states, and the mechanisms by which systems come into equilibrium are not usually grist for the thermodynamic mill. Yet, our goal is to teach students about a small corner of the physical world, the atmosphere, rather than to adhere rigidly to some sacred canon. As a consequence, we believe that it is appropriate even in a course labeled *thermodynamics* to teach a bit about the transport coefficients for energy (thermal conductivity), momentum (viscosity), and mass (diffusion coefficient) of gases. Our experience is that students either know almost nothing about these transport coefficients or what they do know is hopelessly wrong. Students do not seem to learn about the viscosity of gases (in particular, its temperature dependence) in their dynamical meteorology courses. And they seem not to learn about transport coefficients in their physics courses. So the last chapter is on transport coefficients. Aside from their inherent usefulness, they help one to understand why systems depart from equilibrium and how they return.

These days meteorology is being more and more married with environmental engineering. This is a marriage of convenience, and the jury is still out on whether the union will be successful. But if meteorologists are being encouraged to get into bed with engineers, they should at least know how to converse with them. Accordingly, Section 7.1 is devoted to what might be called engineering heat transfer. The essentials can be readily taught without drowning in a sea of empirical relations for heat transfer coefficients. We note that one of us has two degrees in engineering and, many years ago, worked as an engineer. As a consequence of having lived for a time in the engineers' camp, he knows their lingo.

In the preface to Volume 1 of *Kinetic Theory*, a collection of historically important papers by Boyle, Newton, Clausius, Maxwell, and others, its editor, Stephen Brush, avers that "Most writers of textbooks in physics seem to believe that the introduction of anecdotes from the history of physics enhances their exposition of the subject matter itself. Unfortunately, much of the 'history' that one finds in textbooks or popularizations of physics is either false or misleading." Alas, this is an understatement. Students should know a bit about the history of the subject they are studying if for no other reason than this shows them that our illustrious predecessors grappled with the same concepts the students are now struggling to understand. We teach conservation of energy in one lecture. Should we be surprised that students don't understand it immediately, given that it evolved over tens if not hundreds of years? Yet much of what passes for history of science in textbooks does not rise above Parson Weems's homilies about George Washington and the cherry tree. Textbook writers are not obliged to include any history, but those who choose to do so should be held to the same standards as those adhered to by historians of science. There is no excuse for doing otherwise given the existence of the *Dictionary of Scientific Biography*, a rich source of scientific biographies by scholars who took the time and trouble to tell true stories instead of passing on myths. One bit of history we do include is that of thermodynamic diagrams in meteorology (Chapter 6). It should be of considerable interest (and surprise) to meteorology students to learn that these diagrams were invented by Heinrich Hertz, whose name appears in every physics textbook without crediting him with any interest in meteorology. To physicists, Hertz's fame rests on his experimental verification of Maxwell's theory of electromagnetism. Although the invention of thermodynamic diagrams by Hertz is eclipsed by his work in electromagnetism, his meteorological investigations show that his interests (and those of his mentor, Helmholtz) were much broader than those of present-day physicists.

Although we often are critical of the now outmoded ideas of our predecessors, we paid them the ultimate compliment of actually reading what they said, rather than what someone else said they said.

Richard Lewontin, a biologist, writing in *The New York Review of Books* (9 January 1997) notes that "to put a correct view of the universe into people's heads we must first get an incorrect view out. People believe a lot of nonsense about the world of phenomena, nonsense that is a consequence of a wrong way of thinking." Getting incorrect views out of people's head is a task of heroic proportions. By the time young people enter college, their mental furniture is firmly in place, and they fiercely resist being required to move it. We do considerable debunking in this book, which some readers will find distasteful. Other readers may think that we have gone overboard in trying to stamp out misconceptions by frequent ridicule and by pointing out how they impede understanding. Yet the sad truth is that no amount of debunking is sufficient to dislodge misconceptions firmly rooted at an early age. We speak from the experience of having taught a combined total of about 40 years at the largest meteorology department in the United States. We know our customers. We know that the heads of our entering students are filled with nonsense, and that after four years we barely have begun to help them flush it out. All misconceptions against which we inveigh, sometimes repeatedly, are not rare singularities: we encounter them every day. The notion that air is a spongy medium, the pores of which expand with temperature so that they can sop up more water vapor, is more firmly rooted in the minds of students than the Ten Commandments.

To our readers we advise: read this book actively, not passively. Argue with us. Make us justify every assertion, every equation. Try to show that we are wrong, and only after failing to do so, accept what we say as at least provisionally true.

Acknowledgments

Although this book is emblazoned with the names of only two authors, behind them in the shadows stands a small army of people who contributed to it, often without their knowledge, perhaps sometimes even contrary to their wishes. In the forefront of this army are past and present colleagues in the Department of Meteorology at Penn State University, who offered criticisms, asked questions, argued, and sometimes even fought with us. Fading memories and lack of space prevent us from acknowledging each person's specific contribution. The best we can do is thank them in alphabetical order: Tom Ackerman, Peter Bannon, John Dutton, Bill Frank, Mike Fritsch, Lee Grenci, Paul Knight, Ray Najjar, John Olivero, Rosa de Pena, Raymond and Mandy Shaw, Nels Shirer, John Spiesberger, Denny Thomson, Bob Wells, and John Wyngaard. Special thanks go to Alistair Fraser, Dennis Lamb, and Hans Verlinde, our fellow teachers of atmospheric thermodynamics. And this reminds us that we are indebted to hundreds of students for their countless questions and for having to endure the class notes that formed the nucleus for this book.

Beyond Penn State lies another phalanx of colleagues to whom we are grateful: Bill Beasley, Duncan Blanchard, Robert Cole, Chris Curran, Jim Fleming, Clayton Gearhart, George Greaves, Harvey Leff, John Lienhard, Carl Ribbing, and the late Bernie Vonnegut.

Terry Faber and Jean Carpenter helped produce some of the figures; Jean also helped with the design of the front cover artwork. Kay Hale and Helen Albertson, librarians at the University of Miami's Rosenstiel School of Marine and Atmospheric Sciences, helped find some of the historical articles used during the writing of this book.

Our experiences with staff members at Oxford University Press support our contention that dry and lifeless textbooks are the product of dry and lifeless authors, not the result of censorious and Pecksniffian editors. Our editor, Joyce Berry, gave us free rein to write an idiosyncratic book, critical, humorous in places, even ribald. Christine Landau and Karen Shapiro expertly shepherded our manuscript through the various stages of production. Jacki Hartt deftly copy edited the manuscript, working as a surgeon, not as a butcher.

Finally, each of us thanks the other. Despite all the trials and tribulations associated with the writing of this book, exacerbated by our unplanned physical separation halfway through its completion, we are better friends now than when we began.

Craig F. Bohren
Bruce A. Albrecht

Part of this book was written while I was a visiting professor in the Department of Physics at Trinity University, where I taught a course in thermodynamics and statistical physics during the fall semester of 1994. I am grateful to Fred Loxsom and Dick Bartels for making this visit possible. Special thanks go to Dick for the many discussions we had about thermodynamics, even though I couldn't convince him that differentials are the work of the Devil.

Thanks go to my Oak Hall neighbor, architect Joe Westrick, for discussions about the turnover time of air in houses.

As usual, I am most grateful to my wife of 33 years, Nanette Malott Bohren, who again had to endure a house littered with papers, books, and manuscripts, despite which she still patiently pored over the final version for hours, ferreting out errors and inconsistencies.

Craig F. Bohren
Tŷ'n y Coed
Oak Hall, Pennsylvania
July 1997

My interest in atmospheric thermodynamics was inspired by Alan Betts and Wayne Schubert while I was a graduate student at Colorado State University. Over the years I benefitted greatly from their vast knowledge of and physical insight into this subject. And finally, special thanks go to my loving wife Nancy for her support and patience during the writing of this book.

Bruce A. Albrecht
Miami, Florida
July 1997

Contents

1   Introduction: Conservation of Energy   1

   1.1   Thermodynamics: A Science of Measurable Quantities   2

   1.2   Conservation of Energy in Mechanics   5

   1.3   Conservation of Energy: A System of Point Molecules   7

   1.4   A Few Examples of Energy Conservation   13

*1.5   Kinetic Energy Exchanges in Molecular Interactions (Collisions)   14

   1.6   Working and Heating   19

       An Example of Working   23

   1.7   Some Necessary Thermodynamic Concepts and Jargon   28

   1.8   Thermodynamic Internal Energy and the First Law   31

       Irreverent Thoughts about Heat   37

       How Does One Measure "Amount of Heat?"   44

       Description and Explanation   47

       A Few Parting Shots   47

   Annotated References and Suggestions for Further Reading   49

   Problems   54

2   Ideal Gas Law: Pressure and Absolute Temperature   60

   2.1   Gas Pressure and Absolute Temperature: What Are They and What Are They Not?   61

       Ideal Gas Law   62

       A Perspective on Units   69

       Pressure Measurement: Barometer and Manometer   73

       Temperature Scales and Thermometers   77

       Atmospheric Temperature Measurements   79

       Interpretations, Operations, and Explanations   82

       The Nature of Statistical Laws   83

       A Brief History of the Gas Law   85

   2.2   Pressure Decrease with Height: Continuum Interpretation   88

       Pressure–Height Relationships: Thickness   93

2.3   Pressure Decrease with Height: Molecular Interpretation   95

2.4   The Maxwell–Boltzmann Distribution of Molecular Speeds   98

    Why Don't Air Molecules Escape to Space?   104

2.5   Intermolecular Separation, Mean Free Path,
and Intermolecular Collision Rate   106

    Mean Free Path   107

    Intermolecular Collision Rate   109

    Local Thermodynamic Equilibrium   110

2.6   Is the Pressure Gradient in a Gas a Fundamental Force of Nature? 112

2.7   Surface Pressure and the Weight of the Atmosphere   113

    Flat Earthers Take Note!   114

    Why Aren't We Crushed by Airplanes Flying
Overhead?   114

2.8   The Atmosphere Is a Mixture of Gases: Dalton's Law   115

    Mean Molecular Weight   116

Annotated References and Suggestions for Further Reading   121

Problems   127

3   Specific Heats and Enthalpy: Adiabatic Processes   144

3.1   A Critique of the Mathematics of Thermodynamics   144

    "Those Accursed Differentials"   148

    Differentials and Infinitesimals   153

    Are Differentials Necessary in Thermodynamics?   154

    Pure and Impure Thermodynamics: The Indicator
Diagram   157

    Impossible Processes   158

3.2   Specific Heats and Enthalpy   159

    Is the Heat Capacity of Liquid Water Extraordinarily
High?   168

    An Incompressibility Paradox: The Perils of
Idealization   168

    Enthalpy of the (Hydrostatic) Atmosphere   170

3.3   Adiabatic Processes: Poisson's Relations   171

3.4   (Dry) Adiabatic Processes in the Atmosphere   174

    Do Pistons and Cylinders Inhabit the Atmosphere?   177

3.5   Stability and Buoyancy   179

    Buoyancy   180

Dry Adiabatic Lapse Rate and Stability 182

Parcel Oscillations 187

3.6 Specific Heats of Gases 188

The Ratio of Working to Heating at Constant Pressure 194

3.7 Heat Capacities of Mixtures of Gases 196

Water Vapor Demystified 197

Isobaric, Adiabatic Mixing of Moist Parcels 198

3.8 Atmospheric Applications of the First Law 201

3.9 Chemical Reactions and Temperature Changes 206

3.10 Residence Time of the Internal Translational Kinetic
Energy of Earth's Atmosphere 210

Annotated References and Suggestions for Further Reading 211

Problems 214

4   Entropy 229

4.1 Entropy of an Ideal Gas 230

Entropy Change in a Free Expansion 232

Entropy Changes upon Heating and Cooling 235

The Second Law and Stability 242

Entropy of Mixtures: Entropy of Mixing and the
"Gibbs Paradox" 242

Entropy Changes upon Mixing Two Gases with
Different Temperatures and Pressures 245

An Entropic Derivation of Joule's Law 246

Entropy and Disorder: A Persistent Swindle 248

Microscopic Interpretation of Entropy 251

4.2 Entropy Changes of Liquids and Isotropic Solids 252

4.3 Atmospheric Applications of the Second Law 255

Maximum Entropy: Arbitrary Temperature
Distribution in a Solid Slab 258

Entropy Maximization in the Atmosphere 261

Thermodynamic Efficiency: The Carnot Cycle 270

Refrigerators and Coefficient of Performance 275

Thermodynamic Efficiencies of Real Engines 277

Lapse Rate in Water 278

Annotated References and Suggestions for Further Reading   279

Problems   281

5   **Water and Its Transformations**   285

5.1   Evaporation and Condensation of Water Vapor   286

5.2   Measures of Water Vapor in Air   289

Dew, Frost, Defrosters, Dehumidifiers, and Swamp Coolers   294

5.3   The Clausius–Clapeyron Equation   298

Other Enthalpy Differences   303

Mantras and Misconceptions about Phase Transitions   304

Entropy and Enthalpy Differences in Phase Changes   305

Temperature Dependence of Enthalpy of Vaporization   306

Temperature Dependence of Saturation Vapor Pressure: A More Accurate Equation   308

Difference between the Saturation Vapor Pressure above Ice and above Subcooled Water at the Same Temperature   310

Dew Point Depression and Human Comfort in Hot, Humid Weather   313

Lapse Rate of the Boiling Point   315

Evaporative Cooling and Condensational Warming   316

5.4   van der Waals Equation of State   318

Must a Liquid Boil in Order to Evaporate?   329

Can a Solid Boil Before It Melts?   330

Departures from Ideality According to the van der Waals Equation   331

* The Maxwell Construction and Saturation Vapor Pressure   332

An Overview of the Many Successes of the van der Waals Equation   334

5.5   Phase Diagrams: Liquid–Vapor, Liquid–Vapor–Solid, Triple Point   334

5.6   Free Energy   340

5.7   Effect of Air Pressure on Saturation Vapor Pressure   343

5.8   Lowering of Vapor Pressure by Dissolution   346

5.9   Air in Water: Henry's Law   351

Change in Saturation Vapor Pressure with Total Pressure   355

5.10   Size Dependence of Vapor Pressure: Water Droplets, Solution Droplets, and Bubbles   356

Droplet and Bubble Vapor Pressure: Physical Interpretation   357

Mechanical Equilibrium of Balloons, Corneas, Droplets, Bubbles: The Young–Laplace Equation and the Road to Surface Tension   358

Equilibrium Vapor Pressure of Droplets and Bubbles: A Physical Interpretation   360

The Kelvin Equation and the Difficult Birth of Cloud Droplets   361

Vapor Pressure of Solution Droplets   364

Boiling Demystified and More Heresy   369

Annotated References and Suggestions for Further Reading   373

Problems   380

6   Moist Air and Clouds   398

6.1   Precipitable Water in the Atmosphere   398

6.2   Lapse Rate of the Dew Point: Level of Cloud Formation   400

6.3   Density of Moist Air: Virtual Temperature   405

6.4   Wet-Bulb Temperature   408

Is the Temperature of a Wet Bulb the Wet-Bulb Temperature?   413

Humidity Measurements   414

6.5   Lapse Rate for Isentropic Ascent of a Saturated Parcel   416

Equivalent Potential Temperature and Wet-Bulb Potential Temperature   421

An Overview of Temperatures, Real and Fictitious   429

6.6   Thermodynamic Diagrams   430

A Smattering of History   431

Skew-T–log p Diagram   432

Tephigram   438

Other Diagrams   440

6.7   Stability and Cloud Formation   443

Entrainment   450

6.8   Mixing Clouds   455

6.9   Cloud Formation on Ascent and Descent   457

Annotated References and Suggestions for Further Reading   460

Problems   464

7   Energy, Momentum, and Mass Transfer   469

7.1   Energy Transfer by Thermal Conduction   469

Fourier Thermal Conduction Law   471

Thermal Resistance   475

Convective Transfer of Energy   477

Conductivity of Gases: A Few Fallacies Dispelled   478

The Effective Conductivity of Porous Materials   483

The Skin Diver's Fallacy   485

Newton's Law of Cooling: A Study in Historical Error
Propagation   487

Freezing of Lakes   492

Radiative Energy Transfer   493

Radiation and Convection Combined: Dew and Frost
Formation   497

To Insulate or Not to Insulate?   500

Radiation in Porous Media   502

Newton's Law of Cooling According to Newton   503

*Thermometers and Soils as Low-pass Filters   505

Chilliness at High Altitudes: Forced Convection   507

Cooling in Air and Water: Free Convection   510

7.2   Momentum Transfer: Viscosity   515

7.3   Mass Transfer: Diffusion   523

Diffusion Coefficient   526

The Nonexistence of Still Air   527

Growth of Cloud Droplets   528

Annotated References and Suggestions for Further Reading   531

Problems   537

Selected Physical Constants   545

Saturation Vapor Pressure over Water   546

Reference   548

Bibliography   549

Index   565

1

Introduction

Conservation of Energy

Thick catalogs of university courses may list thermodynamics offered by departments of physics, chemistry, mechanical engineering, chemical engineering, and meteorology. Each finds its needs sufficiently different to warrant separate courses. That taken by physicists may emphasize fundamental principles, and applications may be sparse. Chemical reactions are a major topic in the thermodynamics studied by chemists. Mechanical engineers learn about refrigerators, engines, and power plants. Chemical engineers are likely to learn what chemists and mechanical engineers learn, seasoned with processes on the industrial scale. What is distinct about atmospheric thermodynamics?

General thermodynamic principles don't change, but specific applications do. Atmospheric thermodynamics emphasizes water and its transformations, especially the water vapor mixed with mostly nitrogen and oxygen in *moist air*, or simply air because no air free of water vapor exists naturally on Earth. It is a minor component distinct from the others in that it *alone* can undergo transitions to its liquid and solid phases at normal terrestrial temperatures. As a consequence, water complicates atmospheric thermodynamics (Chapter 6) but makes terrestrial life possible. The fraction of water vapor in Earth's atmosphere is about that of argon ($\approx 1\%$), which we could live without, although it does have its uses (Section 7.1).

In atmospheric dynamics courses students learn about atmospheric motions, partly with an eye toward forecasting. But they are often of less interest than what they bring with them. What will the high and low temperatures be? Will it be cloudy or clear? Will it rain or snow? Temperature is a central concept in thermodynamics. Cloud formation, a precursor to rain or snow, is a phase transition from water vapor to liquid water, or liquid to ice. For these reasons atmospheric thermodynamics is taught as a separate subject even though it could be, and to an extent may be, woven into other courses.

A major difference between atmospheric thermodynamics and other brands is that planetary atmospheres cannot be pent up in a laboratory and submit to controlled experiments. Atmospheric states and processes are observed in the wild. Although the density of air decreases with altitude, most of Earth's atmosphere, per

Atmospheric Thermodynamics. Second Edition. Craig F. Bohren and Bruce A. Albrecht, Oxford University Press.
© Craig Bohren and Bruce Albrecht (2023). DOI: 10.1093/oso/9780198872702.003.0001

unit surface area, lies within volumes thousands of times taller than those depicted in engineering, chemistry, and physics textbooks. And gravity may not even be mentioned, whereas gravitational potential energy changes of molecules in Earth's atmosphere are *not* negligible. Without gravity there would be no atmospheric thermodynamics because there would be no atmosphere. And gravity drives convection, waves on water, and atmospheric waves.

1.1 Thermodynamics: A Science of Measurable Quantities

The principles of thermodynamics are firmly rooted in experiments. It is a science of measurable quantities rather than unobservable constructs. It seeks no explanation at a level below what we can observe directly with our coarse senses and low-resolution instruments. It treats phenomena on a *macroscopic*, as opposed to a *microscopic*, scale, that is, the atomic or molecular scale.

These days the existence of invisible molecules is not in doubt. Although deniers of the germ theory of disease still roam the earth, most deniers of molecular reality are in their graves. Only around the first decade of the last century were molecules finally promoted from fictitious to real. But if we were to decide that molecules don't exist, the essentials of thermodynamics would not change. It does not *explicitly* invoke an underlying structure to matter but does not deny it.

Measurements can settle disputes, which distinguishes science from religion and politics. If we were to say, "it feels colder today than yesterday," and you respond, "it feels warmer," we can settle the issue with a thermometer. Whatever it reads, we still may *feel* colder and you, warmer, but at least we can establish objectively that temperatures today are either higher or lower than yesterday. Measurements are evidence in a court of appeals with an unbiased judge (e.g. a thermometer) on the bench. But thermometers can't reveal what temperature *is*. They may be inaccurate but not, unlike the human sensory nervous system, sources of "paradoxical" perceptions of hot and cold. We do not need to know, however, what temperature is in order to investigate the many observable consequences of its *changes*: thermal phenomena such as expansion, contraction, melting, freezing, evaporation, boiling, combustion, etc. As Oliver Heaviside sardonically noted, "Shall I refuse my meal because I do not fully understand the process of digestion?"

Thermodynamics is sometimes said to be *phenomenological*, a science that describes phenomena but avoids interpretation and explanation. We supplement thermodynamics with kinetic theory, which explicitly acknowledges that atoms and molecules exist, move, and interact. Thermodynamics could be discussed without them, which would be like writing a 50,000-word novel using only words without the letter e (it has been done), the most common letter in many languages. If our aim is to understand the physical world, whatever helps is welcome. Ever since atoms were let out of the bag, physical scientists have gravitated more toward atomistic interpretations of macroscopic phenomena. Atoms have become part of everyday language. An atom is not necessarily miniscule but cannot be divided *without* destroying its essence. We are atoms, as are our words. No wonder we find atomistic explanations to our liking.

Thermodynamics is plagued by confusion between the colloquial and scientific meanings of words. *Energy* and *pressure* are most often used without scientific meaning. *Entropy* plays a major role in thermodynamics and is becoming an everyday word, not as much as energy but growing in (mis)use in nonscientific contexts. *Enthalpy* and *free energy* are not nor likely to become everyday words.

Thermodynamics can be distilled into a few general statements called *laws*. A physical law is not something that must be obeyed (by whom?) under threat of punishment, nor is it enforceable. It is a concise generalization from experiments and observations about *regular* physical phenomena, often in mathematical form. It minimizes the effort to grasp a large class of such phenomena, apparently disparate, but is not explanatory. The ideal gas *law* describes an experimentally determined regularity; kinetic *theory* explains it (Section 2.1).

We discuss mostly the first and second laws of thermodynamics (with passing mention of the zeroth and third), which *may* be expressed as follows:

> 1st Law: The energy of an *isolated* system is conserved.
> 2nd Law: The entropy of an *isolated* system cannot decrease.

Ignoring the qualifier "isolated" (Section 1.7) underlies arguments that life violates the second law. John Donne's famous phrase "no man is an island" translated into thermodynamic jargon is, "no animal is an isolated system." Low entropy of an animal is paid for by a higher entropy of its surroundings. An isolated system is an unrealizable idealization. We cast the first law into a more useful, but less general, form in Section 1.6.

Despite its name, thermodynamics is *not* dynamic. It cannot determine how long processes take or how, only their constraints. Nevertheless, some results from thermodynamics play important roles in atmospheric processes.

According to the first law energy is conserved. What about mass? Within classical physics, energy and mass are *separately* conserved. Classical thermodynamics preceded relativity. We now know that separate conservation laws are not rigorous. An example is positron–electron annihilation. Both elementary particles possess mass, but when they interact the most probable result is two massless photons. Energy is conserved but mass is not. We can say that mass has been converted into energy. This has no relevance to atmospheric processes but underlies *positron emission spectroscopy*, an imaging technique in medical diagnosis. For our purposes we can assume that mass is conserved.

The first law is mostly free of controversy. Not so the second: because it leads to predictions about the beginning and end of the world, it intrudes upon the domain of religions. The second law of thermodynamics was a "major battlefield in the heated ideological discussions of the late nineteenth century." Many combatants had "only a perfunctory knowledge of thermodynamics, and some not even that, but lack of knowledge did not always deter them from offering their view on the nature and consequences of the law of entropy increase and doing so with amazing self-confidence."

The first law asserts what *must* occur, the second what *must not*. They demonstrate the greatest strength of thermodynamics: its generality. They also demonstrate its greatest weakness: its generality. Are you confused by this seeming contradiction? Read on.

If you were to memorize the law of conservation of energy, would you then understand the physical world? Not likely. This law is too general to evoke concrete images. Consider, in contrast, the statement that the acceleration g of a body *falling* near the surface of Earth is 9.81 ms^{-2} (*not* the same as saying that g is the gravitational *force* per unit mass). This is *not* a fundamental law. It isn't true on Earth because of its atmosphere. Even if we could eliminate it and stop Earth's rotation, precise measurements would show that g decreases with altitude in the troposphere by a few parts per thousand. At a given location it varies diurnally by a few parts per ten million. It varies with location because Earth is not a homogeneous perfect sphere. Yet our statement about accelerating falling bodies, neither universal nor strictly true, is more concrete than the true statement that energy is conserved.

Conservation of energy is the largest zero-sum game in our solar system, with many players of sizes ranging from microscopic to macroscopic to astronomical. If the energy of some of them increases, that of others must decrease by the same amount.

General laws don't mean much until by using them in different ways we gradually come to better understand them. Memorizing them is not enough. If thermodynamics were only its laws, a thermodynamics textbook could be written on a business card. The enfolding of these laws, the examination of their many consequences, puts flesh on otherwise bare bones.

In stating the two laws of thermodynamics, we did not define energy and entropy. We could define energy as something that is conserved, whereas entropy is something that cannot decrease. This would be logically defensible, but not enlightening.

Concepts such as energy and entropy are *not* understood by defining them. Easy to digest but nonnutritive definitions may give the illusion of knowledge but cannot engender understanding, which itself is elusive. The more fruitful a concept, the more difficult to define, and energy is a remarkably fruitful concept.

If you are disconcerted because energy and entropy cannot be neatly defined, consider words without scientific connotations. Could you define love so comprehensively as to allow for every possible interpretation? Love is defined in dictionaries by strings of words, each defined by other strings, and so on, *ad infinitum*. Library shelves are filled with books on love by poets, philosophers, and novelists, despite which, its definition remains elusive. Love takes its many meanings from the many ways it is used—you can never understand it fully because you can never experience it in all its manifestations. And so it is with energy and entropy. Don't expect to completely understand them from the outset. Be content with gradual and progressive understanding through use. Be prepared for a journey, but don't expect a destination. The modern concept of energy and its conservation evolved along a circuitous path over a few decades in the mid-nineteenth century (preceded by a much longer prehistory), the fruit of several investigators independently coming to similar conclusions expressed in different ways. Is it realistic to expect complete understanding after hearing one lecture or reading one chapter of a book, especially because the disordered multiple origins of thermodynamics are reflected in how it continues to be taught today?

To understand energy, entropy, and many other concepts (scientific or otherwise), we must use them in as many contexts as possible, proceeding from the familiar to the unfamiliar. We are often exhorted to conserve energy because of an impending energy crisis. We have no choice. The salient characteristic of energy is that it *is* conserved. If there is a crisis, it is an entropy crisis (Chapter 4).

In his 1884 Baltimore Lectures Lord Kelvin opined that "I can never satisfy myself until I can make a mechanical model of a thing." Because of the remarkable success of Newtonian mechanics at precisely predicting planetary orbits, it became the model for other physical sciences. But it could treat only bodies moving in free space and acted upon by only inverse-square attractive forces. Thermodynamics borrowed some concepts such as energy but to progress had to go beyond Newtonian mechanics and define an average *internal* energy of objects composed of $N \gg 1$ molecules interacting by unknown forces. With a hypothesis unique to thermodynamics (inapplicable to planetary orbits), the huge number of variables for fluids is reduced to two—temperature and pressure—without having to know an intermolecular force law or solving simultaneously a set of N vector equations of motion. For this reason, we begin our journey with the concept of energy conservation in mechanics, the embryo for further growth.

1.2 Conservation of Energy in Mechanics

Mechanics, the analysis of where things are, where they are going, and how fast, is subdivided into kinematics and dynamics. Kinematics—geometry changing in time—describes motions independent of their causes; dynamics includes them.

As an example of a dynamical law we take the modern form of Newton's second law of motion for a fictitious point molecule with mass m and a velocity $\mathbf{v}$ in a uniform gravitational field $\mathbf{g}$:

$$m\frac{d\mathbf{v}}{dt} = m\mathbf{g}. \tag{1.1}$$

Molecules are often approximated as point masses even though they have internal structure that sometimes cannot be ignored. For the moment, we do. The gravitational field $\mathbf{g}$ near Earth's surface is approximately $-g\mathbf{e}_z$, where $\mathbf{e}_z$ is a unit vector normal to a locally flat spherical Earth. A coordinate system fixed relative to it is not an inertial coordinate system, and hence $\mathbf{v}$ does not strictly satisfy Eq. (1.1). We neglect Earth's rotation and banish (temporarily) its atmosphere.

The scalar product of Eq. (1.1) with $\mathbf{v}$ is

$$m\mathbf{v} \cdot \frac{d\mathbf{v}}{dt} = -mg\mathbf{v} \cdot \mathbf{e}_z, \tag{1.2}$$

which can be written

$$\frac{d}{dt}\frac{1}{2}m\mathbf{v}\cdot\mathbf{v} = \frac{d}{dt}\frac{1}{2}mv^2 = -mg\frac{dz}{dt},$$ (1.3)

where v is the speed of m. By regrouping terms, Eq. (1.3) can be written

$$\frac{d}{dt}(K+P) = 0,$$ (1.4)

where $K = mv^2/2$ is the *kinetic energy* of m and $P = mgz$ is its (gravitational) *potential energy*. Because of their form, kinetic energy can be said to be *energy of motion*, potential energy can be said to be *energy of position*. The total energy of m, the sum of its kinetic and potential energies, is conserved:

$$E = K + P = \text{const.}$$ (1.5)

This equation seems to satisfy a deep-seated human need for constancy. We couldn't survive in a world of incessant change. Tomorrow's rules for survival would be different from those of today. Our brains impose constancy on the world we perceive. An example is *size constancy*. The size of images on our retinas varies inversely with distance, a *geometrical*, but not a *perceptual*, law. Within limits, we perceive people to be about the same size, regardless of distance. If we perceived objects according to geometry, as people moved away from us, they would appear to shrink to the size of dolls. This does not happen because we have learned unconsciously from experience to organize our world perceptually so that it appears approximately constant. Size constancy is only one example among several (e.g. shape, color, brightness). The prominent role that conservation laws play in physics may be partly rooted in our need for constancy.

Conservation of energy, Eq. (1.5), contains no information not in Eq. (1.1). Indeed, it contains *less*: Eq. (1.5) follows from Eq. (1.1) but not conversely. E is the *first integral* of the motion, and integration always is accompanied by loss of details. But the gain is changelessness embedded within change. No matter the trajectory of m, one of its properties, energy, is unchanged. There are other integrals of the motion: force integrated over *time* is a *momentum* change; force integrated over *distance* is a *kinetic energy* change.

From this specific example more general results follow. We acquire at least one handle on energy, its *dimensions*: the product of mass and speed squared (or force and length). *All* energies have these dimensions (not to be confused with units). Another general result is that there are two distinct *forms* of energy: *motional* and *positional*.

Physical dimensions—length, mass, time—are pre-scientific concepts. We cannot compare a length with a mass or a time, but we can compare different lengths with any one of many standard lengths (units), all interconvertible. And similarly for mass and time. Problems can often be solved approximately by dimensional arguments. For example, the time for an object dropped from a height h to reach the ground must be proportional to $\sqrt{h/g}$ (if drag is negligible), which is physically plausible, dimensionally correct, and need not be obtained by solving a second-order differential equation. This kind of dimensional argument is as much an art as a science. Dimensional arguments permeate this book, especially in Chapter 7.

We often use the vague term negligible, by which we mean below the limits of detection by our instruments or, even if not, does not change our conclusions. Two important questions should be asked about all measurable quantities: How large? Who cares?

If P goes down (up), K goes up (down), and hence we can say that potential energy is *transformed* into kinetic energy, or vice versa—a way of summarizing a conservation law. The transformation is one of *form*, not of substance. Potential and kinetic energies have different forms. Energy can only be transformed, neither created nor destroyed, neither generated nor consumed. How do we reconcile this with frequent exhortations to "create" more energy, a literal violation of the first law? What is meant is *transform* more energy to forms more suited to the needs and wants of humans.

The potential and kinetic energies of an arcing golf ball could be measured by taking photographs of it in quick succession with suitably positioned cameras. Its potential energy at its highest point is greatest, its kinetic energy is least. As it descends, its potential energy decreases continuously while its kinetic energy increases correspondingly—until the ball reaches the ground and stops. To do so it must interact with the ground and the state of both must change if only fleetingly and undetectably. But $K = P = E = 0$ and energy isn't conserved—or so it appears. The predecessor to kinetic energy was mv^2, the *vis viva* (living force) of Leibniz. Efforts to understand why it *apparently* can be destroyed led to recognizing forms of energy other than mass-motion kinetic energy. To preserve the comforting principle of energy conservation, we search for energy forms in the ball *and* the ground *not* directly detectable. To reveal these *hidden* or *internal* forms of energy we must go beyond a single point mass to a system of many interacting point molecules.

Molecules, atoms, and electrons are often called particles. But Earth's atmosphere contains what *we* mean by them: discrete, coherent aggregations of many atoms and molecules, which can be collected and observed with a microscope (e.g. dust). Atmospheric particles are not debris of no consequence but serve as cloud condensation nuclei (Section 5.11) and affect climate, visual range, and human health in ways that molecules do not. Because of this we are careful to distinguish between molecules and particles. In the Lagrangian description of fluid flow, an unobservable, *fictitious* particle (also called a parcel) is a collection of molecules with properties (e.g. velocity and density) that may change with time.

1.3 Conservation of Energy: A System of Point Molecules

Consider $N \gg 1$ point molecules, each with *fixed* mass $m_i (i = 1, 2, \ldots, N)$ and position $\mathbf{x}_i$ relative to a reference point. For them to be considered a single macroscopic system they *must* interact, if only weakly and sporadically. To be confined in a container they must interact strongly with its impenetrable walls. We assume that these molecules neither combine nor dissociate. The jth molecule exerts a force $\mathbf{F}_{ji}$ on the ith molecule, and an *external* force $\mathbf{F}_i^e$, originating from outside the system, may act on it. The equations of motion

$$m_i \frac{d\mathbf{v}_i}{dt} = \sum_{j \neq i} \mathbf{F}_{ji} + \mathbf{F}_i^e \quad (i = 1, 2, \ldots, N) \tag{1.6}$$

describe the *individual* trajectories of molecules. By considering idealized *point* molecules we ignore that forces between real nonspherical molecules depend on their relative orientations. According to Newton's third law of motion, the force exerted on the ith molecule by the jth molecule is equal and opposite the force exerted on the jth molecule by the ith molecule: $\mathbf{F}_{ji} = -\mathbf{F}_{ij}$. If we sum Eq. (1.6) over i, the sum of *all* internal forces vanishes, leaving only the sum of the external forces:

$$\sum_i m_i \frac{d\mathbf{v}_i}{dt} = \frac{d}{dt} \sum_i m_i \mathbf{v}_i = \sum_i \mathbf{F}_i^e = \mathbf{F}^e. \tag{1.7}$$

This equation describes the trajectory of the entire system of molecules considered as a *single* entity. The position of its *center of mass* is defined as

$$\mathbf{X} = \frac{\Sigma_i m_i \mathbf{x}_i}{M}, \tag{1.8}$$

where M is the total mass of the system. With this definition, the equation of motion Eq. (1.7) can be written

$$\frac{d\mathbf{P}}{dt} = \mathbf{F}^e, \tag{1.9}$$

where the total linear momentum $\mathbf{P} = M d\mathbf{X}/dt$ of the system is conserved if $\mathbf{F}^e = 0$.

We assume that each *pair* of molecules interacts independently of all the others (called two-body or *pairwise* or *binary* interactions). The failure of textbooks to even mention the possibility of other than binary molecular interactions is probably because gravitational interactions between planets in our solar system are binary. But it does not necessarily follow that *all* molecular interactions are. Consider three molecules, A, B, and C. Interaction of C with B can change its structure, and hence its interaction with A. Gravitational interactions between molecules are binary because they depend only on mass but are often negligible. Intermolecular interactions are almost always implicitly taken to be binary in gases and liquids composed of real atoms and molecules, which may be a very good approximation, but is not rigorously true. Assuming two-body interactions makes analysis much easier. Three-body, four-body, five-body, and so on, interactions do not violate conservation of energy, but they do make its mathematical form more complicated. Sometimes they cannot be ignored. And no law requires all forces to be *central*, directed along the line between two molecules.

The position $\mathbf{x}'_i$ of the ith molecule relative to the center of mass, which in general is not fixed, is

$$\mathbf{x}'_i = \mathbf{x}_i - \mathbf{X}. \tag{1.10}$$

The corresponding relative velocity is

$$\mathbf{v}'_i = \mathbf{v}_i - \mathbf{V}, \tag{1.11}$$

where $\mathbf{V} = d\mathbf{X}/dt$ is the velocity of the center of mass. The mass-weighted average velocity relative to the center of mass vanishes:

$$\langle \mathbf{v}' \rangle = \frac{\Sigma_i m_i \mathbf{v}'_i}{M} = 0. \tag{1.12}$$

The average velocity $\langle \mathbf{v} \rangle$ of the system of molecules is therefore $\mathbf{V}$. Motion relative to (or about) the center of mass can be said to be motion that, on average, goes nowhere. It may be unorganized or random, in contrast with motion of the center of mass, which may be organized. All motion about the center of mass is not necessarily random, rotation of a wheel, for example. A fictitious rigid body is one for which $|\mathbf{x}'_i - \mathbf{x}'_j| = c_{ij}$, where $i \neq j$ and c_{ij} is a constant.

The velocity $\mathbf{V}$ is related to but not the same as the *velocity field* in fluid mechanics. To specify it at any time t we imagine a small volume in the fluid surrounding a point specified by a position vector $\mathbf{x}$. By "small" we mean compared with a probe but large enough to contain sufficiently many molecules that an average velocity is meaningful. This velocity is that of the center of mass of all the molecules in the volume, often written as $\mathbf{v}(\mathbf{x}, t)$ to emphasize that it depends on a point *in* the fluid and time.

Return to Eq. (1.6) and take the scalar product of both sides with $\mathbf{v}_i$ and sum over all i:

$$\sum_i m_i \mathbf{v}_i \cdot \frac{d\mathbf{v}_i}{dt} = \sum_j \sum_{j \neq i} \mathbf{v}_i \cdot \mathbf{F}_{ji} + \sum_i \mathbf{v}_i \cdot \mathbf{F}^e_i. \tag{1.13}$$

The left side of this equation is the time derivative of the total kinetic energy K:

$$K = \sum_i \frac{1}{2} m_i \mathbf{v}_i \cdot \mathbf{v}_i = \sum_i \frac{1}{2} m_i v_i^2. \tag{1.14}$$

Because of Eq. (1.11),

$$K = \frac{1}{2} M V^2 + \sum_i \frac{1}{2} m_i v_i'^2 + \mathbf{V} \cdot \sum_i m_i \mathbf{v}'_i. \tag{1.15}$$

The third sum vanishes because the average velocity [Eq. (1.12)] is a *vector* sum with zero resultant. The average square of the *speed* about the center of mass does not vanish, even though the average *velocity* does. Because speed squared is nonnegative, the sum of speeds squared cannot be zero unless each is zero. The total kinetic energy can be written as the sum of K_{cm}, the kinetic energy of a fictitious body with mass equal to that of the entire system with its center-of-mass speed, and the (hidden) internal kinetic energy about the center of mass K_{int}:

$$K = K_{cm} + K_{int}. \tag{1.16}$$

K_{int} is *invariant*, the same in all coordinate systems moving relative to each other at a constant velocity. Invariant is *not* the same as constant. K_{cm}, however, is different in different coordinate systems. Internal kinetic energy about the center of mass is often manifested macroscopically by absolute temperature (Section 2.1). An implicit assumption underlying our interpretation of K_{int} as internal, unorganized kinetic energy is that the system of molecules does not rotate as a whole. A rotating body has rotational kinetic energy, which can be included in Eq. (1.16) so that K_{int} becomes entirely unorganized kinetic energy.

Simplifying the right side of Eq. (1.13) requires a bit more effort, which we make gradually. We began this section with a simple dynamical system, a point molecule acted on by a uniform, constant force,

$$\mathbf{F} = -mg\mathbf{e}_z, \tag{1.17}$$

which led to the potential energy

$$P = mgz. \tag{1.18}$$

$\mathbf{F}$ is the negative gradient of this potential energy:

$$\mathbf{F} = -\nabla P, \tag{1.19}$$

and therefore is said to be *derivable* from a potential energy. P is defined only to within an additive constant because its gradient is zero. Underlying Eq. (1.17) is the assumption that $z \ll R_e$, where R_e is Earth's radius. From the inverse-square law of gravitation, we obtain (approximately) Eq. (1.18) with an additive constant. Both forms of the potential energy give the same force. But why is it defined as the *negative* gradient? Why not positive? It is rare in our experience for anyone to explain or question this *convention*. Its apparent function is to make conservation of energy [Eq. (1.30)] more pleasing to the eye: potential energies are *added* to kinetic energies. And why the qualifier "potential," which implies that it is not the real thing? The reality of potential energy is discussed in a historical-critical account by Eugene Hecht, who concludes that potential energy is real but superfluous. But it is *convenient*, which is why we use it. Perhaps the simplest way to understand why positional energy is qualified as potential is that it has the *potential* of being transformed into kinetic energy by mechanisms and over times not necessarily known or even knowable. The intermolecular potential energy of a lump of coal can stably exist for hundreds of millions of years, then abruptly be transformed into kinetic energy by combustion. The gravitational potential energy of water behind a dam can be transformed slowly into kinetic energy and put to good use, but not if the dam fails suddenly.

If all external forces are derivable from a potential energy,

$$\mathbf{F}_i^e = -\nabla_i P_i^e(\mathbf{x}_i), \tag{1.20}$$

where the gradient operator in rectangular Cartesian coordinates is

$$\nabla_i = \mathbf{e}_x \frac{\partial}{\partial x_i} + \mathbf{e}_y \frac{\partial}{\partial y_i} + \mathbf{e}_z \frac{\partial}{\partial z_i}. \tag{1.21}$$

Each term in the second sum on the right side of Eq. (1.13) therefore has the form:

$$\mathbf{v}_i \cdot \mathbf{F}_i^e = -\mathbf{v}_i \cdot \nabla_i P_i^e. \tag{1.22}$$

The *external* potential energy (e.g. gravity) of the ith molecule is a function only of its coordinates x_i, y_i, z_i. From the chain rule of calculus

$$\frac{dP_i^e}{dt} = \frac{\partial P_i^e}{\partial x_i}\frac{dx_i}{dt} + \frac{\partial P_i^e}{\partial y_i}\frac{dy_i}{dt} + \frac{\partial P_i^e}{\partial z_i}\frac{dz_i}{dt} = \mathbf{v}_i \cdot \nabla_i P_i^e. \tag{1.23}$$

The two preceding equations lead to the following result for the second sum on the right side of Eq. (1.13):

$$\sum_i \mathbf{v}_i \cdot \mathbf{F}_i^e = -\frac{d}{dt}\sum_i P_i^e = -\frac{dP^e}{dt}. \tag{1.24}$$

From this result it is plausible that, subject to assumptions, the first term on the right side of Eq. (1.13) will have a form similar to that of Eq. (1.24):

$$\sum_j \sum_{j \neq i} \mathbf{v}_i \cdot \mathbf{F}_{ji} = -\frac{d}{dt}\frac{1}{2}\sum_i \sum_{j \neq i} P_{ij}, \tag{1.25}$$

where P_{ij} is the potential energy of interaction between the ith and jth molecules. In the left side of Eq. (1.25) each term $(\mathbf{v}_i - \mathbf{v}_j) \cdot \mathbf{F}_{ij}$ appears only once because of Eq. (1.28). The sum on the right side is over all pairs of molecules so each interaction is counted twice, hence the factor $1/2$. If you are content with a plausible guess, you can skip the following derivation of Eq. (1.25).

First, we assume that the force on the ith molecule because of the jth molecule is derivable from a potential energy $P_{ij} = P_{ji}$:

$$\mathbf{F}_{ji} = -\nabla_i P_{ji}. \tag{1.26}$$

Although this potential energy of interaction could depend on the absolute positions of the two molecules, it is more plausible—and much simpler—to assume that it depends only on their relative positions. With this restriction, Newton's third law is necessarily satisfied. If the potential energy depends only on the distance between molecules,

$$P_{ij}(\mathbf{x}_i, \mathbf{x}_j) = P_{ij}(|\mathbf{x}_i - \mathbf{x}_j|), \tag{1.27}$$

it follows from Eqs. (1.26) and (1.27) that

$$\mathbf{F}_{ji} = -\nabla_i P_{ij} = \nabla_j P_{ij} = -\mathbf{F}_{ij}. \tag{1.28}$$

Because of Eq. (1.28) the first summation on the right side of Eq. (1.13) is a sum over all terms of the form

$$(\mathbf{v}_i - \mathbf{v}_j) \cdot \mathbf{F}_{ji} = -\nabla_i P_{ij}(|\mathbf{x}_i - \mathbf{x}_j|) \cdot \frac{d}{dt}(\mathbf{x}_i - \mathbf{x}_j) = -\frac{d}{dt}P_{ij}, \tag{1.29}$$

where we again invoked the chain rule. From this equation, Eq. (1.25) follows. The derivation is easier if we write $R = |\mathbf{x}_i - \mathbf{x}_j|$ and express $\mathbf{x}_i - \mathbf{x}_j$ and $\mathbf{v}_i - \mathbf{v}_j$ in rectangular Cartesian coordinates; $\mathbf{x}_i - \mathbf{x}_j = \mathbf{x}'_i - \mathbf{x}'_j$ and $\mathbf{v}_i - \mathbf{v}_j = \mathbf{v}'_i - \mathbf{v}'_j$.

All the ingredients are now at hand to blend into a conservation law similar to Eq. (1.5) for a single point molecule extended to a system of a great many molecules:

$$K_{cm} + K_{int} + P_{int} + P^e = \text{const.}, \tag{1.30}$$

where the internal potential energy P_{int} is the sum of all the potential energies of mutual interaction:

$$P_{int} = \frac{1}{2}\sum_i \sum_{j \neq i} P_{ij}. \tag{1.31}$$

Because we write P_{ij} does not imply that we or anyone knows an analytical expression for it. No universal, rigorous, unique intermolecular potential energy exists, even for binary interactions. Various potential energies are chosen for *convenience*, subject to the constraints that they correspond to repulsion at small distances and weaker attraction at large distances (on the molecular scale) and do not predict macroscopic quantities markedly at odds with measurements.

Although the analogy may seem strange, a ball dropped from rest at some height above Earth is like a chemical reaction between atoms and molecules. The internal potential energy of the system of two "atoms," ball and Earth, decreases, accompanied by a corresponding increase in kinetic energy relative to the center of mass. The only difference is the vastly greater size of these macroscopic atoms and that the always attractive force between them is much longer range than interatomic forces. For more about chemical reactions, see Section 3.9.

Potential energy is transformed when you walk at constant speed up a flight of stairs. Your gravitational potential energy increases in a macroscopic vertical displacement at the expense of decreased intermolecular potential energy in your body because of microscopic displacements (chemical reactions). The totality of all such reactions within living organisms is called *metabolism*, a term derived from a Greek word meaning change.

The total kinetic energy Eq. (1.15) is a sum over N molecules, and hence divided by N is an average per molecule *independent* of N. But P_{int} is a sum of $N(N-1)/2 \approx N^2/2$ terms, and if divided by N is not necessarily such an average.

It is *if* intermolecular forces drop off with distance sufficiently rapidly that to good approximation mostly nearest neighbors interact and the sum is dominated by terms such that for each i and at any instant P_{ij} is negligible except for a small range of j. We return to this point in Section 3.2. Because potential energies are defined only to within a constant, intermolecular potential energies are often *defined* as zero at indefinitely large molecular separation. The contribution of gravitational inter-molecular forces to the *internal* potential energy P_{int} is usually negligible, but the *external* potential energy P^e is often Mgz, changes of which may *not* be negligible. Earth may be looked upon as a single massive molecule; average g within tens of kilometers of the surface is proportional to Earth's mass and inversely proportional to its radius squared.

No matter how complicated the motion of this system, no matter how apparently patternless, there is one small island of certainty in a sea of uncertainty: total energy is conserved, and its four components can be interconverted. The more molecules in a system, the more possible conversions that leave the total energy unchanged. We may not witness nor understand them nor be able to predict how long they take. They may be as hidden and inscrutable as conversions of dollars and euros and pesos in banks, which take a percentage of each conversion, whereas energy accounts are balanced at no charge. Conservation of energy is at heart bookkeeping by a cosmic bookkeeper who never cooks the books. Similarity between fiscal budgets and energy conservation is reflected in the term *energy budget*. The related term *energy expenditure* implies *irreversible* energy conversions, often those required to maintain life.

K_{int} and P_{int} are for an *entire* system regardless of its spatial nonuniformity. But the average energies of two or more separated *subsystems* containing large numbers of molecules are not necessarily the same at all times, Earth's atmosphere, for example. We address nonuniformity in Sections 1.8 and 2.5.

All the energies in Eq. (1.30) are of *palpable* matter. To make energy conservation complete we must somehow include the energy of *impalpable* electromagnetic fields (radiation). Earth and everything on it are embedded in electromagnetic fields varying in space and time, which originate from matter. We return to this point in Section 1.8.

1.4   A Few Examples of Energy Conservation

The stage is now set to rethink the trajectory of a golf ball rising to a maximum height, falling to the ground, and stopping abruptly (Section 1.2). The initial total energy of the ball was not zero, yet soon after it stopped it *apparently* had neither kinetic nor potential energy. But it follows from Eq. (1.30) that in the process of stopping the maximum kinetic energy of its center of mass—its *external* or *mass-motion* or *directed* kinetic energy—is transformed into hidden *internal* or *random* forms of energy of the ball *and* the ground. This internal energy is the lifeblood of thermodynamics. Although invisible, the consequences of its changes are in principle, and sometimes in practice, detectable. The internal energy of a fictitious rigid body is *unchangeable* by definition and hence lies outside the domain of thermodynamics as does celestial mechanics, a branch of astronomy, because it analyzes *individual* trajectories of celestial objects like planets and comets.

This transformation of gravitational potential energy into mass-motion kinetic energy and then into internal energy is concrete: we can see and hold a golf ball. And it sheds light on the more abstract descent and ascent of air parcels, which we can neither see nor hold (see the end of Section 3.4).

Consider two examples of energy transformation. A whirring fan is a source of directed motion—a breeze you can feel but not see. You can make it visible with a puff of baby powder. Turn the fan on, then off. The air soon will be still again. What happened to the mass-motion kinetic energy of the breeze? It must have been transformed into internal energy of the air. The notion that energy is "lost" reflects the narrow view that only mass-motion kinetic energy is the genuine article.

Turn off a hot plate when its temperature is constant. Transient air currents swirl above it, observable in whiffs of baby powder illuminated by a flashlight in a dark room. As the plate cools, its internal energy is partly transformed into mass-motion kinetic energy of surrounding air (convection) until the plate and air temperatures are equal. A temperature gradient with a component along the gravitational field drives the convection. This is the origin of winds at all scales. Temperature gradients over Earth's surface are partly transformed into mass-motion kinetic energy. Conservation of energy cannot tell us how or how fast.

In the subsection on free expansion of an ideal gas (Section 4.1) we describe a simple experiment demonstrating conservation of *total* kinetic energy. An increase of *directed* kinetic energy is balanced by a decrease of *random* kinetic energy manifested by an appreciable temperature drop measured with a fast-response thermocouple.

*1.5   Kinetic Energy Exchanges in Molecular Interactions (Collisions)

Equation (1.30) tells us that total energy is conserved, but not how kinetic energy is transferred between different parts of a system of molecules or between different systems of molecules. To go beyond this equation we must consider in detail interactions between molecules. This would be a Herculean task for a system of many molecules, just the kind of system of interest, but we can retreat with honor to two molecules for insights into what happens, on average, when molecules of unequal kinetic energy interact (collide).

Real molecules have internal structure and therefore intramolecular kinetic and potential energies and rotational kinetic energies. Although all these energies may change in collisions, for the moment we treat molecules as having only translational kinetic energy. In Section 3.6 we abandon this fiction. We can envision a collision between two real molecules as that between two bags of atoms jostling around within the bags. Before collision they have different total energies. The bags remain intact and the sum of their total energies, but not necessarily their translational kinetic energies, is conserved in collisions. An atom itself is a bag of negatively charged electrons corralled by a positively charged relatively massive nucleus. Changes in translational kinetic energies of molecules can take a *continuous* set of values. Quantum mechanics restricts changes in rotational and vibrational kinetic energies to a *discrete* set.

Imagine two molecules moving toward each other in such a way that they collide. Because of the short-range forces they exert on each other, their trajectories are altered from what they would have been had they not collided (Fig. 1.1). But before and after collision they are so far apart (many molecular diameters) that we can take the total intermolecular potential energy change to be zero. Their velocities before collision relative to a fixed coordinate system are denoted by $\mathbf{v}_1$ and $\mathbf{v}_2$, and their respective masses by m_1 and m_2. We take molecule 2 as having the higher kinetic energy before collision. Does the kinetic energy of molecule 1 increase, decrease, or remain the same after the collision? In any single collision, any of these outcomes is possible. But what is the average over many collisions?

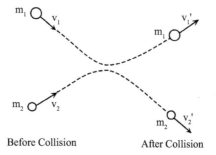

Before Collision   After Collision

Fig. 1.1: A collision occurs between two molecules with different masses and initial velocities if they become close enough to exert short-range forces sufficiently large to change their trajectories.

Regardless of the details of the collision, total momentum must be conserved (absent external forces):

$$m_1\mathbf{v}_1 + m_2\mathbf{v}_2 = m_1\mathbf{v}_1' + m_2\mathbf{v}_2', \tag{1.32}$$

where the primes indicate velocities after the collision. The collision is a black box into which two molecules enter and from which they exit. What goes on in the box is a mystery.

An *elastic collision* is defined as one in which translational kinetic energy is conserved. We *assume* that the molecules collide elastically:

$$\frac{1}{2}m_1v_1^2 + \frac{1}{2}m_2v_2^2 = \frac{1}{2}m_1v_1'^2 + \frac{1}{2}m_2v_2'^2, \tag{1.33}$$

where $\nu = |\mathbf{v}|$. We include the common factor $1/2$, even though it cancels from both sides, to remind us that it is needed if we add potential energy to the total.

There are six unknowns in this problem: the six components of the velocity vectors $\mathbf{v}'_1$ and $\mathbf{v}'_2$ We have only three momentum equations (Eq. 1.32) and one kinetic energy equation (Eq. 1.33), and hence the task of finding the velocity vectors after collision is, in general, impossible. That is, $\mathbf{v}_1$ and $\mathbf{v}_2$ do not uniquely determine $\mathbf{v}'_1$ and $\mathbf{v}'_1$, which is obvious to anyone who has ever played pool (the sign of a misspent youth). Fortunately, we are interested only in the average kinetic energy change, which, as we shall see, is attainable even though all the details of the collision are not.

If we square both sides of Eq. (1.32), we obtain

$$m_1 \Delta K_1 + m_2 \Delta K_2 = m_1 m_2 (\mathbf{v}_1 \cdot \mathbf{v}_2 - \mathbf{v}'_1 \cdot \mathbf{v}'_2), \tag{1.34}$$

here ΔK with appropriate subscripts is the difference in kinetic energies before and after collision. The quantity of interest is the kinetic energy change of molecule 1, ΔK_1, which, from Eq. (1.33) is $-\Delta K_2$, and hence

$$\Delta K_1 = \frac{m_1 m_2}{m_2 - m_1} (\mathbf{v}'_1 \cdot \mathbf{v}'_2 - \mathbf{v}_1 \cdot \mathbf{v}_2). \tag{1.35}$$

From Eq. (1.35), $m_1 = m_2$ implies that $\mathbf{v}'_1 \cdot \mathbf{v}'_2 - \mathbf{v}_1 \cdot \mathbf{v}_2 = 0$, so we assume that $m_1 \neq m_2$ and if necessary take the limit $m_1 \to m_2$.

The velocity of the center of mass

$$\mathbf{v} = \frac{m_1 \mathbf{v}_1 + m_2 \mathbf{v}_2}{m_1 + m_2}, \tag{1.36}$$

is the same before and after collision ($\mathbf{v} = \mathbf{v}'$) because of momentum conservation. The *relative velocities* are

$$\mathbf{v}_r = \mathbf{v}_1 - \mathbf{v}_2, \quad \mathbf{v}'_r = \mathbf{v}'_1 - \mathbf{v}'_2. \tag{1.37}$$

The preceding two equations can be solved to obtain

$$\mathbf{v}_1 = \mathbf{v} + \frac{m_2}{m_1 + m_2} \mathbf{v}_r, \quad \mathbf{v}_2 = \mathbf{v} - \frac{m_1}{m_1 + m_2} \mathbf{v}_r. \tag{1.38}$$

The velocities after collision follow from Eq. (1.38) by attaching primes to all velocities. Equation (1.38) yields

$$\mathbf{v}'_1 \cdot \mathbf{v}'_2 - \mathbf{v}_1 \cdot \mathbf{v}_2 = \frac{m_2 - m_1}{m_1 + m_2} (\mathbf{v}' \cdot \mathbf{v}'_r - \mathbf{v} \cdot \mathbf{v}_r) + \frac{m_1 m_2}{(m_1 + m_2)} (v_r^2 - v_r'^2). \tag{1.39}$$

Equation (1.32) can be rewritten as

$$m_1 (\mathbf{v}_1 - \mathbf{v}'_1) = m_2 (\mathbf{v}'_2 - \mathbf{v}_2), \tag{1.40}$$

and hence Eq. (1.33) rewritten as

$$m_1 (\mathbf{v}_1 - \mathbf{v}'_1) \cdot (\mathbf{v}_1 + \mathbf{v}'_1) = m_2 (\mathbf{v}_2 - \mathbf{v}'_2) \cdot (\mathbf{v}_2 + \mathbf{v}'_2). \tag{1.41}$$

These two equations when combined yield

$$(\mathbf{v}_2' - \mathbf{v}_2) \cdot (\mathbf{v}_1 + \mathbf{v}_1') = (\mathbf{v}_2 - \mathbf{v}_2') \cdot (\mathbf{v}_2 + \mathbf{v}_2'). \tag{1.42}$$

From Eq. (1.38) we have

$$\mathbf{v}_2' - \mathbf{v}_2 = \frac{m_1}{m_1 + m_2}(\mathbf{v}_r' - \mathbf{v}_r). \tag{1.43}$$

With the two preceding equations and the definition of the relative velocity, we obtain

$$(\mathbf{v}_r' - \mathbf{v}_r) \cdot (\mathbf{v}_r + \mathbf{v}_r') = v_r^2 - v_r'^2 = 0, \tag{1.44}$$

from which, together with Eqs. (1.35) and (1.39), it follows that

$$\Delta K_1 = \frac{m_1 m_2}{m_1 + m_2}(\mathbf{v} \cdot \mathbf{v}_r' - \mathbf{v} \cdot \mathbf{v}_r). \tag{1.45}$$

The second scalar product on the right side of Eq. (1.45) is

$$\mathbf{v} \cdot \mathbf{v}_r = \frac{2K_1 - 2K_2 + (m_2 - m_1)\mathbf{v}_1 \cdot \mathbf{v}_2}{m_1 + m_2}. \tag{1.46}$$

Rather than tackling a general collision, we can acquire insight by considering a special collision: $\mathbf{v}_1$ and $\mathbf{v}_2$ are *parallel*. With this assumption, the center-of-mass and relative velocities before collision also are parallel, differing only by a scalar factor:

$$\mathbf{v} = \frac{m_1 v_1 + m_2 v_2 \cos \theta}{(v_1 - v_2 \cos \theta)(m_1 + m_2)} \mathbf{v}_r, \tag{1.47}$$

where θ, the angle between $\mathbf{v}_1$ and $\mathbf{v}_2$, is either 0 or π. There is a difference between collisions with $\theta = 0$ and $\theta = \pi$, as exemplified macroscopically by the different outcomes of head-on and rear-end collisions of automobiles. Equation (1.46) follows from Eqs. (1.36) and (1.37). The scalar product of $\mathbf{v}$ and $\mathbf{v}_r'$ is

$$\mathbf{v} \cdot \mathbf{v}_r' = v_r^2 \cos \Theta \frac{m_1 v_1 + m_2 v_2 \cos \theta}{(v_1 - v_2 \cos \theta)(m_1 + m_2)}, \tag{1.48}$$

where Θ is the angle between $\mathbf{v}_r$ and $\mathbf{v}_r'$. By using

$$v_r^2 = (v_1 - v_2 \cos \theta)^2 \tag{1.49}$$

it follows that

$$\mathbf{v} \cdot \mathbf{v}_r' = \frac{2K_1 - 2K_2 + (m_2 - m_1)\mathbf{v}_1 \cdot \mathbf{v}_2}{m_1 + m_2} \cos \Theta. \tag{1.50}$$

By combining Eqs. (1.45) and (1.50), we obtain

$$\Delta K_1 = \frac{m_1 m_2}{(m_1 + m_2)^2}[2K_2 - 2K_1 + (m_1 - m_2)\mathbf{v}_1 \cdot \mathbf{v}_2](1 - \cos \Theta). \tag{1.51}$$

If $m_1 = m_2$, $\Delta K_1 > 0$ in all collisions (i.e. for all values of Θ and θ) if $K_2 > K_1$. For this special case, the molecule with initially lower kinetic energy gains kinetic energy upon collision. If the two masses are not equal, however, we cannot determine what happens in an *individual* collision: ΔK_1 could be positive or negative. What about the *average* kinetic energy change in all possible collisions?

For speeds v_1 and v_2 the two possibilities for the scalar product $\mathbf{v}_1 \cdot \mathbf{v}_2$ are $\pm v_1 v_2 \cos \theta$, the plus sign for a rear-end collision, the minus sign for a head-on collision. The velocities $\mathbf{v}_1$ and $\mathbf{v}_2$ do not uniquely determine the outcome of the collision, namely, the angle Θ, which lies between 0 and π, although not all angles in this range are necessarily equally likely. The extreme example is provided by two molecules with parallel velocity vectors, which nevertheless do not collide unless they come sufficiently close to each other (see Fig. 1.2). If the molecules are spherically symmetric, the distribution of angles Θ is independent of the angle between $\mathbf{v}_1$ and $\mathbf{v}_2$. With this assumption,

$$\langle \mathbf{v}_1 \cdot \mathbf{v}_2 (1 - \cos \Theta) \rangle = \langle \mathbf{v}_1 \cdot \mathbf{v}_2 \rangle \langle 1 - \cos \Theta \rangle, \tag{1.52}$$

where angle brackets indicate averages over all collisions. If the two possible values for θ (0 and π) are equally likely, $\langle v_1 v_2 \cos \theta \rangle = v_1 v_2 \langle \cos \theta \rangle = 0$, from which it follows that

$$\langle \Delta K_1 \rangle = \frac{2 m_1 m_2}{(m_1 + m_2)^2} (K_2 - K_1)(1 - \langle \cos \Theta \rangle). \tag{1.53}$$

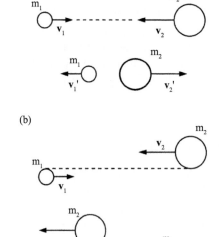

Fig. 1.2: The velocities of two molecules before a collision are not sufficient to determine its outcome. In (a) and (b), the velocities before collision are the same, yet, in (a), the velocities are reversed by collision, whereas in (b), because the initial trajectories are sufficiently displaced from each other, they are unaffected.

On average, therefore, the molecule with the initially lower kinetic energy gains kinetic energy upon collision provided that both types of collision, head-on and rear-end, are equally likely. This is one of the few examples in which the poor get richer (in kinetic energy) and the rich get poorer.

We can take a further step forward by supposing that v_1 and v_2 are not fixed but can vary over all possible values. With the assumption that the two values of the angle between $\mathbf{v}_1$ and $\mathbf{v}_2$ are independent of the magnitudes of these vectors, we have

$$\mathbf{v}_1 \cdot \mathbf{v}_2 = \langle v_1 v_2 \rangle \langle \cos \theta \rangle = 0, \tag{1.54}$$

from which it follows that

$$\langle \Delta K_1 \rangle = \frac{m_1 m_2}{(m_1 + m_2)^2} \left(2 \langle K_2 \rangle - 2 \langle K_1 \rangle \right) (1 - \langle \cos \Theta \rangle). \tag{1.55}$$

The absolute temperature of an ideal gas is proportional to the average kinetic energy of its molecules (Section 2.1). Thus, Eq. (1.55) suggests that if two gases at different temperatures interact (i.e. their molecules collide), the temperature of the initially colder gas will increase, whereas that of the initially hotter gas will decrease. Although we obtained this result by considering elastic collisions of molecules with initially parallel velocity vectors, it holds for molecules whizzing about in three dimensions.

If you find this lengthy derivation tedious, consider a simpler example you can verify on a pool table. All balls have about the same mass. A cue ball struck on its center by a cue stick parallel to the table acquires only kinetic energy (no spin), which is completely transferred to a stationary ball hit head on.

Love, we are told, makes the world go round. It might be more accurate to say that Eq. (1.55) makes the world go round. At least this equation is at the heart of why the atmosphere goes round. Random collisions between groups of molecules with different average kinetic energies result in kinetic energy transfers.

1.6   Working and Heating

Interaction between a system of molecules and its surroundings sometimes may be described by a potential energy P^e (Section 1.3). But often it is not known, and even if it were, it might be unwieldy for systems composed of a great many molecules. Macroscopic forces—averages over many microscopic forces—are not necessarily derivable from a potential, aerodynamic drag, for example, which is ≈ 0 on the Moon but not on Earth. We therefore turn to another way of describing the interaction between a macroscopic system and its surroundings.

The equation of motion of a mass m acted on by *any* force $\mathbf{F}$ is

$$m \frac{d\mathbf{v}}{dt} = \mathbf{F}. \tag{1.56}$$

Take the scalar product of both sides of this equation with $\mathbf{v}$,

$$m\mathbf{v} \cdot \frac{d\mathbf{v}}{dt} = \frac{dK}{dt} = \mathbf{v} \cdot \mathbf{F}. \tag{1.57}$$

Work is defined as a force acting through a distance, and hence the scalar product of $\mathbf{v}$ and $\mathbf{F}$ is the *rate* W at which work is done on m:

$$\frac{dK}{dt} = W. \tag{1.58}$$

Although the term work (*travail*) for a force times a distance is sometimes attributed to Gaspard-Gustave Coriolis (1830s), a name familiar to meteorologists, he had predecessors who have vanished into the mists of history. Given the other meanings of work, effort, for example, it was not the best choice of terms, but at least is used fairly consistently and, unlike heat, was not thought to have a material existence. From the outset, work was a specific way of doing something rather than a thing.

   This equation is another way of stating conservation of energy: the kinetic energy of m is constant *unless* work—positive or negative—is done on it by an external force, which requires that the energy of its source must change accordingly. Gravitational work done *on* m is easier to visualize (and measure) than *by* m. But if work is done on m by Earth, m must reciprocate according to Newton's third law. Given that the mass of Earth is about 10^6 times greater than the mass of its *entire* atmosphere more or less symmetrically distributed, the acceleration of Earth because of interaction with its atmosphere is immeasurably small.

   The foundations were laid in Section 1.3 to extend Eq. (1.58) to a system of point molecules,

$$\frac{d}{dt}(K_{cm} + K_{int} + P_{int}) = \Sigma_i \mathbf{v}_i \cdot \mathbf{F}_i^e, \tag{1.59}$$

which is Eq. (1.30) written differently by using Eq. (1.24). The internal potential energy is defined only to within a constant. All external forces are not necessarily derivable from a potential, but some could be (e.g. gravitational). $\Sigma_i \mathbf{v}_i \cdot \mathbf{F}_i^e$ is the rate of energy transfer because of interaction of the system with its surroundings; dP_{int}/dt is the rate of change of potential energy because of interactions *within* the system. Equation (1.59) is symbolic. We have no hope of following changes of the three energies. The smallest cloud droplet contains about 10^{12} molecules. To make this equation useful we have to approximate it in terms of variables that can be measured by mortals.

   Because air and liquid water are molecular substances (Section 1.7), interaction between them is the simplest example that entails only molecules. Air molecules above water have a *distribution* of energies (Section 2.4), as do the water molecules. Sometimes an air molecule transfers energy to a water molecule, sometimes the reverse. This is always a *two-way* exchange. But if the average energy of the air molecules is greater than that of the water molecules, the water gains energy, a random process. Molecules have no brains and are not intentional entities. They cannot exchange energy unless they exert forces on each other no matter how weakly or briefly (Section 1.5). Take water to be the system and air its surroundings. Every so often a water molecule experiences an exceedingly brief force that changes its trajectory. A moment later it experiences another brief force with a different direction and magnitude. This happens again and again. The trajectory of a molecule is punctuated by a series of hard knocks, which are zero on average but not without consequences. And so it is with all the other molecules in the system. Even though the sum of all these forces

$$\Sigma_i \mathbf{F}_i^e \tag{1.60}$$

vanishes,

$$\Sigma_i \mathbf{v}_i \cdot \mathbf{F}_i^e \tag{1.61}$$

may not vanish, and indeed does not if the average energy of the air molecules is different from that of the water molecules. Molecules also may be acted upon by external forces the sum of which does *not* vanish (e.g. gravity).

The previous paragraphs suggest that we can, in principle, decompose the sum on the right side of Eq. (1.59) into two components. One component, denoted by Q and called the *rate of heating* (or heating rate or simply heating), is the contribution to the sum from forces that vanish on average. The other component, denoted by W and called the *rate of working* (or working rate or simply working), is the contribution from forces that do not vanish on average. Adding "ing" to heat and work emphasizes that heating and working are *processes*, and hence Q and W may be and usually are time dependent. As it stands, Q is only a different kind of working. We often must include a radiant energy component. How best to do so is discussed in Section 1.8.

If we define $U = K_{int} + P_{int}$, we can write Eq. (1.59) as

$$\frac{dU}{dt} + \frac{dK_{cm}}{dt} = Q + W. \tag{1.62}$$

By historical accident, convention, and for brevity, Q, usually is called the rate of heating, more accurately, rate of heating or *cooling* because Q can be negative. If so, it can be called cooling to leave no room for doubt. Beer in a bottle placed in a refrigerator cools ($Q < 0$) without any obvious work being done ($W \approx 0$), whereas air compressed in a bicycle pump is warmed ($W > 0$) and yet $Q \approx 0$. In both processes the positions and velocities of a great many molecules change but the only detectable evidence is temperature (and possibly pressure) changes, a drop in temperature (cooling) or an increase in temperature (warming).

If the system evolves in such a way that the kinetic energy of its center of mass is constant,

$$\frac{dU}{dt} = Q + W. \tag{1.63}$$

Equations (1.62) and (1.63) are general even though we may not know Q, W, and U. Physical understanding and quantitative results follow from these equations only by making assumptions and approximations.

U is the *internal energy* of a system of molecules—the sum of a potential energy arising solely from forces exerted by molecules of the system on each other and a kinetic energy of motion about its center of mass. U may change with time because of interactions of the system with its surroundings. These interactions may be those for which the net force vanishes on average and yet the energy of the system changes because of random collisions with surrounding molecules having a different average energy, and those for which the net force does not vanish on average. No matter how many times it has been asserted that "heat is a form of energy," we do not agree (and we are not alone). *Internal energy* is the kinetic and potential energy of a body, *invisible* but with possibly detectable consequences, added to its possible *visible* mass-motion kinetic energy and potential energy in an external force field (e.g. gravity). Heating (cooling) are macroscopic ways of increasing (decreasing) the internal energy of macroscopic systems.

The dimensions of Q and W are power (force times speed). Because we can determine only energy differences, literal generation of energy contradicts the first law, but generation of power does not. Many possible combinations of Q and W can result in the same dU/dt. Equation (1.62) also allows for the possibility that the kinetic energy of the center of mass can be transformed into internal energy and conversely. Winds at all scales are transformations of internal energy into mass-motion kinetic energy.

As it stands, Eq. (1.63) is still a *deterministic* dynamical equation of little use because U is a function of a great many positions and velocities of molecules, microscopic variables that cannot be calculated or measured. Even worse, Q and W also depend on unknown forces. To leap the great distance from mechanics to thermodynamics, from deterministic to statistical, from discrete to continuous, from many variables to few, we *hypothesize* that to good approximation U for a system of a great many molecules is a function of only a few measurable macroscopic variables such as temperature and pressure. They could vary with time but at any instant would have to be approximately uniform or else the temperature and pressure of the system would be meaningless. Many *different* microscopic states (*microstates*) correspond to almost the *same* macroscopic state (*macrostate*). Although a thermometer inserted into a gas in equilibrium may indicate a constant macrostate, the microstates of the gas, as well as the positions and velocities of *every* molecule, rapidly and ceaselessly change. A thermometer can sample only that part of the gas in its *proximate* (or *immediate*) surroundings. If it were sufficiently sensitive, it could detect tiny local fluctuations. If the temperature and pressure of two equal volumes containing the same number of identical gas molecules are the same, their measured macrostates are (almost) *always* the same, whereas their microstates at any instant *never* are. As a macroscopic process, Q is a consequence of temperature differences between a system and its surroundings, but W is easier to visualize, quantify, and measure.

Everyday observations (cooling of beer in a refrigerator, warming and an increase in pressure of air compressed in a bicycle pump) support our hypothesis but do not prove it. Nor do they prove that temperature and pressure are the *only* measurable properties that uniquely characterize macroscopic matter.

The transition from a microscopic to a macroscopic energy equation is fruitful if it helps us understand some physical phenomena and do calculations that agree acceptably with measurements. But having hypothesized the *existence* of U, Q, and W we need not give them a microscopic interpretation, although we are not forbidden to by thermodynamic thought police. Anything that enhances understanding, short of fairy tales, is permitted. In principle, a result from macroscopic (thermodynamic or continuum) theory should follow from microscopic (molecular or kinetic) theory, and *when feasible* can give a better understanding of the former (e.g. Sections 2.2 and 2.3). But sometimes the cost in mental effort would greatly exceed the benefits, and the fallback position is macroscopic. The two approaches are not always compatible. Thermodynamic reasoning within a restricted but large domain is valid only because of "a meta-law of physics," according to which "physical laws at one length scale are not very sensitive to the details of what happens at much smaller scales."

The first law of thermodynamics in the form of a single scalar equation, Eq. (1.63) is a giant step away from a huge number of coupled vector equations governing unobservable motions impelled by unobservable forces [Eq. (1.6)]. Unlike forces, internal energy is (or can be) conserved, and we have a way of determining its dependence on only two observable properties: temperature and pressure. To go further we need to know how W *might* be determined from theory and measurements. The following example is perhaps the simplest.

An Example of Working

Consider a gas enclosed in a vertical cylinder fitted with a movable, frictionless, and massless piston girded by rings so that no gas can escape (Fig. 1.3). A whimsical definition of thermodynamics is the science of gases in impossible cylinders fitted with impossible pistons, a relic of its rapid growth in the era of steam engines (Section 3.1). Initially, the net force on the piston is zero. The upward force on the piston is pA, where p is the instantaneous *average* pressure over the piston area A. A downward external force F a tad bit *larger* than pA causes the piston to move downward a distance Δx in a time interval Δt thereby compressing the gas. The unnamed agent of F is invisible offstage, but if it does positive work on the gas, the gas must do negative work on it. Work is done *by* the surroundings *on* the gas and $\Delta x / \Delta t < 0$. The average rate of (positive) work done on the gas in time Δt is

$$W = -pA\frac{\Delta x}{\Delta t}. \tag{1.64}$$

The downward force F must continuously increase because the pressure increases. In an expansion, the upward force is a tad bit *smaller* than pA, $\Delta x / \Delta t > 0$, and work is done by the gas *on* its surroundings. A dimensional quantity such as Δx cannot be said to be small except compared with another

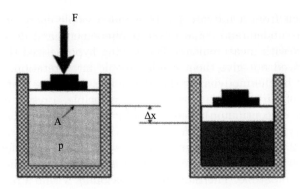

Fig. 1.3: Work is done on a gas in a cylinder fitted with a piston subject to a force F sufficient to move it down a distance Δx in time Δt. p is the average gas pressure over the face of the piston with area A.

quantity with the same dimensions, which we take to be the gas volume V divided by A. In the limit $\Delta t \to 0$, $A\Delta x/\Delta t \to \mp dV/dt$, and we have an expression for the instantaneous working, positive in a compression, negative in an expansion, in terms of measurable (in principle) variables:

$$W = -p\frac{dV}{dt} . \tag{1.65}$$

If we could measure the instantaneous average pressure *on* the piston face and its instantaneous speed, W would follow. But usually we can measure only a pressure somewhere *within* the gas and assume (or hope) that it is close to the average on the piston face. The validity of this equation is determined by how well its testable *predictions* agree with *measurements* and *observations*, not by *a priori* arguments. W sometimes is called *reversible working*. It follows from Eq. (1.65) that the sign of W is reversed if time is reversed ($t \to -t$). Work done in the forward process is equal and opposite that done in the reverse process. For more about reversibility see Section 3.3.

If the gas is compressed dV/dt is negative, the working rate is positive, and *by this process alone* the internal energy of the gas increases. A pressure increase as the gas is compressed requires an increasingly greater applied force to compress it more, what Robert Boyle called "the spring of the air," as distinct from its weight. It could be almost constant as in thermal expansion or contraction of a gas in a cylinder fitted with a freely moving piston.

As a macroscopic quantity, W is divorced from its microscopic origins. But if the total external force $\mathbf{F}^e$ is the sum of a force that is derivable from a potential energy P^e and one that is not, the working rate associated with the former is $W^e = -dP^e/dt$, which can be included in W in Eq. (1.59) or dP^e/dt can be added to the left side of Eq. (1.62). W is then the rate of working that is *not* the negative of the time derivative of a potential energy. The total work done (the time integral of W) in general depends on the path the system follows between two *macroscopic* states (Section 3.1). If the left side of Eq. (1.62) contains only time derivatives (including dP^e/dt), its time integral depends only on the initial and final states. If P^e and K_{cm} are constant and $Q = 0$ (an idealized *adiabatic process*), $W = dU/dt$, and hence its integral depends only on the initial and final states. But we may not know how U depends on macroscopic state variables such as temperature and pressure. Often our aim is to determine ΔU as a function of such variables by measuring integrated W. But adiabatic processes exist only on paper except in the trivial limit of *no* temperature difference between system and surroundings. No matter how deep your pockets, you cannot buy "perfect" insulation (see Section 7.1). All *real* processes are *diabatic* ($Q \neq 0$), but some may be *approximately* adiabatic if $|Q| \ll W$ or the integral of Q over the process time is much smaller than the integral of W.

Adiabatic, meaning "not passing through," was coined by the English engineer William Rankine (1858), but the concept predates him. He also defined energy as the "capacity for doing work," a mantra still extant. This was adequate for his purposes but does not come to grips with the essence of energy as a conserved quantity. His statement about the "waste of heat or energy" reflects a failure to distinguish between them, which persists to this day.

We begin this book on atmospheric thermodynamics (Section 1.2) by banishing Earth's atmosphere. Now we restore it. A solid or liquid body moving in air does work *on* it, which we account for approximately by *nonconservative*—not derivable from a potential energy—forces (drag). $E_b = K_{cm} + P^e$ (P^e is gravitational potential energy) is *not* conserved. As the body moves through air it imparts mass-motion to it, and hence the decrease of E_b is balanced by an increase in the kinetic and potential energy of the air with which the body interacts (if $\Delta U \ll |\Delta E_b|$). The energy budget must be balanced, but exactly how may be difficult to determine.

Drag does not always impede motion. A *wind force* component perpendicular to the gravitational field and acting on small mineral particles can transport Saharan *dust* (*not* sand) thousands of kilometers across the Atlantic Ocean, eventually deposited in South and North America.

A few hidden assumptions underlie Eq. (1.65). Compression requires pressure differences. To move the piston from an equilibrium position, requires a net force on the piston. For a moment the total force on the piston will be greater than pA on the other side. This disequilibrium is necessary if the piston is to move downward and do work on the gas. We assume that compression is sufficiently slow that the pressure changes almost instantaneously from one uniform value to another. Such a process is sometimes called *quasistatic*. But without an approximate criterion, "sufficiently slow" is vague. Slow and fast, like small and big, have no meaning without *quantitative* standards for comparison, if only implicit.

To understand the conditions under which a process *might* be taken to be quasistatic, consider the following example. The piston depicted in Fig. 1.3 is suddenly pulled upward a distance Δx. For a brief moment, the number density of molecules near the piston face is less than in the bulk of the gas. As a consequence, there is a net migration of molecules from the gas toward the piston until the density is again everywhere uniform. We assume that they cannot migrate faster (on average) than their mean speed, which at ordinary temperatures is approximately the speed of sound in air v_s ($\approx 340\,\mathrm{ms}^{-1}$ at $20\,^\circ\mathrm{C}$). The *relaxation time* τ_r, the characteristic time for the gas to return to equilibrium after it is perturbed, is defined by

$$v_s \tau_r = \Delta x. \tag{1.66}$$

The *process time* τ_p is the characteristic time for the process (expansion) to occur, which for this example is

$$v_p \tau_p = \Delta x, \tag{1.67}$$

where v_p is the speed of the process (the speed at which the piston moves). If Eqs. (1.66) and (1.67) are combined,

$$\frac{v_p}{v_s} = \frac{\tau_r}{\tau_p}. \tag{1.68}$$

For a process to be quasistatic requires that

$$\tau_r \ll \tau_p, \tag{1.69}$$

which, from Eq. (1.68), implies that

$$v_p \ll v_s. \tag{1.70}$$

Expansion (or compression) is therefore quasistatic if the speed of the piston is much less than the speed of sound in the gas. We take this to be a *necessary* condition. Whether or not it is *sufficient* is difficult to say with certainty. The only test is acceptable agreement between measurements and theory based on the quasistatic approximation. What is "acceptable" is not absolute.

We snuck in another idealization and approximation. We implicitly assumed that the piston speed is constant, always has been, and will be forever, whereas at some point a *real* piston, and hence the gas, had to be accelerated from rest to a constant speed, then decelerated. Acceleration and deceleration are not counted as contributing to the work. And we implicitly assumed that the only external force on the gas is a consequence of interaction with a piston. What about gravity? If the center of mass of the gas moves downward, work must be done on it by gravity, no different from a ball dropped from rest. This work is negligible (see Problem 3.88).

Fuel-air mixtures are compressed and expanded in the cylinders of engines. If the rotational speed is 60 revolutions per second (3600 rpm), the total distance traveled completely downward or upward by a piston (the stroke) is of order 0.1 m, and hence the process speed v_p is about $10\,\mathrm{ms}^{-1}$, Usain Bolt's average speed in his world record 100-meter dash, more than 30 times smaller than the speed of sound in air. Even the seemingly rapid compressions and expansions in the cylinder of an engine are quasistatic (we support this bald assertion in Section 3.3).

An apparently quasistatic process is blowing up a latex party balloon (see Chapters 2 and 3 for several balloon problems). A rough estimate of the process speed is $0.1\,\mathrm{ms}^{-1}$ (see also Problem 2.48).

Thermodynamicists can be agnostic about the existence of molecules and need not sweat over the microscopic details of how the energy of a gas increases because of interaction with a moving piston. We swap a microscopic description involving myriads of collisions of gas molecules with a moving piston for a macroscopic description in which only the macroscopic variables p and V appear. But without much sweat we can interpret compressional work microscopically.

Consider a gas compressed in an insulated cylinder fitted with a piston. Molecules colliding with its surface are *adsorbed* (Section 2.1)—reside on it for a very short time—then are *desorbed* at velocities uncorrelated with their incident velocities. Except in textbooks, *individual* molecules do not obey the law of specular (mirror) reflection nor are their kinetic energies conserved (see Fig. 2.3).

Denote by $\mathbf{v}_a$ the incident velocity of an adsorbed molecule and by $\mathbf{v}_d$ its desorbed velocity. If $\mathbf{v}_p$ is the piston velocity, the desorbed velocity relative to the cylinder is

$$\mathbf{v}_c = \mathbf{v}_d + \mathbf{v}_p. \tag{1.71}$$

The average square of $\mathbf{v}_c$ is

$$\langle v_c^2 \rangle = \langle v_d^2 \rangle + 2\langle \mathbf{v}_d \rangle \cdot \mathbf{v}_p + v_p^2 \approx \langle v_d^2 \rangle + 2\langle \mathbf{v}_d \rangle \cdot \mathbf{v}_p \tag{1.72}$$

because $v_p \ll v_d$. If we assume that the temperature of colliding molecules is equal to the piston surface temperature and that desorbed molecules are in thermal equilibrium with it, $\langle v_d^2 \rangle = \langle v_a^2 \rangle$, and hence

$$\langle v_c^2 \rangle > \langle v_a^2 \rangle, \tag{1.73}$$

because $0 < \langle \mathbf{v}_d \rangle \cdot \mathbf{v}_p \ll \langle v_d^2 \rangle$.

Although the random kinetic energy of the colliding molecules does not change, their *total* kinetic energy increases because of collisions with a *moving* piston ($v_p > 0$), which is transformed by hidden processes into increased internal energy of the gas. For the piston to transfer kinetic energy it *must* be moving with a finite speed. Even if low, this only increases the time during which the energy of the gas is increased. If the cylinder (and piston) are (ideally) insulated, any increase in the kinetic energy of the gas is *cumulative*. If $\langle \mathbf{v}_d \rangle \cdot \mathbf{v}_p < 0$ then $\langle v_c^2 \rangle < \langle v_a^2 \rangle$, which corresponds to an expansion and a decrease in internal energy. We discuss in more detail in Section 2.4 what is meant by $\langle v^2 \rangle$, even small changes of which result in readily measurable temperature changes with profound consequences for life.

Each successive group of molecules that collides with a moving piston has its average kinetic energy increased a bit. The slower the piston, the smaller the increase but the larger the number of groups. As a consequence, the increase of internal energy is proportional to the distance traveled by the piston if it is small compared with the greatest possible distance (Section 3.3).

All analogies are imperfect, some more imperfect than others, even to the point of being laughable. But analogies may make Eq. (1.72) clearer. You are standing at the edge of a road and as a car passes, you are able to grab and hang onto a door handle, thereby acquiring a velocity relative to the road. Eventually you let go and fly off into a road sign, demolishing it because of your increased kinetic energy. But if you run to catch up with a car moving more slowly than you, hop on, then off, your kinetic energy will have decreased. These processes are mediated by forces difficult to quantify but must exist.

Cessation of piston motion and vanishing of mass-motion translational kinetic energy does not necessarily imply vanishing of *all* nonrandom motion. We can extend the analysis in Section 1.3 to deriving the angular momentum $\mathbf{L}$ about the center of mass of a system of point masses

$$\mathbf{L} = \sum_i (\mathbf{x}_i - \mathbf{X}) \times m_i(\mathbf{v}_i - \mathbf{V}) = \sum_i \mathbf{x}_i \times m_i \mathbf{v}_i - \mathbf{X} \times M\mathbf{V}. \tag{1.74}$$

$\mathbf{V} = 0$ does not necessarily imply that $\mathbf{L} = 0$ or is conserved. A disc rotating at a constant angular speed has angular momentum, no mass-motion translational kinetic energy, but mass-motion rotational kinetic energy *and* internal kinetic energy. Disc brakes transform rotational energy into internal energy. Even in closed air spaces there may be circulating currents. Another example is in our bodies. If we are standing still our translational kinetic energy (relative to the ground) is zero but the mass-motion kinetic energy of our circulating blood is not. Unless we spring a leak, this mass motion goes nowhere.

Absolutely still air is yet another idealization, never realized in Earth's atmosphere or even unventilated, unheated empty rooms (Section 7.3). The surface temperature of humans is usually higher than air temperature, resulting in plumes driven by temperature gradients, and hence density gradients, in air acted on by gravity (Section 3.5). We all are ensheathed in imperceptible buoyant plumes from which we cannot escape.

In many processes Q or W *predominates*. For example, the temperature of an iron bar in a flame increases because $Q > 0$ and decreases because the bar expands slightly at constant pressure ($W < 0$), but $Q \gg |W|$. The temperature of a gas compressed in real cylinders increases because $W > 0$ and decreases because $Q < 0$; W predominates if $W \gg |Q|$. What is observed is always a consequence of $Q + W$.

1.7 Some Necessary Thermodynamic Concepts and Jargon

An air parcel is an example of a *system,* a term used frequently in previous sections. A system is anything on which we focus attention. Sometimes, but not always, it is defined by a bounding surface. The boundary of a parcel (a fixed set of molecules) becomes progressively less distinct with time. There may be no well-defined boundary between system and surroundings. Yet *all* boundaries are fictitious, existing only on paper, not in nature. Although a block of ice may *appear* to have a definite surface, this is only because of the coarseness of our senses. A tiny fraction of a millimeter into the block is definitely ice, and this same distance above the block is definitely moist air. But along the path from inside to outside the ice there is a more or less continuous change from ice to a liquid-like or quasi-liquid or pre-melted layer or film, neither as solid as ice, even well below its nominal melting point, nor as fluid as water, to moist air. The thickness of these layers is of order tens of nanometers (nm) and decreases sharply with decreasing temperature below zero. They are not mere curiosities but play an important role in a surprising array of large-scale phenomena in Earth's atmosphere and cryosphere. Even though no absolutely sharp boundaries exist in nature, this shouldn't restrain us from imagining them as long as we realize that they are mathematical devices to aid our thinking.

Everything in the universe exclusive of a system is its *surroundings*, which usually means only the part that affects the system, its proximate surroundings. What is meant by system and surroundings is a matter of choice. Butt an iron slab A at 50 °C against an iron slab B at 10 °C. If A is the system and B its surroundings, Q is negative. If B is the system and A its surroundings, Q is positive. If $A + B$ is the system, Q is zero. Each choice is equally valid, but one may make analysis easier and more comprehensible.

Changes of a system are so often of interest that we may forget that the surroundings also must change. Q for a system is $-Q$ for its surroundings. If sufficiently large, its changes may be highly localized. The temperature of hot water cooling in air may be approximately uniform at any instant. What about the air? We measured the temperature near a coffee cup filled with water at $\approx 65\,°C$ using a thermocouple. We wrapped the cup with aluminum foil to minimize radiation (see Section 7.1). The thermocouple probe ($\approx 1mm$) at the surface recorded 59 °C. Three millimeters from the surface, the air temperature was $\approx 35\,°C$. At ten millimeters, it was close to room temperature. Air temperature changed only over a thin boundary layer. The cup was immersed in a large, *almost* constant temperature bath, and hence *the* temperature of the surroundings is ambiguous. What is usually meant is the temperature sufficiently far from the system that it is almost uniform. From experience we can predict that this is the final temperature of the water. How long this takes is determined by the boundary layer thickness, thermophysical properties of the water, and size and composition of the cup (Section 7.1).

If no mass is transferred across the boundary of a system, it is said to be *closed* (or confined). But clear-cut boundaries don't always exist. Where, for example, is the boundary between a system consisting of all the nitrogen molecules in a given mass of air, the surroundings being all the other molecules? A system that is not closed is, not surprisingly, said to be *open*. For example, the mass of liquid water in an open vessel may not be constant because of net evaporation (Chapter 5). An *isolated* system exchanges neither energy in *any* form nor mass with its surroundings. A truly isolated system is yet another idealization, useful but existing only on paper. Open systems are difficult to treat rigorously by thermodynamic reasoning, so we steer clear of them except briefly in Section 5.10.

Because of the possibility of chemical reactions, the composition of a closed system is not necessarily fixed (Section 3.9); because of the possibility of phase changes (Section 5.3) neither is its structure.

A live human is *never* a closed system. Even excluding ingestion of food and water, excretion, and perspiration, we continuously inhale more oxygen than we exhale, and exhale more carbon dioxide and water vapor than we inhale. We cannot survive for more than a few minutes deprived of oxygen, evidence that continuous and rapid oxidation of organic compounds in our bodies drives the molecular motors that keep our internal organs humming and maintains a nearly constant core temperature. Death by hypothermia, a decrease of several degrees of temperature, is much slower (Section 7.1).

Although some "not very large" systems are relevant to atmospheric processes (Section 5.10), in most of this book we implicitly assume "very large" systems, which has no absolute meaning. Finite systems—the only kind other than, possibly, our universe—often lie within closed surfaces with finite areas enclosing finite volumes. If the ratio of area to volume is very small, the system can be considered infinite. But this ratio has no absolute meaning. The ratio of the area of a sphere to its volume is proportional to the inverse of its diameter, which is a *dimensional* quantity and hence its numerical value can be anything. Calculate your height in nanometers and you'll understand our point. Molecules near an interface are in an atypical environment, *not* surrounded only by molecules of like kind. A system is large if the number of molecules within several molecular diameters from an interface is a small *fraction* of the total number of a *great many* system molecules provided that the forces between them decrease rapidly with separation (Section 3.2).

The physical state of a gas might be specified by the positions and velocities of all its molecules. If we know where all the molecules are, where they are going, how fast, and the forces between them, we might think that we know everything possible to know *mechanically* about the gas. Could we obtain such detailed knowledge? Not likely. But even if we could, would we be enlightened or overwhelmed? We say mechanically because gases have nonmechanical properties (e.g. odor, toxicity), and individual molecules have internal states that can change.

At or near sea level each cubic centimeter of air contains about 10^{19} molecules, the state of each (excluding internal states) specified by three space coordinates and three velocity components. We would need about 10^{20} numbers to specify the microstate of this system at any instant. Even massive computers would not be up to the task.

A microstate is well beyond direct experience. It contains vastly more information than we can understand, keep track of, or use. Thermodynamics makes the best of a seemingly bad situation and rescues us from this crushing burden of information by reducing the huge number of variables needed to specify a microstate to only a few specifying a macrostate. These few thermodynamic state variables, such as density, pressure, and temperature, are measurable not only in principle but in practice. This enormous reduction in the number of state variables is possible because measuring instruments are slow to respond (compared with characteristic times at the molecular level) and respond to many molecules. The 1 mm diameter probe of our thermocouple is bombarded by about 10^{21} air molecules in one second. We can think of a thermometer as a probe (or sensor) that samples the energies of sufficiently many air molecules to yield a meaningful average whereas a pressure gauge samples their momenta (see Section 2.1 for more details).

A macrostate is an average over a great many microstates. But all averages are accompanied by fluctuations (deviations from average), which are ignorable if they cannot be detected with ordinary macroscopic instruments (e.g. thermometers).

Microscale fluctuations exist and sometimes lead to observable consequences (e.g. Brownian motion). A price must be paid for being satisfied with much less than the most detailed possible description of a system, but often worth it. By fluctuations at the microscopic scale we do not mean the irregular behavior of turbulent flows, which are macroscopic.

We refer so often to molecules because air and water are the primary raw materials of atmospheric thermodynamics. Air is composed of weakly interacting molecules. Liquid water is composed of more strongly interacting molecules. Ice is a *molecular solid,* composed of discrete molecules that interact more strongly than in the liquid. Even in ice a water molecule retains its identity to an extent. But not all solids are molecular. Most are not. Salt, for example, is *not* composed of NaCl molecules, but rather sodium and chlorine *ions* arranged on a regular crystal lattice. Although NaCl molecules exist as vapor above solid or molten salt, their equilibrium number density (Chapter 5) at terrestrial temperatures is immeasurable. Salt bears no resemblance to its constituents. If it did, eating it would *lower* your blood pressure—to zero. But you can slake your thirst by eating ice. Most solids are crystalline or polycrystalline. Glasses are an exception.

The kinetic part of the internal energy of solids is mostly vibrational, unlike gas-phase nitrogen and oxygen in which it is mostly translational. Atoms in solids are more or less fixed in place whereas molecules in gases are ceaselessly moving. Molecules in liquids migrate but not nearly as frenetically as those in gases.

1.8   Thermodynamic Internal Energy and the First Law

Unless stated otherwise, systems of interest have constant mass-motion kinetic and external potential energies (K_{cm} and P^e), their sum sometimes called mechanical energy. Energies (strictly, energy *changes*) of interest—potential and kinetic—are internal or hidden. This internal energy U is the province of thermodynamics, which extends the principle of conservation of energy in mechanics to hidden forms of energy not specified by macroscopic kinematic variables. If there is ambiguity about what is meant by internal it can be qualified as thermodynamic.

If we were thermodynamic purists, we would pretend that we don't know the origins of U. But we know about molecules, and to pretend that we don't would be dishonest and foolish. For a system of molecules, the thermodynamic internal energy is the sum of the kinetic energy about the center of mass and the potential energy arising from forces between them (Section 1.6). Molecules are not point masses, so have other forms of energy (Section 3.6).

Because atmospheric thermodynamics is dominated by properties of and processes in air (the *internal* potential energy changes of which are negligible), it is easy to forget that they determine processes in Earth's atmosphere we could not do without. Internal potential energy changes are the essence of freezing and melting of water, condensation and evaporation of water, and deposition and sublimation of ice (Chapter 5). The hydrologic cycle is driven by internal potential energy changes of water.

By deleting letters we may shorten *therm*[odynamic intern]*al* energy to *thermal* energy, which is not only molecular translational kinetic energy but may also include contributions from motions and interactions *within* molecules as well as from interactions *between* them. But even if we were to look upon thermal energy as shorthand, we prefer the term *internal energy* because thermal energy has become disreputable by misuse. The late Mark Zemansky, a noted teacher of thermodynamics, asserted that "the concept of thermal energy is by all odds the most ambiguous term employed by writers of elementary physics and by chemists." Although we tend to agree, thermal energy at least is better than "heat energy," especially to denote the random *translational* kinetic energy of an ideal gas. In liquids and solids internal potential energy is not negligible. The best term for U is internal energy because it has not been corrupted by the persistent conflation of heat, energy, heat energy, temperature, and motion. The notion, common in textbooks, that "heat is motion," a relic of the nineteenth century, exemplifies this. Although Eq. (2.10) shows that absolute temperature is proportional to the average translational kinetic energy of an *ideal* gas, this is not the only contribution to U because molecules and even atoms have structure. Moreover, a change in U is not always accompanied by a change in temperature, for example, the isothermal freezing of water (Section 5.3). What is sometimes called chemical energy is potential energy *changes* because of spatial rearrangements of atoms and molecules in chemical reactions, which even in processes where $Q = 0$ is accompanied by changes in kinetic energy (Section 3.9) manifested by temperature changes. U of an isolated system is constant, but not necessarily its composition or phase (solid, liquid, gas) or temperature.

There are two macroscopic ways that the internal energy of a *closed* system can be changed in nonzero time: (1) by interaction with its surroundings at a different temperature (heating or cooling); and (2) by a force on the system acting over a distance (working). Heating, cooling, and working are formed from the verbs to heat, to cool, and to work. They denote *processes* or *activities*, not things. To grasp the difference between a process and a thing requires contemplation. Children have difficulty with abstractions, so are told that heat is a thing. Of their own accord they abandon their belief in Santa Claus and the Easter Bunny by around age nine, but some may never abandon material heat.

We admit time into thermodynamics even though it is absent from most textbooks. It is sometimes said that thermodynamics deals only with equilibrium states of matter. What is equilibrium? Some necessary conditions for *strict* thermodynamic equilibrium are no change with time of thermodynamic variables, no gradients of these variables, and no dependence of the system on its history. No change with time means no macroscopic change. At the molecular or microscopic level change is rapid and incessant. Strict equilibrium is an unrealizable idealization. Because of its dendritic shape, a snowflake, even at a uniform temperature below 0 °C, is not in an equilibrium state and will evolve slowly into an unaesthetic blob, losing its status as a Christmas card icon (Section 5.10).

We are often interested in real processes, and all such processes occur in nonzero time. Weather is a process, not a thing. In his provocative book *Rational Thermodynamics*, Clifford Truesdell asserts that the "range of intended applications of classical thermodynamics is to homogeneous systems, bodies that can be described sufficiently by functions of time only." Thus, we do not pretend that processes do not occur in time, and we write time derivatives of thermodynamic variables with neither fear nor apology. But we do not derive equations to determine explicit time dependences of temperature until Section 7.1. The adjective dynamic is applied to processes or systems characterized by changes in time. If time-dependent variables are taboo in thermodynamics, truth in advertising would demand that it be relabeled as thermostatics so as to not mislead consumers. The *dynamical* Equations (1.1) and (1.6) contain explicit derivatives with respect to time.

For a system to be described by functions of time only implies a limit on its size, which would seem to rule out the entire atmosphere. But because of *local thermodynamic equilibrium* (Section 2.5), we can consider the atmosphere to be composed of many small subsystems in each of which thermodynamic variables such as temperature, pressure, and density depend on time only. If this were not true, measuring atmospheric temperature and pressure profiles would be a fool's errand.

If a closed system interacts with its surroundings, its internal energy changes with time:

$$\frac{dU}{dt} = Q + W, \tag{1.75}$$

where Q is the rate of change of internal energy of the system because of temperature differences between it and its surroundings and W is the rate of change of internal energy because of external forces exerted through a distance, reflecting its origins in mechanics; Q and U are unique to thermodynamics. If we know how U depends on temperature and pressure from independent measurements or theory, then in *any* process between two different equilibrium states the integral of Q follows from the mechanical work done, which is usually much easier to measure.

We do not write Q and W with dots over them, as is sometimes done, implying that they are derivatives with respect to time, which in general they are not. Equation (1.75) is the first law of thermodynamics for a system that interacts with its surroundings. In engineering thermodynamics, the right side of this equation usually is written $Q - W$ because engineers (as their name implies) are often interested in engines, systems that do work *on* their surroundings. To engineers, W is positive if work is done *by* the system. Because this positive work entails a decrease in internal energy, where we write W engineers write $-W$.

For simplicity we refer to *the* temperature of the surroundings, implying that it is uniform and constant, sometimes neither. Although this makes no difference to the first law, it does mean that the definite *symbol* Q may not, and usually does not, have a definite mathematical form. Q could be the sum of $Q_h > 0$, a heating rate more or less constant in time, and $Q_c < 0$, a variable cooling rate. $|Q_c|$ decreases as the temperature of an object approaches that of its surroundings; Eq. (7.52) is a simple example.

For an isolated system $Q = W = 0$, and hence

$$U = \text{const.,} \qquad (1.76)$$

which is the thermodynamic version of Eq. (1.5). But U can be constant for a system that is *not* isolated if $Q = -W$ even if both are nonzero.

Equation (1.75) is deceptively simple, perhaps even misleading. It is a symbolic form of the statement that there are indefinitely many *different* macroscopic ways of changing the internal energy of a closed system by the *same* amount. Two particular (idealized) ways are $Q = 0$, $W \neq 0$ and $Q \neq 0$, $W = 0$. As it stands, this equation has limited use. It cannot be solved to obtain functional relations among measurable thermodynamic variables. Writing *symbols* for heating and working does not imply that these processes can be expressed in terms of such variables and their time derivatives; an exception is Eq. (1.65) for W. The first law is not restricted by any microscopic model of matter. We approached this law by way of a system of interacting point molecules interacting by pairwise forces solely to make it plausible. But it holds equally well for gases, liquids, crystalline solids, and amorphous solids. How U depends on measurable quantities such as temperature and pressure requires either measurements or theories based on microscopic models. Such theories must go beyond classical mechanics even for ideal gases (see Section 3.6).

If we integrate Eq. (1.75) over time:

$$\Delta U = U_2 - U_1 = \int_{t_1}^{t_2} Q dt + \int_{t_1}^{t_2} W dt = W_{tot} + Q_{tot}, \qquad (1.77)$$

where

$$Q_{tot} = \int_{t_1}^{t_2} Q dt, W_{tot} = \int_{t_1}^{t_2} W dt. \qquad (1.78)$$

If $U = U(p, V)$, any curve in the pV-plane represents a process, specified parametrically by $p(t)$ and $V(t)$. In the language of calculus dU/dt is the *antiderivative* of $Q + W$. Its time-integral over the curve depends *only* on its end points, and the time it takes is irrelevant—mathematically but *not* physically. In general, the antiderivatives of Q and W *separately* are not known, but if the antiderivative of one is known, that of the other follows.

We are more specific about the meaning of these integrals in Section 3.1. In a *cyclic* process, a system proceeds from an initial state and returns to it. In such a process $U_1 = U_2$ and

$$W_{tot} = -Q_{tot}. \qquad (1.79)$$

Thus, the first law allows for the possibility of *thermodynamic engines*, systems that can undergo a cycle in which $Q_{tot} > 0$, resulting in work being done *by* the system *on* its surroundings ($W_{tot} < 0$ by our convention). Such engines are powered by temperature differences. Q may be both positive and negative during the cycle as long as its integral is positive (it *must* be both to have $\Delta U = 0$). Equation (1.79) is the origin of statements that heat is "converted" into work, which obscures a deeper physical interpretation. Combustion of a fuel results in a decrease in its internal potential energy balanced by an increase in its internal kinetic energy, manifested by a temperature increase. Temperature is associated with random motion (Section 2.1). W embodies directed motion impelled by a directed force. Thus, Eq. (1.79) expresses transformation of random motion into directed motion. For example, molecules impinging on a piston have *random* velocities in a hemisphere of directions, but the combined result of many collisions is a force (pressure times area) in a *single* direction, the inward normal to the piston face. Its surface imposes order. If the average pressure is greater in the heating part of the cycle than in the cooling part, the net result is work done by what is often called, appropriately, the working fluid, an intermediary that does not change in a cycle. To attribute this work to temperature or pressure differences or a conversion of "heat" is a matter of taste, although only the first two are readily measurable (see Section 4.4 for more about thermodynamic engines). The adjective "useful" modifying work has a meaning only relative to human wants and needs. A hurricane is a thermodynamic engine that does *harmful* work.

The validity of Eq. (1.75) does *not* require the system to be uniform at the beginning or end of or during the processes of heating and working. Moreover, only the *difference* in internal energies can be determined. But to determine how ΔU depends on, say, pressure and temperature, we would have to measure them for initial and final quiescent states in which gradients are negligible and mass-motion, translational and rotational, had ceased.

To this point we have considered Q to originate from *apparently* direct, material interactions (molecular collisions) between a system and its surroundings at a different temperature. The qualifier "apparently" signals that the concept of molecules touching in the same sense that our fingers touch the pages of a book is an invalid extrapolation from the macroscopic to the microscopic (see Intermolecular Collision Rate in Section 2.4). As noted at the end of Section 1.3, the energies of complicated electromagnetic fields, and hence electromagnetic radiation, should somehow be included in the first law applied to matter. The simplest way to do this without explicitly considering fields is to add a radiative contribution to Q, in principle and in practice measurable, interpreted as energy transfer because of temperature differences but without direct interaction. For example, the temperature of the Sun is higher than terrestrial temperatures; the temperature of the walls within a house is usually lower than that of bare human skin. To extend this to radiation from microwave ovens and lasers is a bit of a stretch, but we can say that heating by such radiation is indistinguishable from heating by sources the radiation from which does depend on their temperature. All palpable matter is a source of radiation (emission) *to* and a sink of radiation (absorption) *from* its surroundings, processes essential for atmospheric circulation and the existence of terrestrial life. *Net* radiative heating of Earth's surface is the norm during the day, whereas *net* radiative *cooling* often occurs at night. Although we may write heating for brevity, heating and cooling are inverse processes and to privilege heating is physically baseless. Thermodynamics is as much the science of "cold" as of "heat." If the development of refrigerators for food preservation had preceded the development of steam engines for draining mines and propelling locomotives, today we might see books and courses on "cold and thermodynamics" and "cold transfer" as well as or instead of "heat and thermodynamics" and "heat transfer."

The phrase "perception of heat" is easily replaced with a neutral term, *thermal perception*, roughly graded as cold, cool, tepid, warm, and hot. Sensation and perception are so closely related that they are often used synonymously. But perception is the interpretation by brains of sensations from sensory receptors. Temperature is not a thing, but rather a quality that can be made quantitative by means of operations using instruments.

Electromagnetic radiation transports energy. According to the photon interpretation of radiation, the energies of all photons, the transporters of energy, are positive—but so are the translational kinetic energies of all gas molecules, which are neither hot nor cold. The extent to which electromagnetic radiation *can* warm (raise the temperature of) an illuminated body, and at what rate, depends on its composition as well as the directional spectral power density of the radiation. *Net* radiative warming or cooling of a body depends on the *difference* between absorption and emission. The internal energy of a body *simultaneously* increases because of absorption and decreases because of emission.

Energy transported (or mediated) by electromagnetic radiation is not obviously kinetic or potential. But ultimately it originates from transitions *between* energy levels of matter, for example, the Sun and, indeed all matter, and transported even over great distances through nearly empty space. This is akin to dropping a rock into a still pond with a cork floating on it many meters away. The kinetic energy of the rock is transferred to a water wave that acquires kinetic and potential energy, some of which is transferred to kinetic and potential energy of the cork. Molecules in the rock do not interact directly with those in the cork. It owes its increase in energy to a decrease in the rock's kinetic energy, which itself is transformed potential energy.

An inexpensive aid to your education is an infrared (or radiative) noncontact thermometer that converts radiation emitted by an object into a temperature based on assumptions often, but not always, valid. Point it at the overhead sky and measure the (radiative) temperature of clear and cloudy areas. Measure the temperature variation over your hand. Measure the temperature of a bare thigh, then the temperature of the same area with a blanket draped over it. Address the widespread notion that you "lose 40% of your heat from your head," which conveys that heads are especially torrid, by measuring face, neck, and crown temperatures and skin temperatures on other parts of your body (Problem 7.32). Measure predawn ground temperatures after clear, still nights and cloudy, windy nights and compare them with air temperature. Infrared thermometers may not be accurate partly because of uncertain emissivities (Section 7.1) of targets. Differences in actual and inferred temperatures of the same target at different times may be more accurate. An infrared thermometer infers a temperature averaged over an area (spot) by detecting radiation in its field of view. The spot area increases with distance from the target but is *not* the area of the aiming laser spot. To resolve spatial temperature differences requires getting as close to the target as possible, but there is a minimum spot area, and hence resolution. The possibilities are endless for learning directly about reality. This book is a map of reality, in part a guide to exploring it on foot.

The first law, Eq. (1.75), is *not* restricted to processes without chemical reactions or phase changes, although by omission textbooks may convey this. Chemical reactions (Section 3.9) and phase changes (Section 5.3) could be included as pseudo-heating terms folded into Q. But we restrict Q to energy transfer mediated by temperature differences between a system and its surroundings. We write "differences" (plural) because the temperature of the system and its surroundings need not be uniform and often is not. To call temperature changes in *adiabatic* processes ($Q = 0$) heating or cooling is contradictory. Chemical reactions and phase changes are *internal* and hence logically belong on the *left* side of the first law. They can result in temperature changes even if $Q = 0$ and $W = 0$. Perhaps the simplest remedy for an inadequate vocabulary is to refer to adiabatic temperature increases and decreases rather that heating or cooling when we must be crystal clear. $Q \neq 0$ can cause changes in the structure or composition or both of a system, not just its temperature and density.

Irreverent Thoughts about Heat

Internal kinetic and potential energy are added to mass-motion kinetic energy and external potential energy to complete the energy conservation principle. Temperature changes are consequences of internal energy changes. But heat is *not* a form of energy. We try to use the word *heat* as infrequently as possible, especially as a noun. Heat transfer, shorthand for energy transfer driven by temperature differences, is now too widespread to avoid completely. Heat was once thought to be a weightless, colorless, odorless, highly elastic fluid called *caloric*. According to the caloric theory, bodies were heated and cooled because of the passage of caloric from one to another. This theory had its successes and hence deserves a bit of respect. Slide rules were once widely used for multiplication and division. Pocket calculators and desktop computers have made slide rules antiques. They had their day, did what was asked of them, but were superseded. And so it is with the caloric theory. It is officially dead, and hence should receive a decent burial and remain below ground. Although textbooks officially acknowledge the death of caloric, they continue to breathe life into its corpse. Vestiges of the caloric theory remain in the unit the *calorie*, in phrases such as the amount of heat added to or absorbed by or contained in a body, in quantities such as heat capacities and heat content. In everyday speech, heat construed as a material is the norm. We are often told that "heat rises" but no one says that "cold falls" even though this is logically equivalent and can be observed (Problem 3.72). Caloric is one of the ghostly imponderable fluids—ether, phlogiston, electrical and magnetic fluids—invented in the eighteenth and nineteenth centuries in physics and chemistry. All are now dead—except caloric, which lurks in the shadows under a new name (heat). Material heat survives probably because caloric fluid, although nonexistent, is more concrete than abstract energy. Frigorific atoms once had adherents but are just as imaginary as calorific atoms.

We do not say that heat is added to a system. Instead we say that the system is heated. There is no such thing as the amount of heat in a system. We have encountered the notion that heat is conserved, which implies that cold is not. Heat added to a system or cold removed from it are logically equivalent. Conservation of energy demands that Q for a system must be balanced by—Q for its surroundings. The energy of a system can increase or decrease if it interacts with its surroundings. Although working can change the energy of a system, we usually aren't told that it has had work added to or removed from it and hence it contains more or less work. Why isn't work thought of as an imponderable fluid (*laboric?*) that can pass between bodies? Why does heat "flow," "leak," or "drain" but work does not? Why does "work transfer" sound strange but "heat transfer" does not? Heating and working are on an equal footing in Eq. (1.75).

The notion that a body "loses heat" is a sloppy metaphor conveying that bodies possess what some textbooks emphasize they do not. Heat sometimes means temperature, sometimes energy. If two bodies with different temperatures interact *only* with each other they eventually come to the same temperature but *not* the same energy or even energy per unit mass (unless their compositions are identical). This is the fundamental difference between temperature and "heat" meaning energy. *Measured* equality of temperature signals equilibrium, the end of a process at the macroscopic level. Fortunately for us, Earth is *not* in temperature equilibrium with the Sun.

A modern version of Joseph Black's assertion (late 1700s) that "no matter has ever been found so cold as to be destitute of heat" is that all matter possesses a measurable temperature. There is no need to refer to the "internal heat" of, for example, Earth. If by internal heat is meant that the temperature deep in Earth is greater than surface temperatures, this only defines something as "hot" if its temperature is higher than an arbitrary standard and "cold" if its temperature is lower. No standard temperature divides "hot" from "cold."

To show that there is no such thing as the amount of heat in a body, we partly filled a small plastic bottle with water and let it come into equilibrium with its surroundings. We measured the water temperature, capped the bottle, and shook it for a while before uncapping it. Then we measured the temperature. We did not insulate the bottle but held it by its cap. We shook 150 ml of water, *erratically*, not *continuously* for ten minutes. The temperature increased by 4.4 °C. We also spun 200 ml of water for 50 seconds in a food processor. The temperature increased by 0.2 °C (see Problem 3.81).

These are examples in which the temperature of water increased yet without any heat being "added" to it. The water did not interact with its surroundings at a different temperature. Work was done on the water, resulting in an increase in its internal energy, which was manifested by a temperature increase, which could have been obtained by bringing the water into contact with an object at a higher temperature. Measurements made on the water could not tell *how* its temperature increased (absent paranormal theories of water having memory). By measuring the temperature increase we could infer ΔU, and hence W_{tot}, the integrated work [Eq. (1.78)] assuming $Q \approx 0$. But the integrand W would be beyond reach.

Desperate measures sometimes are taken to keep the notion of heat as a substance alive. For example, the notion that "heat is energy in motion." Those who assert this may acknowledge that it is meaningless to talk about the amount of heat in a body. Yet although body A has no heat in it, nor does body B at a higher temperature, when they interact, the heat transferred from B to A exists during, and only during, its transit. This is the reverse of the paranormal phenomenon of teleportation by which an object supposedly is transported between two points but has no existence on its journey between them. Paranormal heat is conceived of as existing neither in A nor in B but only on the journey between them. This view of heat transfer is an example of adhering to a comforting description long after its expiration date. Because so many people have talked and continue to talk about heat as if it were a substance, a corporeal form for it must be invented. Using work as a noun is at least consistent with its definition as a force acting through a distance, but heat cannot be given such a tidy definition except by resorting to paranormal arguments.

For sake of a concrete image, consider a large chunk of concrete cooling in air. We can define the (average) heat flux density from the chunk as Q/A, where A is its surface area. The quantity so defined has the dimensions of a flux *density*, something (in this instance energy) per unit *area* and time. So far so good. We stumble and fall, however, if we try to identify this flux density with the flow of something tangible. To demolish the notion of a heat flux density as a flow of a substance, we consider flux densities in general. We try to consistently distinguish between flux and flux density, which have *different* dimensions, those of energy flux are energy per unit time (power). Mass, mass *density*, and weight are often confused even though their dimensions are different. How often have you heard that "lead is heavier than water" or that "air is lighter than water?"

Consider a space more or less uniformly populated with n identical objects per unit volume. Take them to be bugs all flying in the *same* direction with speed v. Each bug carries a property, mass of bug juice, m_{bj}, say. The axis of an imaginary cylinder in this swarm of bugs is parallel to their flow. The cross-sectional area of the cylinder is A and its length is vt, where t is a time interval during which all bugs in the cylinder have passed through one face of the cylinder. Because the volume of the cylinder is Avt, the number of bugs in it is $Avtn$ and the total amount of bug juice is $Avtnm_{bj}$. The flux density of bug juice is defined as the amount crossing a unit area perpendicular to the flow of bugs in unit time: nvm_{bj}. This is a general result for flux densities of quantities to which physical reality can be assigned. The total flux density of a property is the product of the number density of carriers of the property, the speed of the carriers, and the amount of the property each one carries (e.g. mass of bug juice). Although we defined Q/A as a heat flux density (which does not convey the same idea as a flux density of heat), we are stymied in our attempts to find the carriers of this flux density. That is, Q/A in a gas cannot be decomposed into the product of the number density of molecules, times their speed, times the quantity of heat that each carries. We can specify a flux density of kinetic energy because each molecule has a definite kinetic energy. But then we have a kinetic energy flux density, and there is no need to relabel it as a heat flux density. Moreover, an energy flux density usually is a *net* flux density [see, e.g. Eqs. (7.43) and (7.46)], an *unequal* two-way *exchange* of energy in opposite directions. We exchange energy with our environment, usually giving more than we get.

Only nonzero net energy flux densities result in measurable changes. Moreover, molecules transport not only translational kinetic energy but rotational kinetic energy as well as vibrational kinetic and potential energies (see Section 3.6). And a zero net *mass* flux density, as in solids, does not imply a zero net *energy* flux density. Wherever the term heat flux density is encountered it can be replaced by energy flux density. For Bridgman's criticism of heat (or energy) flux density as a "pure invention without physical reality," see Section 7.1.

Heat sometimes is said to be absorbed or emitted, which has engendered the misconception that heat is "really" radiation. What about the statement that hot bodies radiate "heat," the implication being that cold bodies do not? Snow radiates about seven times *more* per unit area than polished aluminum at 100 °C. All bodies emit radiation in varying amounts, be they hot or cold (ambiguous designations absent definite criteria of hotness and coldness). Radiant heat and heat radiation belong on the scrap heap of misleading and ambiguous terms. Emission spectra depend on the temperature, composition, size, and physical state of emitting bodies. Spectral absorption by bodies depends on their inherent properties *and* the spectral flux density of radiation *incident* on them. The long-lived and widespread notion that the infrared part ($\approx 0.7-2.5$ μm) of direct solar radiation far exceeds in heating power the visible part ($\approx 0.4 - 0.7$ μm) reflects a failure to scrutinize spectra of direct solar radiation at Earth's surface (which depend on many factors, especially zenith angle) *and* the spectral reflectivity of what it illuminates (human skin, soils, the leaves of plants, etc.). The "heating power" of radiation is contingent. Indeed, photon for photon, infrared radiation carries *less* energy than visible radiation. Radiation of *any* wavelength is capable of measurable heating given sufficient power and a suitably chosen illuminated body. A supposed distinction between light and heat is a relic of an era before quantitative spectral radiometry, now about 150 years old. But there *is* a distinction between visible (to humans) and nonvisible radiation. Radiant emission by a wood fire (≈ 900 K) is more than a million times greater in the infrared than in the visible, and yet the fire is visible. What is true for the Sun is not true for a fire, even though we can see and are warmed by both. Huddled around a campfire in winter we are warmed directly by radiation and indirectly by air warmed by the fire.

Statements that heat cannot spontaneously flow from a cold body to a hot body may be replaced by something more concrete: if two bodies with different temperatures are placed in contact, and together form an *isolated* system, the temperature of the hotter body always decreases whereas that of the colder body always increases. But the hotter body is not the active partner, the colder body the passive partner. Energy transfer is a *two-way transaction* in which the hotter body gives more than it gets *on average* whereas the colder body gets more than it gives *on average*; unequal energy *exchange* rather than transfer more accurately conveys this. Transfer is the *net* result of an exchange. This is analogous to (net) evaporation (Section 5.1) of water. Evaporation (outgoing) is accompanied by condensation (incoming). In Section 7.1 we derive an expression for energy transfer in a gas because of a temperature gradient by considering energy flux densities in two *opposite* directions. The *difference* between them is the *net* energy flux density. Another example is *net* radiation, where a system and its surroundings, not in direct contact, are at different temperatures. We continuously emit infrared radiation ($\approx 5 - 15$ µm) to a room that absorbs most of it. But only a fraction of the radiation emitted by the room in approximately the same wavelength interval is absorbed by us despite the larger wall surface area relative to ours because most of the radiation emitted by the room (walls, floor, and ceiling) is absorbed by it. The exchange of radiant energy between Earth and Sun favors Earth (per unit area) by a factor of more than a billion. The Sun receives about as much of its own radiation reflected by Earth as emitted by Earth.

Infrared means beyond the red, a spectral region from about 0.7 µm to 1000 µm, but often not qualified. The qualifiers near, middle, and far are not much help because they have different meanings in different fields. "Infrared" unaccompanied by wavelengths is incomplete information. Unless stated otherwise, by infrared we mean $5 - 15$ µm. *Shortwave* is often used for solar radiation, *longwave* for terrestrial infrared radiation, which is ambiguous because roughly half of the solar spectrum outside Earth's atmosphere is in the infrared. Black-and-white "infrared" photographic film is sensitive in the wavelength region 700–900 nm.

Energy exchange between "cold" air and a "hot" surface is a two-way transaction. *Some* gas molecules in the air have energies appreciably greater than average (Fig. 2.9), and hence *individually* can transfer energy to *some* of the surface atoms or molecules. But on average it is the other way around, consistent with thermodynamic laws and observations. Nonradiative thermal interaction between liquids, solids, and gases is always two-way.

Why have we taken so much trouble to purge heat from your scientific vocabulary? An adiabatic process is *defined* as one in which $Q = 0$, often said to be a process in which "no heat is absorbed." An example is combustion, catalyzed by platinum black, of hydrogen and oxygen in an insulated, sealed container. The temperature of the products of this reaction is higher than that of the initial gas mixture. Yet this process is often described as one in which "heat is absorbed" (from where is anyone's guess) or "generated" or "evolved." Contradictions such as this, rife in thermodynamics, are an impediment to understanding. Because the energy of the contents of the container is constant a better description is that the *form* of this energy changes. Energy *transformations* are essential to the existence of Earth and life on it.

Consider two alternative ways of describing adiabatic combustion of hydrogen in oxygen: (1) the temperature is higher following combustion; and (2) heat is generated. The first is a concrete statement that can be verified quickly and directly with a thermometer. The second is abstract, invoking a hypothetical quantity that can at best be inferred indirectly only from temperature measurements. Which do you prefer?

Saying that a temperature increase is a consequence of "heat generation" is not a physical explanation. The total energy of the contents of the container is constant, but rearrangements of molecules in combustion may result in a decrease of internal potential energy (associated with bound configurations of atoms) balanced by an increase of internal kinetic energy, manifested macroscopically by a temperature increase. But temperature can *decrease* in chemical reactions (Section 3.9). Combustion and corrosion are similar, differing by greatly different reaction rates. Combustion results in large temperature increases in short times. Corrosion is so slow that it is nearly adiabatic and isothermal; on a cold morning you can't warm your hands over a pile of rusting nails.

The origins of the frequently encountered assertion that heat is motion go back to well before 1857, the date of publication of Rudolf Clausius's famous paper entitled "The Nature of the Motion which we call Heat," which begins with a reference to a previous paper "in which heat is assumed to be a motion." For almost 200 years this has been the dogma faithfully passed on even by those who argue that there is no such thing as the amount of heat in a body.

According to kinetic theory, gas molecules are in incessant motion at speeds greater than the fastest trains. This theory has been verified by many experiments. In fact, verification is so conclusive that rival theories have been swept into oblivion, and we forget—or, more likely, never were told—that well into the nineteenth century the prevailing microscopic model of a gas, at least in Britain and France, was an assemblage of molecules held *motionless* in place by mutually repulsive forces. The triumph of the kinetic theory is not that it established heat as motion but that invisible motion in gases at the microscopic scale is ceaseless and breathtakingly fast by the standards of sluggish humans.

In Chapter 2 we show that the temperature of an ideal gas is proportional to the mean kinetic energy of its molecules. But if heat is motion and temperature is motion, it would follow that heat and temperature are essentially the same, a return to the state of confusion prevailing a century or two ago when the distinction between heat and temperature was vague.

It is easy to deal with the definition of heat: it does not exist, so why try to define it? Heating (cooling), however, denotes macroscopic *processes* in which the internal energy of a system changes by virtue of a temperature difference between it and its surroundings. Why "macroscopic?" Because we invoked temperature. Only a collection of a great many molecules can be said to have a temperature.

We know of no examples in which invoking a substance called heat leads to increased physical understanding. Consider the common experience of rubbing your hands together to warm them on a cold morning, an example of *frictional working*. We could describe this as the generation of heat (even though $Q \approx 0$). Or we could simply say that the temperature of the skin on our hands increased. The second statement can be verified quickly and easily (e.g. with an infrared thermometer); the first cannot. The mechanism for the temperature increase (a fleeting 1–2 °C) is beyond our understanding based on possible macroscopic measurements. Calling this mechanism the generation of heat adds nothing to our understanding. At the macroscopic level, one hand exerts a force on the other through a distance, resulting in working, which raises the internal energy of both, manifested macroscopically by a *measurable* temperature increase. You might retort by saying that generation of heat is just a short way of summarizing this process. An equally short way is to say that skin temperature increases.

Authors blithely switch back and forth between heat, energy, heat energy, thermal energy, heat content, temperature, and other terms to mean the same thing. We have often seen the assertion that "heat is a form of energy." But it is rare to see "work is a form of energy," even though this follows logically from the first law. This jumble of inconsistent terms, aside from indicating unsureness, violates a fundamental rule of technical writing: avoid ambiguity and promote clarity. In other words, once you define a wiffle, it must remain a wiffle and not, on a whim, be relabeled as a waffle. Because the concept of energy, which is *conserved*, whereas heat is not, is so central to thermodynamics, wherever heat can be replaced by energy we gain in clarity and economy.

A single term—*internal energy*—is physically descriptive, neutral, and applicable to all matter, which, regardless of its temperature has internal energy, even Antarctic glaciers in the depths of winter. The internal energy of a fixed mass decreases with decreasing temperature. But at constant temperature, the internal energy of most liquids and solids increases with increasing pressure.

The essence of thermodynamics is not that it is the study of heat but rather of phenomena we may qualify as thermal to signify that temperature is a *central* concept, unlike in mechanics, and electromagnetism, in which it is *peripheral*. Heating ($Q > 0$), cooling ($Q < 0$), and working are distinct macroscopic energy transfer *processes*, more like the transfer of ideas than of some kind of tangible stuff. If we tell you something, your brain is changed as is ours. You and we remember what we said, possibly for a long time. There must have been physical *changes* in our brains even though nothing physical was transferred. Changes in the *internal* energy of a system and its surroundings are analogous. Nothing material is transferred. Where this analogy breaks down—as analogies usually do—is that knowledge is not conserved. Your gain is not equal to our loss. If we throw a ball to you and you catch it, a material form of energy (mass–motion kinetic energy) *is* transferred, as is mass.

Hot and cold describe *subjective perceptions* of sentient beings, but the temperature dividing hot from cold is *indefinite*. We learn the meanings of these two words from childhood experiences. Internal energy is a *quantity*, but only *differences* can be inferred from measurements of other quantities. Temperature is a measurable *absolute variable*, often related to internal energy but different from it. For example, liquid water in equilibrium with ice at the *same temperature* 0 °C has a greater internal energy (*per unit mass*) than ice (Section 5.4). The molecular interpretation of U is a sum of motional (kinetic) and positional (potential) internal energies. Motional energies include rotational and vibrational energies of *single* molecules. The intermolecular potential energy contribution in liquids and solids is not negligible. In 1931 John Lennard-Jones wrote that "Temperature is a manifestation of kinetic energy, cohesion of potential energy, and the interplay between these two forms of energy is responsible for many of the observed physical properties of matter." Heat and work are *not* "forms of energy"; kinetic and potential energy are. Temperature and internal energy are unique to thermodynamics. The qualifier internal conveys that this kind of energy is not *directly* perceptible to our senses. But we may be able to perceive its changes *indirectly* with a thermometer.

Unfortunately, heating and cooling often denote temperature changes (positive or negative) even if $Q = 0$. If $W \neq 0$, ΔU can be positive or negative, and so can the temperature change. $\Delta T > 0$ is often denoted as "heat generation," thereby perpetuating the confusion between temperature and heat. In chemical reactions with $Q = W = 0$, ΔT can be positive *or* negative. Calling these "exothermic" and "endothermic" reactions, respectively, is dressing up a temperature increase or decrease (all that can be measured readily) in jargon. If an exothermic reaction is the "release" of heat, it follows that an endothermic reaction is the "capture" of heat or perhaps the "release" of cold. The temperature of an isolated system can increase *or* decrease, its energy conserved but possibly *transformed.* Quantity of heat is *not* a fundamental concept. It is not conserved. Heat carries the connotation of temperatures that are (indefinitely) "high."

Despite the official demise of the caloric theory, heat (but rarely cold) in everyday discourse still connotes an ethereal fluid. But as noted by Gilbert N. Lewis, "science demands some refinement of common ideas, and often a more limited and technical use of language." This is especially true of thermodynamics, the two most important concepts of which, energy and entropy, are often used in confusing ways.

If criticism in this and other chapters nettles you, contemplate the words of John Roche:

> History can suggest that certain technical terms, invented a century ago, perhaps, are now obsolete or misleading ... Historical study can reveal ... that certain redundant explanations still linger on in modern textbooks. It can show that valid insights have been lost sight of, or that errors of interpretation introduced long ago into the foundations of a theory may survive unnoticed into the present."

We sometimes see "heat" written in scare quotes, implying skepticism or disagreement without saying why. We are not so restrained. We are doing only some necessary, but long-deferred, janitorial work. Faulty and misleading explanations, confusing and inconsistent terminology, have escaped the broom for too long. Don't shoot the janitors!

How Does One Measure "Amount of Heat?"

Without knowing how physical quantities are measured, at least in principle if not in detail, equations are black marks on white paper. One definition of a unit of heat, the calorie (gram or "small" calorie), is the "amount of heat" necessary to increase the temperature of one gram of pure water at standard atmospheric pressure from 14.5 °C to 15.5 °C. This seems simple enough until you impertinently ask, how can one measure the "amount" of such a ghostly entity? One can't, at least not by collecting it as it flows into a bucket. All that can be measured *directly* is work done in different processes and resulting temperature changes. For example, measure the work done (approximately) adiabatically on a fixed mass of water such that its temperature increases from 14.5 °C to 15.5 °C. Then *postulate* that the same increase could have been obtained by heating ($Q > 0$) without *any* work being done, the essence of the first law. This yields the "mechanical equivalent of heat," an archaic term now downgraded to a mere conversion factor between different units for the *same* quantity, namely, *energy*. Thus, the specific heat capacity of water (Section 3.2) can be measured by working rather than heating. The SI unit for energy is the joule (J) to honor James Prescott Joule for his precise measurements of the mechanical equivalent of heat by measuring temperature changes corresponding to measured amounts of work. He did not take pains to insulate his system but corrected for (integrated) $Q \neq 0$, which was small relative to total work done because temperature differences were ≈ 1 °C. He believed on religious grounds that energy is conserved: it is "absurd to suppose that the powers with which God has endowed matter can be destroyed any more than that they can be created by man's agency." He was a science hobbyist whose day job, managing a brewery, informed his precise temperature measurements. He determined to within 0.5% what became the BTU (British Thermal Unit), 778.2 foot-pounds, about 1055 J. More important, he obtained similar values for different liquids and solids, evidence of the consistency of the first law and that it holds for several materials and ways of doing work on them. He showed that the same temperature increase could be obtained by mechanical means (working) as by heating. Keep in mind that the "unit of heat" was defined as a specific temperature increase of a specific mass of pure water. His experiments were a milestone in understanding the energy concept, even though he did not use it. He confused "force" with what is now called energy, and "heat" with temperature. This understandable state of confusion more than 170 years ago should have vanished completely by now but hasn't.

The *thermochemical calorie* is now *defined* as exactly 4.184 J and is independent of any material. Other "calories," not substantially different, are in use. The SI unit for power (energy per unit time) is the watt (W), the units of Q and W. Different units for energy changes because of observably different macroscopic processes are relics from an era when energy and its conservation were not well understood.

One way to increase the temperature of water is to put it in a rigid, insulated container. An electrical resistor connected to an external power supply (e.g. a battery) is immersed in the water. Everything within the container is the system, which interacts with its surroundings only by way of the power supply. The temperature increase is often called Joule heating, but it would be more accurate to call it *Joule working*: an electric field exerts forces on charges (usually electrons) in a resistor. These accelerated mobile charges transfer kinetic energy to relatively immobile ions, atoms, and molecules. Underlying *all* measurements are theories, sometimes kept under wraps in the hope that no one will notice. The *theoretical* total *work* done on the resistor in a definite nonzero—not an indefinite "infinitesimal"—time t is $I^2 Rt$, where I is the steady current in it and R its resistance. The details of this *process* are buried in R, which depends on the composition of the resistor and its geometry. Its temperature increases because of this work, and water in contact with it is warmed because of a temperature difference *within* the system. But it is not necessary to measure this difference in order to determine Q. Overall, this is approximately adiabatic. If it were not, we could not determine ΔU.

Consider the reverse process. Take the system to be a battery attached by a wire to its surroundings, a light bulb. ΔU of the battery is negative, its magnitude $I^2 Rt$. The internal potential energy of the battery decreases because of chemical reactions within it. Chemists would likely call this chemical energy. What about "electrical energy?" The potential energy of interaction within and between atoms and molecules is ultimately electromagnetic even though they may be electrically neutral. If electrons and protons carried no charge, the world as we know it would not exist. Nuclear potential energy changes result from rearrangements of neutrons and protons at the scale of atomic nuclei, which can break apart (fission) or join together (fusion), the energy budget balanced by the greater kinetic energy of the products of these reactions than of the reactants. Potential energy changes in nuclear reactions are much greater than in molecular reactions because the forces between neutrons and protons are much greater than those between molecules even though nuclear forces are of much shorter range ($\approx 10^{-15}$ m). Radioactive decay is a transformation of nuclear potential energy into kinetic energy of emitted alpha and beta particles and energy of radiation (gamma rays). The many qualified "energies" (thermal, heat, chemical, internal, mechanical, electrical, atomic, nuclear, ...) obscure an underlying unity: all are kinetic and potential. But kinetic energy changes in nuclear reactions are often interpreted as conversion of mass to kinetic energy, which can be calculated from the difference between measured nuclear masses of reactants and products. Kinetic energy changes in chemical reactions also could be interpreted as mass conversion but rarely are because mass changes are undetectable. For example, the mass of a water molecule is not measurably different from the sum of twice the mass of a hydrogen atom plus that of an oxygen atom.

Electric power lines carry currents that can be directed to resistors, for example, heating elements of electric stoves, coils in water or space heaters, and filaments of incandescent light bulbs, and do work resulting in temperature increases. Or current can be directed to motors in refrigerators, fans, and pumps. Electric bills are based on time-integrated electrical working (current times voltage) expressed in energy units. The entire power grid from beginning to end is a sequence of energy transformations. The potential energy of water falling over a dam is transformed into rotational kinetic energy of turbines that drive electrical generators that produce currents. Literal "production" and "generation" of energy used in everyday parlance violate the first law.

How does electromagnetic radiation fit into this? A fire, for example, is a reaction between oxygen and a fuel (wood, oil, etc.) resulting in a temperature of the products higher than that of the reactants and electromagnetic radiation, partly visible, that can propagate even through empty space and increase the energy of what it illuminates. Radiation is an intermediary. According to classical theory, the electric field of radiation does work on what it illuminates (absorption). The reverse process, emission, is a spontaneous decrease of the energy of the emitter, which does not violate conservation of energy. Emission sometimes is qualified as "thermal" to distinguish it from luminescence (fluorescence and phosphorescence) nearly independent of the temperature of the source. For example, the fluorescence spectrum of light emitted by fireflies is similar to visible light from the Sun at a much higher temperature than the abdomen of a firefly could endure.

Description and Explanation

In a letter to the editor of *Philosophy* dated January 1936, Norman Campbell suggested "that the distinction between description and explanation is that between values concerning which men agree and values concerning which they differ." The ratio of specific heat capacities γ for different gases over a range of temperatures (Section 3.3) can be measured with the help of macroscopic theory; the result is a table of numbers. This is strictly a description, an account of facts unaccompanied by an explanation of *why* γ has the values it does and *why* they are different for different gases. We need not explain why if our only aim is to use γ for some purpose. If no one ever asked for the causes of physical phenomena, there would never be any explanations. We are severely critical of ones that can be shown to be illogical, contradictory, or downright idiotic, especially if they have been known to be for 100–200 years. Although many explanations are demonstrably incorrect, none are absolutely correct or complete. The test of a good *provisional* explanation is that it does not have to be unlearned; it is correct subject to stated caveats and may be subsumed in more complete explanations. Microscopic quantities may follow from macroscopic quantities given a microscopic theory, and vice versa. But two or more microscopic theories may predict the same macroscopic quantities *to within experimental error*. In the real world, measurements are delimited by error bars often airbrushed out of textbooks.

A Few Parting Shots

Equation (1.30) is the first integral of a differential equation but does not prove conservation of energy. All mathematical theorems are exact by definition. Proven correctly once, they are proven forever. The only test they must pass is logical consistency with underlying axioms and definitions. Equations purported to be about the real world face a more severe, but less definitive, test: acceptable agreement with measurements within experimental error. And this is always *provisional*. To date no one has shown by experiment that energy is not conserved. If you crave fame, perform one that can be repeated by others showing that energy *sometimes* is not conserved. In the early decades of the twentieth century, highly competent physicists argued that energy might be conserved only on average, not necessarily in individual atomic processes. That idea was shot down by experiment. But the increase of entropy is now thought to hold only on average. A corollary is that the entropy of a *single* molecule (except possibly a macromolecule composed of hundreds or thousands of atoms), like its temperature but not its energy, is meaningless.

Temperature is inherently a statistical concept applicable only to ensembles of a great many molecules (Section 2.1).

Energy conservation follows from the symmetry of the fundamental equations of classical mechanics and quantum mechanics, which make no distinction between past and future. They are time-reversal symmetric. But this does not imply that the real world *must* be. These equations are models of reality, and we cannot expect exact correspondence with reality. The only exact model of the universe is the universe itself. The final chapter on energy conservation may never be written.

A slightly cynical view of conservation of energy is that if it appears to be violated, a new form has to be invented to save the law. For example, the electrically neutral neutrino, now known to have a mass a miniscule fraction of the electron mass, was postulated by Wolfgang Pauli in 1930 to account for an apparent violation of energy conservation in the decay of atomic nuclei by electron emission (beta decay). A quarter of a century later Frederick Reines and Clyde Cowan announced the direct detection of the elusive neutrino, a remarkable experimental accomplishment (for a first-hand account, see Reines's 1995 Nobel Prize lecture, "The Neutrino, from Poltergeist to Particle"). Bridgman, not one to shrink from criticism, including of himself, asserted that conservation of energy is "much more than a convention, and it enables us to make predictions in a much wider range of circumstances than those that compelled us to recognize the existence of new forms of energy."

No experimental science is exact, thermodynamics not excepted. But the laws of thermodynamics as they stand are too abstract and are not well-suited to help us grapple with physical reality. To extract anything useful or interesting from them requires idealizations and approximations, which we shamelessly make in the rest of this book, as we must to obtain testable relations between measurable macroscopic variables such as temperature, pressure, and density. The justification for this rough-and-ready approach is *good enough* agreement with a necessarily *finite* set of experiments and observations. Experiment connotes something done in a laboratory under *controlled* conditions. By this criterion, meteorologists don't do experiments, nor do astronomers and geologists. The term "numerical experiment" is misleading if not dishonest. The difference between *simulating* and experimenting is that experimenters have calluses on their hands and dirt under their fingernails.

Throughout this book we try to follow Henri Poincaré's example and "clearly distinguish between what is experiment, what is mathematical reasoning, what is convention, and what is hypothesis."

If you are irked by the many caveats and qualifiers in this chapter (with more to come), keep in mind that repetition does not guarantee correctness. An example is the unqualified textbook parabolic trajectory of an object propelled obliquely upward from Earth's surface. Drag is often neglected, not because it is necessarily negligible but because it is messy, depending on the size, shape, orientation, and speed of the object and air density. Absent drag, the trajectory is *elliptical* not parabolic. NASA rocket scientists know this. Some—but not all—elliptical trajectories that intersect Earth are *approximately* parabolic.

Annotated References and Suggestions for Further Reading

For two additional thermodynamic "laws" (the 4th Law: No piece of experimental apparatus works the first time it is set up; and the 5th Law: No experiment gives quite the expected numerical result), see Thomas S. Kuhn, 1961: Measurement in modern physical science. *Isis*, Vol. 52, pp. 61–93. This is worth reading in its entirety. He notes that there are no consistently applied or agreed-upon criteria for "reasonable" agreement between theory and experiment, which varies between fields and over time.

We assume that students understand the concept of mass, but to be sure, we recommend Herbert L. Jackson (1959) "Presentation of the concept of mass to beginning physics students," *American Journal of Physics*, 27, pp. 278–80. Jackson is a man after our own hearts in that he explains mass by discussing how it behaves rather than by way of a largely meaningless definition (such as "quantity of matter"). But there is no consensus on how it should be defined. And there has been plenty of hand-wringing over this. Even without a completely adequate definition and rock-solid understanding of the concepts force and mass, they nevertheless can be used to calculate the orbits of planets and satellites with remarkable accuracy.

For insights into conservation of energy, see Milton A. Rothman (1988) *A physicist's guide to skepticism*. Buffalo, NY: Prometheus, 1988, chs. 1 and 4. He notes (p. 24) that "Energy, as an abstract concept, is truly a human invention, as opposed to the things that really exist in nature."

If you agonize over the meaning of energy, take solace in accepting that "It is important to realize that in physics today, we have no knowledge of what energy *is*." Richard P. Feynman, Robert B. Leighton, and Matthew Sands (1963) *The Feynman lectures on physics*, Vol. I. Boston: Addison-Wesley, 1963, section 4.1. "Energy is the capacity for doing work" is a mantra aimed at placating students who insist on tidy definitions of physical concepts.

For a discussion of various kinds of perceptual constancy (size, color, brightness, and shape), see Herschel W. Leibowitz (1965) *Visual perception*. New York: Macmillan.

Section 1.5 was inspired by Problem 1.9 in Frederick Reif (1967) *Statistical physics*. New York: McGraw-Hill.

For more about pairwise additive intermolecular forces, see Jacob N. Israelachvili (2011) *Intermolecular and surface forces*. 3rd edn. New York: Academic. He writes (p. 128) that "Unlike gravitational and Coulomb forces, van der Waals forces are not generally pairwise additive ... [the] effect is usually small."

For a brief discussion of the consequences of departures from pairwise additivity of molecular forces, see chapter 10 in Anthony J. Stone (2013) *The theory of intermolecular forces*. 2nd edn. Oxford: Oxford University Press.

In Section 1.6 we give a classical interpretation of the difference between heating and working. For a quantum-mechanical interpretation see Frederick Reif (1965) *Fundamentals of statistical and thermal physics*. New York: McGraw-Hill, secs. 2.6 and 2.7. Briefly stated, he interprets heating as a consequence of the change of occupancy of fixed energy levels, whereas working is a consequence of the change of energy levels of fixed occupancy. For an elementary discussion of quantized energy levels, see Section 3.6 of this book.

For a thorough treatment of the pre-melting of ice, see J. G. Dash, A. W. Rempel, and J. S. Wettlaufer (2006) "The physics of pre-melted ice and its geophysical consequences," *Reviews of Modern Physics*, 78, pp. 695–741.

For critical discussions of the confusing and contradictory uses of heat, see Mark W. Zemansky (1970) "The use and misuse of the word 'heat' in teaching physics," *The Physics Teacher*, 8, pp. 295–300; Myron Tribus (1968) "Generalizing the meaning of 'heat'," *International Journal of Heat and Mass Transfer*, 11, pp. 9–14; Richard A. Gagglioli (1969) "More on generalizing the definitions of 'heat' and 'entropy'," *International Journal of Heat and Mass Transfer*, 12, pp. 656–60.

For a concise history of the caloric theory, see Sanborn C. Brown (1950) "The caloric theory of heat," *American Journal of Physics*, 18, pp. 367–73.

The story of the hard-won battle to establish *Molecular reality* is rarely told. Yet we are fortunate that one of the leading generals, Jean Perrin, has left a beautifully written popular account of his experimental investigations in his book *Atoms*, published in 1920 by Constable. Perrin's contributions are placed in their historical perspective by Mary Jo Nye in her 1972 *Molecular reality* published by American Elsevier. Perrin's technical paper on Brownian movement and molecular reality is reprinted in Mary Jo Nye (ed.) (1984) *The question of the atom*. Los Angeles: Tomash—a superb collection of original papers. For an even thicker collection, see the two volumes edited and annotated by Henry A. Boorse and Lloyd Motz (1966) *The world of the atom* published by Basic Books. If you take the time to read the seminal papers by our illustrious predecessors, you will be moved to tears by the clarity and individuality of their writing, both of which largely have been banished from modern scientific papers.

The impossibility of analyzing "the flow of heat into the product of a density and a velocity, something which we can always do when we are dealing with the flow of matter" is noted by Percy W. Bridgman (1961) *The nature of thermodynamics*. New York: Harper, p. 23. Clifford Truesdell's assertion about the "range of intended application of classical thermodynamics" is in his *Rational thermodynamics* published by McGraw-Hill in 1969, p. 4. The first five pages of this book are great fun to read. He doesn't mince words.

Medicine is another field in which myths and bogus explanations are widespread. For debunking the myth that you "lose 40% of your heat from your head," see Rachel C. Vreeman and Aaron E. Carroll (2008) "Festive medical myths,"

BMJ, 337: a2769 (December 17). If you are wearing a thick, down-filled, hooded parka, but are barelegged, then what?

For a critique of potential energy see Eugene Hecht (2003) "An historical-critical account of potential energy: Is PE really real?" *The Physics Teacher*, 41, pp. 486–92. He describes clearly and succinctly the evolution over many years of the terms kinetic and potential energy and the principle of conservation of energy and summarizes arguments that potential energy is not a real physical quantity. He concludes that it is real but superfluous because there is only kinetic energy and mass.

For a good critical treatment of how to deal with the "challenges to physics teachers associated with the terms *heat* and *heating*" see Harvey S. Leff and Carl E. Mungan (2018) "Isothermal heating: purist and utilitarian views," *European Journal of Physics*, 39, 045 103 (11 pp).

Our irreverent thoughts about heat are shared by one of the pillars of physics education, Robert Romer, longtime editor of the oldest and best-known journal devoted to physics education at the college level, who published a strongly worded condemnation of heat as a noun. And he cites us as allies. See his editorial: Robert Romer (2001) "Heat is not a noun," *American Journal of Physics*, 69, pp. 107–09.

Garrett Hardin, a biologist writing in of all places a psychiatric journal, 1957: "The threat of clarity," *American Journal of Psychiatry*, 114, pp. 392–96, displays a better understanding of the difference between a thing and a process than is the norm in textbooks. He notes that heat is a false substantive. It is "like the singing of a bird, an activity, process—and not a substantive or object like apples or oranges. Its category has rather more to do with verbs than with nouns." We tend to *reify* concepts in physical theories, to make a thing out of what is not (e.g. the reification of heat). This is understandable given the limitations of our minds and of the languages they have created. But it is nonetheless a source of confusion and ambiguity.

In the Refences for Chapter 4 we cite a series of papers by Harvey Leff in *The Physics Teacher*. In the first of these he notes that "largely because of remnants of the erroneous and obsolete caloric theory, the history of heat and work and the language used for it has been a tortuous one."

Subjective perceptions of hot and cold are created by the brain as a consequence of signals sent along nerve fibers from localized hot and cold sensors on skin. Although it can be anesthetized, thermometers cannot, which is one reason they were invented. Skin is the main natural human temperature sensor, although it is imprecise, limited, and unreliable. For more about thermal sensation, a fascinating subject in itself, see Herbert Hensel (1982) *Thermal sensations and thermal receptors*. Springfield, IL: Charles C. Thomas.

For details about thermal plumes around humans, see Brent A. Craven and Gary S. Settles (2006) "A computational and experimental investigation of the human thermal plume," *Journal of Fluids Engineering*, 28, pp. 1251–57.

For more about measurements of the "mechanical equivalent of heat," see James Prescott Joule (1884) *The scientific papers of James Prescott Joule*. London: The Physical Society, pp. 298–328.

The "thermodynamics reference most frequently cited in physics research literature" is the first edition of Herbert B. Callen (1985) *Thermodynamics and an introduction to thermostatistics*. 2nd edn. Hoboken: John Wiley & Sons. Many physics, chemistry, and engineering students (undergraduate and graduate) were taught thermodynamics using this book. Another widely used textbook with many editions is Mark W. Zemansky and Richard H. Dittman (1997) *Heat and thermodynamics: an intermediate textbook*. 7th edn. New York: McGraw-Hill. Both books use the term *internal energy* for U, *not* heat energy.

Cold is often left out in the cold, in contrast to its counterpart, heat (a hot item that basks in the warm glow of its admirers). To bring cold in from the cold, see the collection of essays on cold edited by Kostas Gavroglu (2014) *History of artificial cold, scientific, technological, and cultural issues*. New York: Springer.

Now largely forgotten, a school of thought called energetics emerged in the late nineteenth century, mostly in Germany. "The energeticists believed that scientists should abandon their efforts to understand the natural world in mechanical terms and should give up atomism as well in favor of a new worldview based entirely on relations among quantities of energy." This was not the view of cranks or pseudo-scientists; scientists of the highest caliber defended it. But their ideas were swept away in the rising tide of atomism. For a brief history see the entry on energetics in John L. Heilbron (ed.) (2003) *The Oxford companion to the history of modern science*. Oxford: Oxford University Press.

For John Roche's comments about misleading terms and errors of interpretation in textbooks, see John Roche (ed.) (1990) *Physicists look back: studies in the history of physics*. Bristol: Adam Hilger, p. 144.

The quotation at the end of this chapter is from Henri Poincaré's (1952) *Science and hypothesis*. Mineola: Dover, p. 89. Now more than 100 years old, this book aimed at a general audience is still in print and worth reading.

If we piqued your interest in Brownian motion, we recommend Phil Pearle, Brian Collett, Kenneth Bart, David Bilderback, Dara Newman, and Scott Samuels (2010) "What Brown saw you can too," *American Journal of Physics*, 78, pp. 1278–89. The authors took pains to get the story right. Contrary to a widespread misconception, Brown did not observe the motion of pollen grains, but the much smaller granules *within* the grains released by them when they burst. But Brown was not the first, and knew it, to observe what is now called Brownian motion.

Our prickly comments about heat echo those by Percy W. Bridgman (1960/1927) *Logic of modern physics* New York: MacMillan, p. 125: "it is not possible in the general case to find anything which we can call heat as such ... The heat concept is in the general case a sort of wastebasket concept, defined negatively in terms of the energy left over when all other forms of energy have been allowed for." Bridgman's fame was for *experimental* thermodynamics, but he thought deeply about its foundations. Most scientists who take a philosophical interest in their science are theoreticians. They have the free time. Experimentalists are too busy trying to get their instruments to work.

We mention briefly in Section 1.6 the long-distance transport of Saharan dust, a topic of unusually wide scope meriting its own book. For just one sample from an extensive literature, see Joseph M. Prospero and Olga L. Mayol-Bracero (2013) "Understanding the transport and impact of African dust on the Caribbean Basin," *Bulletin of the American Meteorological Society*, 94, pp. 1329–37.

The fabled slaying of the material heat dragon by Count Rumford, "a figure beloved of the writers of textbooks," is debunked by the masterly work of Robert Fox (1971) *The caloric theory of gases*. Oxford: Oxford University Press, pp. 4, 99. Rumford's observation of the boring of cannons did not spell the immediate rejection of the caloric theory and was (at best) a footnote. Fox treats rival caloric theories and their decline in detail.

For Bridgman's (1959) assertion about conservation of energy (Section 1.8), the result of long reflection that caused him to change and refine some of his previous views, see P. W. Bridgman's "'The Logic of Modern Physics' after Thirty Years," *Daedalus*, 88, pp. 518–26.

Thomas Kuhn notes that "The history of science offers no more striking instance of the phenomenon known as simultaneous discovery. Already we have named twelve men, who, within a short period of time, grasped for themselves the essential parts of energy and its conservation. Their number could be increased but not fruitfully." See Thomas Kuhn (1977) *The essential tension: selected studies in scientific tradition and change* Chicago, IL: University of Chicago Press, pp. 66–104. We agree. Thermodynamics is a soup that has been spoiled by too many cooks.

For a proof that trajectories are approximately parabolic (absent drag) if their maximum curvature is much greater than Earth's curvature, see Lior M. Burko and Richard H. Price (2005) "Ballistic trajectory: parabola, ellipse, or what?" *American Journal of Physics*, 73, pp. 516–20.

To learn the full story about the long and circuitous path to what is now called Newtonian gravity, we highly recommend Eugene Hecht (2021) "The true story of Newtonian gravity," *American Journal of Physics*, 89, pp. 683–92. This is the kind of history that students should learn, instead of the potted histories in most textbooks. Newton was not a lone genius. He was influenced by and acknowledged others, some now mostly forgotten.

The statement about a meta-law of physics (Section 1.6) is in Benny Lautrup (2011). *Physics of continuous matter: exotic and everyday phenomena in the macroscopic world*. 2nd edn. Boca Raton: CRC Press, p. 1. He conveys many good physical insights and uses theory to explain everyday phenomena rather than wallow in solutions to differential equations. Although he is a theoretical physicist, he is not oblivious to the world around him.

The concept of a fluid particle in fluid mechanics is not as straightforward as implied in textbooks, which often dodge exactly what is meant by this. For a critical review of it see Ricardo P. Pecanha (2015) "Fluid particles: a review," *Journal of Chemical Engineering and Process Technology*, 6, p. 238.

The quotations in Section 1.1 about the second law of thermodynamics being a "battlefield" are in Helge Kragh (2008) *Entropic creation: religious contexts of thermodynamics and cosmology.* Farnham: Ashgate, p. 1, 2.

Problems

1. The prototype for elastic collisions is often taken to be those between billiard balls. But if you were to walk blindfolded into a pool hall, you would know immediately that collisions between billiard balls cannot be perfectly elastic. How?

2. In the late 1970s a California inventor invented a machine (called SLX) that split ordinary tap water into hydrogen and oxygen. The hydrogen then could be burned in oxygen and used to run cars. A U.S. Senator's science advisor told reporters that SLX "works ... It is such a profound breakthrough in thermonuclear physics that it will astound the world by providing an unlimited energy source." Suppose that a friend of yours who is seriously considering investing in SLX comes to you for advice. What advice would you give? It must be accompanied by arguments that your friend, who is not a scientist, can understand.

3. Hydroelectric power is generated when falling water turns turbines. Many years ago, so the story goes, a member of the U.S. Congress came up with a scheme to increase the power output of hydroelectric power plants. Water generates power, he reasoned, by falling from a high to a low elevation. Therefore, it is obvious that more power can be produced by pumping water after it has passed through the turbines to its previous high elevation and allowing it to again flow downward through the turbines. What do you think of this scheme? Does your answer depend not on how *much* power can be produced but how it is *distributed* over a 24-hour period?

4. We began this chapter with the assertion that no "air free of water vapor exists naturally on Earth." Try to refute this assertion by scanning climatological records for the driest places on Earth. Try to find a station that has reported a relative humidity of exactly zero.

5. Rainfall rate, expressed as a depth per unit time, is a volumetric flux density of water: volume of water per unit area and time. How is the mass flux density of rain related to the rainfall rate R? You may assume that all raindrops have the same size and terminal velocity. Generalize this result to real rain, the drops of which are distributed in size (diameter), and hence also in terminal velocity.

6. Hydroelectric power generation exploits the kinetic energy of falling river water. Rain is also falling water, and hence transports kinetic energy. Estimate the power (in watts) that could be generated in a typical rainstorm. To do this problem you need the result from Problem 5 and also have to know, guess, or look up typical values for terminal velocities of raindrops (see References,

Chapter 3), rainfall rates, and the horizontal extent of rainstorms. From having observed rain you should be able to make reasonable guesses for all three quantities. If you do your calculations correctly, you should obtain a fairly large number for the power in a rainstorm. Why, then, do we not directly use falling rain to generate electrical power?

7. As a continuation of the previous problem, estimate the rate at which kinetic energy of rain falls on the entire planet. You need the globally averaged yearly rainfall, which, if you don't know, you should be able to guess within a factor of two. The number you obtain is a reasonable estimate of the maximum hydroelectric power generation possible. Shared equally, how much of this power would be allocated to each inhabitant of our planet? On the basis of your result, what can you say about the feasibility of worldwide hydroelectric power generation?

8. In Section 1.5 we considered the collision between two rigid, impenetrable molecules. In so doing we assumed that total momentum of the center of mass is conserved in the collision if there are no external forces on the molecules. But this leads to an apparent paradox. Consider, for example, the head-on collision between two molecules with unequal (in magnitude) momenta (i.e. the total momentum of the system is not zero). Although the total momentum before and after collision may be equal, if the trajectory of one molecule is reversed, it must at some instant come to rest. And if this molecule comes to rest, so must the other which implies that the total momentum is zero. How do you explain this? No mathematics is necessary.

9. A 0.22 caliber rifle and a 0.22 caliber pistol can use the same ammunition (caliber indicates the diameter of the bullet). Yet the muzzle velocity of a rifle bullet is appreciably greater than that of a pistol bullet (muzzle velocity is the velocity of the bullet as it emerges from the barrel of the rifle or pistol). Why? Estimate the ratio of muzzle velocities of the rifle and pistol. Assume that propulsion of a bullet is a constant-pressure process.

   The principle behind shotguns, rifles, and pistols is essentially the same. When fired, an explosion occurs in the shell. This explosion, which occurs behind the projectiles (bullets or shot), propels them down the barrel.

   After you have made your estimate, you might try to find data on muzzle velocities of rifles and pistols. If there are any discrepancies between measured values and what you calculated, try to explain them.

10. Estimate how fuel efficiency is affected by increasing the maximum speed limit from 85 to 100 kph (assuming that drivers obey these limits) on highways. Assume that at these two speeds engine efficiency is the same. Also assume that the rolling resistance of a car (friction of tires against asphalt, bearing friction, etc.) is negligible compared with wind resistance (drag) at highway speeds, which increases as speed squared. If you are really ambitious, you can do an experiment to determine if your estimate is correct. At least try to design such an experiment. That is, give all the necessary steps you would have to take in order to show a difference in fuel efficiency at the two different speeds.

11. High jumpers can go over a bar in such a way that their center of mass goes *under* the bar even though every part of a jumper's body must go *over* the bar. Please explain. A simple sketch should help. Discuss why it might be advantageous for a high jumper to jump in this way.

12. What power is required to maintain an automobile at a cruising speed of 100 kph on a level road? Neglect all drag except that due to wind resistance. The drag force F_D on an object moving at constant speed v through a fluid is $F_D = \rho v^2 \, AC_D \, / \, 2$, where ρ is the density of the fluid (air in this example), A is the frontal area of the object, and C_D is its (dimensionless) drag coefficient. A good estimate for the drag coefficient of an automobile is 0.25, and you can assume that it is approximately independent of speed. Estimate the frontal area of automobiles by examining a few. Your result (unless you make an error) should come somewhat as a surprise. Why is the horsepower (1 hp $\approx$ 746 W) rating of many automobiles so much higher than needed for cruising even at 100 kph? *Hint*: Calculate the average power needed to accelerate an automobile (you pick the make and model) from 0 to 100 kph in, say, 10 seconds. Neglect drag during the acceleration.

13. What power is required for an automobile to climb a 10% grade at a constant speed of 100 kph? Grade is the vertical distance divided by the horizontal distance times 100%. Pick the make and model of automobile. Compare the power you obtain with the result of the previous problem. What do you conclude on the basis of your calculations?

14. The horsepower of the diesel engines in large trucks that haul freight long distances is not much greater than the horsepower of the engines in ordinary automobiles, and yet these trucks weigh about ten times more. Why can trucks travel at fairly high speeds (often faster than automobiles), but with engines less powerful per unit total mass than those of many automobiles? What price do trucks pay for their comparatively low power-to-weight ratio?

15. Suppose you take a long trip by automobile in which you experience tail winds as frequently as head winds. That is, the average wind velocity is zero. Equivalently, suppose that you drive 100 km in one direction and then 100 km in the opposite direction during which time the wind velocity is constant, and hence the average wind velocity for the round trip is zero. You might think that under these conditions, tail winds cancel head winds with no net effect on fuel efficiency. Show that when the average wind velocity is zero, wind still reduces fuel efficiency. *Hint*: Keep in mind that the speed of a car is measured relative to the ground, as is the speed of the wind, whereas the wind speed v that determines drag (see Problem 12) is that relative to the automobile.

16. Extraction of kinetic energy of wind is among the many schemes for power generation without burning fossil fuels. You don't need to know much about the details of wind energy to estimate how the maximum power generation by a wind turbine depends on wind speed. Please do so. By making a judicious guess as to the area of the blades on a wind turbine, you can estimate the minimum average wind speed required such that a wind turbine of realistic dimensions

could supply an appreciable fraction of the power needs of an average household. *Hint*: This problem is related to Problems 10 and 12.

17. Potential energy is sometimes said to be energy that can be stored more or less indefinitely. What is the greatest length of time you can think of that potential energy has been stored and yet readily transformed into kinetic energy almost instantaneously. Devise schemes for storing kinetic energy and discuss why it would be difficult to store indefinitely. Can you think of examples in which kinetic energy can be stored and what limits its storage time?

18. If 1 kg of, say, gasoline, were to be lifted from the surface of Earth to a distance where its gravitational field is negligible (many Earth radii), the change in gravitational potential energy of the gasoline would be comparable to the change in molecular potential energy upon combustion of the gasoline. What does this result tell you about the magnitude of molecular forces relative to the force of gravitation?

19. If you did Problem 10 correctly, you will have concluded that fuel efficiency decreases with increasing speed. Yet if you wander through the lot of an automobile dealer and peek at stickers on the windows of cars, you'll find two average efficiencies displayed, one for city driving, the other for highway driving. The efficiency for city driving, which is usually slower than highway driving, is invariably the smaller of the two figures. This seems to contradict the result of Problem 10. Resolve this apparent contradiction.

20. A newspaper article on heat pumps informs us that "the electric heat pump is flameless and doesn't use any energy to create heat. It simply transfers heat from areas where it is not wanted into areas where it is." Please discuss. See Problems 11–13 in Chapter 4 for more about heat pumps.

21. By thinking carefully about Eq. (1.56) you should be able to determine that it cannot be correct without qualification. This does not require knowing any physics beyond what was known to Newton. This simple problem is an example of something students should do, namely, interpret equations and probe their limitations, rather than just memorize them.

22. We ended this chapter with Joule heating, our system a resistor immersed in water in an insulated container, the surroundings a power supply. If we take the resistor as the system, how would the process it undergoes be described?

23. One of the authors regularly hikes at a steady pace straight up a "mountain" with an elevation change of about 160 m. His mass is 65 kg. What is the minimum work done? Keep in mind that Q and W are negative. Express your result in kilocalories ("food calories"). What do you conclude? According to the first law, this work must be accompanied by a *decrease* in the internal energy of the "system." What was the nature of this energy decrease? What is the average power output for a 20-min ascent time? Express it in horsepower and compare it with that of small gasoline engines (e.g. walk- behind lawn mowers). A human "engine" is sometimes said to be 25% efficient. How does this change your results? Express your result in plain-donut calories. (See Section 3.9).

If you think about this problem carefully you may learn a lot. Most of the systems we consider in this book are extremely simple. It is a huge jump from gas in a cylinder to the human body, within which there are complex *macroscopic* internal energies of motions that, fortunately, usually don't go anywhere on average. As an aside, the current consensus is that exercise is not especially effective at reducing weight but is at maintaining a weight loss by eating less food. And there are many other health benefits of regular exercise.

24. Equation (1.27) is the assumption that the force between two *point* masses depends only on the distance between them. What about the force between two *real* molecules? Can you envision that it might not be dependent solely on this distance and why? This problem exemplifies something we encourage students to do. When they encounter an assumption in a derivation, implicit or explicit, try to find an example in which it may not be strictly valid. Then try to assess the importance of making an invalid assumption.

25. Aerodynamic drag is often assumed to be "negligible" (without proof) in treatments of the acceleration of bodies falling from rest in a gravitational field. What is really meant is that it is a nuisance. But neglecting drag is an unnecessary dodge. Using the expression for drag in Problem 13, show under what conditions it is negligible. Set up the equation for the acceleration. Assume a sphere sufficiently large that its drag coefficient is constant. First, derive an expression for the terminal speed of a falling body, the speed at which its acceleration is zero, and ask yourself if it makes physical sense. Analysis is simplified by writing the speed, time, and distance in dimensionless form using combinations of g and the terminal velocity. The result is a first-order nonlinear differential equation within the capabilities of students with a year of calculus under their belts. From the solution for the speed one can determine the time over which gt is a good approximation. Pick a few real bodies (density and size) and determine their terminal velocities. Critically examine the assertion that *all* bodies, regardless of size and shape, fall at the same speed. From the speed follows the distance.

    For more about this problem see Gerald Feinberg (1965) "Fall of bodies near the Earth," *American Journal of Physics*, 33, pp. 501–02.

26. Do individual water molecules in air fall in Earth's gravitational field? If so, why don't they accumulate on the ground? Do cloud droplets fall to the ground (in the absence of updrafts)? What about raindrops? What is the difference? This problem requires ideas from Chapter 2 and some careful thinking.

27. As an example of irreversibility consider a mass m dropped from rest from a height h on a flat Earth but with a real atmosphere. Suppose that when m reaches z = 0, time is reversed, and the velocity of m is reversed. All you have to know about drag is that it is opposite the velocity. Show that the maximum height of m is less than h.

28. An object at rest dropped from some height in air experiences drag, a macroscopic force directed opposite to the object's velocity, and hence *negative* work

is done on it by the air. But this means that the object does *positive* work on the air. What might be observable consequences of this? You may have seen something related to this problem. You are driving on a highway behind a large truck moving at a fairly high speed. The shoulder of the road is dirt. What do you observe?

29. Two billiard balls of the same mass approach each other at equal speeds and collide. Assume that the collision is elastic in that the total kinetic energy of each ball is the same before and after collision. But at the moment of collision both balls completely stop and hence their total mass-motion kinetic energy briefly is zero. There must have been a compensating brief increase of some form of energy. What is it?

30. Hydroelectric dams and wind turbines are examples of engines that are *not* thermodynamic in that they do not depend *directly* on temperature differences. Why do we emphasize directly?

31. A useful exercise is to consider in (qualitative) detail the process of inserting the bulb of a liquid-in-glass thermometer into air, say, at a higher temperature. Can you come up with a simple explanation of thermal expansion? *Hint*: Consider a simple pendulum.

32. Estimate the number of molecules in the solid Earth relative to that in its atmosphere. Try to do this without looking up any numbers or even using a calculator or, better yet, with your hands in your pockets. Make plausible guesses. This is an exercise in simple reasoning about an *order of magnitude*.

33. At the end of Section 1.5 we argue that a cue ball on a pool table shot at a stationary ball loses all its kinetic energy, which is transferred to the ball it hits. Show that this follows from Eqs. (1.32) and (1.33).

2

Ideal Gas Law

Pressure and Absolute Temperature

Before diving headfirst into the details of gas pressure and temperature, we digress on the history of the term *gas*. We owe it, and more, to Johannes Baptista van Helmont, born in Brussels in 1579, died there in 1644. Although he entertained notions that today would be judged nonsensical, such as spontaneous generation of scorpions from herbs crushed between bricks, belief in alchemy, and the philosophers' stone, this is true about many of our illustrious predecessors. We can hardly be scornful of him for not knowing 400 years ago what is known today. Because of him and others like him, our understanding of the natural world has progressed beyond superstition and mysticism.

English speakers mangle the Dutch word gouda so that its first syllable becomes *goo*. A more accurate representation of the pronunciation of this delicious cheese might be *howdah*, where the initial consonant is a fricative, produced by forcing breath through a constricted passage. The best evidence is that *gas* comes to us from the Greek by way of the Dutch (Flemish). The initial g is meant to represent, to a speaker of Dutch, the Greek χ, the first letter of *chaos* (χάος), "any vast gulf or chasm, the nether abyss, empty space, the first state of the universe." Van Helmont did not only coin a term, but he also recognized the existence of gases other than air and water vapor. According to Partington, van Helmont "gave the name *gas sylvestre* to the wild spirit (*sylvestris*, of the wood; sylva or silva, a wood or forest), the untameable [sic] gas which breaks vessels and escapes into the air." He observed that if certain chemical reactions occur in a closed vessel, it may be burst by the resulting expansion of gases thereby produced. But he wandered beyond the concrete realm of observation into the dreamland of speculation, considering different volatile products to be the essence of the objects from which they were obtained. We have retained the chemical qualities of gases but abandoned their spiritual qualities, although in uncharitable moments we sometimes associate the gaseous emanations of some of our more odious fellow humans with their character.

Given that chaos originally meant "empty space," it has had an illustrious career engendering the term *gas* and is now attached to a branch of science, one of the pioneers of which is the meteorologist Edward Lorenz, who used chaos "to refer

Atmospheric Thermodynamics. Second Edition. Craig F. Bohren and Bruce A. Albrecht, Oxford University Press.
© Craig Bohren and Bruce Albrecht (2023). DOI: 10.1093/oso/9780198872702.003.0002

collectively to processes ... that appear to *proceed according to chance* even though their behavior is in fact determined by precise laws." Although the motions of individual molecules in a gas in equilibrium may be governed by precise laws, their motion as a whole is chaotic in the sense of complete *microscopic* unpredictability. Yet, out of this chaos comes *macroscopic* predictability in the form of the ideal gas law.

2.1 Gas Pressure and Absolute Temperature: What Are They and What Are They Not?

Volume is usually a clear geometrical concept, whereas pressure and temperature, central concepts in meteorology, are clouded in misconceptions and half-truths. Consider the following assertions:

1. Air pressure is just the weight of the atmosphere above us.
2. The decrease of air pressure with altitude is a consequence of fewer air molecules pushing on those below.
3. Absolute zero is the temperature at which all motion ceases.
4. As air temperature increases so does pressure, and vice versa.
5. Cold air is denser than hot air.

None of these can withstand critical scrutiny. Two are false, and three are false in general, but can be made true with caveats.

We tackle the first by a demonstration. Add a spoonful of water to a beverage can, heat it until the water boils, then quickly recap the can. It will be crushed as it cools, which can be hastened by running water over it. This demonstrates *air* pressure. But sometimes we are told that the *entire* atmosphere above the can crushes it. How can air molecules 10 km, 10 m, or even 1 m from the can interact with it? Would this demonstration have failed if done in a tightly sealed room because the weight per unit area of air in the room is thousands of times smaller than that of the entire atmosphere?

If Earth were flat and in the absence of vertical accelerations (caveats rarely stated) air pressure at the surface would be equal to the weight (per unit area) of the air from the surface upwards (Section 2.7). But the can is crushed *only* because of its *proximate* surroundings, not the entire atmosphere.

Imagine taking an insulated, rigid container of air 1000 Earth radii beyond Earth. The pressure of the air would not change even though its weight would be a million times smaller. Pressure and weight (and mass) are different and should be so treated, but often are not.

Balance a four-meter aluminum rod on a knife edge and gently push the tip of one end down a centimeter. All parts of the rod rotate, apparently simultaneously, by a predictable amount. A line through it intersects $\approx 10^{12}$ atoms *linked* by cohesive forces. Make the same motion in air without the rod. Molecules four meters away do not move. A line through the air intersects $\approx 10^9$ molecules *not linked* to each other. Aluminum is *cohesive*, air is not. Arguments that are valid for solids and liquids are not for gases. A 3000-meter-high cylinder of the strongest concrete would be crushed by its own weight, and its diameter would have to be greater than about 200 m or it would buckle.

Ideal Gas Law

Newton argued that the structure of a gas is crystalline: gas molecules on a lattice are held in (almost) fixed positions by repulsive forces. Because of his authority, this model, accepted by many prominent scientists up to about 1860, the only possible interpretation of "heat (temperature) is motion" is *vibrational* rather than translational. We now know that gas molecules are in rapid, straight-line motion between sporadic collisions. To tackle the remaining assertions we need the ideal gas law, which we derive by molecules-in-motion arguments. The molecules in a truly ideal gas do not interact with each other *at all*. A real gas approximates an ideal gas to the extent that potential energy *changes* are negligible. A better qualifier is *idealized*, signifying the product of human imagination. A brutally honest qualifier is *fictitious*. An ideal gas is sometimes called a perfect gas, but perfection is meaningless without knowing by what criteria. To add to the confusion, by a perfect fluid is sometimes meant one with zero viscosity and thermal conductivity (Chapter 7) and characterized solely by a mass density and isotropic pressure. But an ideal gas (a fluid) has a *finite* viscosity and conductivity and so in this sense is *not* a perfect fluid. Why should something be called "perfect" because it makes theories simpler?

Consider N identical gas molecules with mass m in a cubical container of volume V, its edges parallel to a rectangular coordinate system xyz; $n = N/V$ is the statistically uniform number density (concentration) of molecules. For our purposes here we assume that *intermolecular* collisions are negligible, whereas collisions with solid or liquid surfaces (*contramural*) are not. Molecules interact with the *solid walls* of this container in collisions with them. One wall is coincident with the yz plane. Before collision with the wall a molecule has an x component of momentum mv_x, where v_x is the component of its velocity normal to the wall. We assume that molecules undergo no kinetic energy change in collisions obeying the law of *specular* reflection (angle of incidence equals angle of reflection). With these assumptions, the x component of momentum after collision is $-mv_x$ (the y and z components are unchanged), and hence, the change in momentum upon collision is $2mv_x$. As we show later, specular reflection is almost *never* true, nor is kinetic energy conserved in *individual* collisions, despite which by dumb luck we get the right answer. Reflection of molecules by real walls is more like *diffuse* reflection of light by white matte paper.

Newton's second law of motion for a molecule with momentum $\mathbf{p}$ acted on by a force $\mathbf{F}$ is

$$\frac{d\mathbf{p}}{dt} = \mathbf{F}. \tag{2.1}$$

A molecule that collides with a wall has its momentum changed, and thus must have experienced a force. But if the wall exerts a force on a molecule, it must exert an equal and opposite force F_x on the wall. From Eq. (2.1), the time integral of this force is the momentum change:

$$\Delta p_x = 2mv_x = \int F_x dt. \tag{2.2}$$

From the bug example in Section 1.8, the flux density of a property is the number density of carriers of that property times their speed and the property of interest. If gas molecules are moving in all directions with equal probability, only half of them at any instant will be moving toward the wall. Thus, the x component of the flux density of molecules impinging on the wall is $nv_x/2$, During the time interval τ, $nv_x A\tau/2$ molecules strike a patch with area A on the wall. The total time-integrated force on this area is

$$\frac{1}{2}nv_x 2mv_x A\tau = \int_0^\tau F_x dt, \tag{2.3}$$

where F_x is the *instantaneous* force on A. The *time-averaged* force during τ is defined as

$$\langle F_x \rangle = \frac{\int_0^\tau F_x dt}{\tau}. \tag{2.4}$$

How large must τ be? In one nanosecond, much shorter than the response time of pressure sensors, $\approx 10^9$ air molecules at sea level collide with 1 mm^2 of wall. The *pressure* on the wall is the average

$$p = \frac{\langle F_x \rangle}{A} = nmv_x^2. \tag{2.5}$$

We implicitly assumed that all molecular speeds are the same, whereas they are distributed about an average, some molecules faster, some slower, but in all directions with equal probability, and so we replace v_x^2 by its average:

$$p = nm\langle v_x^2 \rangle, \tag{2.6}$$

where $\langle\,\rangle$ denotes an average over the *equilibrium distribution* of molecular speeds (Section 2.4). If molecules are moving in all directions with equal probability, averages of the velocity components in three mutually perpendicular directions must be the same:

$$\langle v_x^2 \rangle = \langle v_y^2 \rangle = \langle v_z^2 \rangle. \tag{2.7}$$

Because

$$\langle v^2 \rangle = \langle v_x^2 + v_y^2 + v_z^2 \rangle = \langle v_x^2 \rangle + \langle v_y^2 \rangle + \langle v_z^2 \rangle, \tag{2.8}$$

we can write Eq. (2.6) as

$$p = \frac{2}{3}\frac{N}{V}\left\langle \frac{mv^2}{2} \right\rangle. \tag{2.9}$$

Thus, the pressure of an ideal gas is two-thirds its internal kinetic energy density. Because of only linear momentum changes in collisions with *walls*, p depends on the average molecular *translational* kinetic energy independently of rotational and vibrational energies (Section 3.6). This holds equally well for monatomic, diatomic, and polyatomic molecules *if* intermolecular interactions are negligible.

We define *absolute temperature* T by

$$kT = \frac{2}{3} \left\langle \frac{mv^2}{2} \right\rangle, \tag{2.10}$$

where k, usually called *Boltzmann's constant* (despite having been introduced by Planck), is $\approx 1.38 \times 10^{-23} \mathrm{JK}^{-1}$. We *assume* that it is the same for all gases. Joseph Black called what we now recognize as absolute temperature "the [unknown] beginning of heat." T is *inherently* a statistical concept. The qualifier absolute is warranted because internal kinetic energy, unlike mass–motion kinetic energy (Section 1.3), is invariant. With this definition (see Section 2.4 for subtleties), the ideal gas law Eq. (2.9) can be written

$$pV = NkT. \tag{2.11}$$

This law applies only to molecules in an ideal world. It is independent of their shape and the number and kinds of atoms per molecule. All real gases disobey this law (often a misdemeanor). It becomes a better approximation as n decreases but not indefinitely: at some point p and T become meaningless. Equation (2.11) is a particular *equation of state*. Its simplicity is only apparent. It describes to good approximation a system of impenetrable complexity. This form of the *theoretical* gas law, from which others can be derived, is (within limits) in accord with the *experimentally* determined law. We can write Eq. (2.11) as

$$p = m \frac{N}{V} \frac{k}{m} T = mn \frac{k}{m} T = \rho RT, \tag{2.12}$$

where $\rho = mn$ is the mass density and k/m is the *specific* (or *individual*) *gas constant R*. Another form is $pv_m = R^*T$, where $v_m = M/\rho$ is the molar specific volume, M is the molecular weight, and R^* is the *universal gas constant* (Section 2.8); $pv_m/R^*T = p/\rho RT$ is sometimes called the *compressibility factor*. It is 1 for ideal gases and so its difference from 1 is a measure of the departure of real gases from ideal. It is very close to 1 for atmospheric nitrogen and oxygen because pressures are much lower than their critical pressures (Section 5.4).

Equation (2.11) is called a law—but *not* a fundamental law—because (within limits) it expresses a *regular* relation between measurable gas properties: p, V, N, and T are not all independent. This is a regularity of *averages* with exceedingly small relative deviations.

$p = \rho RT$ usually is the appropriate form of the ideal gas law for atmospheric applications because pressure, temperature, and mass density vary from point to point in an atmosphere without a definite volume. Nevertheless, Eq. (2.11) is the more general form. An implicit assumption underlying this equation is that p and T

are uniform in V (yet another idealization). In Section 2.5 we show how to extend the ideal gas law to a nonuniform gas (e.g. Earth's atmosphere).

The form of Eq. (2.11) is independent of any specific molecule. Although all ideal gases are created equal, every liquid is unlike every other. Liquids are outlaws. There is no "ideal liquid law," a universal relation between pressure, density, and temperature. Because of the ubiquity of liquid water, its relevance to so much science and technology, and its "anomalous" behavior, much experimental and theoretical effort has been directed at determining its equation of state. The results are far from simple. Gases and liquids are fluids, their flows described by the Navier–Stokes equation. But because of nonnegligible interactions between molecules in liquids separated by average distances of order molecular sizes, a liquid is *not* simply a very dense gas.

Equation (2.11) implies that equal volumes of *all* gases at the same temperature and pressure have exactly the same number of molecules. This holds only for *ideal* gases. For two different *real* gases—the only kind—the number of molecules for given p, V, and T is *not* exactly the same.

Boltzmann's constant is sometimes said to be "small." However, it has dimensions, and hence, indefinitely many *numerical* values depending on the choice of units. A dimensional quantity can be said to be small or large only *relative* to a standard. Compared with the energy of macroscopic bodies, kT is indeed small, even for $T = 4000$ K. At 298 K, kT is about 10^{13} times smaller than the kinetic energy of a golf ball whacked off a tee.

Because Eq. (2.11) is a relation between four thermodynamic variables, p, V, N, and T, there is *not* a unique relation between any two of them *unless* the other two or their ratio is constant. Equation (2.13) is a relation between *three* thermodynamic variables: p, ρ, and T. And yet how often have you heard that cold air is denser than warm air? From this unqualified premise a *logically* unassailable conclusion follows: if temperature decreases with altitude, density must *increase*, contrary to experience. Air density goes up as temperature goes down *provided that pressure is constant*, which it is not, except approximately in shallow atmospherical layers.

The ideal gas law is not a relation between any two variables without a constraint on a third. Figure 2.1 shows a continuous trace of air temperature and pressure during a week. There is no obvious relation between them. Sometimes temperature goes up and pressure goes down. Sometimes temperature goes up and pressure goes up.

Two other ways of writing the ideal gas law follow from Eq. (2.11):

$$pv = RT, \qquad (2.13)$$

$$p = nkT, \qquad (2.14)$$

where the *specific volume* $v = 1/\rho$ is the volume per unit mass (V/Nm). Meteorologists often use α for specific volume because v denotes a horizontal component of the wind velocity. We also use v for molecular speed but in contexts where it is unlikely to be confused with specific volume.

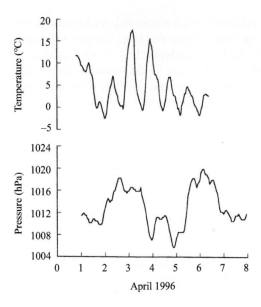

Fig. 2.1: Pressure and temperature in State College, Pennsylvania during a week in April 1996. Note the lack of any obvious relation between pressure and temperature.

We can rewrite Eq. (2.10) so that it expresses a relation between absolute temperature and mean molecular kinetic energy:

$$\left\langle \frac{1}{2}mv^2 \right\rangle = \frac{3}{2}kT. \tag{2.15}$$

This equation is behind well-meaning (but misguided) attempts to explain absolute zero—admittedly a difficult concept—by saying that it is the temperature at which all motion stops. To show the danger of extrapolating the ideal gas law to absolute zero, we appeal to Eq. (2.11). A fixed number N of gas molecules is in a cylinder fitted with a movable piston so that the pressure within the cylinder is constant. According to Eq. (2.11), as T goes to zero the volume of the gas goes to zero until poof!—it vanishes.

Equation (2.15) implies, erroneously, that all motion comes grinding to a halt at $T = 0$. But this equation applies only to *ideal* gases. At sufficiently low temperatures, all *real* gases liquefy, almost all solidify (helium at normal pressures is the exception) and are far from behaving ideally. Nitrogen at standard pressure becomes liquid at 77 K and solid at 63 K. Moreover, according to the simplest and least-controversial form of the third law of thermodynamics, absolute zero is unattainable. We say no more about this law because it is not relevant to atmospheric thermodynamics. Nor is *zero-point energy*: in the limit $T \to 0$, crystalline matter still has vibrational kinetic energy, which cannot be explained by classical physics. Although this also is not relevant to atmospheric thermodynamics, throughout this book we do not tell you what we know to be untrue even if it may not matter. Equation (2.15) can neither prove nor disprove that all motion ceases at absolute zero.

The definition of absolute temperature as proportional to the average translational kinetic energy of an ideal gas in equilibrium provides a mechanical (or kinetic) interpretation of temperature. But this does not mean that temperature is "just" motion. Like many concepts worth understanding, temperature has many facets. And wouldn't it be jarring to call the random motion of molecules in a gas at $-100\,^{\circ}$C "heat?"

Although no gas is strictly ideal, the air in Earth's atmosphere is to good approximation. We know of no terrestrial atmospheric phenomenon that is a consequence of air departing from ideality. If it did, atmospheric thermodynamics, dynamical meteorology, and aerodynamics would be more complicated (Section 5.4). If you are tempted to invoke departures from ideality to explain the behavior of the gaseous atmosphere, think again. We must qualify this by restricting ourselves to altitudes where air follows simple laws, which is true up to 100 km and probably well beyond. But we eventually reach the domain of rarefied gas dynamics and thermodynamics where laws lower down no longer are in force.

Before continuing to grapple with the pressure concept, we address a disturbing point dropped in the previous section: we got the right answer by wrong reasoning.

Almost invariably, the ideal gas law is derived under the assumption that the angle between the velocity of an incident molecule and the normal to the surface with which it collides is the same as the angle between the velocity of the reflected molecule and this normal. This assumption of specular (mirrorlike) reflection seems plausible until we realize that on a molecular scale even highly polished surfaces are exceedingly rough. The normal to an *ideal* planar surface is the same everywhere over the surface, whereas the normal to *real* surfaces changes erratically from point to point (Fig. 2.2). The average distance between peaks and valleys (roughness height) of a slate chalkboard is about 1000 times greater than the diameter of an

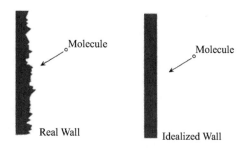

Fig. 2.2: Reflection of a gas molecule by an idealized perfectly smooth wall and by a real wall with surface roughness larger than molecular scale.

air molecule. To expect it to obey the law of specular reflection upon collision with a chalkboard is like expecting a soccer ball fired at the Alps to rebound at exactly the same angle it made initially with the mean normal to the surface.

What really happens is that a gas molecule impinging on a surface is incorporated on it and rattles around for an exceedingly brief time but long enough that, if it returns to the gas, it does so with equal probability in any direction. A molecule retains no memory of its incoming trajectory, and hence, its collision with the surface is the opposite of perfectly specular. If so, why did we get the correct ideal gas law? We were lucky. Pressure is an average over *many* collisions, so the precise details of *individual* collisions are irrelevant. Each molecule is not specularly reflected, but gas molecules arrive at and leave the surface with equal probability in all directions. To every incoming molecule with a trajectory making a particular angle to the (average) normal to the surface, there is an outgoing molecule (but not, in general, the same molecule) with a trajectory making that same angle with the normal (Fig. 2.3). It is *as if* each collision were specular, whereas only the ensemble of collisions is.

In writing this second edition we discovered that in 1857 Rudolf Clausius (see References, Chapter 5) recognized that specular reflection of molecules and equal probability of angle of reflection (as depicted in Fig. 2.3) yield the same gas pressure, which is an average over a distribution of velocities. He also recognized that the *average* speed of the incident and reflected molecules must be the same, even though *individual* incident and reflected speeds need not be (he implicitly assumed that the wall and gas temperature are the same). An incident molecule does not necessarily interact with only a single molecule on a surface densely packed with molecules, but rather with many. The interaction is not binary and even if it were, the energy of the molecule on the surface would have to be unchanged for the kinetic energy of the incident molecule to be unchanged, which follows from Eq. (1.33).

The brief stay of gas molecules on surfaces after colliding with them is called *adsorption*. In the indexes of ten textbooks on college-level physics, some old, some new, some famous, two in double-digit editions, some difficult to lift without a hoist, adsorption is absent. We conclude from this that, to physicists, adsorption does not exist, which if true, neither would they. Adsorption is akin to condensation without a phase change; desorption is akin to evaporation (Section 5.1).

Adsorption is not an obscure phenomenon of little consequence. Without it, energy exchange between gases and solids would not be possible. We would live in an environment of mostly radiant energy exchanges. Although the ground and atmosphere are directly heated by radiation, no matter how hot the ground, there could be no energy exchange between it and air by convection. Radiative heating or cooling of air would be nonuniform, which would lead to temperature gradients, and hence, density gradients. But nothing on the ground, including us, could be warmed or cooled by air. Adsorption is essential for human existence. We could not exchange energy with the air around us if molecules did not reside very briefly on our skin, giving or taking energy before leaving in random directions.

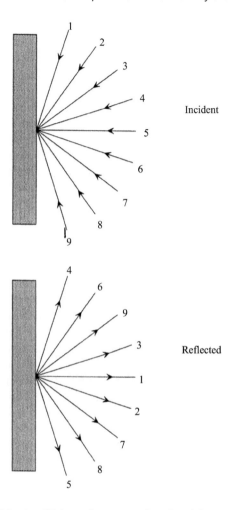

Fig. 2.3: In each individual collision of a gas molecule with a wall, the law of specular reflection (angle of incidence equals angle of reflection) is *not* necessarily obeyed but is by an ensemble of many collisions if all angles of incidence and reflection are equally likely.

A Perspective on Units

Physical quantities are in SI units unless noted otherwise. MKS (meter-kilogram-second) is *part* of SI (Système international), which has seven base units. Although you are likely to be familiar with the SI units of force, energy, and power, your sense of the magnitudes of these quantities may be hazy. Here's a little test to check your SI awareness. Lift this book and estimate its weight in pounds (if you live in the United States) or its mass in kilograms. This should be easy. But approximately how many newtons (abbreviated as N) must you exert to lift the book against the force of gravity (if you estimated the mass in kilograms, you have an edge on this one)? If you estimated 10 N without pause, you not only are SI literate, but also you have the right attitude to be a good scientist because you rounded off to an even number. If you estimated 8 N, you might do well guessing weights at carnivals. If you estimated 7.6 N, you cheated and weighed the book. For rough estimates, round to numbers easy to use in calculations. Giving a physical quantity to many digits does not necessarily indicate precision. For example, a broomstick diameter of 2.1047576 cm looks impressive to those who are easily impressed by the appearance, rather than the substance, of science, but it is either dishonest or silly or both: the diameter of a broomstick need not and cannot be known to eight digits—or even four. Unfortunately, the widespread use of pocket calculators has resulted in digit inflation. We have often asked students to *estimate* a quantity that under the best of circumstances can be measured to at most two digits, and yet have been given a four-digit (or more) answer. If every superfluous digit resulted in an equal number of lashes with a cane, digit inflation would soon come to a halt. But we have to content ourselves with tongue lashings.

Early in their education, science and engineering students should have been given a brief lecture on *significant digits* in *measured* as distinct from *defined* quantities. Yet we are amazed at how many mature scientists must have been asleep or absent during that lecture. We have seen "estimates" varying by a factor of ten, but the range presented to three or four digits. Modelers unconstrained by reality may report values to *four* (or more) digits for computed quantities that can be known to *one* at best. A typical example of digit inflation is in a paper in a dermatology journal, the authors of which are probably safely dead, too old to care, or unlikely to read this book. Their paper reports to *six* digits the number of skin cells (corneo-cytes or squames), of order 100,000 and with dimensions of order micrometers, shed per cm^2 from various parts of the body. Here, six-digit precision is not only impossible, but it is also useless. To kick the habit of excessive digit addiction, imagine how you would react to a forecast high temperature of 34.1788 °C. And this is as good a spot as any to criticize "normal" air temperatures. They are *averages* over specified time intervals. For example, the daily maximum temperature at a given station is a time series with moving (or running) averages that depend on duration: one, two, three ... decades. The antonym of normal is abnormal, which connotes something undesirable or worse. This is especially true of normal applied to core body temperature (see Section 7.1, where we discuss hyperthermia and hypothermia).

To reinforce your sense of what a newton means, each time you lift an empty glass or coffee cup, think a few newtons. Soon you'll acquire a familiarity with this unit of force.

For the next part of the SI literacy test, lift this book from your waist to above your head. Roughly how much energy was required? If you estimated 10 joules (1 J = 1 Nm) without hesitation, you pass. If you looked for a ruler to measure the distance the book was lifted, you missed the point. By the end of this book you should have learned that physical understanding often requires only rough estimates. For some purposes, g can be taken to be 10 m s^{-2} and π approximated as 3 or even 1. Although precision may be needed in some calculations, even then we should be impressed by neither the number of digits displayed, nor the size of the computer used to produce them.

In nutritional science, the SI unit for energy is the kilojoule (kJ), 4184 "small" calories (cal) or 4.184 "large" calories (kcal). The dietary Calorie or Cal is one kcal.

If you now have a better sense of force and energy in SI units, move on to power. If you lift this book over your head in one second, you do work at the rate of about 10 watts (1 W = 1 J s^{-1}). Imagine powering a 100 W light bulb using a hand-driven generator. This is not so easy, roughly equivalent to lifting this book over your head ten times each second, about seven times faster than we could do—and we couldn't do it for long.

Horsepower (hp) is still used for rating internal combustion engines and electrical motors. One mechanical horsepower is 745.7 W; one metric horsepower is 735.5 W (75 kg × 9.8066 m s^{-2}). The metric ton (tonne) is 1000 kg, corresponding to a weight of about 9800 N. Probably the most familiar energy unit (non-SI) because it appears on electrical utility bills is the kilowatt-hour (kWh), 3600 kJ. In the United States, the average annual energy usage per household is about 10,000 kWh, approximately ten 100 W light bulbs burning day and night. The units of energy flux density are J s^{-1} m^{-2} or W m^{-2}. To put this in perspective, the direct energy flux density (*irradiance*) of solar radiation at the top of Earth's atmosphere is about 1365 W m^{-2}. A typical household appliance (e.g. toaster, iron, hair drier) is rated at about 1000–1500 W. For solar panels to generate this much electric power at 20% efficiency from this irradiance would require a collection area of *at least* four square meters. Because solar radiation is the ultimate source of the atmosphere's energy, an upper limit for *any* atmospheric energy flux density is ≈ 1000 W m^{-2}, both positive *and* negative.

The SI unit of pressure is the Nm^{-2} or pascal (Pa), which is a bit more difficult to comprehend. If you place your hand on a flat surface and lay this book on your hand, the pressure you experience is roughly 1000 Pa (0.15 lb in^{-2}). If you were to drive a nail through the book and support it on the nail, the pressure exerted would increase to 10^6 Pa or more, but highly localized. One pascal is small compared with pressures we normally experience. For example, we estimate the average pressure on our feet when we are standing still to be approximately 15 kPa. If this seems like a high pressure, it is because we have small feet.

The newton, pascal, joule, and watt are SI units *derived* from the three *base units* meter, kilogram, and second. The SI unit for frequency is the hertz (Hz), one cycle per second. Authors who still use cycles per second were asleep under a tree for 20 years.

Atmospheric pressure at Earth's surface is about 100 kPa. Unfortunately, this doesn't help with our perception of pressure because we rarely sense this magnitude. We may, however, become aware of pressure *differences* as our ears pop when we change altitude quickly. A painful demonstration of the force exerted by pressure differences is what we sometimes experience flying in unpressurized aircraft if we have a head cold. Blocked eustachian tubes can prevent the pressure on both sides of the tympanic membrane (eardrum) to equalize as the altitude of the aircraft changes. But even flying in pressurized aircraft can be painful for someone with a bad cold because the cabin is pressurized to only about 85 kPa, so pressure changes as large as 15 kPa are typical during landing and takeoff. If you have ever flown on commercial aircraft with young children as passengers, you know that landings can sometimes be accompanied by screams and cries of pain. The eustachian tubes of children are relatively small and do not allow the pressure across the eardrum to equalize easily, especially when the external pressure increases during landing.

Although the SI unit of pressure is the pascal, meteorologists often use a holdover from the cgs system, the *millibar* (mb), where 1 mb = 100 Pa (hence, 1 bar = 10^5 Pa). Even though the SI system is used for calculations, upper-air constant-pressure contour lines (isopleths) are labeled in millibars (e.g. 850, 700, and 500 mb), as are sea-level isopleths on surface weather maps. A millibar is identical to the SI unit the *hectopascal* (hPa) = 100 Pa; the decimal unit prefix *hecto*, the inverse of *centi*, is flanked by its more familiar prefixes *kilo* (×1000) and *deca* (×10). To be SI-compliant requires only a word change.

Other units of pressure are still used. Atmospheric pressure is often measured with mercury barometers and expressed in inches or millimeters of mercury. Tire and other pressure gauges in the United States are read in pounds per square inch (psi), whereas in neighboring Canada, gauges indicate kPa. The *atmosphere* is still a widely used unit of pressure. It was originally defined as the pressure exerted by a 760-mm column of mercury at 0 °C and an acceleration due to gravity of 9.80665 m s^{-2}. In SI units the atmosphere is *defined* as exactly 101.325 kPa (1013.25 mb), which is close to the original definition. This does not imply that we know, can know, or need to know the average global sea-level pressure to six digits. The global sea-level pressure is difficult to determine precisely. Many of the observations needed for a global average are made above sea level, which requires estimating sea-level pressure from surface pressure measurements. This cannot be done without approximations because the measured surface pressure must be extrapolated to sea level. Furthermore, the approximations needed to make this extrapolation vary from country to country. Observations are not made uniformly over Earth's surface. Many areas over oceans and unpopulated land are not included in estimates of the global average. Finally, global pressure is not constant because of variations in water vapor. The seasonal variation in sea-level pressure is about 0.5 mb. Estimates by Kevin Trenberth (1981) give a global sea-level pressure of 1011.0 mb with an uncertainty of about 0.1 mb. This is close to the 1012 mb estimate made in 1887 by Kleiber using a much more limited data set. For many purposes the approximate average sea-level atmospheric pressure is 1000 mb.

Pressures vary substantially from this average on several time and space scales. For large-scale (synoptic-scale) weather systems (cyclones and anticyclones) with spatial scales of order 1000 km and time scales of order days, there can be variations of more than 40 mb. Extreme variations on the synoptic scale can cover a range of over 200 mb. The highest observed sea-level pressure was 1083 mb at Agata, Siberia in 1961 and the lowest was 870 mb associated with Typhoon Tip over the western Pacific Ocean in 1979. For medium-scale (mesoscale) weather events with spatial scales of one to hundreds of kilometers and time scales less than a day, pressure can vary as much as 10 mb. Some changes within mesoscale systems can occur relatively rapidly, for example, as great as 10 mb at a fixed location over several minutes in association with outflow boundaries (traveling gravity waves) generated by mesoscale severe storms. Superimposed on the synoptic-scale and mesoscale pressure variations are the diurnal and semidiurnal oscillations associated with atmospheric tides. Because the atmosphere, like the ocean, is a fluid, it experiences tides manifested by small pressure oscillations. But atmospheric tides are both thermally and gravitationally excited, the thermal component dominant. Amplitudes of tidal oscillations are less than 1 mb in mid-latitudes and about 3 mb in the tropics. The smaller mid-latitude amplitude is even more difficult to extract from time series of observed pressures because the synoptic variability in pressure is much larger than in the tropics.

SI base units evolved over about 200 years. The second was originally 1/86400 of a day, then a fraction of a particular year, and finally the time for a definite number of periods of radiation from a particular transition of a cesium-133 atom. The meter was originally 1/10,000,000 of the length of a meridian through Paris from the equator to the North Pole, then the length of a platinum-iridium bar at 0 °C, then a certain number of wavelengths, and finally the distance traveled by light in free space in 1/299,792,458 s. The kilogram evolved from the mass of one cubic meter of pure water at 0 °C to the mass of a platinum–iridium cylinder to being defined in terms of Planck's constant and the definitions of the second and meter. These units have been divorced from all physical objects other than the cesium atom. But measuring lengths by counting wavelengths is usually not practical or necessary. A *mise en pratique* (putting into practice) is a set of instructions that allows a definition to be realized at the highest level.

Pressure Measurement: Barometer and Manometer

Many years ago one of the authors and a colleague fell into casual conversation with a graduate student in meteorology. We discovered by accident that he did not know how a barometer works. We expressed our surprise so emphatically that he became enraged and cursed us roundly. We probably deserved some of the invective he hurled at us. We did needle him. But this was not entirely an act of mischief: we were genuinely shocked that someone could go so far in the formal study of meteorology and yet not know the principle of the barometer, an instrument of almost Stone Age simplicity. The memory of that long-ago shouting match compels us to include the following brief discussion of pressure-measuring instruments.

A barometer is a long tube made of glass so that you can see into it, one end of which is sealed. The tube is filled with a liquid, usually mercury, and then upended into an open pool of the same liquid (Fig. 2.4). Most of the liquid remains in the tube, although there will be a space between the liquid surface and the sealed end. What holds up the column of liquid in the tube?

The pressure in the space above the liquid column is assumed to be zero. It isn't, for reasons discussed in Chapter 5, but usually so small that it can be taken to be zero. The pool in which the tube is inserted is exposed to air at the

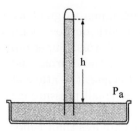

Fig. 2.4: A barometer measures absolute pressure. The pressure p_a balances the weight per unit area $\rho g h$ of the liquid column, where ρ is the density of the liquid and g is the acceleration due to gravity.

pressure p_a. Continuity of pressure requires that the pressure at the surface of the liquid in the pool also be p_a. If this liquid is in equilibrium, there is no horizontal gradient of pressure in the pool, and hence, the pressure in the tube at the height of its intersection with the liquid surface is also p_a. The upward force acting on the column of liquid is therefore $p_a A$, where A is the cross-sectional area of the tube. For this column to be in equilibrium, the sum of all forces on it must be zero. Gravity provides the downward force, the weight of the column, $\rho A h g$, where ρ is the density of the liquid, h is the height of the column, and g is the acceleration due to gravity. A force balance yields $p_a = \rho g h$. By measuring the height of a column of liquid with known density, we can determine the pressure of the air above the pool. Because mercury is about 13.6 times denser than water, to measure atmospheric pressure near sea level with a water-filled barometer would require it to be $\approx 10\,\mathrm{m}$ (≈ 33 ft) long. For this reason, mercury, not water, is the liquid used in barometers, even though water is cheaper and nontoxic; mercury also has a much lower freezing point $\approx -38\,°\mathrm{C}$.

Barometers measure ambient pressure p_a—the pressure of the air in contact with the open pool. Except for the barometer, pressure gauges we encounter in our daily lives measure the *difference* between absolute pressure and ambient pressure, called *gauge* pressure. Examples of gauges are tire pressure gauges and gauges on pumps of various kinds. The simplest gauge that measures pressure differences is a *manometer*, a U-shaped glass tube (Fig. 2.5), different from a barometer only in that one end of a manometer tube is open to the atmosphere, the other end exposed to a fluid at the pressure p to be measured. The height of the column of liquid in a manometer tube times the liquid density and g is the pressure difference $p - p_a$. To determine p requires barometric pressure. Blood pressure is a gauge pressure.

Measurements such as those depicted in Figs. 2.4 and 2.5 depend on theory, which predicts that pressure differences are $\rho g h$, where h is measured. This is (provisionally) valid only if the product ρh is the same in the same environment for liquids with appreciably different densities. Measured pressure is independent of A if it is large enough that capillarity is negligible.

Evangelista Torricelli is credited with the invention of the mercury barometer in 1643 although Vincenzo Viviani made the actual measurements. The beginnings of

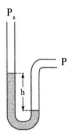

Fig. 2.5: A manometer measures pressure differences. $p - p_a$ balances the weight per unit area $\rho g h$ of the liquid column, where ρ is the density of the liquid and g is the acceleration due to gravity.

the development of the barometer were based on observations that a siphon would not work for heights over about ten meters. An experiment by Gasparo Berti around 1640 was made to study the production of a "vacuum." He attached a glass flask to the top of a pipe filled with water with its bottom submerged in an open reservoir of water. When a valve at the bottom of the pipe was opened, it was found that the water column in the pipe descended a fixed amount with a final height of the water column of about "18 cubits" (about 10.5 m). Because the purpose of this experiment was to determine if there was a vacuum in the evacuated flask, there was no attempt to relate the height of the water column to atmospheric pressure. Torricelli learned of this experiment and understood the mechanism by which the water column was supported. With this understanding he used much denser mercury to provide a more practical way to measure atmospheric pressure. Later experiments suggested by Blaise Pascal (1648) compared the heights of a mercury column at the base of a lava dome and at about 1000 m above its base, which showed that atmospheric pressure decreases with height. To honor Torricelli, a pressure unit widely used in high-vacuum work is the *torr*, defined as 1/760 of standard atmospheric pressure (1.333 hPa). The term vacuum is used carelessly to denote *any* pressure below local atmospheric pressure (negative gauge pressure, sometime called tension) or identically zero pressure (unattainable). Because the partial pressure of atmospheric water vapor is less than (total) atmospheric pressure, vapor pressure is sometimes called vapor tension (positive), a misleading term because liquid water can exist in a state of negative pressure called tension (Section 5.4).

Pascal's barrel is a demonstration attributed to him even though it cannot be found in his collected works. Spectacular videos have been made of it. A long (≈ 50 m), thin vertical tube is tightly fitted to a sealed barrel filled with water. Water is added to the tube until the barrel bursts. And yet the weight of water in the tube is a tiny fraction of the weight of water in the barrel. This dramatically shows the difference between weight and pressure. It also demonstrates the (relative) incompressibility of liquids because of very strong repulsive forces at small intermolecular distances. Gravity also plays a role by its unrelenting downward force on molecules, but intermolecular repulsion is what bursts the barrel. No magic is involved (see An Incompressibility Paradox in Section 3.20).

Fortin (around 1800) developed a mercury barometer that could be transported more easily than the earlier mercury barometers. Measurements using a Fortin barometer corrected for the change in density of mercury with temperature and the variation of gravity with altitude and latitude can be accurate to about 0.3 hPa. Fortin barometers are still in some use, although discontinued by the United States National Weather Service in 1999 when new instruments became available for accurate automated measurements of atmospheric pressure.

An *aneroid* (without water) *barometer* (invented by French scientist Vidie in 1844) provides a less accurate means for measuring atmospheric pressure. The original aneroid barometers were constructed using a sealed (usually metallic) container from which air had been removed to lower the internal pressure. A spring in the container resists compression and prevents it from collapsing. External pressure changes cause deformations of the container, which can be amplified with mechanical levers. In the past, aneroid barometers were used to construct *barographs* for recording pressure by a pen attached to the barometer lever and in contact with a drum wrapped in paper and rotated to give a time trace of pressure. Barometers of this type also were constructed to be very sensitive to small changes of pressure to make *microbarographs*, which can resolve pressures to microbars but with an accuracy of only millibars. They can be used to track pressure changes associated with gravity waves, frontal passages, gust fronts, and other small-scale weather events. They also can be used to track the diurnal and semidiurnal pressure changes associated with atmospheric tides.

Aneroid barometers have been used in aircraft altimeters and rawinsondes— balloon-carried packages of instruments to measure upper-level pressure, temperature, humidity, and winds. Aneroid barometers with mechanically driven readouts have been replaced by aneroid barometers that produce electrical signals proportional to the pressure. At present, two main types of electronic pressure sensors are used. One is based on capacitance sensitive to pressure and the other on resistance sensitive to pressure. The capacitive sensor has an evacuated cavity with two electrode plates separated by an air gap. One of the electrodes is solid, while the other is flexible so that the gap between the electrodes will change with pressure changes and thus change the capacitance. The other, a resistive pressure sensor, uses strain sensors attached to an evacuated cavity that changes shape with pressure.

Micromechanical sensors were developed in the 1970s and can provide accurate and precise ambient pressure measurements. Capacitance pressure sensors were used routinely in the 1980s for radiosonde pressure measurements and in automated weather stations. With today's technology, signals from these sensors can be digitized easily for further processing and storing and dissemination of observations. The development of accurate electronic barometers resulted in the replacement in 1999 of the mercury barometers used by the United States National Weather Service. The new barometers are much easier for routinely and automatically recording and reporting pressures than mercury barometers, which also have inherent safety issues because of the large amounts of mercury they use.

Capacitive and resistive pressure sensors have been miniaturized as Micro Electro Mechanical Systems with sizes from 1 to 100 μm. They can be combined with signal processing electronics in packages with dimensions of about 2mm × 2mm × 1mm for integration into smartphones and other mobile devices. Although these sensors may have limited accuracy (about 2 hPa), they can have good precision (0.1 hPa) that allows for resolution of small pressure changes, which can be used to estimate small changes in altitude. Pressure observations from smartphones also have the potential to provide dense and widespread surface pressure observations that may be used as input in short-term weather forecasting models.

Temperature Scales and Thermometers

Although the concept of temperature originated in human sensations of hot and cold, they are not reliable measures of temperature. As Ernst Mach noted, our "sensory organs ... have been biologically adapted not for the advancement of science, but for the maintenance of favorable conditions of life." We often estimate the temperature of the air around us by our sense of how hot or cold we feel. Yet our feelings are affected not only by air temperature, but also by wind speed as well. And those of us who live where summers can be hot and humid know that on two days determined to be equally hot by a thermometer, we are likely to feel less uncomfortable on the drier of the two days (Section 5.3).

You can demonstrate the subjectivity of your perception of temperature by stepping with a bare foot on a coin that has been lying on a carpeted floor. Although they are at about the same temperature, the coin will feel colder than the carpet because energy is transferred from your skin to the coin more rapidly. For an explanation, see References at the end of Chapter 7.

Thermometers yield objective temperatures. The basis for measuring them is the zeroth law of thermodynamics, which states that two systems at the same temperature as a third are at the same temperature as each other. Although this appears to be a trivial statement, it allows the temperature of a system to be obtained without direct comparison with a standard. A similar law, rarely explicitly stated, is applied when a ruler is used to measure distance or when almost any quantity is measured. If we measure the length of two objects with a ruler, and if the ruler readings are the same, we take it for granted without having to invoke anything so grand as a law that the two objects have the same length.

In the simplest application of the zeroth law a *thermoscope* can be used to determine if two systems are at the *same* temperature. A thermoscope can be constructed by using a liquid sealed in a glass tube. If the level of the liquid in the tube is the same when the tube is separately in equilibrium with two different systems, they are at the same temperature. No scale is needed. A scale on the glass tube indicates a numerical temperature.

Thermometers are based on materials with properties sensitive to temperature. For example, in a liquid-in-glass thermometer, both the liquid and the glass expand and contract with temperature changes by different amounts. Or one could confine a gas to a fixed volume and use pressure as a thermometric property. And various electrical devices (e.g. resistors) are sensitive to temperature changes.

Regardless of the material used to make a thermometer, calibration is required. It is simplified if the thermometric property varies linearly with temperature so that $T = ax + b$, where x is the thermometric property and a and b are constants. They can be determined by measuring x at two known temperatures. Although the *ice point* (defined in Section 5.5) and the *boiling point* (Section 5.4) are two common reference points, there are subtle issues, discussed in Chapter 5, that must be addressed when trying to use these points for accurate calibrations.

Two common temperature scales are those developed by Fahrenheit and by Celsius. The logic behind the Celsius scale, which has an ice point of 0 °C and a boiling point of 100 °C, is obvious. But the logic is less obvious for the Fahrenheit scale, which has an ice point of 32 °F and a boiling point of 212 °F. Fahrenheit claimed to have used a mixture of salt water and ice to obtain a 0 °F calibration point. But this would have been unreliable because the depression of temperature below the ice point depends on the concentration of salt in the solution. Middleton addresses this issue in a discussion of the controversial origin of the Fahrenheit scale. Regardless of its origin, those living in the United States appear to be stuck with it for the foreseeable future, even though it has been abandoned everywhere else in the world, including the countries where Fahrenheit was born, lived, and died. Although upper-air temperatures in the United States are reported and recorded in degrees Celsius, surface air temperatures are still reported in degrees Fahrenheit. Perhaps the current trend in the United States of using both Celsius and Fahrenheit scales for public displays of temperature will give Americans a better sense of temperatures in degrees Celsius.

To please your Swedish friends, express temperatures in degrees Celsius, not degrees centigrade. Anders Celsius was born (1701) and died (1744) in Uppsala, Sweden, following his father as professor of astronomy in the university there, one of the most ancient and venerable institutions of higher learning in Europe. Celsius's contribution to thermometry was to divide the temperature scale into 100 degrees between the boiling and freezing points of water, although his ordering was the *reverse* of what we now use: 100 °C corresponded to the ice point; 0 °C corresponded to the boiling point.

The conversions between Fahrenheit and Celsius temperatures are typically taught at the secondary school level. But those who use these conversions infrequently may not be able to recall them quickly. It is easy to remember that 1 °C equals (9/5) °F, but where in the calculation 32 is added or subtracted can tax the memory. An alternative method, based on equality of the two temperature scales at −40 degrees, may be easier to remember:

1. Add 40 to the initial temperature regardless of whether you are going from °C to °F, or vice versa.
2. Multiply by 5/9 to go from °F to °C or 9/5 to go from °C to °F.
3. Subtract 40 to obtain the final temperature.

Try it.

For most applications in atmospheric thermodynamics we use the absolute or Kelvin temperature scale with an ice point T_0 of 273.15 K (unlike °F and °C, the degree sign is omitted). The original basis for this scale was measuring pressures exerted by a real gas in a vessel of fixed volume at both the boiling point and the ice point using different amounts of gas in the vessel. The ratio of these pressures (p_b/p_0) is tabulated as a function of p_0. If extrapolated to $p_0 = 0$, this ratio is found to be 1.36609 for any gas. Thus, it is possible to define

$$\frac{T_b}{T_0} = \lim_{p_0 \to 0} \frac{p_b}{p_0} = 1.36609, \tag{2.16}$$

where T_b is the boiling point. For $T_b - T_0 = 100$ K, it follows that $T_b = 373.15$ K. The Kelvin scale is determined experimentally and requires no assumptions about the behavior of gases near absolute zero. Furthermore, for an ideal gas no extrapolation is needed to show that p_b/p_0 is a constant independent of the gas.

Absolute temperatures on the Fahrenheit scale are expressed in degrees Rankine (°R or R) by adding 459.67 to degrees Fahrenheit. Some engineers in the United States still use degrees Rankine. To convert to Kelvin, multiply by 5/9.

Although the basis for the Kelvin scale originates from gas thermometry, it is difficult to apply to defining thermodynamic temperature. Consequently, a practical temperature scale was adopted in 1927, with revisions made in 1948, 1960, and 1968. The most recent revision resulted in the International Practical Kelvin Temperature Scale of 1968 (IPTS-68). Kelvin (K), the unit of temperature in the IPTS-68, is defined as *exactly* 1/273.16 of the thermodynamic temperature at the triple point of water, which is therefore 273.16 K (0.01 °C). Defining the Kelvin scale in this way allows for the possibility that the interval between the ice point and the boiling point may not be exactly 100 K. More recent measurements indicate that the boiling point may be closer to 99.975 °C rather than 100 °C assigned in IPTS-68. The SI temperature unit, the kelvin, is now *defined* in terms of a fixed value of Boltzmann's constant and the definitions of the second and meter.

The ice point and the boiling point are *not* intrinsic properties of pure water. The lowest temperature at which liquid water exists *stably* is the triple point; the highest temperature at which it is distinguishable from the vapor is the critical temperature, 647.1 K (see Section 5.5). Both temperatures are the same on Mars and on Earth, unlike the boiling point (see Sections 5.3 and 5.10).

From the point of view of human comfort (or even survival), an air temperature decrease from 20 °C to −20 °C is drastic. But this is only a 14% decrease in *absolute temperature*, the temperature of scientific relevance. The degree size of the Celsius (or Fahrenheit) scale is arbitrary, as is its zero point. *Relative* differences in absolute temperature are independent of scale.

Atmospheric Temperature Measurements

Atmospheric temperatures are used for applications ranging from public displays of air temperature to measurements of small-scale temperature fluctuations associated with turbulence. The choice of a particular thermometer depends on various factors, including accuracy, response time, durability, cost, and the type of display and data recording used with the temperature sensor. For example, liquid-in-glass thermometers are used when a fast response is not required, and manual recording of data is acceptable. Sensors with thermometric properties related to the electrical properties of a material provide signals that can be digitized and stored. For simple experiments inexpensive electronic thermometers that are moderately accurate and provide a digital display are widely available.

The temporal response of a temperature sensor often satisfies

$$\frac{dT}{dt} = -\frac{T - T_a}{\tau}, \tag{2.17}$$

where T_a is the actual temperature, T is the temperature measured by the sensor, and τ is the sensor's time constant (see Section 7.1). If a thermometer at an initial temperature T_i is immersed in a fluid at a constant temperature T_a, the solution to Eq. (2.17) is

$$T = T_a + (T_i - T_a)e^{-t/\tau}. \tag{2.18}$$

Measured temperature exponentially approaches T_a, and after time τ the difference between T and T_a decreases by $1/e$. In many atmospheric applications T_a changes with time, and hence it is necessary to choose a thermometer with τ adequate to resolve the temperature changes of interest. Sensors used for public displays of temperature do not require nearly as short response times as sensors used to measure the temperature fluctuations associated with atmospheric turbulence. The response time of a typical mercury-in-glass thermometer is about 60 s if exposed to a wind of about $5 \, \mathrm{m \, s^{-1}}$. Fast-response thermometers on research aircraft may have response times in the range 0.1–0.01 s.

In general, the time constant of a sensor can be decreased by decreasing its size (Section 7.1). How small the sensor can be, however, may depend on the stresses it will encounter. To make temperature measurements from aircraft, for example, sensors must withstand stresses due to high wind speeds and the impact of cloud and precipitation particles. Yet a relatively fast response is required because the flow of air past a sensor may be greater than $100 \, \mathrm{m \, s^{-1}}$. For protection, these sensors are sometimes placed in housings designed to remove water droplets from the air before it impinges on sensors. Such housings also can reduce the cooling due to wetting of sensors. Unfortunately, they also increase the response time and give a response function more complicated than Eq. (2.18). Atmospheric temperatures measured from aircraft also must be corrected for aerodynamic heating of the probe due to friction and adiabatic compression. For an aircraft flying at $100 \, \mathrm{m \, s^{-1}}$ (≈ 225 mph) the temperature increase can be as high as $5\,^\circ\mathrm{C}$, about $3\,^\circ\mathrm{C}$ being more likely (Problem 3.18).

Although liquid-in-glass thermometers were used for measurements of air temperatures for many years, they have been replaced by electrical sensors that provide outputs that can be digitized, displayed, and saved. The most popular sensors are thermocouples, resistive elements, and semiconductor sensors. Thermocouples are constructed by making two junctions using two dissimilar metals. If the two junctions are at different temperatures, a voltage proportional to the temperature difference is produced. Thermocouples are practical for measuring temperature differences and can be used when a fast response is required. If thermocouples are used to make absolute temperature measurements, the temperature of one of the junctions, called the reference junction, must be known. Although it can be immersed in an ice bath for laboratory temperature measurements, this is not usually practical for atmospheric measurements. Other electrical temperature sensors can be used to provide the temperature of the reference junction embedded in a material with a large total heat capacity, and the other junction of the thermocouple can provide relatively fast response temperature measurements. Their accuracy depends on how accurately the reference temperature can be measured. Thermocouples for meteorological applications are usually copper–constantan (constantan is an alloy of copper and nickel).

Platinum wire resistors are used to make fast-response thermometers. The response time is minimized by using very thin wire. The resistance of the wire increases as the temperature increases. Because the change in resistance with temperature is not large, a long wire is needed to obtain resistance changes that can be resolved.

Thermistors are semiconductor devices whose resistance decreases as temperature increases. The change in resistance for a given temperature change is much larger than that for simple resistors and can be used for accurate temperature measurements. Thermistors and other semiconductor temperature sensors are used in integrated circuits to develop compact packages for temperature measurements. Inexpensive electronic thermometers using integrated circuits provide accuracies of about $2\,°C$ with precisions of about $0.2\,°C$ and a resolution of $0.1\,°C$. Accurate electrical thermometers have been used for atmospheric research since the early 1970s and in the 1990s replaced the mercury-in-glass thermometers used by the United States National Weather Service.

Infrared thermometers (radiometers) can be used to measure temperatures, the basis for which is that radiation emitted by an object varies with its temperature. They respond quickly and are useful for mapping surface temperatures from aircraft or determining temperatures of the bases or tops of thick clouds. These thermometers are inexpensive, readily available, and used for a variety of purposes: measuring temperatures of human skin and of surfaces inside and outside buildings or for classroom demonstrations. They operate in a spectral region, $8 - 13$ µm, where there is little absorption by atmospheric gases. For more applications, see Section 1.8. Some radiometers make use of emission by atmospheric CO_2 at 4.3 µm and 15 µm to infer air temperatures. Using a radiometer on aircraft eliminates wetting problems when making measurements in clouds. In addition, the volume of air sampled is larger than that sampled by an *in situ* probe.

The temperature of air or water (or anything) is yet another idealization. Consider a fixed volume containing a constant number of air molecules. Although the air may be macroscopically quiescent, any smaller *subvolume* is not isolated from surrounding air. Thus, the number of molecules in a subvolume varies because molecules randomly flit into and out of it. The relative fluctuation is of order $1/\sqrt{N_{av}}$, where N_{av} is the average number of molecules in the subvolume. A volume of air 0.01 mm on a side contains about 3×10^{10} molecules at sea level, the inverse square root of which is $\approx 10^{-5}$. We expect relative fluctuations of absolute temperature in this subvolume to be *at least* this large. Actual air temperature fluctuations are usually much larger. Within an unheated, unventilated room, over times of order seconds, temperatures can vary by $0.1-1\,°C$. "Room temperature" is often *defined* as $20\,°C$. All objects in real rooms, including their walls, do not come to exactly the same temperature. Fluctuations are even greater in the free atmosphere, and measuring small-scale fluctuations is critical for characterizing the turbulence structure within the atmospheric boundary layer.

Defining absolute temperature of an ideal gas (e.g. air) by Eq. (2.16) is easier than measuring it accurately. The thermometer must be in equilibrium with air and only air. Indoors, this usually is not a problem because the radiative temperature of walls, floors, and ceilings is likely to be close to air temperature. But a thermocouple probe (1 mm) in air at $18\,°C$ registered a temperature increase of $3\,°C$ about 1 cm from a hand at about $27\,°C$. Even if an outdoor thermometer is shaded from direct solar radiation, it is unavoidably illuminated by infrared radiation from its surroundings. But radiation does not contribute to temperatures taken with an oral thermometer.

Digital thermometers *record* fluctuations in temperature, even when they do not exist, because electronic circuits are inherently noisy (unlike liquid-in-glass thermometers). To estimate how noisy a thermocouple is, embed it deeply in a large granular mass (sugar, sand, etc.) that has been isolated for a long time. We used sugar and measured $\pm 0.1\,°C$ at around $17\,°C$ with our inexpensive (\$30) digital thermometer. Also keep in mind that the *operating* temperature of a thermometer is *not* the same as its temperature *range*. We discovered this when our thermometer was wildly incorrect in subfreezing weather.

Interpretations, Operations, and Explanations

Although we begin this section with molecular interpretations of temperature and pressure, this was not strictly necessary. It would have sufficed to define them by specifying a set of *operations* for measuring them consistently and reproducibly. Whoever takes your temperature needs to know only that a number on a thermometer scale should fall within a specified range. Temperature is what a thermometer measures, pressure is what a barometer or manometer measures. For many purposes we don't have to understand what they *are*. But molecular interpretations give us the feeling that we *do*.

The operational philosophy expounded by P. W. Bridgman in 1927 in his *Logic of Modern Physics*, according to which we do not know the meaning of a physical concept unless we have a set of physical operations for measuring it, has fallen out of fashion. Nevertheless, if a physical concept *can* be defined operationally, all to the good, but is not required of *all* concepts. If *none* within a particular theory can be, it is not a physical theory. It cannot be tested.

If authors of textbooks *choose* to give physical explanations, they must not be *demonstrably* false. We begin this section with a few examples. More follow. The supply is large. Our criterion for debunking an error is that it must have appeared several times where it should not have. Faulty physics in comic books may be amusing but is not our concern.

The Nature of Statistical Laws

We obtained the correct form of the ideal gas law without knowing the angle at which each gas molecule rebounds after collision with a wall. This kind of snatching knowledge from the jaws of ignorance exemplifies the statistical approach to scientific laws. Although Eqs. (2.11) and (2.13) do not explicitly betray their statistical origins, they apply only to equilibrium properties of ensembles of a great many molecules. By equilibrium we always mean statistical equilibrium. Pressure, temperature, and density are *not* properties of single molecules.

The earliest known occurrence of the term *statistics* was in the seventeenth century, when it was used in the sense of "pertaining to statists or to statecraft." The origins of statistics are more in the social and biological than in the physical sciences. The term *statistics* was not coined until the rise of nation states with large populations and the desire to govern these states for the greater good. Thus, records began to be kept of statistics: births and deaths, incomes, infant mortality, death by various diseases, and so on.

As an example of the statistical approach in epidemiology, consider a man who is 40 years old and has been smoking at least a pack of cigarettes daily since age 20. Will he die of lung cancer before age 60? This is impossible to answer. We cannot make any measurements—cholesterol, blood pressure, pulse rate, whatever—that will enable us to state with certainty that he will die or not of lung cancer in the next 20 years. For this *individual* we are ignorant, but we can make accurate predictions about a large *population* of similar individuals. For a population of people who share the common characteristic of being male, 40 years old, and having smoked at least a pack of cigarettes a day for 20 years, we can predict fairly accurately what *fraction* of this population will die of lung cancer in the next 20 years. This is statistical knowledge.

Statistical thinking pervades many fields: medicine, agriculture, public health, biology, psychology. There is an important difference between statistical thinking in these fields and in physics. In the biological sciences it seems that in principle it is not possible to predict the fate of a single individual drawn from a population. Now and probably forever, it is not possible to make physiological and chemical measurements of humans and then predict with certainty their fates.

Within the framework of Newtonian mechanics, the trajectory of an individual gas molecule is (in principle) knowable *if* we know the dynamical law $\mathbf{F} = m\mathbf{a}$ for each molecule in a gas. Why don't we just integrate the equations of motion to determine the trajectory of every molecule? There are at least three reasons. One is computational complexity. Given the vast number of molecules in even a small volume of gas, computing molecular trajectories, equipped with perfect knowledge of the dynamical laws and the initial state of each molecule, would be a heroic task. And, as noted in Section 1.3, the intermolecular force law is *not* known. Even if it were and we could do the computations, the results would likely overwhelm us. Would you be enlightened to know the trajectory of every gas molecule that entered your lungs? Would you feel that this helped you to understand respiration? This is more than you want or need to know or could know.

For these reasons, statistical methods are used in physics, especially the physics of matter in bulk (i.e. gases, liquids, and solids). We gain in simplicity at the expense of ignoring details. Wherever there are averages there are fluctuations, deviations from averages. They do not disappear just because we ignore them; they are still present and may be negligible, a nuisance, or useful.

Suppose that we had a perfect computer that could exactly solve the dynamical equations of motion for the atmosphere—sometimes called the primitive equations. And suppose that we could specify with perfect accuracy the initial state of the atmosphere at every point. Could we then predict the weather forever? Alas, no. The equations of motion, seemingly exact, apply only to *averages*. Wind velocity, just like pressure, density, and temperature, is an average. Fluctuations about these averages eventually would make their presence felt.

But some statistical knowledge about physical systems (such as gases) is more reliable than that about biological systems (such as cigarette smokers). All gas molecules of a given kind (e.g. nitrogen) are identical, unlike all 40-year-old, male smokers, who have many more characteristics they do not share than those they do. And then there is the matter of sample size. Even the smallest volume of likely interest in a gas contains vastly more individuals than an epidemiologist could dream of studying. And gas molecules, unlike humans, cannot read predictions about their behavior and change it accordingly.

Equation (2.12) is a statistical law with robust predictive accuracy within the domain of its validity. Given p and T at a particular time around a particular point in air, ρ can be predicted with near certainty. But predictions of *future p* and *T* at that point entail more than the ideal gas law and become increasingly uncertain with increasing time. The distinction between these two types of "prediction" is not always recognized.

We have used the term *macroscopic*, the antonym of which is *microscopic*. These terms do not mean big and small, which have no meaning until we specify a measuring stick. You will sometimes encounter the terms *visible* and *invisible*, which are misleading synonyms for macroscopic and microscopic. A gas is invisible but macroscopic. A single gas molecule, also invisible, is microscopic. The essential distinction between macroscopic and microscopic is not that between large and small linear dimensions, but that between a sample with a large number $N(\gg 1)$ of individuals and a sample with a small number. A macroscopic system contains sufficiently many individuals that averages have meaning, and fluctuations about these averages are usually negligible. A microscopic system is one for which this is not true. But there is no sharp boundary between macroscopic and microscopic. How small does a particle have to be (how few molecules) before it ceases being macroscopic but is not yet microscopic? Fortunately, even the smallest particles in the atmosphere contain so many molecules that usually we do not have to worry about the shadowy mesoscopic region lying between microscopic and macroscopic.

Sharp boundaries between physical categories are in our minds, not in nature. An example is the liquid–vapor interface (see Section 1.7 for another example). We can talk about such an interface, but it does not truly exist. Between a liquid and its vapor lies a shallow intermediate region neither as dense as the liquid nor as dilute as the vapor. We may often ignore this intermediate region, but we cannot deny its existence.

A Brief History of the Gas Law

The first ascent of a major mountain is preceded by tragic deaths and failed expeditions. Mt. Everest was not conquered on the first attempt. Climbers and sherpas died, expeditions turned back, before the first footprints of mortals were imprinted in snow on the summit of the world's highest mountain. Once the route was established, however, it became almost a stroll, subsequently reached by more than 4000 climbers. Similarly, the evolution of the ideal gas law did not proceed in such an orderly and logical fashion as that presented in a previous section. Scientific papers, and even more so textbooks, for the most part propagate false history. Once a law is known, it is easy to obtain it in a much tidier fashion than actually occurred. The real evolution was not a straight line from ignorance to truth, an exercise in logic, the unfolding of syllogisms—but rather a tortuous, zigzag path with many dead ends and controversies along the way.

To trace the evolution of the gas law, we rewrite Eq. (2.11) as

$$V = \frac{NkT}{p}. \tag{2.19}$$

For the same pressure but at a reference temperature T_0, the ice point, the corresponding volume is

$$V_0 = \frac{NkT_0}{p}. \tag{2.20}$$

The previous two equations can be combined to yield

$$V = V_0 \frac{T}{T_0}. \tag{2.21}$$

Now write T as $T_0 + t$, where t is in degrees Celsius, to obtain

$$V = V_0 \left(1 + \frac{t}{T_0} \right) = V_0(1 + \alpha_0 t), \tag{2.22}$$

where α_0 is the (isobaric) volume coefficient of thermal expansion

$$\alpha = \frac{1}{V} \frac{\partial V}{\partial T} = \frac{1}{T} \tag{2.23}$$

at $T = T_0$. Sometimes α_V is used to distinguish between volume and linear expansion. We may also write Eq. (2.23) as

$$\frac{V - V_0}{V_0} = \alpha_0 t. \qquad (2.24)$$

If $t = 100\,°C$ and $\alpha_0 = 1/273$, $V/V_0 = 1.366$. Gay-Lussac obtained 1.375 for several real gases. His aim was "to investigate the expansion of gases and vapors for *one definite rise of temperature* [e.a.] and to make it clear that this is the same for all these fluids." This rise was from the temperature of melting ice to that of boiling water, which he did *not* measure with a thermometer because he rightly considered existing thermometers to be inaccurate. Equation (2.22) or (2.24) is often called the law of Charles and Gay-Lussac or, worse yet, Charles's law. *The Dictionary of Scientific Biography* delivers a harsh verdict on the scientific accomplishments of Jacques-Alexander-César Charles (b. Beaugency, France, 1746; d. Paris, 1823): "Charles published almost nothing of significance … [His] achievements in science were relatively minor." On the bright side, however, "Charles developed nearly all the essentials of modern balloon design." If you are a balloonist, take your hat off and observe a moment of silence to honor the memory of Charles—but don't credit him with a gas law he did not publish or even discover. Joseph Louis Gay-Lussac (b. St. Léonard, France, 1778; d. Paris, 1850) deserves the credit for Eq. (2.23) for $t = 100\,°C$.

Gay-Lussac was, like Charles, a balloonist. Perhaps this stimulated a correspondence by means of which Gay-Lussac learned about Charles's unpublished work. In 1804 Gay-Lussac set a new altitude record for a balloon ascent of 7016 m, which was not exceeded, and by only 44 m, until 1850.

He was not the first to investigate the thermal expansion of gases. He had several predecessors, whom he graciously acknowledged—perhaps too graciously. It is only because of his published work that we know about the unpublished work of Charles, which was not undertaken with sufficient care to establish a general law. Two paragraphs in Gay-Lussac's 1802 paper cost him having to share the stage with Charles or to be pushed off it altogether. Gay-Lussac's great achievement, supported by careful experiments, was to establish a universal law: "All gases in general expand equally between the same degrees of heat provided that they are all brought under the same conditions." Because he considered only two "degrees of heat," he cannot be credited with Eq. (2.25) for *any* temperature t, which he understood. At the end of his 1802 paper he states that "to complete this research it remains for me to determine the law of the expansion of gases and vapors … for any degree of heat … I am at work upon this new research." We found no evidence that he published the results of this research. He may have been distracted by measuring the composition of air. In 1804 he published measurements made at 6636 m from a balloon showing an oxygen fraction the same as at sea level (21.49%) (Section 2.8). In 1805 he and Alexander Humboldt showed that water is composed of two parts hydrogen and one part oxygen.

Although Gay-Lussac apparently did not publish his measurements of the expansion of gases at any temperature between the melting point of ice and the boiling point of water, Jean-Baptiste Biot's *Treatise on Physics* (1816) describes them in detail.

According to Eq. (2.23), the temperature t_{abs} at which $V = 0$ is

$$t_{abs} = -\frac{1}{\alpha_0}. \tag{2.25}$$

This is *not* the temperature at which all motion ceases but the temperature at which the volume of an ideal gas (at constant pressure) would vanish *if* such a gas existed at very low temperatures, which it does *not*. Absolute zero, at least within the bounds established by the ideal gas law, is an *extrapolated* temperature. A measurement of the coefficient of thermal expansion of a gas at the ice point gives the value of absolute zero in degrees Celsius.

Equation (2.22), is only part of the ideal gas law. To complete it we need another relation between two of the three thermodynamic variables—pressure, temperature, and volume—one of them being held constant. Robert Boyle performed the definitive experiments establishing

$$pV = \text{const.} \tag{2.26}$$

for constant temperature (b. Lismore, Ireland, 1627; d. London, England, 1691) and he published the results in 1660 in *New Experiments Physico-Mechanicall, Touching the Spring of the Air and Its Effects*. In the second edition (1662) he gives the inverse relation between pressure and volume at constant temperature and (implicitly), constant number of molecules or mass, which is rightly called Boyle's law, dubiously called the Boyle–Mariotte law, and in France called Mariotte's law.

Two different characters than Boyle and Edme Mariotte (d. Paris, France, 1684) can hardly be imagined. Boyle was an aristocrat with a long pedigree, the son of the Earl of Cork, a member of a prominent Anglo–Irish family, educated at Eton and by private tutors. Mariotte's antecedents are a mystery. That he was born is indisputable, but where and when are uncertain. He surely was educated, but where and by whom is another mystery. Again, *The Dictionary of Scientific Biography* delivers a harsh verdict: "Mariotte's works ... rest in large part on fundamental results achieved by others." It wasn't until 1679, 15 years after Boyle published his results, that Mariotte published "a review and reconfirmation of what was already known." Yet even though Mariotte relied on previous work by Boyle, he did not acknowledge this intellectual debt. Nor was this an isolated example, a rare lapse from good manners. Huygens also accused Mariotte (after his death) of plagiarism. Although the law was first published by Boyle, and by this criterion should bear his name, it was "discovered by Power and Towneley, accurately verified by Hooke, accurately verified again by Boyle (aided in some degree by Hooke), first published by Boyle but chiefly publicized by Mariotte, in short, the 'law of Power and Towneley, and of Hooke and Boyle, and—to some degree—of Mariotte'."

A theoretical derivation of the ideal gas law chronologically straddles the experimental determination of Boyle's law (1662) and of Gay-Lussac's law (1802). Daniel Bernoulli (b. Groningen, The Netherlands, 1700; d. Basel, Switzerland, 1782) exemplifies the maxim that the best avenue to success is to choose your parents carefully.

Daniel's father was John Bernoulli. *The Dictionary of Scientific Biography* lists eight Bernoullis, possibly a record for a family.

In Daniel Bernoulli's *Hydrodynamica* (1738) we find a successful application of the kinetic theory of matter, its apparent calm roiled by ceaseless, rapid, and chaotic motion of its invisible constituent molecules. He adopted the now-nearly universal heuristic device of imagining a gas enclosed in a cylinder fitted with a weighted piston: "The compressing weights are almost in the inverse ratio of the spaces which air occupies when compressed by different amounts." He obtained this result under the assumption that "the velocity of the particles [molecules] is the same in both conditions of the air, the natural condition as well as the condensed." Bernoulli acknowledged that this "law has been proved by many experiments," although he cites neither Boyle nor Mariotte. Bernoulli then went on to assert that by heating air a greater weight would be needed to keep the air at constant volume because of the "more intense motion in the particles of air." The weight of the piston "should be in the duplicate ratio" of the velocity of the "particles" of air. This appears to be consistent with Eqs. (2.10) and (2.11), from which $V \propto \langle v^2 \rangle / p$ follows, if we interpret him to mean that the weight of the piston (p) must be proportional to the mean speed squared of the air molecules.

Presumably on the basis of Eq. (2.15) or similar equations, it is often said that "heat" is merely motion or a mode of motion, which was established by the kinetic theory of matter. But Daniel Bernoulli knew better more than 250 years ago: "Heat may be considered as an *increasing* [e.a.] internal motion of the particles." The interpretation of Eq. (2.15) is that the *temperature* (of an ideal gas) is a macroscopic manifestation of microscopic motion. All bodies are composed of molecules in motion, regardless of temperature, even down to absolute zero. The kinetic theory established that apparently stationary gases are composed of vast numbers of rapidly moving molecules that on average do not go anywhere. This was a remarkable discovery, which was by no means obvious, and even contrary to common sense. To complete the *experimentally* determined ideal gas law in the form of the *theoretical* gas law Eq. (2.11), we need Avogadro's *hypothesis* that equal volumes of all gases at the same temperature and pressure contain the same number of molecules (Section 2.8).

2.2 Pressure Decrease with Height: Continuum Interpretation

Atmospheric pressure decreases with height, as anyone who has climbed a high mountain knows. Why and how does pressure decrease? We can answer this from two points of view: the continuum and the molecular. The mathematical results are identical but are interpreted quite differently.

From the continuum point of view a gas is a continuous medium, filling all space and leaving no gaps or pores. We know better. But let us pretend that a gas is continuous and see what follows.

Consider all the air within a thin planar region with cross-sectional area A between z and $z + \Delta z$, where z is the height above the surface of Earth. We may pretend that it is flat because the vertical extent of its atmosphere is small compared with Earth's radius. In the absence of vertical accelerations, this slab of air is in equilibrium (or, as sometimes said, hydrostatic equilibrium), and hence, the sum of the forces on it must be zero. One force is its weight, Mg, which acts downward. The total mass of air M within the slab is $\rho A \, \Delta z$, where ρ is the average air density in the slab. Because we are pretending that air is a continuum, we have little choice but to say that all the air below z exerts a force $p(z)A$ upward on the air above it and all the air above $z + \Delta z$ exerts a force $p(z + \Delta z)A$ downward on the air below it. With these assumptions, the force–balance equation is

$$p(z)A - p(z + \Delta z)A - \rho A \, \Delta z \, g = 0. \tag{2.27}$$

Divide both sides of Eq. (2.27) by $A \, \Delta z$ and take the limit as $\Delta z \to 0$:

$$\lim_{\Delta z \to 0} \frac{p(z + \Delta z) - p(z)}{\Delta z} = \frac{dp}{dz} = -\rho g, \tag{2.28}$$

which is called the *hydrostatic equation*. With a sign change, it accurately describes the *increase* of pressure of *water* with depth because the density of water is almost independent of pressure and temperature (Fig. 3.9). But if Eq. (2.28) is combined with the ideal gas law Eq. (2.12), we obtain

$$\frac{1}{p}\frac{dp}{dz} = -\frac{g}{RT}, \tag{2.29}$$

with the solution

$$\frac{p}{p_0} = \exp\left(-\int_0^z \frac{g}{RT}dz \right), \tag{2.30}$$

often called the *barometric formula*. The pressure p_0 at $z = 0$ is determined by how the atmosphere was formed and subsequently evolved. g, R, and T determine only how pressure *decreases* with altitude. But does it always decrease? Underlying Eq. (2.29) is the *assumption* of hydrostatic equilibrium (or balance). Could pressure increase with altitude, at least over short distances and for short times? We don't know.

Because temperature varies with altitude, we cannot evaluate the integral in Eq. (2.30) without knowing the atmospheric temperature profile. But because *absolute* temperature does not vary greatly ($\approx 20\%$) from the surface to the tropopause, we may take T to be constant and equal to some average value T_{av} without appreciable error. With this approximation, Eq. (2.30) becomes

$$p \approx p_0 \exp\left(-\frac{gz}{RT_{av}} \right). \tag{2.31}$$

The quantity RT_{av}/g must have the dimensions of length, so we call it the *scale height*, and denote it by H:

$$p \approx p_0 \exp\left(-\frac{z}{H}\right).$$ (2.32)

This is a good approximation, as evidenced by Fig. 2.6, which compares an exponential fit to pressure versus height in the troposphere with the pressure decrease for the U. S. Standard Atmosphere (a composite atmosphere designed by a committee, not a column of air preserved in a 10-km glass tube in Boulder, Colorado).

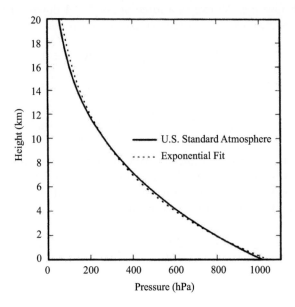

Fig. 2.6: Pressure versus height in the troposphere. The solid curve is for the U.S. Standard Atmosphere; the dashed curve is an exponential fit, with scale height 7.29 km, to this pressure distribution.

Equation (2.32) tells us that pressure (and hence, density) decreases nonuniformly with height. Unlike the oceans, the atmosphere does not end abruptly at some height below which there is atmosphere and above which there is none. It just fades away. Yet, although the atmosphere does not have a definite height, it does have a definite characteristic scale height, the vertical distance over which atmospheric pressure diminishes by about $1/e$. Subject to the same approximations underlying Eq. (2.32), density also decreases exponentially with height:

$$\rho \approx \rho_0 \exp\left(-\frac{z}{H}\right).$$ (2.33)

This is also a good approximation, as evidenced by Fig. 2.7, which compares an exponential fit of density versus height in the troposphere with the density decrease for the U.S. Standard Atmosphere. On the summit of Mount Everest (8849 m),

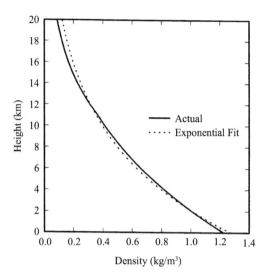

Fig. 2.7: Density versus height in the troposphere. The solid curve is for the U.S. Standard Atmosphere; the dashed curve is an exponential fit, with scale height 8.55 km, to this density distribution.

the number density of air molecules is about 35% that at sea level; at the absolute ceiling of commercial aircraft (12,800 m), about 22%. Mount Everest has been climbed many times without the use of supplemental oxygen. The first to do so, Reinhold Messner, was still alive at 76. It appears that the altitude ($\approx$ 5100 m) of the highest permanent human habitation is La Rinconada, a mining village in the Peruvian Andes with a population in the thousands.

Although the atmosphere is not isothermal, and hence, does not have a unique scale height, we still may define the scale height at each altitude as

$$-\frac{1}{p}\frac{dp}{dz} = \frac{1}{H_p} = \frac{g}{RT},$$ (2.34)

where the subscript p is appended to indicate that the scale height for pressure is not necessarily that for density H_ρ:

$$-\frac{1}{\rho}\frac{d\rho}{dz} = \frac{1}{H_\rho}.$$ (2.35)

The two scale heights are equal if the temperature is locally constant, which follows from the ideal gas law:

$$\frac{1}{H_\rho} = \frac{1}{H_p} + \frac{1}{T}\frac{dT}{dz}.$$ (2.36)

At altitudes where the temperature is decreasing, $H_p < H_\rho$. To be strictly correct, we should consider the variation of R with altitude, but this becomes appreciable only above about 100 km (Fig. 2.13). Because the arithmetic average of these two scale heights is close to 8 km, this easy-to-remember height is usually adequate for estimating vertical gradients of pressure and density, especially because the U. S. Standard Atmosphere exists only on paper. The average temperatures corresponding to these two scale heights are about 250 K for pressure, 290 K for density. The standard temperature at sea level is 15 °C and the standard rate at which temperature decreases in the troposphere (*lapse rate*) is 6.5 °C km^{-1} (for more about lapse rates see Section 3.4).

Although $H_p > 0$, the temperature gradient can be negative and usually is except close to surfaces. If $dT/dz < -g/R$, $H_\rho < 0$ and density would *increase* with height if $dT/dz < -34$ °C km^{-1}. This is contrary to experience, and so we conclude that density almost always *decreases* with height. But from the mantra that fluids in Earth's gravitational field are stable if density decreases with height, the atmosphere (a fluid) is almost always stable, which is not true, but is for liquids. For more about atmospheric stability see Section 3.5.

We haven't finished after having derived Eq. (2.31). Derivations are a means, not an end. We are left with the most important task, namely, interpreting equations. This is constrained by our (false) assumption that matter is continuous. In fluid mechanics, the pressure gradient is a "fundamental" force, and yet in physics there are only four fundamental interactions (also called forces): strong, weak, electromagnetic, and gravitational. Two of them have been unified as the electroweak interaction. Atmospheric scientists need not know the details of *all* sciences, but they should know enough to not say anything that is wildly incorrect. See Section 2.6 for a further discussion of the pressure gradient.

Gases are sometimes said to be *compressible*, in contrast with liquids, which are said to be *incompressible*. Because nothing is truly incompressible, it would be more accurate to say that gases are compressed much more easily than liquids (and solids). To give some meaning to this vague notion, we define the isothermal coefficient of compressibility (or simply isothermal compressibility or unqualified compressibility) as the fractional change in volume for a given change in pressure:

$$\kappa_T = -\frac{1}{V}\frac{\partial V}{\partial p} = \frac{1}{\rho}\frac{\partial \rho}{\partial p}. \tag{2.37}$$

The inverse of κ_T is the *bulk modulus*. It follows from Eq. (2.11) that the isothermal compressibility of an ideal gas is $1/p$. For such a gas we can approximate Eq. (2.37) in finite form as

$$\frac{\Delta V}{V} \approx -\frac{\Delta p}{p}. \tag{2.38}$$

The fractional change of volume of an ideal gas upon compression (or expansion) is approximately the fractional change in pressure. You can demonstrate that the fractional change in volume of a liquid is much less than the fractional change in pressure: try to compress water in a closed syringe. No matter how red in the face you become from pushing on the plunger, the change in volume of the water will be imperceptible. The compressibility of air $(1/p)$ is about 10^{-5} Pa^{-1}; that of liquid water is about 10^5 times smaller. For this reason, the density of water in the oceans is nearly constant with depth, whereas the density of air usually decreases with height (see Section 2.3 for a physical explanation).

If interactions between molecules in an ideal gas are negligible, why isn't its isothermal compressibility infinite and why does it decrease with increasing pressure? To compress a gas it must be *confined* in a *solid* cylinder fitted with a *solid* piston. As the piston is pushed downward, the pressure on it steadily increases because of increased n, and hence, molecular collisions with the piston. In a liquid, forces between tightly packed molecules are not negligible, and the piston must overcome strong intermolecular repulsion.

Because we pretend that air is continuous and we know that it is compressible, our interpretation of the decrease of air density with height is almost forced to be the following. All the air above a given height "presses" on all the air below it, which is compressible. At each successive height there is less air above to compress that below, and hence, density is greatest at the surface and decreases with height. We envision layers of air pressing on each other, encouraged all the more by the term pressure itself. This interpretation is unobjectionable provided that we don't switch languages. But we are not innocent of knowledge about molecules and can hardly resist translating from continuum language into molecular language. And here is where we stumble and fall. It is almost inevitable to interpret pressure in a gas as resulting from collisions between gas molecules. After all, if one part of a gas presses on another part (continuum language), this must be the result of gas molecules colliding with each other (molecular language). This is a bad translation. One way of demonstrating this is by way of our molecular derivation of the ideal gas law. Pressure was obtained as a momentum transfer between gas molecules and a *solid* wall. A molecule headed toward the wall had to collide with it. But the number density of molecules in a solid wall is about 1000 times greater than that of air molecules near sea level. Thus, the rate of collision between gas molecules cannot be the same as the rate of collision between gas molecules and the walls of their container.

The expression for pressure, Eq. (2.5), was derived by considering only one of its aspects: measurability. We measure pressure almost always by means of a solid or liquid surface. An ordinary barometer is a tube filled with mercury, a metallic liquid. A manometer is filled with water or perhaps alcohol. Other pressure gauges have metallic surfaces. To grasp the physical interpretation of pressure, rather than how it is measured, we have to consider what happens *within* a gas, rather than at a wall. Before doing this in the following section, we briefly give some meteorological applications of Eq. (2.30).

Pressure–Height Relationships: Thickness

The hydrostatic equation, Eq. (2.29), and the ideal gas law yield pressure–height relationships for several applications. With the assumption that R and g are independent of altitude (good approximations in the troposphere), we may integrate Eq. (2.29) between two altitudes z_1 and $z_2 > z_1$ on a flat Earth to obtain

$$z_2 - z_1 = -\frac{R}{g} \int_{z_1}^{z_2} \frac{T}{p} \frac{dp}{dz} dz. \tag{2.39}$$

The left side of Eq. (2.39) is an altitude difference, and hence, logically can be called a *thickness*. By transforming the variable of integration from z to $\ln p$, this equation can be written

$$z_2 - z_1 = \frac{R}{g} \int_{p_2}^{p_1} T d(\ln p) = \frac{R}{g} \langle T \rangle \ln \left(\frac{p_1}{p_2} \right), \tag{2.40}$$

where p_j is the pressure at z_j and $\langle T \rangle$ is an average temperature weighted by $1/\ln(p_1/p_2)$. Equation (2.40) can be written as

$$z_2 - z_1 = \langle H \rangle \ln \left(\frac{p_1}{p_2} \right), \tag{2.41}$$

where $\langle H \rangle$ is the scale height calculated from $\langle T \rangle$; this is sometimes called the *hypsometric equation*. Barometric hypsometry is the measurement of topography; bathymetry is its underwater equivalent. The *thickness* $z_2 - z_1$ represents an average temperature of a region between two constant-pressure surfaces expressed as a length. The temperature should be the virtual temperature (see Section 6.3) to account for the dependence of the density of air on its variable water vapor concentration.

To obtain some sense for thickness magnitudes, consider a 100-mb layer near the surface ($p_1/p_2 = 1000/900$). The average temperature in this layer is around 285 K, and hence the corresponding thickness is about 850 m (≈ 1 km). Near the tropopause, however, the thickness of a 100-mb layer ($p_1/p_2 = 200/100$) is approximately five times greater. The thickness of the 500-mb surface is that of the layer between the surface ($p_1 \approx 1000$ mb) and the altitude at which the pressure is 500 mb. The thickness of the 500-mb surface is around 5000 m and varies from about 4700 m to 5800 m between the tropics and polar regions.

A common map used by meteorologists for forecasting has 500–1000-mb thickness isopleths (lines of constant thickness) plotted with sea-level pressure isopleths (isobars). This map can be used to help identify surface fronts and to identify the snow–rain line associated with mid-latitude cyclone precipitation. A rule of thumb used by forecasters puts the snow–rain line along the 5400-m thickness line. Thus, snow might be expected in areas with thicknesses lower than 5400 m and rain would be expected in the areas with thicknesses higher than 5400 m. But since the actual snow–rain line thickness varies with altitude and season, some additional knowledge of local conditions is needed to refine forecasts.

Another application of Eq. (2.41) converts surface pressure observations to a sea-level pressure by writing

$$z_g = H_g \ln \left(\frac{p_0}{p_g} \right), \tag{2.42}$$

where z_g is the elevation where p_g is measured and p_0 is the sea-level pressure. The scale height H_g is usually approximated from the near-surface air temperature. From this equation,

$$p_0 = p_g \exp\left(\frac{z_g}{H_g}\right). \tag{2.43}$$

From the thickness estimate, Eq. (2.42), for a layer near the surface, a 1-mb pressure change is associated with 8.5-m change in height. For example, at Denver, Colorado (elevation ≈ 1600 m) the sea-level pressure would be about 180 mb higher than the observed surface pressure, whereas at Kansas City, Missouri (elevation ≈ 300 m) the sea-level pressure would be about 35 mb higher. To produce a sea-level pressure map, most land station pressure observations are converted to sea-level pressure.

Another critical application of a pressure–height relationship derived from the hydrostatic equation is in the design and fabrication of aircraft altimeters. They indicate altitudes used by pilots and are constructed using mechanical or electrical aneroid barometers. Thus, the altimeter provides a pressure altitude based on application of the hydrostatic equation to a standard atmosphere. That used for aircraft altimeters and adopted by the International Civil Aviation Organization (ICAO) is identical to the U.S. Standard Atmosphere. For this atmosphere, the mean sea-level temperature is 15 °C and the surface pressure is 1013.25 mb. From the surface to 11 km, the temperature decreases at the rate of 6.5 K km^{-1} and from 11 km to 20 km, the temperature is constant at -56.5 °C. The temperature structure defined by this standard atmosphere when used in the hydrostatic equation provides a fixed p-z relationship used for the design and calibration of aircraft altimeters used worldwide. During flight, the altimeter is set at the pressure where the height is zero. For aircraft (in the U.S. and Canada) flying 18,000 feet above Mean Sea Level (MSL) the altimeter is set at standard atmospheric pressure. When aircraft are assigned to a given flight level, this is a pressure altitude. The geometric height of the aircraft above the surface will not be constant, but the pressure at an assigned level will be. By using the ICAO standard atmosphere worldwide, there is no confusion about the pressure altitude at which aircraft are flying and they can be assigned flight levels to maintain safe separations. Below 18,000 ft (called the Transition Altitude) pilots set their altimeters to the local surface pressure to ensure that that flight levels at these altitudes are closer to the geometric heights, which are also critical at landing. Modern aircraft are equipped with radar altimeters that provide accurate heights above the surface at altitudes below 2500 feet during landings and takeoffs. Pressure altitudes (flight levels) are in feet except for flights over Russia, China, and North Korea, where they are in meters.

2.3 Pressure Decrease with Height: Molecular Interpretation

The continuum interpretation of the decrease of pressure with height (Section 2.2) is a reversion to Boyle's model of air as "a heap of little bodies, lying one upon another, as may be resembled to a fleece of wool." He was a prolific experimenter and a keen observer, but not informed by theories that did not yet exist. Almost 400 years later, in light of molecular reality and the kinetic theory of gases, we can do better.

The dimensions of pressure are those of momentum flux density: momentum per unit area and time. This is a clue, not a coincidence. Consider a *mathematical* yz plane within a gas. Molecules with positive x momentum transport positive momentum into the region to the right of this surface. Molecules with negative x momentum transport negative momentum out of this region. Transport of positive momentum into a region is equivalent to transport of negative momentum out, just as an increase in salary is equivalent to a decrease in indebtedness. The net momentum transported into the region to the right of the imaginary surface is

$$\frac{1}{2}nv_xmv_x + \frac{1}{2}nv_xmv_x. \tag{2.44}$$

The first term is the flux density of positive momentum to the right; the second (identical) term is the flux density of negative momentum to the left. The momentum flux density, Eq. (2.44), is the pressure p in Eq. (2.5). Thus, we interpret pressure, construed as a *property* of a gas, as a momentum flux density, whereas we interpret *measured* pressure as a transfer of momentum to a wall by collisions. Although the pressure in a gas is positive, the force on a solid or liquid surface imparted by colliding gas molecules is $-\mathbf{n}p$, where $\mathbf{n}$ is the unit outward-directed (into the gas) normal to the surface.

An ideal gas is one for which intermolecular forces are negligible, and hence, the pressure of such a gas is interpreted solely as a momentum flux density. For liquids, the interpretation of pressure is not so distinct. Molecules in a liquid transport momentum, but they also are sufficiently close that they exert nonnegligible forces on each other. Both momentum flux density (kinetic) and intermolecular forces (dynamic) contribute to the pressure in liquids.

To demonstrate the consistency of the momentum–flux interpretation of gas pressure, we derive the decrease of pressure with height from a molecular point of view. As before, the system of interest is the air between z and $z+\Delta z$. This system is not closed, however, because molecules (hence, momentum) cross its upper and lower boundaries. The two momentum flux densities are not equal. At the lower boundary $p(z)A$ is the net rate at which positive z momentum enters the system. At the upper boundary $p(z+\Delta z)A$ is the net rate at which positive z momentum leaves the system. Because of the force of gravity (which acts downward), positive z momentum within the system is decreased at the rate $nA\,\Delta z\,mg$, where n is the mean number density of molecules, each with mass m, in the system. For it to be in equilibrium, its momentum must be constant, which requires that the rate at which momentum enters the system minus the rate at which it leaves minus the rate at which momentum is decreased by gravity be zero:

$$p(z)A - p(z+\Delta z)A - nmgA\,\Delta z = 0. \tag{2.45}$$

Divide both sides of Eq. (2.45) by $A\,\Delta z$ and take the limit as $\Delta z \to 0$ to obtain

$$\frac{dp}{dz} = -nmg = -\rho g, \tag{2.46}$$

which from Eq. (2.14), can be written

$$\frac{1}{p}\frac{dp}{dz} = -\frac{mg}{kT}.$$

(2.47)

Because $m/k = 1/R$, this is no different from Eq. (2.29), but it is interpreted differently. The force on molecules is gravitational. Pressure within a gas is a momentum flux density, not an intermolecular force per unit area.

Equation (2.46) is equivalent to Newton's third law for a *point mass* in a uniform gravitational field, $dp_z/dt = -mg$, extended to a gas of weakly interacting molecules in such a field. The vertical gradient of the momentum flux density (pressure) is equivalent to the time rate of change of the z-component of momentum p_z; ρg, the gravitational force per unit volume, is equivalent to the force mg.

With the assumption of an approximately constant absolute temperature, Eq. (2.47) can be integrated to obtain

$$p = p_0 \exp\left(-\frac{mgz}{kT}\right).$$

(2.48)

The numerator of the quotient in the exponential function is the potential energy of a molecule; the denominator, except for a constant factor, is its mean kinetic energy. We can interpret the decrease of pressure with height as a struggle between gravity pulling inexorably downward and the helter-skelter motion of molecules in all directions (including upward). To show further the plausibility of this interpretation, we can write the scale height in molecular terms as

$$H = \frac{kT}{mg}.$$

(2.49)

In the limit of infinite kT/mg, the scale height becomes infinite, and pressure is uniform throughout the atmosphere. Does this make sense? Finite gravity cannot restrain highly energetic molecules from spreading uniformly throughout space. And in the limit of zero kT/mg, the scale height vanishes because all molecules are irresistibly pulled to the surface.

The explanation of the decrease of atmospheric pressure with altitude because of fewer air molecules pressing down on the molecules below them is, like a cockroach, difficult to kill. It has survived for more than 300 years. Students learn it at an early age and may cling to it for the rest of their lives. Although wrong, its popularity is understandable. No thought is necessary to teach or learn it and it *is* correct for solids and liquids. For example, if you place a set of metal plates, one on top of the other, on a bathroom scale and remove the top plate, the scale will indicate a lower weight. The weight of the top plate was balanced by the upward short-range molecular forces, ultimately of electrical origin, exerted on it by its immediate neighbor. By Newton's third law the top plate must exert a downward force on its neighbor. When the top plate is lifted even a smidgeon, it no longer exerts a downward force on what is now the top plate. This is a consequence of the short-range forces between plates in contact. Each plate can be looked upon as a giant molecule composed of a great many linked atoms. Contrast this with air. Even at sea level the average separation between air molecules is around ten molecular diameters. The force between two molecules as a function of distance between them is sometimes described by the easy-to-remember but approximate Lennard–Jones (or 6–12) potential. Its slope is positive for distances greater than the molecular diameter with a $1/r^6$ dependence, and negative for smaller diameters with a $1/r^{12}$ dependence. The force at distances beyond about a few molecule diameters is weakly attractive, whereas at shorter distances is strongly repulsive (hence, the low compressibility of water). The attractive force decreases as the ratio of molecular diameter to separation to the *seventh* power. Thus, over a distance equal to the average separation, the average attractive force decreases by a factor of 10^7. Only during relatively rare encounters do air molecules exert brief repulsive forces on each other. Arguments applied to solids and liquids, which we can touch, do not apply to gases. If the upper 5 km of the atmosphere were to be removed suddenly, its weight per unit area would decrease accordingly. But the decrease of surface pressure would take a much longer time (see Problem 7.32) because the average number density would decrease by a net upward diffusion of molecules, not because they had suddenly shed a burden.

Stephen Brush, a physicist-historian who has written extensively on the history of thermodynamics and kinetic theory, avers that "we now realize that the pressure of a gas at ordinary density is almost entirely due to impacts with walls rather than to continuously acting interatomic forces. But the amazing historical fact is that it took almost two centuries for the majority of physicists to accept that conclusion." Is this really true? There have been many studies of student misconceptions. We need more studies of their *sources*: teachers and textbooks.

The *collision cross-section* (σ) is an approximate way of describing the fleeting forces that deflect straight-line trajectories of real molecules (Section 2.5). We need not know the details of these forces. σ does not appear in Eq. (2.47), even implicitly. There is no necessary relation between the mass of a molecule and its collision cross-section. The gravitational force on a molecule depends only on its mass, not its structure, whereas σ ultimately originates from electromagnetic forces between molecules and depends on their structure. Thus, the decrease of pressure with height cannot be the result of molecules pressing down on each other.

In Martin Knudsen's list of four assumptions that "form the very foundations" of the kinetic theory, the third is "The pressure caused by the movement of the molecules is the only one existing in a gas, when it is in the ideal state." An ideal gas is *defined* as one in which changes in its internal energy at constant temperature are negligible [see Eq. (3.52)], *not* that it is collisionless. Indeed, in Section 2.5 we show that the (volumetric) collision rate in an ideal gas is staggeringly high.

2.4 The Maxwell–Boltzmann Distribution of Molecular Speeds

In Section 2.1 we defined absolute temperature in terms of the *average* molecular speed squared. This implies that all molecules in a gas do not have the same speed, some speeds being more likely than others. This is hardly a surprise, given the role of collisions in kinetic energy transfers (Section 1.5). Even if all the molecules in a uniform gas initially had the same speed but different directions, collisions would soon cause them to be distributed in speed. Purely by chance, a molecule might undergo a series of collisions in each of which it gained kinetic energy. Such a molecule would have a kinetic energy appreciably higher than the average, which means that there must be other molecules with energies appreciably lower. We assume that the kinetic energies evolve to a definite equilibrium distribution. But it is not obvious that a *gas* of molecules should exhibit such a statistical regularity despite the highly irregular trajectories of *individual* molecules.

Collisions play the same role in driving a system of molecules to equilibrium as a mixer does in making cake batter. All its ingredients—flour, salt, sugar, eggs, shortening—are dumped into a bowl and mixed into a homogeneous batter, the equilibrium state. Once thoroughly mixed, the batter does not spontaneously unmix itself, nor does further mixing change the batter. This is not impossible, merely exceedingly unlikely. And so it is with a gas. Any initially nonequilibrium distribution will evolve spontaneously to an equilibrium distribution. Given that T was defined by $\langle mv^2 \rangle$, we expect this distribution to depend on T.

For an ideal gas at temperature T the distribution of molecular speeds v is given by the *Maxwell–Boltzmann distribution function*:

$$f(v) = A4\pi v^2 \exp\left(-\frac{mv^2}{2kT}\right),$$ (2.50)

where

$$A = \left(\frac{m}{2\pi kT}\right)^{3/2}.$$ (2.51)

Without the exponential factor, $f(v)$ integrated over all speeds would be indefinitely large. Like all continuous distribution functions, f is *defined* by an integral, rather than by its value at each point. That is,

$$\int_{\nu_1}^{\nu_2} f(v)dv$$ (2.52)

is the fraction of molecules with speeds within any limits ν_1 and $\nu_2 > \nu_1$. Although $f(\nu) \neq 0$, its integral approaches zero if $\nu_2 \to \nu_1$. A distribution function is to be distinguished from a point function such as $E(v) = mv^2/2$. The distinction is that if Eq. (2.52) is transformed to energy E, the transformed distribution function is *not* simply $f[v(E)]$ but its product with dv/dE. This is a *theorem* about the transformation of variables of integration, which like all theorems must be proven by logical arguments, *not* by multiplying dv by dE/dE (see Section 3.1).

Because of the mean value theorem of integral calculus, we can write

$$\int_{v}^{v+\Delta v} f(v)dv = f(\bar{v})\Delta v = A4\pi \bar{v}^2 \Delta v \exp\left(\frac{-m\bar{v}^2}{2kT}\right),$$ (2.53)

where $v < \bar{v} < v + \Delta v$. The smaller the interval Δv, the more closely $\bar{v}$ is approximated by v. The constant A is such that f is normalized:

$$\int_0^\infty f(v)dv = 1. \tag{2.54}$$

The infinite upper limit of integration is *symbolic*, meaning much larger than $(2kT/m)^{1/2}$. Note that f does not specify *which* molecules have a speed v but what *fraction* of a large population of molecules has this speed within a small range. Because of frequent collisions, any given molecule will undergo a series of rapid promotions to the highest energy levels followed by equally rapid demotions to the lowest. Molecules have no feelings, so are not disturbed by their ever-changing status in the energy hierarchy.

The definition Eq. (2.10) is not dishonest, but it is incomplete. Because the distribution of molecular speeds in a gas can be *anything* in principle, so can its average kinetic energy. For theoretical T to be well defined it must be determined by a *particular* distribution, Eq. (2.51), resulting from molecules interacting and exchanging energy in such a way that any arbitrary initial distribution evolves to a statistically unchanging distribution. Because it *contains* temperature as a parameter, we have gone around in a circle. We cannot define temperature in terms of an average kinetic energy without temperature in the distribution function determining that average. We see only one way out of this loop. Measure the velocity distribution in a gas at a uniform temperature (easier said than done, but it has been done). Determine the parameter kT in Eq. (2.50) that best fits these measurements and compare it with kT obtained from Eq. (2.11) using measured p, V, and N. If they agree to within experimental error, we have a self-consistent theory.

The term "average" is used loosely—including by us—but there are indefinitely many averages and distribution functions (or weight functions) for the same variable. All averages are *not* equally-weighted arithmetic averages.

The Maxwell–Boltzmann distribution can be derived without *explicitly* considering molecular collisions even though they are necessary for its existence. They mix any arbitrary nonequilibrium distribution to an equilibrium distribution. Except for extremely rarefied gases, the collision rate is so high that the time this takes is a tiny fraction of a second (Section 2.5). Although we shall not derive Eq. (2.50), we can interpret it physically and try to make it plausible.

At any instant, the velocity components of a molecule are (v_x, v_y, v_z), which we can display graphically as a point in a three-dimensional velocity space (Fig. 2.8). The velocities of each molecule in a gas of N molecules are represented by a set of N points in this space. The volume between v and $v + \Delta v$ is approximately $4\pi v^2 \Delta v$ (if $\Delta v/v \ll 1$). For this reason, $4\pi v^2$ is often called the *density of states*. Because kinetic energy depends on v, this volume of velocity space is proportional to the number of ways of obtaining a particular kinetic energy (within some small range). All the points within a thin spherical shell of fixed thickness Δv in velocity space correspond to approximately the same kinetic energy, and the greater the radius of this shell, the more points it encloses. Note the appearance of the quantity $4\pi v^2$ in the distribution function Eq. (2.50). All else being equal, therefore, we

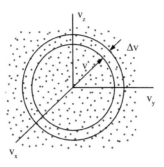

Fig. 2.8: The velocity of a molecule is a set of three numbers, its velocity components, which can be plotted as points in a three-dimensional velocity space. Each point in this space corresponds to the velocity of a molecule. For a given thickness Δv of a thin spherical shell in velocity space, more points are enclosed by the shell the greater its radius v.

expect more molecules to have speeds with a range between v and $v + \Delta v$, where Δv is fixed, the greater the value of v simply because there are more ways of obtaining speeds within this range. But all else is not equal. The factor $\exp(-mv^2/2kT)$ accounts for the decreased likelihood that a molecule will have a given energy. Insight into the origin of the exponential factor comes from Eq. (2.48) for an isothermal atmosphere in equilibrium in a gravitational field, according to which the decrease of number density n with height is

$$\frac{n}{n_0} = \exp\left(-\frac{mgz}{kT}\right). \tag{2.55}$$

Because mgz is the potential energy of a molecule in a gravitational field, we may interpret Eq. (2.55) as the probability that a molecule has a potential energy mgz. From Chapter 1 we know that potential and kinetic energies are interconvertible. If we had to guess the probability of a molecule having a kinetic energy $mv^2/2$, we would be on safe ground assuming that it has the form of Eq. (2.55) with potential energy replaced by kinetic energy, which indeed gives us the exponential factor in Eq. (2.50).

Although this equation is adequate for our limited purposes, it is not universally valid. It breaks down for high concentrations, for example, an electron "gas" in a metal, even at room temperature, with a number density a thousand times that of sea-level air, and at very low temperatures. The Maxwell–Boltzmann distribution of molecular speeds had to wait half a century for experimental verification when the *existence* of molecules was (mostly) beyond controversy.

The *most probable speed* v_p is that for which f is a maximum. By setting its derivative to zero, we obtain

$$v_p = \sqrt{2kT/m}. \tag{2.56}$$

With this result we can write the distribution function, Eq. (2.50), as

$$f(v) = A4\pi v^2 \exp\left(-\frac{v^2}{v_p^2}\right). \tag{2.57}$$

To determine how v_p depends on temperature, differentiate Eq. (2.56) with respect to temperature

$$\frac{dv_p}{dT} = \frac{1}{2}\frac{v_p}{T}, \tag{2.58}$$

and approximate the derivative by a ratio of differences $\Delta v_p / \Delta T$ to obtain

$$\frac{\Delta v_p}{v_p} \approx \frac{1}{2}\frac{\Delta T}{T}. \tag{2.59}$$

If the temperature drops from 20 °C to −20 °C, the most probable molecular speed drops by only about 7%. On average, therefore, air molecules in summer move only slightly faster than in winter; Fig. 2.9 shows the shift in the distribution of speeds with temperature. We must dress differently to be comfortable outdoors in summer and winter (not to mention stave off hypothermia and hyperthermia) because the many gas molecules with energies around the average don't drive chemical reactions but those (relatively few molecules in the high-energy tail of the distribution do (see Problem 2.63 and the discussion at the end of Section 7.1).

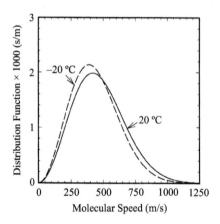

Fig. 2.9: Maxwell–Boltzmann distribution of molecular speeds in nitrogen at −20 and 20 °C.

The approximation underlying Eq. (2.59) can be applied to estimates of how the relative change of one physical variable depends on the relative change of another without doing detailed calculations or even worrying about units. The quantity $\Delta v_p / v_p$ is dimensionless, as is $\Delta T / T$.

The most probable speed is only one of several possible speeds that can be used to characterize a distribution of molecular speeds. We also can define the *mean speed*, denoted by $\langle v \rangle$, by

$$\langle v \rangle = \int_0^\infty v f(v) dv = \frac{2}{\sqrt{\pi}} v_p. \tag{2.60}$$

The *root-mean-square speed* is

$$v_{ms} = \sqrt{\langle v^2 \rangle} = \sqrt{\frac{3}{2}} v_p, \tag{2.61}$$

where

$$\langle v^2 \rangle = \int_0^\infty v^2 f(v) dv. \tag{2.62}$$

These three speeds, v_p, $\langle v \rangle$, and v_{ms} (see Fig. 2.10), are not greatly different (they stand in the ratio 1:1.13:1.22), and any one of them can be used to characterize the "average" speed of a collection of molecules distributed in speed. When you need to be precise, specify which average is meant. The Maxwell–Boltzmann distribution is broad. Its relative standard deviation $\sqrt{\langle v^2 \rangle - \langle v \rangle^2}/\langle v \rangle$ is 0.42 and its full width at half maximum is $1.2 v_p$.

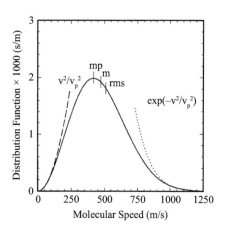

Fig. 2.10: Because the Maxwell–Boltzmann distribution function is a product of two functions, one of which increases with speed (dashed curve), the other of which decreases (dotted curve), it has a maximum at v_p, the most probable speed. The three commonly encountered average speeds—most probable (mp), mean (m), and root-mean-square (rms)—are indicated by short vertical lines. The gas (nitrogen) temperature is 20 °C.

The temperatures in our everyday lives are in Celsius or Fahrenheit degrees, which are convenient and tell us what we need to know. But as evidenced by Eqs. (2.50) and (2.55) and similar equations, when we take off our baseball caps and put on our science caps, we switch to absolute temperature in kelvins.

Why Don't Air Molecules Escape to Space?

They *do*! Earth is not encased in a gigantic impermeable bubble. At issue is not whether air molecules escape to space, but at what rate. To determine it, we must first obtain an expression for the *escape velocity*.

We cannot pretend that Earth is flat for this problem. Consider an air molecule so high in the atmosphere that it is unlikely to interact with other molecules because of the greatly reduced molecular number density. The potential energy of this molecule in Earth's gravitational field at a distance r from the center of Earth is

$$P = -\frac{GM_e m}{r}, \tag{2.63}$$

where m is the mass of the molecule, G is the universal gravitational constant, and M_e is the mass of Earth. The latter two quantities are related to g, the acceleration due to gravity at the surface:

$$g = \frac{GM_e}{R_e^2}, \tag{2.64}$$

where R_e is the radius of Earth. The total energy of the molecule is conserved:

$$\frac{1}{2}mv^2 - \frac{mgR_e^2}{r} = \text{const}. \tag{2.65}$$

From this we can determine what speed, the *escape velocity* v_{esc}, a molecule must have at the surface $(r = R_e)$ such that its speed in the limit $R_e/r \to 0$ is zero:

$$v_{esc} = \sqrt{2gR_e}. \tag{2.66}$$

Only molecules with speeds greater than v_{esc} can escape Earth's gravitational downward pull. By the "surface" we mean sufficiently high that molecular collisions are unlikely, but small compared with Earth's radius (≈ 6400 km). The escape velocity is the velocity an object would have at the surface if dropped from a height R_e in a uniform gravitational field (with no air resistance). For $R_e = 6400$ km and $g = 9.8 \, \text{m s}^{-2}$, the corresponding escape velocity is $11{,}200 \, \text{m s}^{-1}$, considerably higher than v_p at ordinary temperatures. For example, the most probable velocity for nitrogen (N_2) is

$$v_p = 24.3\sqrt{T} \, \text{m s}^{-1}, \tag{2.67}$$

which yields $v_p = 401 \, \text{m s}^{-1}$ at $0 \, °C$; v_p for oxygen (O_2) is about 7% smaller.

To proceed, we need the fraction f_{esc} of molecules with speeds greater than the escape velocity:

$$f_{esc} = 4\pi A \int_{v_{esc}}^{\infty} v^2 \exp\left(\frac{-mv^2}{2kT}\right) dv. \tag{2.68}$$

By transforming the variable of integration, we can write this as

$$f_{esc} = \frac{4}{\sqrt{\pi}} \int_x^\infty u^2 e^{-u^2} \, du, \tag{2.69}$$

where $x = v_{esc}/v_p$. Repeated integration by parts gives

$$\int_x^\infty u^2 e^{-u^2} \, du = \frac{1}{2} e^{-x^2} \left(x + \frac{1}{2x} \right) - \frac{1}{4} \int_x^\infty \frac{e^{-u^2}}{u^2} \, du. \tag{2.70}$$

The integral on the right side of Eq. (2.70) satisfies the inequality

$$\int_x^\infty \frac{e^{-u^2}}{u^2} \, du < e^{-x^2} \int_x^\infty u^{-2} \, du = \frac{e^{-x^2}}{x}. \tag{2.71}$$

Because average molecular speeds (at ordinary temperatures) are several hundred meters per second, whereas the escape velocity is more than ten thousand meters per second, $x \gg 1$, and hence,

$$\int_x^\infty u^2 e^{-u^2} \, du \approx \frac{x}{2} e^{-x^2}. \tag{2.72}$$

Equation (2.72) in Eq. (2.68) gives

$$f_{esc} \approx x 10^{-x^2/2.3}, \tag{2.73}$$

where we transformed from a power of e to a power of 10.

For nitrogen at a temperature around 0 °C, the ratio of escape velocity to most probable speed x is about 27.8, which corresponds to $f_{esc} \approx 10^{-335}$. To put this exceedingly small number in perspective, we estimate the total number of molecules in the atmosphere. We showed in Section 2.2 that mass density, and hence, number density n, decreases approximately exponentially with height:

$$n = n_0 \exp\left(-\frac{z}{H} \right), \tag{2.74}$$

where H is the scale height and n_0 is the number density at the surface. The total number N_{tot} of molecules, most of which are nitrogen, in the entire atmosphere is obtained by integrating Eq. (2.74) from the surface to infinity ($R_e \gg z \gg H$) and multiplying the result by the surface area of Earth A_e:

$$N_{tot} = A_e \int_0^\infty n_0 \exp\left(-\frac{z}{H} \right) dz = A_e H n_0. \tag{2.75}$$

For a scale height of 8 km, a surface number density of $10^{25} \ \mathrm{m}^{-3}$, and $R_e = 6400 \, \mathrm{km}$, $N_{tot} \approx 4 \times 10^{43}$. Only one nitrogen molecule out of every 10^{335} has the requisite speed to escape Earth's gravitational attraction, yet there are fewer than 10^{44} nitrogen molecules of *all* speeds in the *entire* atmosphere. For this reason we need not lose sleep worrying about nitrogen (and oxygen) escaping from our atmosphere. Gravity provides an effective, although imperfect, seal.

Low-molecular-weight gases are much less tightly bound to Earth than nitrogen, oxygen, and argon. According to Eq. (2.56), the most probable velocity of a molecule is inversely related to its mass. At a temperature for which the most probable speed of a nitrogen molecule is 400 m s^{-1}, the most probable speed of a hydrogen molecule is 1500 m s^{-1}. For molecular hydrogen, $v_{esc}/v_p \approx 7.4$, and from Eq. (2.73) $f_{esc} \approx 10^{-23}$. We conclude that molecular hydrogen could leak from Earth at a substantial rate, which explains why its atmosphere does not contain much hydrogen (some is produced by photodissociation of water vapor at high altitudes). If Earth's atmosphere eons ago had contained appreciable molecular hydrogen, by now it would have escaped to space. The most probable speed of atomic hydrogen is 2100 m s^{-1}, $v_{esc}/v_p \approx 5.3$, and $f_{esc} \approx 10^{-11.7}$.

We treat in this section perhaps the simplest escape mechanism, called *Jeans escape*, one type of thermal escape. There are other thermal and nonthermal escape mechanisms, some more important on Earth. Collisions with asteroids or comets could dwarf the otherwise feeble wasting away of atmospheric gases. Absent such catastrophes, Earth is predicted to be desiccated in a few billion years because of the escape of hydrogen, without which there is no water. It is unlikely that anyone will be around to care.

2.5 Intermolecular Separation, Mean Free Path, and Intermolecular Collision Rate

Nonsense about the alleged limited capacity of air to hold water vapor can be demolished readily by showing that air is mostly empty space. From the ideal gas law, Eq. (2.11), the number of molecules per unit volume is

$$n = \frac{N}{V} = \frac{p}{kT}. \tag{2.76}$$

The inverse of n is the average volume of space allocated to each molecule, and hence, the cube root of $1/n$ is the average separation d between molecules:

$$d = n^{-1/3} = \left(\frac{kT}{p}\right)^{1/3}. \tag{2.77}$$

For $p = 10^5$ Pa and $T = 293$ K, $d = 3.4$ nm (1 nm $= 10^{-9}$ m). Although molecules are fuzzy (but not lovable) without sharp boundaries, the approximate diameter d_m of air molecules is about 0.3 nm. The volume fraction of air occupied by matter is the ratio of the molecular volume to the volume allocated to a molecule, approximately $(d_m/d)^3 \approx 10^{-3}$. Only about one part (by volume) per thousand of air at sea level is occupied by matter, which is roughly equivalent to a few cows grazing on an acre of pasture, not to flank-rubbing cows in a feedlot.

Mean Free Path

The intermolecular separation d is *not* the average distance a molecule must travel before interacting with another molecule. To estimate this distance, consider an imaginary cylindrical volume with cross-sectional area A and length x in a gas (Fig. 2.11). The total number of molecules in the cylindrical volume is nAx.

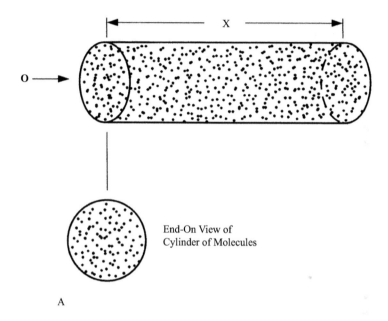

Fig. 2.11: An imaginary cylindrical volume in a gas. The number of molecules in the cylinder is proportional to its length x and cross-sectional area A. The total target area encountered by a molecule incident on the end of this cylinder is proportional to the number of molecules in it.

We pretend that molecules are tiny impenetrable spheres of diameter d_m. Two such spheres interact if their centers approach each other within a distance d_m. For determining collisions between two molecules, we can treat a projectile molecule as a point and a target molecule as a sphere with projected cross-sectional area πd_m^2, called the *collision cross-section* and denoted by σ. A point molecule is deflected if and only if its trajectory intersects a (finite) target molecule. Suppose that a point molecule traveling parallel to the axis of the cylinder is incident on one end. The position on A at which it enters the cylinder is random. How far can a point molecule go before it encounters (collides with) a target molecule? Some point molecules may suffer a collision in a very short distance, while others may travel much longer distances. If we ignore shadowing of one target molecule by another, the total target area presented to a point molecule over a distance x is $nAx\sigma$. The total projected target cross-sectional area per cylinder cross-sectional area is therefore

$n x \sigma$. We may roughly interpret this as the probability that a point molecule suffers a collision in a distance x. If this quantity is 1, the molecule is unlikely to escape a collision. The corresponding distance λ is called the *mean free path for collision* (or simply the mean free path), and is given by

$$\lambda = \frac{1}{n\sigma}. \tag{2.78}$$

Does this expression make physical sense? As n increases, the mean free path decreases, which could hardly be otherwise. And when the collision cross-section decreases, the mean free path increases, which also conforms to expectations. And because molecular number density decreases exponentially (approximately) with height, mean free path increases exponentially (Fig. 2.12).

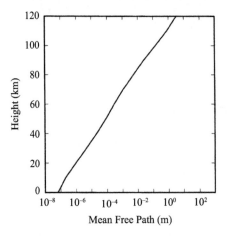

Fig. 2.12: Mean free path between collisions of air molecules versus height for U.S. Standard Atmosphere.

If we were to take more care in determining the average distance between collisions made by a molecule, accounting for its relative motion as it collides with one moving target after another, we would obtain an expression similar to Eq. (2.78), differing only by a numerical constant not appreciably different from one. An implicit assumption underlying this entire section is that a discrete collision exists. That is, the trajectory of a molecule is a continuous set of straight line segments with discontinuous slopes. This occurs only approximately in a gas in which the average distance between molecules is much greater than the distance over which a collision occurs (approximately a molecular diameter).

The *geometrical* collision cross-section σ of an air molecule, based on its fuzzy diameter, is $\approx 3 \times 10^{-19}$ m^2. Near sea level, the number density of air molecules is about 3×10^{25} m^{-3}. These two values in Eq. (2.78) yield a mean free path of 0.1 μm (1 μm $= 10^{-6}$ m), 30 times greater than the average separation between molecules. This makes physical sense. Molecules have no brains. Only a thinking molecule (in a gas, which is mostly a void) could arrange its itinerary so that the distance between successive collisions would be approximately equal to the (average) intermolecular separation. After each collision, a molecule would have to draw a bead on the nearest molecule and head for it. In a liquid, the concept of a collision becomes vague because all molecules are continuously in the force fields of their neighbors. Such molecules are never truly free.

At a height of 110 km, the number density is 10^{20} m^{-3}, but the average separation is only about 200 nm. The vastly greater one meter mean free path is because the molecules are so small that they do not "see" each other, so to speak. Although the pressure at this height is low, it barely makes it into the category of high vacuum. This adds weight to the argument that pressure in gases is *not* a consequence of intermolecular collisions but is *measured* by collisions with solid or liquid surfaces.

The concept of mean free path was forced upon Rudolph Clausius (Section 4.1) by the Dutch meteorologist Christophe Buys-Ballot's criticism of a flaw that kinetic theorists had overlooked: the high speeds of molecules calculated by Clausius were inconsistent with the slow mixing of gases. He had to go back to the drawing board and account for the finite size of molecules and the consequent short distance a molecule travels before colliding with another. But the story does not end there, because dispersion of odors in apparently still air is much *faster* than predicted by molecular diffusion (Section 7.3).

In a strictly ideal (collisionless) gas the mean free path would be infinite for finite number density. We can postulate an ideal gas of negligibly interacting molecules and derive its equation of state to be compared with measurements. But it depends on the equilibrium properties temperature and pressure, and in a strictly ideal gas there is no mechanism for attaining equilibrium. Each molecule goes its way independently of all the others. If we pretend that molecules do not interact, we get some correct results, but also some that are wildly incorrect. For sound waves to propagate in a gas, molecules must interact cooperatively to some degree. Idealization is always a deal with the devil.

Intermolecular Collision Rate

To drive home the message that pressure is not a direct manifestation of collisions of gas molecules *with each other*, we consider these collisions in more detail. Assume that all the molecules are fixed except one. We can take all the fixed molecules to be points, and the moving molecule to be a sphere with projected area σ. In a time interval Δt, the moving molecule with speed v sweeps out a volume $v \, \Delta t \, \sigma$. All the fixed molecules in this volume collide with the moving molecule. Thus, the total number of collisions in the time Δt is the number density n of molecules times the volume $v \, \Delta t \, \sigma$; the collision rate $nv\sigma$ follows from dividing by Δt. This is the collision rate per molecule. To obtain the collision rate per unit volume we multiply by the number of molecules per unit volume: $n^2 v\sigma$.

Because molecules are distributed in speed, we should interpret v as a mean relative speed. We know from Section 2.4 that average speeds, be they mean, most probable, or root-mean-square, are proportional to the square root of absolute temperature T and inversely proportional to the square root of molecular mass m. From this and the ideal gas law, we can write the rate of intermolecular collision per unit volume within a gas as

$$C\frac{p^2\sigma}{\sqrt{m(kT)^3}}, \tag{2.79}$$

where C is a constant not appreciably different from one. A more careful derivation would give the precise value of C, but it is of no importance here. In a strictly ideal gas, $\sigma = 0$.

Consider two different gases, both at the same temperature and pressure, but with different collision cross-sections (σ) and masses (m). According to Eq. (2.79) the rates of intermolecular collision are different. Neither pressure nor temperature is determined uniquely by intermolecular collision rates. None of the thermodynamic properties such as pressure, temperature, and density depends explicitly on collisions, but they play a role off stage, anonymously, so to speak, like stage workers for a play. Collisions are the mechanism by which energy and momentum are transferred from molecule to molecule, and they also are the randomizers or stirrers of molecular motion. How rapidly does this stirring occur in a gas? For a number density $n = 10^{25}$ m^{-3}, a collision cross-section of 10^{-19} m^2, and a mean speed of 400 m s^{-1}, the collision rate is 4×10^{33} m^{-3}s^{-1} = $4 \times 10^{15}\mu$m^{-3}s^{-1}. This high collision rate in a tiny volume is at the root of *local thermodynamic equilibrium* (see the following subsection). Even if the collision cross-section were a million times smaller, the collision rate would be more than a billion per second.

The collision cross-section σ is conspicuous by its absence from the Maxwell–Boltzmann distribution, Eq. (2.50), even though if it were zero, this equilibrium distribution could not evolve from an initially nonequilibrium distribution. Causal agents sometimes are invisible.

We cannot extrapolate the concept of collision from the macroscopic to the microscopic level. We say that billiard balls collide, we hear them click, and from a macroscopic point of view they touch, a meaningless concept at the microscopic level. The balls exert strong forces on each other over distances so small and for times so short that we cannot observe that they do not touch. A collision between molecules is an interaction at distances on the molecular scale in which one or more of their properties before collision—kinetic energy and vector linear momentum—are different after. Our perception of touch originates from receptors on our skin that send electrical signals to our brains to be interpreted. We touch, smell, taste, hear, and see with our brains, which molecules don't have.

Local Thermodynamic Equilibrium

To understand local thermodynamic equilibrium we first must understand what is meant by pressure, temperature, and density *at a point*, the existence of which is implicit in Sections 2.2 and 2.3. To determine the local density in a fluid, for example, we imagine a volume V in the fluid centered on the point $\mathbf{x}$. The total mass within this volume is M. The density of this volume of fluid is defined as

$$\rho = \frac{M}{V}. \tag{2.80}$$

What happens to this ratio as V shrinks to zero? Does the limit

$$\lim_{V \to 0} \frac{M}{V} \tag{2.81}$$

exist? Not in a strict mathematical sense. As long as V contains many molecules, M/V does not vary appreciably. But this cannot hold for arbitrarily small V. Suppose, for example, that V is comparable to d^3 for a gas. This volume is so small that at one instant it might contain only a single molecule, which might then migrate out of the volume, leaving it empty. For such a small volume, the density would fluctuate wildly. By density at a point we mean density in the *neighborhood* of a point—a volume large enough to contain many molecules but small enough to be representative of the neighborhood. In a gas at normal temperature and pressure, a cube only 1 μm on a side contains 10^7 molecules. If we do not probe gases on a scale finer than this, fluctuations in the ratio M/V arising from migration of molecules into and out of V do not have readily observable consequences. We further assume that in the neighborhood of a point, molecules have an equilibrium distribution of speeds, which then determines the temperature in that neighborhood.

For equilibrium, the internal energy U of a fluid (gas or liquid) is a function of, say, pressure and temperature. We assume that if we double its mass, all else being equal (e.g. density and temperature), its internal energy will double, and hence, is an example of an *extensive* thermodynamic variable. It is proportional to the mass of the system. See Section 3.2 for why this is a valid assumption. Pressure and temperature, however, are *intensive* variables, independent of the size of the system. An extensive variable can be converted into an intensive variable by dividing by mass or volume. For example, mass and volume are themselves extensive variables, whereas their ratio, density, is an intensive variable. The ideal gas law, Eq. (2.12), is a relation between only intensive variables. We often denote by lower-case letters intensive variables formed from extensive variables. For example, u denotes the *specific* internal energy or internal energy per unit mass, an intensive variable. Although we dutifully pass on the definitions of extensive and intensive thermodynamic variables, all thermodynamic variables do not lie in either one or the other category. A better term for intensive is *local* and a better term for extensive is *global*.

The specific internal energy of a fluid in equilibrium is, say, a function of temperature and pressure (or density):

$$u = u(p, T). \tag{2.82}$$

According to the postulate of local thermodynamic equilibrium, if $p(\mathbf{x}, t)$ and $T(\mathbf{x}, t)$ are the pressure and temperature at the point $\mathbf{x}$ at time t, the corresponding local specific internal energy is

$$u(\mathbf{x}, t) = u[p(\mathbf{x}, t), T(\mathbf{x}, t)], \tag{2.83}$$

where the function u is for strict thermodynamic equilibrium. That is, the local specific internal energy of a system *not* in equilibrium is the same function of (local) pressure and temperature as for equilibrium. The integral of $\rho(\mathbf{x}, t)u(\mathbf{x}, t)$, the specific internal energy per unit volume, over a volume is its total internal energy, where $\rho(\mathbf{x}, t) = \rho[p(\mathbf{x}, t), T(\mathbf{x}, t)]$.

The postulate of local thermodynamic equilibrium is just that: a postulate, not a law. It expresses a hope. For many meteorological applications, especially in the troposphere, local thermodynamic equilibrium is a good approximation because of the exceedingly rapid rate of stirring by collisions in even tiny volumes. But we shouldn't be surprised if this postulate sometimes fails.

Local thermodynamic equilibrium, exemplified by Eq. (2.83), provides one motivation for meteorologists to study equilibrium thermodynamics, which at first glance might seem inapplicable to the atmosphere. It is never in a state of entire thermodynamic equilibrium, as evidenced by gradients of temperature, pressure, and density, winds, clouds, and ever-changing weather. Yet, because of local thermodynamic equilibrium, the equilibrium relations among thermodynamic variables are applicable to a dynamic atmosphere. In deriving Eq. (2.31) we implicitly assumed that $p(\mathbf{x}, t) = \rho(\mathbf{x}, t)RT(\mathbf{x}, t)$, even though the ideal gas law was derived and has been experimentally verified only for uniform gases in equilibrium. The result is a barometric equation in accord with the observed decrease of pressure with altitude, which is what matters.

2.6   Is the Pressure Gradient in a Gas a Fundamental Force of Nature?

The pressure gradient in air is sometimes taken to be a fundamental force of nature. This is acceptable if you don't believe in atoms, if you treat air as if it were a continuous medium, and if your knowledge of physics ends with what was current in the mid-nineteenth century. But if you accept the existence of atoms, the strict interpretation of a pressure gradient in a gas as a force is incorrect. We argued previously that the physical meaning of pressure in a gas is a momentum flux density. To show why we can pretend that pressure gradient is a force, consider Newton's second law

$$\frac{d\mathbf{p}}{dt} = \mathbf{F}, \tag{2.84}$$

where $\mathbf{p}$ is the momentum of a point mass acted on by a force $\mathbf{F}$. Everything on the right side of this equation is a force; everything on the left side is a rate of change of momentum. But we can just as easily write it as

$$0 = -\frac{d\mathbf{p}}{dt} + \mathbf{F}. \tag{2.85}$$

We can interpret $-d\mathbf{p}/dt$ as a force because it appears on the right side of the equal sign. A mere juggling of terms can change the interpretation. This is similar to what happens with the pressure gradient in a *gas*. It is not a fundamental force, but rather a momentum flux density gradient masquerading as a force. Wherever there is a pressure (i.e. momentum flux density) gradient, momentum increases or decreases within a region. Our second derivation of the hydrostatic equation is by way of a momentum balance, rather than a force balance. In *liquids*, electromagnetic forces between molecules contribute to the pressure.

Only two kinds of fundamental forces are relevant to atmospheric processes: gravitational and electromagnetic. The atmospheric engine is not propelled by nuclear forces except indirectly by way of nuclear reactions that are the source of solar radiation—the ultimate power source for the atmosphere. Gravitational forces depend on mass; electrical forces depend on charge. A force is fundamental if it exists at atomic and subatomic levels. The pressure gradient force fails this test.

Gravity pulls us down even when standing on solid ground. Why don't we accelerate downward? The net force on us must be zero, but what is the *nature* of the force that balances gravity? When we ask students, their responses are usually inadequate: the normal force, the pressure force, the reaction force, and so on. Few of them seem to have been taught that forces of electrical origin can balance gravity. They have a much shorter range than gravitational forces because charges can be both positive and negative, whereas mass is only positive. But electrical forces are much stronger than gravitational forces. It takes the entire Earth to pull you downward by gravity, whereas forces acting only on a thin layer on the soles of your shoes resists your downward acceleration. But these repulsive *interactions* (not "true" forces) between molecules that are electrically neutral but composed of charges do not have a simple classical explanation. It is not just electrostatic repulsion between charges of like sign, although negative electrons and positive nuclei are the agents. When molecules get very close together, classical laws go out the window. See Section 3.2 for a brief discussion of repulsion.

The role of gravity in atmospheric processes is fairly transparent. Many of the equations of meteorology contain g explicitly. But electrical forces are hidden from sight: charge does not appear explicitly in the equations of fluid flow. But electrical forces determine properties such as viscosity and collision cross-sections. Make no mistake about it, electromagnetic forces determine atmospheric processes.

2.7   Surface Pressure and the Weight of the Atmosphere

With a better understanding of pressure, it is safe to show how surface pressure and the weight of the atmosphere are related. The total weight per unit area of a column of air extending from the surface to infinity is

$$\int_0^\infty \rho g \, dz, \qquad (2.86)$$

where both ρ and g depend on altitude z. Again, the upper limit is symbolic, meaning sufficiently high to include most of the atmosphere. Because of Eq. (2.29), this weight is

$$\int_0^\infty \rho g\,dz = \int_0^\infty -\frac{dp}{dz}dz = p_0, \qquad (2.87)$$

where the pressure at infinity (wherever that is) is negligible. This is the infamous result that surface pressure is "just" the weight (per unit area) of a column of air. But underlying it is the requirement, not always satisfied, that there be no vertical accelerations and Earth be flat. More important, pressure and weight are independent concepts. This distinction, often forgotten today, was made by Robert Boyle more than 300 years ago, as evidenced by the title of one of his treatises on the expansion of gases: *A Continuation of New Experiments Physico-Mechanical Touching the Spring and Weight of the Air, and their Effects* (1669). By the "spring" of air is meant its pressure, as distinguished from its "weight."

Flat Earthers Take Note!

An assumption smuggled into Eq. (2.86), probably without anyone noticing, is that Earth is flat. Yet according to reliable reports, it is decidedly spherical, although because of social media modern flat Earth societies are on the rise. This sphericity together with the finite vertical extent of the atmosphere combine to yield an average surface pressure that is *not* "just" the weight of the atmosphere per unit surface area of Earth, even for strict hydrostatic equilibrium. Showing this and estimating the magnitude of the discrepancy is Problem 16.

Why Aren't We Crushed by Airplanes Flying Overhead?

We sometimes have been asked to explain what happens when an airplane is flying overhead if air pressure at the surface is merely the weight of everything above us? Does the airplane contribute to the surface pressure? If so, why doesn't a heavy commercial airliner leave a wake of death and destruction along its flight path?

To answer these questions, consider a vertical path upward from the surface to where it intersects the airplane. The weight per unit area of all the air along this path is

$$\int_0^b \rho g\,dz, \qquad (2.88)$$

where the subscript b denotes the bottom of the airplane. The weight per unit area of all the air above the airplane is

$$\int_t^\infty \rho g\,dz, \qquad (2.89)$$

where t denotes the top of the airplane. From the hydrostatic approximation, Eq. (2.29), the sum of the previous two integrals becomes

$$\int_0^b \rho g dz + \int_t^\infty \rho g dz = p_0 - p_b + p_t. \qquad (2.90)$$

This can be written

$$\int_0^b \rho g \, dz + \int_t^\infty \rho g \, dz + p_b - p_t = p_0. \qquad (2.91)$$

The pressure difference $p_b - p_t$ must equal the weight per unit area of the airplane (airplanes for which this is not true either are on the ground or soon will be). Thus, the surface pressure p_0 is indeed the sum of the weight per unit area of *everything* overhead, air and airplane. The hydrostatic approximation underlies this result, which is not strictly valid in the immediate vicinity of the airplane. But if we ignore departures from a hydrostatic atmosphere near the airplane, our conclusion holds.

The surface pressure with no airplane present is

$$p_0' = \int_0^\infty \rho g \, dz. \qquad (2.92)$$

If we subtract Eq. (2.91) from Eq. (2.92) we obtain

$$\Delta p_0 = p_0' - p_0 = p_0 - p_t - (p_b' - p_t'), \qquad (2.93)$$

where $p_b' - p_t'$ is the pressure difference between the altitudes of the bottom and top of the airplane (with no airplane present).

A Boeing 747 is flying at 10 km, its overall height is 19.3 m, which corresponds to a pressure difference $p_b' - p_t'$ of about 70 Pa. Its gross wing area is about 500 m^2; its mass is about 350,000 kg, its weight about 3,500,000 N. The pressure difference $p_b - p_t$ is about 7000 Pa, from which we subtract 70 Pa to obtain a pressure increase at the surface $\Delta p_0 = 6900$ Pa, about 7%. This is substantial. A pressure increase (or decrease) of this magnitude inside a house would likely shatter its windows. But this is the *maximum* possible. A local increase in pressure immediately below the airplane will set up horizontal pressure gradients, and hence, airflow that will destroy them. Similarly, when a fish is added to a tank of water, the weight of the fish contributes to the pressure on the bottom of the tank, but this weight is distributed over the entire area of the bottom, not just on the region below the fish.

2.8   The Atmosphere Is a Mixture of Gases: Dalton's Law

To this point we have assumed, usually implicitly, that all molecules in the gas of interest are identical. But the atmosphere is a mixture of gases, mostly nitrogen and oxygen with a smidgeon of argon, variable amounts of water vapor, and seasoned with trace gases such as carbon dioxide, ozone, and methane.

Because air is to very good approximation an ideal gas, the fact that it is a mixture poses no special problems. According to *Dalton's law of partial pressures*, the total pressure of a mixture of (ideal) gases is the sum of the pressures of each one as if it alone occupied the volume V:

$$p = p_1 + p_2 + p_3 + \cdots, \tag{2.94}$$

where the pressure p_j of the jth component is

$$p_j V = N_j kT. \tag{2.95}$$

The theoretical basis for this law is the assumption that the components are ideal and that in equilibrium their mean kinetic energies are the same.

Although John Dalton (b. Eaglesfield, England, 1766; d. Manchester, 1844) is often claimed by chemists as one of their own, meteorologists have an equal right to claim him. His first scientific interests were in meteorology, which he never lost; he kept weather records for 57 years up to his death. We owe to Dalton the earliest definition of the dew point. His interest in meteorology led him to ponder the state of water vapor in air. The prevailing view in Dalton's time was that air is a kind of solvent for vapor, attracting it by chemical affinity. Dalton rejected this view, arguing on the basis of his measurements that the amount of water in air does not depend on its pressure. That is, air and water vapor are independent components of a mixture, a conclusion he generalized to other gaseous mixtures. Although we now know that Dalton's law of partial pressures is an approximation, it is nevertheless a very good one, especially for meteorological purposes.

Those who claim that there is nothing new under the Sun are fond of pointing out that a finite atom as the ultimate limit of divisibility of matter was invented by Greek philosophers thousands of years ago. Yet, these ancient speculations about atoms were just one example among many about the structure of matter, all equally unverifiable, all equally good (or bad). What Dalton did, on which most of his fame rests, was to develop a theory of atoms subject to experimental confirmation (or refutation). He transformed atomism from speculation to a quantitative science, a natural progression from his interest in weather to experimental investigations of moist air to "a powerful and wide-ranging new approach to the whole of chemistry."

Dalton's law of partial pressures is correct, even though it was obtained by incorrect reasoning. It was based on his assumption that Newton had *proved* that an atom of one kind repels an atom of the same kind but not of a different kind. Each kind of gas in a mixture "acts like a vacuum to the others ... as if it were the only gas present." Because he did not believe that two atoms of the same kind could form a molecule, he got the weights of molecular oxygen and nitrogen wrong.

Mean Molecular Weight

The density of a mixture of gases can be written

$$\rho = \Sigma_j n_j m_j, \tag{2.96}$$

where n_j is the number density of the jth component. Because n, the total number density of molecules in a mixture, is the sum of the number densities of all its components, we can write the ideal gas equation as

$$p = kT\Sigma_j n_j. \tag{2.97}$$

These two equations can be combined to yield

$$p = \rho \frac{k}{\langle m \rangle} T, \tag{2.98}$$

where the mean molecular mass of the mixture is

$$\langle m \rangle = \frac{\Sigma_j n_j m_j}{\Sigma_j n_j} = \Sigma_j f_j m_j, \tag{2.99}$$

where $f_j = n_j/n$ is the number fraction (also mole fraction) of the jth component.

Until 2019, the *mole* (abbreviated mol) was defined as the number of ^{12}C atoms in 12 g of this isotope of carbon. *Avogadro's constant*, N_a, now is *defined* as $6.02314076 \times 10^{23}$ mol^{-1}. Its magnitude, within 0.05%, is the number of molecules in 22.4 liters of an ideal gas at one atmosphere and 0°C. Avogadro's constant is a link between microscopic and macroscopic scales. *Loschmidt's constant*, 2.687×10^{25} m^{-3}, is the number density of ideal gas molecules at 0 °C and 1 atmosphere.

The terms molecular weight, molecular mass, and molar mass, denoted by M, are used interchangeably. For our purposes we take the molecular weight to be the total mass *in grams* of one mole of identical molecules, and hence, the mass m of a molecule is M/N_a and we can write Eq. (2.98) as

$$p = \rho \frac{R^*}{\langle M \rangle} T, \tag{2.100}$$

where $R^* = N_a k$, now *defined* as 8.314462618 J mol^{-1} K^{-1}, is called the *universal gas constant* because it does not depend on any particular gas. The mean molecular weight $\langle M \rangle$ of the mixture is the number-weighted average of the molecular weights of its components:

$$\langle M \rangle = \frac{\Sigma_j n_j M_j}{\Sigma_j n_j} = \Sigma_j \frac{n_j}{n} M_j = \Sigma_j f_j M_j. \tag{2.101}$$

The gas constant R^*/M for any specific gas is called the specific gas constant; the quantity $R^*/\langle M \rangle = \langle R \rangle$ is the mean gas constant of a mixture (the brackets around $\langle R \rangle$, indicating an average, are often omitted), the mass-weighted average of the specific gas constants of its components:

$$\langle R \rangle = \frac{\Sigma_j n_j M_j R_j}{\Sigma_j n_j M_j}. \tag{2.102}$$

Because of the definition of the mole, M expressed as a mass has cgs units, whereas these days all right-thinking folks use the SI system. But the mole is an SI base unit. Sometimes a distinction is made between a mole and a kilogram mole, which is 1000 times larger. But if the term *mole* is not modified, it is likely to mean the gram mole, rather than the kilogram mole. For calculations in which *all* quantities are given in SI units, be sure to use the appropriate value for R^*, 8341 J kmol^{-1} K^{-1}.

Because the molecular weight of water is less than that of nitrogen and oxygen, the density of moist air is always less than that of dry air for the *same p/T*. The reduction, however, is small, a factor of $1 - 0.38f$, where f is the number fraction of water molecules (of order 0.02). We return to this point in Section 6.3.

A useful expression for the number density of molecules of a single kind with mass $m = M/N_a$ composing a homogeneous material with mass density ρ is $n = \rho N_a/M$, where ρ is in g cm^{-3}; $10^3 n$ is the number density in m^{-3}. That of liquid water is 3.3×10^{28} m^{-3}, about a thousand times greater than the number density of air molecules at sea level. Mass density is often a proxy for number density, which is more fundamental. With one exception, molecules do not interact by way of forces dependent on their masses. Mass-dependent gravitational forces are not negligible on Earth, even though the gravitational field of each of its molecules is miniscule, because there are about 10^{50} of them and their fields are additive.

Moist air is a mixture of dry air and water vapor. The major components of dry air are nitrogen, oxygen, and argon. The table shows the fractions of these components, up to about 25 km, taken from the *Smithsonian Meteorological Tables*.

Component	Number Fraction	Molecular Weight
Nitrogen (N$_2$)	0.7809	28.01
Oxygen (O$_2$)	0.2095	32
Argon (A)	0.00934	39.95

The mean molecular weight of dry air is therefore 28.94, only a few percent greater than that of nitrogen, and the mean gas constant for dry air is 287 J kg^{-1} K^{-1}. Because the lower 100 km of the atmosphere is well mixed, the mean molecular weight of air is almost constant up to this height, then decreases (see Fig. 2.13). But as we show in Section 6.3, even in the troposphere R is not strictly constant because of the variable water vapor fraction of air.

Nitrogen, oxygen, and argon are (relatively) *permanent* and well mixed in the troposphere. Because of their structure (diatomic, monatomic) they are not *infrared-active*, unlike polyatomic water vapor, carbon dioxide, and methane. Water vapor in the lower troposphere varies considerably in time and space, whereas concentrations of the other two are much smaller but increasing slowly because of human activities. How often have you seen the unqualified assertion that methane is much more "potent" at "trapping heat" than is carbon dioxide. This is a sloppy, if not misleading, metaphor. What is meant is that infrared-active gases emit infrared radiation downward, the result being higher surface temperatures. How do we fairly compare CO_2 and CH_4 as emitters? For $T \approx 300$ K and infrared wavelengths around 10 μm, emission by a methane molecule is appreciably greater than that by a carbon dioxide molecule in the same environment. But emission per unit volume is the *product* of concentration and emission per molecule. The concentration of methane is

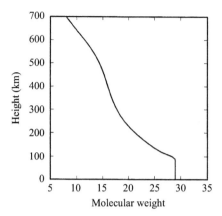

Fig. 2.13: Mean molecular weight versus height for U.S. Standard Atmosphere.

200 times *smaller* than that of carbon dioxide but is increasing at a faster rate, and thus, methane is getting more attention. The common thermophysical properties of air are unaffected by trace gases. Emission of infrared radiation per unit volume is *not* such a property.

We do not use the term "greenhouse gas," a misleading metaphor aimed at the general public that originates from a faulty understanding of real greenhouses. *Any* house suppresses mixing of interior air with colder outside air. Real houses *leak*, as required by building codes. The essential difference between a glass-walled greenhouse is that the furnace is outside (solar radiation) whereas in an ordinary house it is inside.

We expressed the relative amounts of the various gases making up air as number fractions. You may find these relative amounts expressed as volume fractions, which may be confusing. The volume V_j of N_j molecules at a pressure p and temperature T is, from Eq. (2.11),

$$V_j = N_j \frac{kT}{p}. \tag{2.103}$$

At a *standard* temperature and pressure, therefore, volume is proportional to number. According to Dalton's law, the volume V occupied by a mixture of gases is, at the same standard temperature and pressure (an important caveat),

$$V = \frac{kT}{p} \Sigma_j N_j = \frac{kT}{p} N, \tag{2.104}$$

where N is the total number of molecules of all kinds. The number fraction of the jth component in a mixture of (ideal) gases follows from Eqs. (2.103) and (2.104),

$$\frac{N_j}{N} = \frac{V_j}{V} = \frac{p_j}{p}. \tag{2.105}$$

Number fraction and volume fraction are equivalent subject to the requirement that the volume of the mixture and of each of its constituents are measured at the same temperature and pressure. This odd way of expressing relative amounts of gases in mixtures (such as air) is a relic of the methods used by early investigators of the properties of gases, who found volume easier and more accurate to measure than mass. Today, concentrations of trace gases in the atmosphere are often expressed in ppm (or ppmv), meaning parts per million (by volume), or even ppb (or ppbv), meaning parts per billion (by volume). These small concentrations surely were not obtained by measuring gas volumes. Expressing concentrations, especially exceedingly small concentrations, as volume fractions is anachronistic. Number fractions or, equivalently, mole fractions, apply equally well to liquid and solid solutions as to mixtures of gases, which are also solutions, although not usually called such.

The dimensionless parts-per notation is used in different ways, sometimes carelessly. In chemistry, *molarity* is defined as the number of moles of solute divided by the number of liters of *solution*—not solvent. One liter of pure water contains *approximately* 55.6 moles. The density of water depends slightly on temperature and pressure and the liter is a *volume* unit. If a *small* amount of solute is dissolved in a solvent (water, say), the volume of the solution is not greatly different from that of the solvent. For example, a one molar (1M) aqueous solution is about 98% water. If two milligrams of solute are dissolved in one liter of water, the volume of the solution will be almost one liter. The concentration 2 mg L^{-1} is not dimensionless but sometimes written as 2 ppm because one liter of water is about 1000 gm, and therefore the ratio of masses is 2×10^{-6}. There are only two unambiguous non-dimensional concentrations: mass fraction and mole fraction.

We emphasized "small" in the previous paragraph. If one liter of alcohol is added to one liter of water, the resulting volume is about 1.9 liters, an example of $1+1 \neq 2$. Water molecules interact differently with alcohol molecules than with each other, and hence, the structure of the solution is different from that of its components.

Because of the interpretation of pressure as a momentum flux density, we can apply a momentum balance to each component of air:

$$\frac{dp_j}{dz} = -\rho_j g, \tag{2.106}$$

where ρ_j is the density of the jth component. The ideal gas law, Eq. (2.13), for each component is

$$p_j = \rho_j R_j T, \tag{2.107}$$

where $R_j = k/m_j$ is the gas constant for a molecule with mass m_j. If T is approximately constant with altitude, Eqs. (2.106) and (2.107) can be combined to yield

$$p_j = p_{0j} \exp\left(-\frac{m_j g z}{kT}\right), \tag{2.108}$$

and the decrease in number density n_j with altitude is

$$n_j = n_{0j} \exp\left(-\frac{z}{H_j}\right), \qquad (2.109)$$

where the scale height of the jth component $H_j = kT/m_j g$ is inversely related to its (molecular) mass. If this held for oxygen and nitrogen with scale heights different by 15%, at 6.6 km the decrease in the concentration of oxygen should be about 12% less than that of nitrogen. But in 1804, Gay-Lussac (Section 2.1) measured at this altitude an oxygen fraction the same as at sea level (21.49%). Up to about 100 km, the *relative* concentration of each of the components of air, except water vapor, is constant—evidence that the atmosphere (especially the lower atmosphere) is well mixed. But even *imperceptible* vertical motions prevent differential settling depending on molecular weight (see Section 7.3). Water vapor, however, is a component of air that comes and goes. Were it not for water vapor, the mean molecular weight of air would be constant, and its density would depend only on temperature and pressure. We discuss in more detail in Section 6.3 how the variable molecular weight of air because of its variable water vapor fraction is accounted for in meteorology.

Annotated References and Suggestions for Further Reading

For biographies of van Helmont, see J. R. Partington (1961) *A history of chemistry*, vol. 2 New York: Macmillan; Walter Pagel (1944) "J. B. Helmont (1579–1644)," *Nature*, 153, pp. 675–76; and the entry for van Helmont by Walter Pagel in *Dictionary of scientific biography*.

For the origins of the word chaos, see *Oxford English dictionary*.

Edward Lorenz's definition of chaos is in his (1993) *The essence of chaos*. Seattle: University of Washington, p. 4.

The ideal gas law is derived in several books on kinetic theory, among them James Jeans (1940) *An introduction to the kinetic theory of gases*. Cambridge: Cambridge University Press; Leonard B. Loeb (1961) *The kinetic theory of gases*. Mineola, NY: Dover Books; Earle H. Kennard (1938) *Kinetic theory of gases*. New York: McGraw-Hill. Although our derivation is similar to ones in these books, few (if any) authors point out that individual molecular collisions with real walls *do not* obey the law of specular reflection, yet despite this incorrect assumption a correct result is obtained. Under *very* restricted conditions, gas molecules incident on a surface *may* obey the law of specular reflection. This requires surfaces much smoother than almost all natural solid surfaces, as cold as possible (to reduce oscillations of the surface molecules), and low molecular weight molecules incident at near-grazing angles. For more about this, see R. G. J. Fraser (1938) *Molecular beams*. New York: Chemical Publishing Company, p. 19, and Loeb, pp. 331–5.

For a brief history of the now forgotten static theory of gases, see Eric Mendoza (1990) "The lattice theory of gases: a neglected episode in the history of chemistry," *Journal of Chemical Education*, 67, pp. 1040–2.

For a good treatment of adsorption, see Jan Hendrik de Boer (1968) *The dynamical character of adsorption*. 2nd edn. Oxford: Oxford University Press.

For a wealth of information on barometry and altimetry, see *Manual of Barometry*, Vol. 1. Washington, DC: U.S. Government Printing Office (1963).

The history of the barometer is treated by William Edgar Knowles Middleton (1964) *The history of the barometer*. Baltimore: Johns Hopkins University Press. See also by the same author: (1969) *Invention of the meteorological instruments*. Baltimore: Johns Hopkins University Press.

For a thorough discussion of sea-level pressure, see Kevin E. Trenberth (1981) "Seasonal variations in the global sea level pressure and the total mass of the atmosphere," *Journal of Geophysical Research*, 86(C6), pp. 5238–46.

For a good, short introduction to thermometry, see Mark W. Zemansky (1964) *Temperatures very low and very high*. Mineola, NY: Dover Books. For a more detailed treatise, see Terry J. Quinn (1983) *Temperature*. London: Academic Press.

For a treasure trove of information about thermometry, see American Institute of Physics (1941) *Temperature, its measurement and control in science and industry*, vols. 1–6. Reinhold Publishing.

For a recent treatise on temperature measurements, see J. V. Nicholas and D. R. White (2001) *Traceable temperatures: an introduction to temperature measurement and calibration*. 2nd edn. John Wiley & Sons.

For a history of thermometry, see William Edgar Knowles Middleton (1966) *A history of the thermometer and its uses in meteorology*. Baltimore: Johns Hopkins University Press.

The zeroth law of thermodynamics is criticized by Otto Redlich (1970) "The so-called zeroth law of thermodynamics," *Journal of Chemical Education*, 47, pp. 740–1. He points out, correctly, that all of physical science hinges on zeroth laws, although rarely explicitly labeled as such. He concludes that the zeroth law of thermodynamics "is neither a law nor a statement of fact but a guideline for checking our description of nature."

See *Dictionary of Scientific Biography* for biographical sketches of Celsius (Sten Lindroth) and Fahrenheit (J. B. Gough).

For a good discussion of the contributions of Celsius to thermometry, see Olof Beckman (1997) "Anders Celsius and the fixed points of the Celsius scale," *European Journal of Physics*, 18, pp. 169–75.

For discussions of meteorological instruments, see Leo J. Fritschen and Lloyd W. Gay (1979) *Environmental instrumentation*. New York: Springer; William Edgar Knowles Middleton and Athelstan F. Spilhaus (1953) *Meteorological instruments*. 3rd rev. edn. Toronto: University of Toronto Press.

For a good history of statistical thinking, see Theodore M. Porter (1986) *The rise of statistical thinking 1820–1900*. Princeton: Princeton University Press.

See *Dictionary of scientific biography* for biographies of Boyle (Marie Boas Hall), Gay-Lussac (M. P. Crossland), Charles (J. B. Gough), Mariotte (Michael S. Mahoney), and Daniel Bernoulli (Hans Straub). We highly recommend the *DSB* for accurate biographies of scientists. Do *not* rely on textbooks, the authors of which usually uncritically repeat the inaccuracies in previous textbooks. To find out what our predecessors themselves said about their work, not what one textbook writer thinks that another textbook writer thinks that these scientists *might* have said, read the original papers. You'll be helped in this undertaking by William Francis Magie (1935) *A source book in physics*. London: McGraw-Hill, an invaluable compendium of biographical sketches together with excerpts from writings by a host of famous scientists: Boyle, Gay-Lussac, Bernoulli, to name just a few. Magie's book contains the passages we quote from Bernoulli's *Hydrodynamica*.

For details about Boyle's contributions to gas theory, see Nora M. Mohler (1939) "The spring and weight of the air," *American Journal of Physics*, 7, pp. 380–9. The history of Boyle's law is unraveled by I. Bernard Cohen (1964) "Newton, Hooke, and 'Boyle's Law' (Discovered by Power and Towneley)," *Nature*, 204, pp. 618–21.

Despite the superb article, now more than eighty years old, by W. James Lyon (1938) "Inaccuracies in the textbook discussions of the ordinary gas laws," *American Journal of Physics*, 6, 256–9, to this day credit is still given to Charles for what is solely the intellectual property of Gay-Lussac. D. S. L. Cardwell, a historian of thermodynamics, unearthed the origins of how the law for the expansion of gases has become wrongly attributed to Charles. Although Cardwell argues than if anyone deserves to share honors with Gay-Lussac it is John Dalton, "Gay-Lussac was more systematic ... He had the new professional touch: he gave full details of his apparatus and the experiments he carried out." See D. S. L. Cardwell (1971) *From Watt to Clausius: the rise of thermodynamics in the early industrial age*. Ithaca: Cornell University Press, pp. 129–31.

Pressure and density versus height for the U.S. Standard Atmosphere were obtained from *U.S. Standard Atmosphere*. Washington, DC: U.S. Government Printing Office (1976).

Recognition that gas pressure is a consequence of molecular motion rather than gas molecules pushing against each other goes back at least to Maxwell (1860s). For example, in his paper entitled "On the dynamical evidence of the molecular constitution of bodies," listed as number 71 in *The Scientific Papers of James Clerk Maxwell* (Dover, 1960), he asserts that "we may ... attribute the pressure of a fluid either to the motion of its particles or to a repulsion between them ... The pressure of a gas cannot ... be explained by assuming repulsive forces between the particles. It must therefore depend, in whole or in part, on the motion of the particles." Martin Knudsen's much later endorsement of this is in his 1950 *The kinetic theory of gases: some modern aspects*. 3rd edn. London: Methuen, p. 2.

The Maxwell–Boltzmann distribution of molecular speeds is derived in, for example, the books on kinetic theory previously cited.

For more on how Earth's atmosphere is leaking, see David C. Catling and Kevin J. Zahnle (2009) "The planetary air leak," *Scientific American,* 300(5), pp. 36–43.

To learn more about atmospheric tides see Frank Le Blancq (2011) "Diurnal pressure variation: the atmospheric tide," *Weather,* 66, pp. 306–07, and Sydney Chapman and Richard S. Lindzen (1970) *Atmospheric tides: thermal and gravitational.* Dordrecht: D. Reidel.

An illustration of how a dense network of barometers can be used to examine the pressure variations associated with different scales of atmospheric motions is provided in Alexander A. Jacques, John D. Horel, Erik T. Crosman, and Frank L. Vernon (2015:) "Central and eastern U.S. surface pressure variations derived from the US Array Network," *Monthly Weather Review,* 143, pp. 1472–93.

Although the official lowest *measured* pressure is 870 mb for Typhoon Tip, 860 mb was *inferred* for *Haiyan.* See Karl Hoarau, Mark Lander, Rosalina De Guzman, Chip Guard, and Rose Barba (2017) "Did typhoon *Haiyan* have a new record minimum pressure?" *Weather,* 72, pp. 291–95.

For a history of blood pressure measurement see Jeremy Booth (1977) "A short history of blood pressure measurement," *Proceedings of the Royal Society of Medicine,* 70, pp. 793–99.

For a history, various derivations, and extensions of the barometric formula see Mário N. Bereberan-Santos, Evgeny N. Bodunov, and Lionello Pogliani (1997) "On the barometric formula," *American Journal of Physics,* 65, pp. 404–11.

Our postulate of local thermodynamic equilibrium is similar to that given by Donald D. Fitts (1962) *Nonequilibrium thermodynamics.* London: McGraw-Hill, ch. 3.

To appreciate the many contributions of John Dalton to science, see the collection of essays in D. S. L. Cardwell (ed.) (1968) *John Dalton & the progress of science.* Manchester: Manchester University Press, especially Gordon Manley's essay on Dalton's accomplishments in meteorology. We also recommend the essays by W. D. Wright and Cyril Hinshelwood. A biographical sketch of Dalton is by Arnold Thackray in the *Dictionary of Scientific Biography.*

For the history of the transition of Avogadro's constant from a hypothesis more than 200 years ago to a quantity measured with increasing precision by different methods to its final status as an SI base unit defined to eight digits beyond the decimal point, see Peter Becker (2001) "History and progress in the accurate determination of the Avogadro constant," *Reports on Progress in Physics,* 64, pp. 1945–2008.

For a clever, simple measurement of pressure inside a balloon during inflation (and deflation) see Julien Vandermarlière (2016) "On the inflation of a rubber balloon," *The Physics Teacher,* 54, pp. 56–7. The author used the barometer of a smartphone inserted into the balloon to measure its pressure history. Theories of the pressure in a balloon are not simple because rubber is not a simple linear elastic material. Not only that, but it also exhibits *hysteresis*: the inflation and deflation histories are not the same

We now take the Fahrenheit and Celsius temperature scales for granted, but they are the end products of evolution over more than a hundred years. Thirty-five different scales in the period 1641–1772 are listed in Henry C. Bolton (1900) *Evolution of the thermometer: 1592–1743.* New York: Chemical Publishing Company.

For criticism of the terms intensive and extensive see Otto Redlich (1970) "Intensive and extensive properties," *Journal of Chemical Education*, 47, pp. 154–6.

Oxygen in Earth's atmosphere is well-mixed horizontally and vertically with an average concentration of 20.95%. To show this required developing methods over about a hundred years beginning in the late eighteenth century by some of the most accomplished scientists of their time. For an excellent history, see Francis Gano Benedict (1912) *The composition of the atmosphere with special reference to its oxygen content.* Publication No. 166. Washington, DC: Carnegie Institution of Washington.

For interesting speculations about how the surface pressure of Earth's atmosphere might have been much higher in the distant past see Octave Levenspiel, Thomas J. Fitzgerald, and Donald Pettit (2000) "Learning from the past: Earth's atmosphere before the age of dinosaurs," *Chemical Innovation*, 30(12), pp. 50–5.

For measurements of temperature and wind fluctuations in ventilated rooms see A. K. Melikov, U. Kröger, G. Zhou, T. L. Madsen, and G. Langkilde (1997) "Air temperature fluctuations in rooms," *Building and Environment*, 32, pp. 101–14, figs. 1–4. For measured temperature fluctuations from 6 m to 200 m above the ground made from a captive balloon, see C. E. Coulman (1969) "A quantitative treatment of solar 'seeing'," *Solar Physics*, 7, p. 122–43, fig. 3.

As an example of how complicated an equation of state for a liquid can be, see C. A. Jeffrey and P. H. Austin (1998) "A new analytic equation of state for water," *Journal of Chemical Physics*, 110, pp. 484–96.

Despite its tenuousness, the lunar atmosphere is quite interesting. See Alan S. Stern (1999) "The lunar atmosphere," *Reviews of Geophysics*, 37, pp. 453–91.

For a discussion of the potential application of pressure observations from smartphones, see Clifford F. Mass and Luke E. Madaus (2014) "Surface pressure observations from smartphones: A potential revolution for high-resolution weather prediction?" *Bulletin of the American Meteorological Society*, 95, pp.1343–9.

For a history of the centuries-long quest for ever-lower pressures see P. A. Redhead (1999) "The ultimate vacuum," *Vacuum*, 53, pp. 137–49.

For a good history of the belief in a flat Earth, see Christine Garwood (2007) *Flat Earth: the history of an infamous idea*. New York: Macmillan. Many modern flat-Earthers are just having fun by being contrary, but some really do believe that Earth is flat.

For more about von Guericke's demonstration with horses and hemispheres (Problem 2.81) see a critical account of a re-creation in Theodore E. Madey and William L. Brown (eds.) (1984) *History of vacuum science and technology*. College Park, MD: American Institute of Physics, pp. 61–3.

For a good treatment of the basic physics of water wheels, see Mark Denny (2007) *Ingenium: five machines that changed the world*. Baltimore: Johns Hopkins University Press, ch. 2. If this whets your appetite for more, see the review by Emanuele Quaranta and Roberto Revelli (2018) "Gravity water wheels as a micro hydropower energy source: a review based on historic data, design methods, efficiencies, and modern optimizations," *Renewable and Sustainable Energy Reviews*, 97, pp. 414–27.

For Stephen Brush's assertions about the molecular interpretation of gas pressure, see his 1970 "Interatomic forces and gas theory from Newton to Lennard-Jones," *Archive for Rational Mechanics and Analysis*, 39, pp. 1–29. This is reprinted in his anthology of classic papers from 2003, *The kinetic theory of gases*. London: Imperial College Press. He also avers that "there is serious doubt as to whether it is worthwhile trying to establish a single 'realistic' force law for the interaction between two atoms or molecules."

For an account of Buys-Ballot's role in the mean free path concept, see Stephen G. Brush (1983) *Statistical physics and the atomic theory of matter, from Boyle and Newton to Landau and Onsager*. Princeton: Princeton University Press, pp. 53–4. For how Dalton got the right result by wrong reasoning, see pp. 31–4. The first chapter is a more or less correct concise history of the kinetic theory of gases, including how Boyle obtained by a simple but clever experiment the gas law that bears his name. As a physicist–historian of physics, most notably kinetic theory and thermodynamics, Brush recognizes that "heat is a form of molecular motion" ... "is not very meaningful" (p. 49) but does express the interpretation of nineteenth-century scientists. You can be certain that Brush's views are informed by a careful reading of the original papers. He does not propagate myths. As just one example, he wrote on page 68 that the system of Copernicus "achieved no significant reduction in the number of circular motions needed to represent planetary motions."

For a delightful account of the discovery of the noble gases, especially argon, see John H. Wolfenden (1969) "The noble gases and the periodic table: telling it like it was," *Journal of Chemical Education*, 46, pp. 569–76. He notes that the true story of how scientific discoveries were made in chemistry tend to be replaced by "what the story *might* have been if the [past] investigators had been infallible rather than human as well as conversant with the intellectual framework of chemistry today. Truth is not always stranger than fiction but it is consistently more interesting." This applies to historical truths replaced by myths in all sciences.

For a well-researched, fascinating, and remarkably clear treatment of the long history of the metric system see John Bemelmans Marciano (2014) *Whatever happened to the metric system? How America kept its feet.* New York: Blooms-bury. Despite its subtitle, this book is not just about America and includes standardization of currency, the prime meridian, and time standards. Not long ago the number of different measures of weight and length was a metrical Tower of Babel.

The unlikely origins of statistical physics, exemplified by the Maxwell–Boltzmann distribution, are in social physics. In 1874, Maxwell wrote "The Modern Atomists have ... adopted a method ... new in ... mathematical physics, though it has long been in use in ... statistics." See Theodore M. Porter (1981) "A statistical survey of gases: Maxwell's social physics," *Historical Studies in the Physical Sciences*, 12, pp. 77–116. Social physics yields averages having deviations of order 1% for large ensembles (e.g. millions of human sex ratios at birth); statistical physics yields averages having exceedingly small deviations for huge ensembles of molecules; ensemble weather forecasting yields averages having appreciable deviations for small ensembles of forecasts. For a history of ensemble forecasting see, John M. Lewis (2005) "Roots of ensemble forecasting," *Monthly Weather Review*, 133, pp. 1865–85.

Atmospheric *thermo*dynamics originated in the early eighteenth century because

> observations multiplied, as did pressure-height rules based on them ... Each investigator, from his unique set of incorrect measurements, adopted his own formula, constants, and theory ... The way out of this chaos lay in the perception of the hypsometric question as a physical or meteorological, not simply a hydrostatic problem...the actual condition of the atmosphere had to be taken into account.

See Theodore S. Feldman (1985) "Applied mathematics and the quantification of experimental physics: the example of barometric hypsometry," *Historical Studies in the Physical Sciences*, 15, pp. 127–95

Problems

1. An aerodynamicist once was interviewed on a radio science program about his theory that millions of years ago the density of the atmosphere was twice what it is today. This conclusion was based on calculations showing that the giant reptilian birds of eons ago could not fly in the present atmosphere. Climatologists, however, criticized this theory on the grounds that the increased density would imply temperatures higher than are believed to have existed in the past. His response to this criticism was as follows: "I thought that it was

rather common knowledge that as temperature increases density *decreases.*"
Discuss this response.

2. The following passage was taken from a high school textbook on earth science:
"A small mass of cool air over a pond is a parcel of air. A small mass of warm
air over a nearby parking lot is another parcel of air. Because the molecules in
the warm-air parcel are more spread out, there are fewer molecules in the warm
air than in the cool air. The weight of the warm air is less than the weight
of the cool air. As a result, the warm air exerts less pressure on the parking
lot than the cool air exerts on the water. The higher the air's temperature,
the lower the pressure." Suppose that you are a consultant to the company
that published this textbook. Your opinion is sought on its revision. What
comments would you have on the preceding paragraph?

3. When a cold house is heated quickly by turning on a furnace or lighting a
fire, why don't the windows blow out because of the increased pressure? The
volume of a house is constant, and everyone knows that pressure increases
with temperature when volume is held constant. If you answer this question,
don't abandon common sense. You have lived in houses, so don't ascribe to
them magical properties they cannot have. Before answering this question,
you might estimate the total force on a large window as a consequence of a
10–20% rise in pressure inside a house above that outside.

4. A large air-tight container is fitted with a manometer and an expansion chamber (see Fig. 2.14). Pressure changes Δp associated with volume changes ΔV of
the expansion chamber can be measured with a manometer. With this device
it is possible to measure a person's volume, and hence, density. How? State all
necessary assumptions. Again, as in the previous problem, don't throw common sense to the wind. Whatever scheme you devise must be plausible. Don't
let the fact that humans breathe bog you down. The solution to this problem
is the same for a human or a large rock in the chamber. *Hint:* The volume
changes of the expansion chamber are small compared with the total volume
of the air-tight container.

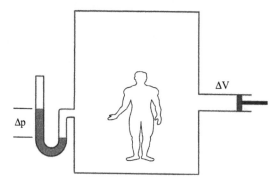

Fig. 2.14: A large air-tight container fitted with a manometer and an expansion
chamber for measuring a person's volume and density.

5. It is common for people to describe hot, humid air as "heavy." How do you interpret this?

6. One way that pressure is regulated in a pressure cooker is to put a weight over a small opening in the lid of the cooker. As the cooker is heated, the internal pressure increases until the weight is lifted slightly and gases exit the cooker, thereby decreasing the internal pressure until the weight again prevents gases from exiting the cooker. If the mass of the weight is 0.1 kg and the diameter of the opening is 2.5 mm, at what internal pressure will gases begin to escape from the cooker? Assume an external pressure of 1000 mb (10^5 Pa). When you obtain your answer, ask yourself if it makes sense on the basis of what you know (or can learn) about pressure cookers.

7. In 1993, a brief note appeared in the August 30 issue of *Chemical & Engineering News* about a column in a Cincinnati newspaper in which an explanation was given for why so many home runs occur in the city's stadium when the weather is hot and humid. A local TV weatherman provided the explanation: "A, hot air molecules are bigger than cold air molecules, so there's less resistance; and B, hot air is always rising, so there's nothing to hold the ball down." Don't bother to criticize this explanation. Instead, provide your own. That is, give a plausible physical explanation on the basis of what you learned in this chapter. You are not being asked to give *the* explanation, but rather, *an* explanation—one that might not be correct, but at least isn't idiotic.

8. If you insert a straw into water, cover the end not submerged with your finger, then remove the straw, some water is retained in it. If you remove your finger from the end of the straw, water flows from it. Please explain. If you just put the straw in water, don't put your finger over the open end, and withdraw the straw, it does not retain water. Is this result independent of the straw (or glass tube) diameter? Can you do simple experiments to determine if it is? This problem is more subtle than you might think.

9. Because the atmosphere is never completely free of particles, air is a mixture of various gases and particles. Show that the presence of particles in air results in only a very small correction to the ideal gas law. *Hint*: Liquid water content of cumulonimbus is around $1\,\text{g m}^{-3}$.

10. A snorkel is a tube used by skin divers for breathing while swimming under water near the surface. Why aren't these tubes longer than about 50 cm? Why is it necessary for divers to carry tanks of air in order to descend well below the surface? Why don't divers just breathe through long tubes that protrude above the surface of the water? *Hint*: The length of snorkels is not limited by engineering considerations. To answer this question requires no more than the ideal gas law and a bit of common sense about the capability of humans.

11. There is no well-defined top to the atmosphere: it just slowly fades away. Yet the real atmosphere is equivalent to (has the same weight per unit area as) a fictitious atmosphere with constant density equal to the surface value up to a finite height and with zero density above this height. Show that the height of this fictitious atmosphere is the scale height for density of the real atmosphere. This *uniform, finite atmosphere model* is often sufficient for simple analysis that can sidestep the vertical distribution of density in an atmosphere with no well-defined upper boundary.

12. Compute what the pressure of liquid water would be (at room temperature) if it obeyed the ideal gas law. The gas constant for water vapor R_v is $461.5 \, \mathrm{J \, kg^{-1} \, K^{-1}}$. On the basis of your calculations, what do you conclude?

13. Suppose that you wake up one morning to discover that a tire on your bicycle is completely flat. You have a pressure gauge, with which you verify that the tire pressure it reads is indeed zero. Do you then deduce that a flat tire is a perfect vacuum?

14. One of the authors once drove from Texas to Iowa during December. The temperatures during this three-day trip ranged from $80\,°\mathrm{F}$ to $20\,°\mathrm{F}$. If the tires of his car did not leak on the trip, how much variation in tire pressure might have been expected? Tire pressure is measured after the car has been sitting for several hours. Assume that a tire gauge indicated 40 psi for all tires at the beginning of the trip. *Hint*: Don't do this problem before understanding the previous one.

15. The acceleration due to gravity, g, decreases with height, yet in computing such quantities as the total mass of the atmosphere, we often take g to be constant. How can we get away with this? Make your answer quantitative.

16. Show that because Earth is a sphere (more or less) and the atmosphere has a finite scale height, surface pressure is not exactly equal to the weight of the atmosphere per unit area of Earth's surface even for strict hydrostatic equilibrium. Estimate the magnitude of the discrepancy between the weight per unit area and surface pressure. *Hints*: Write the hydrostatic equation in spherical polar coordinates and assume that atmospheric variables depend only on the radial coordinate. Or, to obtain a quick-and-dirty answer to this problem, you can pretend that the atmosphere has a finite thickness (the scale height) and uniform density equal to the sea-level value (see Problem 11).

17. Find a set of about 50 measurements of any variable with an appreciable range (a factor of two or more). Examples are the price of 50 common stocks; the populations of 50 cities and towns; the weekly maximum or minimum temperatures where you live. For this set of data calculate the following averages: arithmetic mean (what is often taken, without qualification, to be *the average*), root mean square, logarithmic mean, and root mean cube. The logarithmic mean is the inverse logarithm of the arithmetic mean logarithm of the values; the root mean cube is the cube root of the arithmetic mean cube of the values. What do you conclude from your analysis?

18. Cold air, we are often told, is denser than hot air. Try to think of an empirically verifiable fact, known to many people, that contradicts this assertion taken at face value (i.e. with no qualification).

19. If you have a taste for precision measurements, you might try to think of as many corrections as possible, and estimate their magnitude, that have to be applied to the measurement of atmospheric pressure with a barometer. For example, to convert a height (the quantity measured) into pressure, one needs the density of mercury, which varies with temperature. Try to find the coefficient of thermal expansion of mercury so that you can estimate the maximum error in pressure obtained by assuming a constant density for mercury One

also needs g, the acceleration due to gravity, which varies over the planet. If you look in a book on geophysics, you may find the horizontal variation of g. What about the vertical variation? That is, how much does the variation in g between Death Valley, California, and the summit of Mt. Everest affect barometer readings? The pressure in the barometer tube above the liquid column is assumed to be zero, another source of error (see Problem 5.13). And what about the effect of surface tension (capillarity)?

You might think that the barometer tube must have a uniform cross-sectional area. Convince yourself that this is not necessary and that a nonuniform tube introduces no error.

20. An alert thermodynamics student, Peter Ernst, brought us an article entitled "Overcoming Altitude" in a mountain biking magazine. To set the stage, the author of this article tells readers that "the reason it's harder to breathe at altitude isn't that there's less oxygen in your air. Rather, there's less barometric pressure forcing air into your lungs." Do you accept this explanation? Please discuss. If you would like to learn more about high-altitude physiology, including its history, we recommend the excellent collection of articles edited by John B. West (1981) *High altitude physiology*. Stroudsburg, PA: Hutchison Ross. This collection contains several classics of mountaineering and also excerpts from the pioneering work of Paul Bert in the previous century.

21. In our everyday lives we encounter mostly temperatures expressed in Celsius or Fahrenheit degrees. Yet, no physical quantity can depend on these temperatures because zero on these scales is arbitrary. The only exception is physical quantities that depend on temperature *differences*.

Our experience, and that of colleagues, is that in solving problems involving temperature, students frequently fall into the trap of interpreting T as temperature in °C or °F. Failure to convert temperatures to degrees Kelvin often leads to absurd results. For example, the mean speed of a gas molecule is proportional to the square root of absolute temperature T. But if you carelessly interpret T as a temperature in degrees Celsius, you are forced to conclude that at temperatures below $0\,°C$ the mean speed of a molecule is the square root of a negative number. To drive home the importance of expressing temperatures in degrees Kelvin in calculations, find as many examples as you can in this chapter and throughout this book of absurd results that follow from failing to remember that T in any expression for a physical quantity is in degrees Kelvin, not Celsius or Fahrenheit.

22. Suppose that you need to measure temperature profiles in the atmosphere to within $0.1\,°C$ using a balloon-borne thermometer. A typical lapse rate is about $5\,°C\,km^{-1}$; the ascent rate of the balloon is about $2\,m\,s^{-1}$. Estimate the greatest response time of the thermometer that will enable you to accomplish your task.

23. If you have a thermometer, or can get your hands on one, devise a simple experiment to measure its response time. Compare the response time of your thermometer in air and in water.

24. A colleague of ours happened upon an article in which it was claimed that Napoleon decided, on the advice of his scientific experts, against building what was eventually the Suez Canal linking the Mediterranean and Red Seas

because of the approximately one-meter elevation difference between Port Said and Suez, a distance of about 180 km. Our colleague was curious about the possibility of measuring this elevation difference with a barometer. How accurately would you have to measure barometric pressure in order to determine an elevation difference of one meter? What problems would you face in trying to determine by barometric measurements the one-meter elevation difference between Port Said and Suez?

25. The elevation of Las Vegas, Nevada, is about 600 m. That of Boulder, Colorado, is more than 1600 m. A few years ago the University of Nevada at Las Vegas (UNLV) basketball team played the University of Colorado team in Boulder. The coach of the UNLV team was interviewed beforehand and asked if he was concerned about how the higher elevation would affect his players. His response was, "I'm not worried. My players won't be affected by the elevation. We're playing indoors." Please discuss the coach's response.

26. One of the authors used to drive a pickup truck older than most of his students. When he was inside the truck and closed its door, the latch frequently didn't catch. Repeated slamming of the door didn't help. So he rolled the window down and slammed the door, which then shut tightly. Please explain.

27. During the 1996 Olympic Games in Atlanta, newspapers and magazines noted *ad nauseam* that certain athletes have only 3% body fat. From this one would gather that there is also an international competition for the lowest body fat. How might this number be measured? An autopsy would be scientifically acceptable, but the athlete might complain. *Hint*: See Problem 4.

28. The kinetic energy density of wind (mass motion) is a product of density, which decreases with height, and wind speed squared, which generally increases with height (in the troposphere). Estimate the height at which the kinetic energy density of wind is a maximum. You may assume that wind speed increases linearly with height and is zero at the surface. The latter assumption is called the *no-slip boundary condition*. For a demonstration of this, liberally sprinkle talcum powder on a sheet of clean, black matte paper and do your best to blow all the powder off the paper. You are unlikely to be successful.

29. One of our former students works at an airport. He noticed that the maximum allowable load of airplanes depends only on air temperature and asked us about this. What would you have told him? How would you expect the maximum load for a given airplane to depend on the location of the airport? For the sake of concreteness, consider the difference between the airports at La Paz, Bolivia, and Buenos Aires, Argentina. For a given airport and airplane, what would you expect for the range (expressed as a percentage) of maximum loads? *Hints*: To answer this question you must know or learn what determines the maximum lift an airplane can generate. And careful examination of the pressure and temperature data in Fig. 2.1 (and similar records) should help you.

30. Many baseball fans know that Coors Field, home of the Colorado Rockies, in Denver, is the best park in the country to hit home runs. In 1996, for example, there were 271 home runs in Coors Field compared with 155 in Pittsburgh's Three Rivers Stadium, similar in size to Coors Field (410–415 feet to the center field fence). Even more telling, during the 1996 season the Rockies had 149

home runs at home compared with 72 on the road. One of our students had heard that the ease of hitting home runs in Denver was due to lower gravity there. Please respond to this assertion with a physical explanation supported by rough estimates of how different factors could affect the distance a well-hit ball might travel. To learn more about the physics of baseball, we highly recommend Robert K. Adair (1994) *The physics of baseball.* 2nd edn. New York: Harper.

31. The escape velocity for Earth is about 11 km s^{-1}. This is the minimum velocity an object must have in order to overcome the gravitational attraction of Earth. Does a rocket to the Moon have to reach the escape velocity? What is the essential difference between a rocket and, say, a gas molecule, or, for that matter, a grand piano propelled by a giant slingshot toward the Moon? You may ignore the presence of the atmosphere.

32. A recent theory of ice ages attributes them to Earth passing through an intensely cold region of space. What do you think of this theory?

33. By assuming constant temperature we were able to integrate the hydrostatic equation combined with the ideal gas law to obtain an expression for the exponential decrease of pressure with height. But suppose that temperature decreases uniformly with height. That is, $-dT/dz = \Gamma$, where Γ is called the *lapse rate.* How does pressure now decrease with height? On physical grounds you should be able to guess first if pressure decreases with height more or less rapidly when temperature decreases with height than it does in an isothermal atmosphere. Check your mathematical result against your physical intuition.

34. A relation between pressure and altitude is the basis for *altimetry*, the measurement of altitude. Airplanes are equipped with altimeters that convert pressure into altitude. If the atmosphere were isothermal, z would be proportional to $\ln(p/p_0)$. But the atmosphere is not isothermal, and hence, this relationship is not good enough. Airplane altimeters are based on the U.S. Standard Atmosphere, for which the surface temperature is 15 °C, the surface pressure is 1013.25 mb at mean sea level, and the lapse rate is 6.5 °C km^{-1}. What error (percent) is made in the difference between any two altitudes by assuming a constant temperature equal to the standard sea-level value instead of a constant standard lapse rate and sea-level temperature? *Hint:* The easiest way to do this problem is to obtain a relation between $\ln(p/p_0)$ and altitude and expand the relation for a nonisothermal atmosphere in a power series, truncating the series at the quadratic term.

35. Obtain an expression for temperature versus pressure altitude in an atmosphere with a constant lapse rate. *Hint:* This is a variation on Problem 33. By pressure altitude is meant the pressure at a given altitude.

36. If the lapse rate of the atmosphere were constant with height, what would be the thickness of the entire atmosphere? Estimate a numerical value for this thickness using a realistic value for surface temperature and lapse rate. *Hint:* You can do this problem by straightforward integration, or you can be crafty and guess the answer based on your knowledge of the dimensions of thickness and of the only physical quantities that enter into this problem. To guide you, also make use of the physical interpretation of thickness as an average pressure. How do you expect thickness to change with surface temperature? How do you expect thickness to change with lapse rate?

37. Obtain an expression for the thickness between any two arbitrary pressures p_1 and p_2 for a constant lapse rate. This is a generalization of the previous problem.

38. In physics textbooks the atmosphere usually does not exist (too messy), and hence, bodies falling in Earth's gravitational field accelerate without limit. In the rare physics textbook that acknowledges the existence of the atmosphere, it may be noted that because of drag (air resistance) a falling body steadily approaches (but never reaches) a *terminal velocity*. But reality is even more complicated. The drag experienced by an object moving in a fluid depends on, among other things, its density. As a consequence, a body falling in Earth's atmosphere is subject to drag that varies with altitude. Although the equation of motion for such a body is messy to solve, one can arrive at some general conclusions without extensive computations. Show that a body falling in air of uniform density does not have a maximum velocity (although it does have a limiting velocity). Then show that when the altitude dependence of air density is accounted for, a falling body can have a maximum velocity greater than the terminal velocity in sea-level air. You may assume that the drag force is $\rho v^2 C_D A/2$, where ρ is the density of air, C_D is the drag coefficient (assumed to be constant) of the body, A is its cross-sectional area, and v is its velocity. Also show (by crude arguments) that the maximum velocity should be greater the higher the altitude from which the body (initially at rest) is dropped. If this problem piques your interest, we recommend Mohazzabi, P. and Shea, J. H. (1996) "High altitude free fall," *American Journal of Physics*, 64, pp. 1242–6, which begins with a fascinating true story of a high-altitude free fall. Reality is often more interesting than fiction.

39. In 1994, the October 24 issue of *El Pais*, Spain's leading newspaper, had an article about a new record for the distance traveled in one hour by a cyclist. The line above the article read "Tony Rominger dethrones Miguel Induráin." Rominger rode his bicycle 53.832 km, 792 m more than Induráin had done on the same track. In the body of the article we find the following (loosely translated): "humidity, a factor which increases air resistance ... was 78%, 18% more than during Induráin's attempt on the same track." This assertion embodies some misconceptions and errors. Please discuss.

40. Ask someone how a drinking straw works. You will likely be told that you "suck water up the straw." But this is a description, not an explanation. Please provide one and make it quantitative. It will help if you carefully observe what you do when you use a straw.

41. What is the kinetic energy density of wind relative to the random kinetic energy density of air? All that is wanted is an approximation. What is the ultimate origin of the kinetic energy of wind? What does this tell you?

42. The concentration of carbon dioxide in exhaled air is about 100 times greater than in inhaled air. Oxygen is consumed in metabolism, slow burning of carbohydrates in the human body, a reaction product of which is carbon dioxide. Suppose you are in a perfectly sealed space. How long before you would

be uncomfortable or even sick? You need the average tidal volume, the normal volume of air inhaled and exhaled, and the resting breathing rate. This is a difficult problem to do rigorously. As a first approximation consider what happens in a single breath, then multiply this by the number of breaths in a particular time. Assume that the exhaled air, depleted slightly of oxygen rapidly mixes with the ambient air. Derive an expression for the decrease of oxygen concentration in one breath and be sure that it makes physical sense. Assume that every carbon dioxide molecule produced is at the expense of one oxygen molecule. Within limits you can multiply this by the number of breaths in a particular time. Eventually, however, the linear equation will cease to be valid. More likely the decrease will be exponential. If you are ambitious, you can write a program that will determine the oxygen fraction in ambient air for each breath. If you have a morbid sense, estimate how long you could live in a well-sealed coffin.

43. The definition of isothermal compressibility suggests processes in which an increase in pressure *causes* a decrease in volume. But compressibility is just as applicable to the reverse process: a decrease in volume *causes* an increase in pressure. The compressibility of liquid water is about $46\text{Pa}^{-1} \times 10^{-11}$. How much of a fractional volume decrease is necessary for a substantial pressure increase, say, of order 10 kPa? Can you find such a process that all normal humans experience every day, although not aware of it and not under conscious control? *Hint*: We should be grateful for this process, without which we'd forever be infants.

44. In Section 2.2 we give scale heights for the decrease of the number density of air molecules. But the scale height for the decrease of number density of water vapor is roughly one-fourth that of air. Why? In Chapter 5 we give another related problem.

45. Given the scale height for air, what is the total number of air molecules in the entire atmosphere per unit area expressed in terms of the surface number density?

46. Estimate the vertical pressure gradient at Earth's surface. How does this compare with maximum horizontal pressure gradients? Do not express these gradients to more than one significant digit. The word "estimate" is a signal to not give results to nine digits, even though your calculator does. You can use judgment, your calculator cannot. The horizontal gradient requires poring over surface pressure maps with a horizontal distance scale or finding measurements.

47. From the ideal gas law, if number density decreases pressure decreases accordingly for constant temperature. Consider a fixed volume containing atoms of a particular kind, for sake of argument, hydrogen. They react to form molecular hydrogen. The number of molecules decreases by half, and hence, so does the pressure. But although the number density has decreased, the mass per molecule has doubled. Why, therefore, does the pressure decrease? *Hint*: Carefully consider and state all assumptions. You need to look at this from a molecular point of view.

48. One of the authors has a hunting lodge attached to his house. A door in the lodge opens into a wood room, the door to which just barely closes securely. The two doors are at right angles about two meters apart. If from outside the lodge its door is quickly opened, the door to the wood room opens (into the lodge) about half a centimeter and then stops even though it is free to swing. The time this takes is difficult to measure with a stopwatch, but it is of order one second. If the lodge door is left open and the wood room door secured, then when the lodge door is closed the wood room door remains secured. Explain. To make this quantitative, the dimensions of the wood room door are 0.9 m × 1.8 m.

49. How does a circular rubber suction cup work? If you lightly place it on the smooth surface of an object, there is essentially no adhesion between the cup and the surface. But if you push downward on the cup and flatten it you can lift the object (if it is not too heavy). Why? What is the maximum weight (per unit area of the cup) that can be lifted? Area is that corresponding to the diameter of the cup. Try to solve this problem with the least amount of effort and verify your prediction experimentally.

50. You probably have read or heard that a blood pressure of 120/80 or lower is "normal." What do these numbers mean? What are their units? Stephen Hales is credited with the discovery of blood pressure in the middle of the eighteenth century. He inserted a brass pipe of diameter 1/6 inch into the artery of a 14-year-old mare tied down on her back. A glass tube nine feet long of about the same diameter was affixed to the brass pipe. The blood rose in the tube to a height of 8 feet, 3 inches. From this, estimate the maximum blood pressure of the mare. You can take the density of whole blood to be 1060 kg m^{-3}. Do a bit of searching to determine if your estimate is realistic.

51. With a small hand-held anemometer, determine the highest wind speed you can attain with your breath. From this estimate the relative pressure difference between your lungs and the ambient air while you are blowing. *Hint*: Apply Bernoulli's principle. From this pressure difference, estimate the corresponding relative volume change. Based on your results, critically examine the story of *The Three Little Pigs*.

52. Strictly, molecules in an ideal gas do not exert forces on each other. If so, why is the isothermal compressibility of such a gas finite? *Hint*: Think about how one would measure compressibility. Ask yourself if a compressibility $1/p$ makes physical sense. Strictly by physical arguments you should be able to guess that the compressibility decreases with increasing pressure. Moreover, the dimensions of compressibility are inverse pressure. If you only memorize a formula, you have no more comprehension than a parrot. Try to understand formulas instead of memorizing them. The ability to understand distinguishes you from a parrot.

53. Under what conditions would density *increase* with height in Earth's atmosphere? At what heights would this most likely occur?

54. Above what height in Earth's atmosphere would you expect the postulate of local thermodynamic equilibrium to be increasingly invalid? All that is wanted

is a rough estimate. How does your calculation depend on assumptions you had to make?

55. We have often seen the assertion that carbon dioxide is heavier than air. Without qualification this is meaningless. Discuss.

56. Air molecules, we are told, are in perpetual rapid (on the human scale) motion. But we are also told that perpetual motion is impossible. Reconcile these two apparently contradictory statements. What is the ultimate source of the motion of air molecules?

57. Show by a simple argument that molecules incident on a surface at a given velocity must reside on it for a finite time before being thrown off at a different velocity. *Hint:* Direct your thoughts to Eq. (2.1).

58. Galileo is credited with the invention of the open air thermoscope, the precursor to the thermometer. He fused a glass bulb about the size of a hen's egg to a long, thin glass tube. He held the bulb in his hand to increase the temperature of the air within the bulb. Then he thrust the open end of the tube into water, which rose to a height of about 0.2 m. From this, estimate the difference between the temperature of his hand and air temperature. Assume that the volume of air in the tube is negligible relative to that in the bulb. If Galileo had used mercury instead of water, how high would the column of mercury risen?

59. This problem is related to the previous one. One drawback to the open air thermoscope is that the height of the liquid in the tube depends on both air temperature and barometric pressure changes. Estimate the error in air temperature because of plausible barometric pressure variations. Assume that the thermoscope is calibrated for some atmosphere pressure p_{a0} and air temperature T_0 so that $n_0 = p_{a0}/kT_0$ in the bulb is fixed.

60. This problem is related to the previous two. A smaller source of error in a thermoscope is often swept under the rug. What is it and how can it be reduced? And there is an even smaller source of error. What is it? This may seem like nitpicking but ferreting out all possible sources of error and either correcting for, eliminating them, or determining that they are negligible for the purpose at hand is the essence of making good measurements.

61. The unqualified assertion that air expands upon heating has appeared in print countless times. Discuss.

62. If $g(x)$ is a point function and we want to transform from x to $x = h(y)$, the transformed point function is $\bar{g}(y) = g[h(y)]$. But this is not valid for transforming distribution functions. Transform the distribution function Eq. (2.51) for v to a distribution for $E = mv^2/2$. The integral of $\bar{f}(E)$ over equivalent limits must be the same as the integral of $f(v)$.

63. If you did the previous problem correctly you will have discovered that when the Maxwell–Boltzmann distribution is transformed from speed to energy, the result is a function that cannot be integrated analytically. The distribution function can be made simpler by assuming that the molecules are confined to a plane and deriving a two-dimensional Maxwell–Boltzmann distribution

using the same kind of arguments underlying the three-dimensional distribution. Then transform the two-dimensional distribution to energy and find the fraction of molecules with energy greater than some specified value E_0. With this result, what can you say about how this fraction changes if the temperature is increased from T_1 to T_2? Why might this have consequences for rates of chemical reactions? Using the two-dimensional approximation to the three-dimensional Maxwell–Boltzmann distribution you can estimate the molecular energy greater than which the concentration of molecules at sea level is negligibly small. You will have to decide what concentration you consider to be negligibly small.

64. Explain the physics of inflating an ordinary toy balloon. *Hint*: The ideal gas law in the form Eq. (2.11) is useful. Is the pressure inside the balloon different from that outside? If so, why, and is the difference positive or negative. What observation supports your conclusion?

65. What is the average separation between liquid water molecules relative to the size of a water molecule? What does this result tell you?

66. This is a problem based on Eq. (2.11) that will help make sense of gas concentrations expressed in parts by volume. Allow the oxygen in a given volume of air to completely react with something at a given T and p. After the reaction bring the air to the same T and p and measure the volume. What is the oxygen concentration in the sample?

67. It seems not to be well known that the concentration of carbon dioxide in the air that humans exhale is 100 times that in inhaled air because of metabolism, the oxidation (slow combustion) of hydrocarbons. Carbon dioxide and water vapor are the primary reaction products. Estimate how much the world's human population contributes to the annual emission of carbon dioxide into Earth's atmosphere. We will give you one number you need, the volume of air exhaled per breath (tidal volume). Take it to be 500 ml.

68. Contrary to popular opinion, the Moon has an atmosphere. Assume a number density of 10^5 cm^{-3}. Estimate the collision mean free path. Figure 2.12 will make this easier. Based on your result, what can you say about the "temperature" of the lunar atmosphere?

69. Based on the discussion of flux densities in this and Chapter 1, *estimate* the rate at which air molecules at standard pressure and temperature collide with a wall per unit area.

70. In Section 2.1 we mention the pressure unit the torr. The lower limit of pressure in high vacuum technology has been about 10^{-14} torr for many years. Estimate the corresponding molecular number density and collision rate with a wall.

71. Estimate the smallest pressure difference that could reasonably be measured using a mercury manometer? What about a water manometer? Why is water or an oil likely to be used as the liquid in a manometer instead of mercury as in a barometer?

72. Estimate roughly the altitude in Earth's atmosphere at which the concept of pressure is almost certainly invalid.

73. Temperatures of race car tires can reach up to 90 °C. If the temperature of a tire when cold is 20 °C, what would the initial inflation pressure have to be so that the gauge pressure of a tire at 90 °C would be 35 psi (apparently, even tire pressures in Formula I cars in Grand Prix racing are expressed in psi)? How sensitive is this calculation to ambient atmospheric pressure? State all assumptions. Since race car pit crews need to underinflate tires used for tire changes during a race, they can use tables that indicate the initial tire pressure needed for a known cold tire temperature or digital pressure inflation gauges that automatically give the cold tire pressure. Write down the equations that would be needed to develop a tire inflation pressure table or to develop an algorithm that would be needed to develop a digital temperature gauge.

74. A shoe manufacturer boasted in an in-store ad that the secret to the "anti-gravity system" of one of their sport shoes was "a sole packed with millions of micro-bubbles, each filled with lighter-than-air nitrogen bubbles." What say you about the truth of this ad? Explain your response in simple terms that an ad producer might understand. What might be the basis for the "anti-gravity" claim?

75. Experiments by Berti ($\approx$ 1640) found the height of a water column in a closed pipe with an open base in a reservoir to be about 18 cubits. If we *assume* that 1 cubit = 18 inches, what surface pressure in mb (hPa) does this height correspond to without accounting for a finite water vapor pressure above the water surface in the closed pipe? At 20 °C, this pressure is about 23 mb (see Section 5.3). How would this vapor pressure affect the height of the water column? Does the 18 cubits estimate seem reasonable? Berti estimated the height of the water columns about one day after he started the experiment. What factors might have contributed to a lower water height than expected? What about the geography of Italy and where Berti lived and worked?

76. An engineer at Portland State University developed the "World's Tallest Barometer." He used vacuum-pump oil in a closed glass tube with the bottom opening submerged in a reservoir of oil. During about one year the mean column height was 12.4 m with a variation of ±0.4 m. The density of the oil is 830 kg m^{-3} at 20 °C, and has a negligible vapor pressure at this temperature. To what surface pressure does this mean height correspond, and what is the pressure variation associated with the variation? If this barometer is used to resolve pressure changes to a precision of 1 mb, how precisely must the height of the oil column be measured? How does this precision compare with the precision needed for a mercury barometer? What is the basis for the claim "world's tallest"? Do you believe it? Do not assume that he is lying.

77. Barometers in smartphones can measure altitude changes more precisely than GPS or in locations where GPS may not be available. To what precision must the pressure change be measured to give a one-meter precision in change in altitude? Use an ambient pressure of 1000 mb and temperature of 20 °C. How precisely will these two parameters have to be measured to achieve less than a 1% uncertainty in the quantity dz/dp needed to calculate an altitude change associated with a measured pressure change?

78. We state in Section 1.8 that the potential energy contribution to U of liquids and solids is not negligible. If it were, they would not be cohesive. What about *changes* in U at constant temperature because of pressure changes? The potential energy contribution to U depends on the average separation between molecules. Express the isothermal compressibility in terms of this separation in, say, water. What do you conclude from this? Take the compressibility to be 0.45 GPa^{-1}. Compare this result with that for an ideal gas.

79. Respiration is an everyday example of how your body exploits the ideal gas law. Without knowing the details of human anatomy, you should be able to explain the basic physics of inhaling, exhaling, and "holding" your breath.

80. In chemistry the *bond enthalpy* is the energy required to break a chemical bond to separate two atoms in a molecule. Bond enthalpies are expressed in units of kJ mol^{-1} and are in the range 100 to 1000 kJ mol^{-1}.

    We can consider a ball, say, resting on the surface of Earth, to be a giant two-atom molecule: one atom the ball, the other Earth. What is the gravitational bond enthalpy of this "molecule"? What does this tell you about the stability of Earth's atmosphere (how strongly it is bound to Earth)? *Hint*: By dimensional analysis you can guess the amount of energy required to completely remove a ball from the surface of Earth (neglect air resistance) to infinity. Or you can obtain this from the gravitational potential energy per unit mass. The ball is composed of molecules with molecular weight M.

    What does this problem tell you about the relative magnitude of the forces holding atoms together and the force of gravity binding objects (including its atmosphere) to Earth?

    What is the gravitational bond enthalpy on the Moon? What does this tell you?

81. In 1654, Otto von Guericke performed a demonstration in Magdeburg (Germany) treated in several textbooks. He made two brass hemispheres (diameter $\approx$ 50 cm) that could be clamped together with a good seal. A valve attached to one of the hemispheres enabled him to pump out much of the air within them. Two teams of eight horses (four pairs), each team attached to opposite ends of the hemispheres and pulling in opposite directions, could not pull them apart until the valve was opened. Determine the force necessary to do so. This story becomes more interesting if you go beyond textbooks. First, why did he use two teams? Would one team have been better or worse? In an ideal world, could the horses have pulled the hemispheres apart? A rule of thumb is that a horse can exert a horizontal force equal to its weight. How might *one* (normal) horse been able to pull the hemispheres apart?

82. The greatest depth in the ocean is about 11 km. Estimate the density at this depth. Assume a uniform temperature. Don't fret about salinity. Assume a constant compressibility for water of 46×10^{-11} Pa^{-1}. From the density determine the average separation between water molecules at 11 km and at the surface. Express your results relative to the diameter of a water molecule (≈ 0.3 nm). What do you conclude? *Hint*: Begin with the hydrostatic equation.

You will have to carry more decimal places than usual and then round off your final result.

83. In Section 2.1 we state that in one nanosecond about 10^9 air molecules at sea level strike 1 mm^2. Verify this. In fact, verify all our calculations. We do make mistakes.

84. The precursor to modern hydroelectric power plants is the gravity water wheel, a major source of power for about 2000 years until the invention of steam engines in the late 1700s. What two parameters determine the power output of a gravity water wheel? What would be required for it to yield a continuous power of 1000 W? Assume an efficiency of 70%. Assume that the inlet water flow (headrace) is near the top of the wheel. Pick reasonable values. To get a better feel for the flow rate convert it to ls^{-1}. Estimate the cross-sectional area of the inlet flow assuming a reasonable flow speed. How much does the kinetic energy of the water supply add? If you have access to a river or stream, try estimating the flow speed.

85. We usually assume constant N in Eq. (2.11). A problem in which it isn't has historical significance. A steam engine is driven by water vapor at temperatures greater than 373 K. To generate steam requires burning fuel. James Watt argued (not in these words) that in an expansion of steam the total work done in a cycle if a *fixed* amount of steam expands at constant temperature is less than if during the expansion steam is continuously added to keep the pressure constant. But the fractional decrease in work could be more than compensated for by using less steam. Show this. Express your result in terms of N_1, the fixed number of molecules, N_2, the total number of molecules in the variable number expansion at its end, and the ratio V_2/V_1 of the final volume to the initial volume.

86. Although Earth's atmosphere has no definite upper boundary, show that about 90% of the mass of the atmosphere lies in the lower 20 km.

87. Is it ever possible for a real container with volume V, but one that does not leak, to have an *exactly* fixed number of gas molecules at every instant? Assume no chemical reactions.

88. When we open a can of evaporated milk, we make two holes ($\approx$ 1 cm) in the top near the rim and opposite each other. Why two? Think of a simple experiment to show why, do it, and explain the result? Use water in the can. Don't waste milk.

89. To convince yourself that the ideal gas law is hopelessly wrong when applied to a liquid such as water (we have seen this done), calculate the pressure of water with a density 1000 kg m^{-3} and temperature 293 K from the ideal gas law.

90. The sharp change at 100 km in the mean molecular weight versus height (Fig. 2.13) cries out for a physical explanation. Try to give one that is plausible. If you seek it from your smart phone, you will have learned a fact but not how to think. An explanation does not consist of spouting jargon, which can only jog the memory of someone who knows the explanation.

91. What we might call "the balloon theory of clouds" is that they do not fall (they do) because they are "less dense than air." The concentrations and size distributions of cloud droplets are highly variable. For a concentration of 490 cm^{-3} and a narrow size distribution centered around a droplet diameter of 10 μm for continental cumuli (Squires, 1958c, Fig. 2.4), what is the density of this cloud relative to that of the air? What is the density of moist air relative to that of dry air for the same p/T?

92. The bond energy of molecular oxygen, 498 kJmole^{-1}, is the energy required to dissociate one mole into atomic oxygen. From the Maxwell–Boltzmann distribution of molecular speeds, why is raising the temperature of, say, molecular oxygen an extremely inefficient way of dissociating it into atoms? You can write a computer program to calculate the fraction of molecules with *energy* above the dissociation energy as a function of temperature. How is oxygen dissociated in Earth's (upper) atmosphere?

93. NASA's Space Simulation Vacuum Chamber is about 23,000 m^3 and can reach a pressure of about 4×10^{-6} torr. This is the largest high-vacuum chamber in the world. What is the number density at this pressure? What is the average separation between molecules? How many molecules does it contain? What is the total mass? How does this number density compare with that in space, which is variable?

94. Lord Rayleigh's discovery of argon (c. 1894) caused considerable controversy because the unpalatable implication was that a 1% component of air had escaped detection for more than a century. Why was argon so difficult to detect? He made precise measurements of the masses of gases in a two-liter flask under standard conditions. Nitrogen prepared from air was slightly denser than from ammonia, suggesting that nitrogen in air was mixed with an unknown "impurity." From the mass of the argon atom relative to that of the nitrogen molecule and the number of argon atoms relative to nitrogen, you can determine the relative masses of Rayleigh's pure and "impure" nitrogen (at the same temperature and pressure). How large a difference in mass did he have to measure? This was a brilliant piece of scientific detective work, but Rayleigh was lucky. Why? *Hint:* Equation (2.11) is the best starting point. This problem requires nothing more. Rayleigh was awarded the Nobel Prize in physics, William Ramsay in chemistry (1904) for their co-discovery of argon. It wasn't enough to detect argon. Its molecular weight and concentration had to be determined.

95. This is related to the previous problem. How might it be possible to isolate argon from a volume of air. This was done in 1895. What would be the relative volume of the isolated argon at standard temperature and pressure?

96. To get a better feeling for the newton, calculate the drag (Problem 1.13) on a car of your choice at 100 kph. Calculate the component of gravity opposite and along the direction of motion as a function of grade, the tangent of the slope angle expressed as a percent. Compare these forces with the weight of the car. What do you conclude? AC_D and masses for many makes and models

of car are easy to find. From the solution to the equation of motion of a car on an absolutely flat road and little wind, ACD can be estimated from the time it takes the speed of a car to decrease speed by a given amount if suddenly put into neutral (neglecting all other forms of resistance). Try the experiment. We measured ACD within a factor of two of the published value. Was this likely an underestimate or overestimate of drag?

97. The Pascal's barrel demonstration is dramatic, discussed in several textbooks, but rarely given a molecular interpretation. It is presented almost as a kind of miracle, or at least a paradox, that a relatively small amount of water added to a closed container of water can cause it to explode. We simply note in passing in Section 2.1 that this is because of strong repulsive molecular forces. For a detailed treatment, see Vigoureux, D. and Vigoreux, J-M. (2018) "An investigation of the effects of molecular forces in Pascal's barrel experiment," *European Journal of Physics*, 39, p. 025003 (12 pp).

98. Newton apparently believed that if all the particles in the solar system were to come together, the volume would be that of a nut. If all the molecules in Earth's atmosphere were stripped of their electrons and jammed together into a ball, how large would it be? The diameter of an atomic nucleus is about 10^{-15} m.

99. Two different types of molecules with different masses react in a fixed volume. Assume that the reaction is complete, and that the molecules react one for one. After the reaction the temperature is restored to its initial temperature. What is the final pressure of the gas? The answer is trivial if you mechanically apply the ideal gas law: one half the original pressure because the number density of molecules is halved. But the mass per molecule has increased. How do you reconcile this with the predicted factor of two decrease of pressure? This is an example of a simple problem in which getting the "right" answer is not enough if you want to understand something, rather than just invoke an equation.

100. Suppose that we mix two hypothetical, absolutely collisionless gases at different temperatures in a container at a fixed temperature. What happens? *Hint*: What is the essential characteristic of a container?

101. How many molecules in liquid water are in close contact with a thermocouple probe of diameter 1 mm? What does this tell you? What if the probe diameter were 10 nm? The diameter of a water molecule is about 0.3 nm.

102. What is the escape velocity on the Moon? What is its mean density relative to that of Earth? What does this tell you about the Moon?

103. Molecules do not have precisely defined diameters, if for no other reason than that they are not spherical. Each experimental method, based on a theory, of determining molecular diameters yields somewhat different values. Estimate the diameter of a nitrogen molecule in air from the density of liquid nitrogen, 808 kg m^{-3}. State your assumptions.

104. We assert in Section 2.1 that "The meter was originally 1/10,000,000 of the length of a meridian through Paris from the equator to the North Pole." We encountered a book by a respectable author who stated that the meter was 1/10,000 of this distance. Which is correct?

3

Specific Heats and Enthalpy

Adiabatic Processes

The first law of thermodynamics in its *symbolic* form Eq. (1.75) is of little help toward understanding atmospheric states and processes. To make it more suited to this aim, we must make approximations and cast it in terms of measurable variables by uniting it with the ideal gas law. Before performing this marriage, we offer some premarital counseling on the mathematics of thermodynamics. We advise readers who have taken courses in thermodynamics to read the following section in order to understand why we sometimes depart from notation hallowed by long use, yet physically, mathematically, and pedagogically indefensible. Percy W. Bridgman, whose contributions to thermodynamics were of the highest order, considered thermodynamics to be "the most difficult branch of physics to teach" in part because "thermodynamics uses an unfamiliar brand of mathematics." Emboldened by these words, we critically examine this brand because this book is aimed at teaching thermodynamics.

3.1 A Critique of the Mathematics of Thermodynamics

The first law of thermodynamics (Section 1.8)

$$\frac{dU}{dt} = Q + W \tag{3.1}$$

applied to a closed system for which the rate of working is

$$W = -p\frac{dV}{dt} \tag{3.2}$$

has the form

$$Q = \frac{dU}{dt} + p\frac{dV}{dt}. \tag{3.3}$$

The internal energy U of a closed *simple system* may be considered a function of only two independent variables—temperature T and volume V, for example—which determine the third variable, pressure p, by an equation of state (e.g. the ideal gas law). A simple system is "very large" in the sense described in Section

Atmospheric Thermodynamics. Second Edition. Craig F. Bohren and Bruce A. Albrecht, Oxford University Press.
© Craig Bohren and Bruce Albrecht (2023). DOI: 10.1093/oso/9780198872702.003.0003

1.7 and *isotropic*, a condition satisfied by gases, liquids, amorphous solids, and polycrystalline materials. We assume that the system is sufficiently uniform that assigning it a *single* pressure and temperature is a good approximation and, unless otherwise stated, no chemical reactions or phase changes occur. Without the former assumption, the following partial derivatives are meaningless. According to the chain rule

$$\frac{dU}{dt} = \frac{\partial U}{\partial T}\frac{dT}{dt} + \frac{\partial U}{\partial V}\frac{dV}{dt}. \tag{3.4}$$

The custom in thermodynamics is to enclose partial derivatives in parentheses and append subscripts to denote the variable or variables that are held constant in the differentiation:

$$\frac{\partial U}{\partial T} = \left(\frac{\partial U}{\partial T}\right)_V. \tag{3.5}$$

We don't follow this custom. Context is often sufficient to establish which variables are held constant. But if you wander beyond the pages of this book, you are likely to find strange-looking partial derivatives.

A peculiarity of thermodynamics is that the mathematical problems it addresses are fairly simple, but its mathematical notation is horrible—almost, it seems, designed to be confusing to compensate for the simple mathematics. Thermodynamics textbooks use bizarre mathematical notation unlike that in any other branch of physics and that mathematicians do not use. Space is wasted on belaboring the issue of whether differentials are perfect or imperfect, exact or inexact. We don't care because we dispense with them altogether. They are unnecessary (and students haven't the foggiest idea what a differential is other than it vaguely means an itsby bitsy difference). Its primary function seems to be to maintain the pretense that thermodynamic processes do not occur in time. In the traditional approach to thermodynamics, variables change, but we are not told with what. Time derivatives have been expelled from thermodynamics, along with time itself. We restore time to its rightful place. If you are familiar with the traditional notation of thermodynamics, and hence are enraged or mystified by our pungent criticism, please read the following detailed arguments that support it. We are not isolated crackpots. Our heretical views are shared by others.

A central mathematical concept in physical theories is that of a *function*—a rule for establishing a correspondence between two sets of quantities called *variables*. We may consider a function to be a hopper into which a variable of one kind is entered and out of which a unique variable of another kind emerges. We may write this correspondence symbolically as

$$y = f(x), \tag{3.6}$$

where x is the input variable, y is the output variable, and f is the symbolic name of the function.

In mathematics, the variables x and y usually are pure numbers. In thermodynamics, they are mostly dimensional quantities, such as pressure, temperature, volume, and internal energy.

Functions can often be specified by algebraic operations on numbers. An example is the function called Square It. If we feed any number x into this function it dutifully expels another number y, which is x squared. We denote this symbolically by

$$y = f(x) = x^2, \tag{3.7}$$

where f is the symbolic name of the function Square It, and y is the value that f assigns to x.

Now consider another function called Square Root It. This function is expressed algebraically as

$$x = h(u) = u^{1/2}. \tag{3.8}$$

If a variable u is first fed into the function h and the resulting x fed into the function f, the result is y. This is equivalent to a composite function or hopper into which we could feed u to obtain y:

$$y = u = f[h(u)] = g(u). \tag{3.9}$$

We give this composite function a different symbol g because it is neither Square It nor Square Root It, but rather Take the First Power of It. Strictly, we should always rename the composite function as a reminder that it is different from both f and h. Because the alphabet is short, we are often sloppy and write

$$y = f(u) = u \tag{3.10}$$

to indicate the functional relationship between u and y, which (appearances to the contrary) is not the same functional relationship as in Eq. (3.7).

Because of this sloppiness in naming functions we run into notational problems in thermodynamics. For example, the internal energy U of a simple system could be a function of T and p. If we were careful mathematicians, we would write this functional relationship symbolically as

$$U = F(p,\, T). \tag{3.11}$$

U is a function of the pair of variables p and T, which can be considered a single variable. F is a hopper into which we feed temperature and pressure pairs and that spits out a single internal energy. But p itself is a function of V and T,

$$p = D(V, T), \tag{3.12}$$

from which we can obtain another functional relationship:

$$U = F[D(V,T), T] = G(V, T). \tag{3.13}$$

Why the different symbols G and F for the two functions that yield the same quantity? The *rule* for obtaining U for a given (p, T) is *not* the same rule for obtaining U for a given (V, T). In thermodynamics, however, we are sloppy. We give the same name to both functions and (worse yet) attach the symbol for the variable U to the function that yields U. That is, we write

$$U = U(p, T) \tag{3.14}$$

and

$$U = U(V, T). \tag{3.15}$$

Here we commit two mathematical sins: we confuse the name of a function with that of a variable, and we use the same symbol for different functions. U is the same physical quantity, internal energy, but $U(p, T)$ and $U(V,T)$ are different functions. A loss of mathematical purity is more than compensated for by a gain in clarity, but this gets us into trouble in the notation for partial derivatives. Suppose we take the partial derivative of F with respect to T holding p constant? If we were careful mathematicians, we might write this as

$$\frac{\partial F}{\partial T}. \tag{3.16}$$

Alternatively, we could write

$$\frac{\partial U(p, T)}{\partial T}. \tag{3.17}$$

We also can take the partial derivative of G with respect to T holding V constant, which can be written as

$$\frac{\partial G}{\partial T} \tag{3.18}$$

or

$$\frac{\partial U(V, T)}{\partial T}. \tag{3.19}$$

What is done instead in thermodynamics—and only in thermodynamics—is to use subscripts to distinguish the two different partial derivatives with respect to T:

$$\left(\frac{\partial U}{\partial T}\right)_p \tag{3.20}$$

and

$$\left(\frac{\partial U}{\partial T}\right)_V. \tag{3.21}$$

This notation signals that the partial derivatives are obtained from two different functional relationships: one between U and (p, T), and another between U and (V, T). But this is at the expense of creating a notation not seen outside of thermodynamics. Moreover, it sometimes is cumbersome. For example, suppose that we consider a system containing n different molecular species. Its internal energy is a function of p, T, and the number N_1, N_2, ..., N_n of each species (which could change because of chemical reactions):

$$U = U(p,\ T, N_1, N_2,\ \ldots,\ N_n). \tag{3.22}$$

The partial derivative of this function with respect to N_j using traditional thermodynamic notation is a monster:

$$\left(\frac{\partial U}{\partial N_j}\right)_{p,T,N_1,N_2,\ldots,N_{j-1},N_{j+1},\ldots,N_n}. \tag{3.23}$$

Accordingly, we do not write partial derivatives with subscripts, but they may be warranted if ambiguity might otherwise result.

"Those Accursed Differentials"

In no other branch of physics do differentials play such a central role—and are so confusing and unnecessary—as in thermodynamics. The first law is often written in differential form as $dU = \mathrm{d}Q + \mathrm{d}W$, the peculiar symbol d indicating that differentials for working and heating are a different species from those usually encountered. In our experience, students are never quite sure what a differential is. Moreover, differentials are not only obscure, they also impede students from understanding calculus. Strong words. Can we back them up?

What do card-carrying mathematicians have to say about differentials? In the early 1950s, a series of articles on differentials appeared in *American Mathematical Monthly*, a journal devoted to the teaching of mathematics. After surveying these articles, the editor, C. B. Allendoerfer, opined that "after reading the numerous papers submitted to Classroom Notes on differentials, and after discussions with other mathematicians, your editor is convinced that there is no commonly accepted definition of differential which fits all uses to which this notation is applied." If mathematicians are uncertain about what a differential is, physicists are even more so, and students of physical sciences are likely to be lost.

Allendoerfer asserts further that "if we wish to make calculus an intellectually honest subject and not a collection of convenient tricks, it is time we made a fresh start." This is what we attempt to do here, more than 60 years after his suggestion. His three-page discussion of differentials is a model of clarity and forcefulness. He doesn't pull any punches: when he encounters nonsense, he labels it as such.

To cut the tangled knot of differentials, our fresh start begins with the concept of the derivative of the function Eq. (3.6) as the *limit* of a quotient

$$\lim_{\Delta x \to 0} \frac{f(x + \Delta x) - f(x)}{\Delta x}. \tag{3.24}$$

The derivative is a difficult concept, and grasping it is one of the major obstacles to be hurdled by students of calculus. Both the numerator and denominator in Eq. (3.24) are zero when $\Delta x = 0$, and the quotient $0/0$ is meaningless, but the *limit* of this quotient *may* have a finite value for *some* x. If so, it is called the derivative of f (or y) with respect to x. It would be handy to have some kind of symbol for the limit Eq. (3.24). The most common notation used today is

$$\frac{df}{dx} \quad or \quad \frac{dy}{dx}, \tag{3.25}$$

which is attributed to Leibniz. This notation *reminds* us that the derivative is the limit of a quotient, although the derivative is *not* a quotient with numerator dy and denominator dx.

The notation Eq. (3.25) also helps us remember certain theorems. For example, suppose that we have a function $y = f(x)$ and another function $x = h(u)$; thus, y is also a function of u. To obtain the derivative of y with respect to u, we multiply two derivatives:

$$\frac{dy}{du} = \frac{dy}{dx}\frac{dx}{du}. \tag{3.26}$$

Like all theorems, this one must be *proven,* but we can remember it by noting that if we naively cancel dx in the denominator of dy/dx with dx in the numerator of dx/du *as if* the two dxs were numbers, we get the correct result. This canceling of dx is a fortunate accident—an attribute of the notation that helps us remember theorems but does not prove them. To drive home this point, we list some of the symbols that have been suggested for derivatives and by whom:

$\dot{y}$	Newton
$\frac{dy}{dx}$, $dy : dx$	Leibniz
$[y \perp x]$	J. Landen
y'	Lagrange
∂y_x	Martin Ohm
Dy	Pierce
y_x	G. S. Carr

Suppose that mathematical history had taken a different turn and we had adopted the symbol ∂y_x for the derivative of y with respect to x? Would it then have been impossible to prove Eq. (3.26) because we would have no dx to cancel?

At this point you may be asking yourself "Why all this quibbling?" After all, if we cancel numerator and denominator in Eq. (3.26), we get the right answer. Correct, but if you fall into the habit of getting right answers by wrong reasoning, you eventually may get wrong answers. Consider, for example, the fraction

$$\frac{16}{64} = \frac{1}{4}.$$

By crossing out the six in the numerator with that in the denominator of this fraction, we reduce it to its correct lowest common denominator. A lucky accident, you might say. To show that this was not an accident, we'll pick another fraction at random

$$\frac{19}{95} = \frac{1}{5}.$$

Again, we get the right answer by cancelling the nines in the numerator and denominator. Since we have proven this result for two fractions, it surely must be true for all fractions. We had better make sure. Pick another fraction at random

$$\frac{26}{65} = \frac{2}{5}.$$

Again, the rule works for the third time, and students know that "what I have told you three times is true." We have established irrefutably that one can cross out numbers in the numerators and denominators of fractions to reduce them to their lowest common denominator. Not so?

Let us move on to integrals for other examples of how playing carelessly with symbols, instead of thinking about what they mean, can lead to confusion. The integral of a function f of x over the interval from $x = a$ to $x = b$ is defined by dividing this interval into n subintervals of width $\Delta x = (b-a)/n$ and taking the limit of the sum

$$\lim_{n\to\infty} \Sigma_1^n f(x_j)\,\Delta x, \tag{3.27}$$

where x_j is some value of x in the ith subinterval. As n approaches infinity, Δx approaches zero, yet the number of terms in the series becomes indefinitely large in such a way that the limit Eq. (3.27) *may* not be zero but *could* be infinite. If this limit is finite, it is called the definite integral of f over the interval (a, b). The symbol

$$\int_a^b f(x)\,dx \tag{3.28}$$

is often used for the limit Eq. (3.27). But many other symbols can be (and have been) used. Symbols cannot be used to prove anything, although some are more helpful than others in aiding our memories. For example, suppose we express x as a function of u : $x = h(u)$. With this transformation of variables the integral Eq. (3.28) becomes

$$\int_a^b f(x)\,dx = \int_\alpha^\beta f[h(u)]\frac{dx}{du}du = \int_\alpha^\beta g(u)\,du, \tag{3.29}$$

where $a = h(\alpha)$, $b = h(\beta)$, and $g(u) = f[h(u)]\,dx/du$. We can remember this theorem, but not prove it, by noting that we can cancel the du in the symbol dx/du with the du in the symbol for integration over the variable u. But the integral symbol Eq. (3.28) is neither sacred nor inevitable. If we had chosen a different symbol, containing no du to cancel, the theorem Eq. (3.29) would still hold.

If we go beyond functions of one variable to functions of two or more variables, we encounter inconsistencies. In the expression

$$df = \frac{\partial f}{\partial x}dx + \frac{\partial f}{\partial y}dy \tag{3.30}$$

for a two-variable function, why can dx be treated as an infinitesimal subject to the operations on finite numbers, but ∂x cannot? They can't be cancelled because extending a trick that works for Eq. (3.29) does not always work. A trick is not a proof. Another example of the breakdown of a trick is naively applying the chain rule to a second derivative. It works for the first derivative, that is, $dy/dt = (dy/dx)(dx/dt)$, but $d^2y/dt^2 \neq d^2y/dx^2(dx/dt)^2$ unless x is a linear function of t.

When we turn to integrals over a planar region, the trick of canceling symbols as if they were numbers fails. Consider, for example, the integral of $f(x,y)$ over a planar region:

$$\iint f(x,y)\,dx\,dy. \tag{3.31}$$

We can transform the variables x and y (rectangular coordinates) to r and θ (polar coordinates):

$$x = r\cos\theta, \qquad y = r\sin\theta. \tag{3.32}$$

With this transformation the integral in Eq. (3.31) becomes

$$\iint f(r\cos\theta,\ r\sin\theta)r\ dr\ d\theta. \tag{3.33}$$

We cannot obtain Eq. (3.33) from Eq. (3.31) by canceling symbols. And a trick for transforming to polar coordinates doesn't work if we want to transform to elliptical or some other coordinates (and why not?).

We are further along in our efforts to get to the root cause of confusion about differentials. The symbol dx appears in Eqs. (3.26) and (3.29). Is this the same quantity? No, of course not: dx is not a quantity. In Eq. (3.26) dx is part of a *symbol* for the derivative; in Eq. (3.27) dx is part of a *symbol* for the integral.

To add to the confusion, we may define the differential dy as a linear function of $f' = dy/dx$ (note the change in notation) and the differential dx by

$$dy = f'\ dx. \tag{3.34}$$

In Eq. (3.34) dy and dx are numerical values. These symbols are not some type of nonexistent infinitesimals, quantities too small to be seen with the naked eye. They are simply *finite* numbers (variables), as is the derivative f'. The dy and dx in Eq. (3.34) are *not* the same dy and dx in Eq. (3.25).

The *finite* difference Δy between two values of a function $f(x)$ for two values of x differing by the *finite* value Δx is *approximately*

$$\Delta y \approx f'\ \Delta x. \tag{3.35}$$

This is a handy approximation we use often, but sometimes it is written in the form of Eq. (3.34). Do you see the difference? Equation (3.34) is exact by definition, whereas Eq. (3.35) is approximate.

We unearthed four ways in which the symbol dx is used: (1) in the derivative of a function; (2) in the integral of a function; (3) in a differential (a linear function); and (4) in an approximation for finite differences. All these symbolic uses of dx are different. Given these four ways the same symbol is used for different mathematical concepts, it is no wonder that students are confused about differentials.

Our experience is that many students think that dx represents a number, and hence, manipulate it in mathematical expressions as if it were. But they draw the line at what can be done with dx. For example, it would be unusual for a student to take the square root of dx. Why not? If dx really is just the symbol for a number, we should be willing to square it, root it, take the sine of it, and so on. But students don't do this. When we turn to partial derivatives, even stronger, unconscious taboos are at work. It is a rare student who will rip apart a partial derivative and manipulate the resulting pieces. Somehow, students can sense (even if they may not know why) that ∂x and dx are subject to different rules, and that even dx can stand only so much mauling. Although derivatives of higher order than the first are not conceptually difficult, it would be unusual to see dx^n ripped from the nth derivative of a function and treated algebraically and d^np as a stand-alone quantity. But woe to the teacher who writes Eq. (3.28) without dx, even though real mathematicians, who are meticulous about rigor, know that it can be omitted without damaging integrals. Its primary value is as a device to help remember Eq. (3.29). Shock, indignation, and rioting in the classroom at seeing no dx reflects the teaching of calculus as a "collection of tricks."

Differentials and Infinitesimals

Further confusion about differentials arises from their connection with the outmoded concept of an infinitesimal—a quantity that is not zero but imagined to be smaller than any positive number. William McGowen Priestley, in his beautifully written *Calculus: An Historical Approach*, asserts that the "notion of the *infinitesimal* is one of the most elusive ideas ever conceived. Attempts to describe it ... bordered upon the comic." Chapter 10 of Priestly's book puts infinitesimals in their proper historical perspective. He treats them sympathetically, but critically. We highly recommend this chapter, to which we cannot do justice here.

Infinitesimals, briefly stated, may be looked upon as training wheels invented by Leibniz to enable him to learn to ride a bicycle called calculus. But even Leibniz reluctantly had to abandon them. We don't need them. The well-defined concept of a limit has replaced the vague and unsatisfactory infinitesimal—but not in thermodynamics, a kind of museum for archaic mathematics.

A review, entitled "The Joy of Infinitesimals," of a book on elementary calculus begins "The history of infinitesimals is surely a case of small quantities generating large controversies." We acknowledge that infinitesimals have been made mathematically rigorous in a branch of mathematical logic called *nonstandard analysis* by means of which the concepts of calculus can be developed in terms of infinitesimals rather than limits. In the 1970s there were claims that nonstandard analysis would take the mathematics world by storm, but it appears to have subsided without having uprooted the usual way of teaching the derivative as the *limit* of a quotient. Nor have we seen any evidence that nonstandard analysis has made great inroads into the physical sciences. The extent to which it is useful in pure mathematics is beyond our competence. We are in favor of rigor but only enough to prevent us from making stupid mistakes in the mathematical analysis of physical problems. If "rigor" has negative connotations, replace it with the gentler word "clarity."

Aside from the vagueness of infinitesimals, they have at least two objectionable features in the teaching of thermodynamics. If an infinitesimal is something so small that no one knows how small, how can one infinitesimal be said to be larger or smaller than another? Yet, this is done in textbooks. Further, consider the so-called infinitesimal process described symbolically by $dU = dQ + dW$. Here, dU is said to be the infinitesimal change in internal energy of the system, dW is the infinitesimal work done in the process, and dQ is the infinitesimal heat added (subtracted).

Yet, it is easy to imagine processes in which the difference between the initial and final internal energies is arbitrarily (i.e. infinitesimally) small but total working and heating are not. Such a process is shown in Fig. 3.1. It is *almost* cyclic (the initial and final states are almost the same). In this process, $U_2 - U_1$ is arbitrarily small—hence, could be represented by the infinitesimal (or differential) dU—but total working and heating are not infinitesimal.

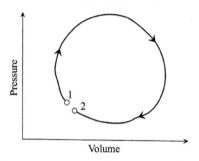

Fig. 3.1: An idealized thermodynamic process considered as a succession of equilibrium states can be depicted by a curve in the pV-plane. The change of internal energy can be made arbitrarily small by making the initial (1) and final (2) states arbitrarily close to each other. Yet the total work done, which is the area enclosed by the (nearly) closed curve, is not vanishingly small.

Are Differentials Necessary in Thermodynamics?

In an expository article on differentials, Mark Kac and J. F. Randolph begin with the assertion that "as easy and desirable as it is to get along without differentials, they seem to be with us to stay." This attitude is one of resignation, of bowing to the inevitable. Karl Menger is more willing to resist. In a superb expository article on the mathematics of thermodynamics, he opines that *"physicists in presenting the elements of the theory* [of thermodynamics] *should refrain from referring to the poorly understood concept of a general differential altogether"* (emphasis in original).

From the resignation of Kac and Randoph to the resistance of Menger, we come to the thundering defiance of Clifford Truesdell: "thermodynamics never grew up ... the unfortunate who reads about thermodynamics even today is made to follow KELVIN's preference for differentials ... and to suffer over again the insecurity CLAUSIUS seems to have felt whenever he used calculus." Truesdell heaps undisguised scorn on the mathematics of thermodynamics, poking fun in particular at its notation: "Not only do differentials replace derivatives, but even derivatives look different," and he provides the following example:

$$\frac{\delta^{rev} Q}{dV} = T\left(\frac{\partial S}{\partial V}\right)_p.$$

What a strange hybrid creature—the mathematical equivalent of the centaur, half man and half horse! And you don't have to search far to find equally strange creatures in thermodynamics, a kind of game refuge for mathematical unicorns, centaurs, satyrs, and sphinxes.

Although Bridgman did not explicitly attribute the mathematical difficulty of thermodynamics to differentials, they are probably what he had in mind. If mathematicians and physicists of great eminence and sagacity have asserted in print that differentials are confusing and unnecessary, why have they not been expunged from thermodynamics books? Alas, the dead hand of tradition still clings to the teaching of thermodynamics. Better to confuse students unto the nth generation than depart from sacred writ.

Differentials are not necessary in thermodynamics. All the mathematics in this book has one aim: to enable you to understand part of the physical world, namely, atmospheres of Earth and other planets. If you acquire this understanding without ever once wrestling with differentials or feeling hamstrung by their absence, they are unnecessary. But differentials are more than a barrier over which successive generations of students must leap. Differentials are a swindle. They convey the paranormal view that thermodynamic processes occur but not in time. All physical processes are states evolving in time: a system is different now from what it was a few moments ago, and it will be different a few moments later. Many applications of atmospheric thermodynamics are describing or predicting the evolution of the thermodynamic state of a system. Return for a moment to Eq. (1.58), which was derived from Eq. (1.57), an equation of *motion* in which time is explicit. It would be unusual to see this written as $dK = dW$ and similarly for Eq. (1.59). With unlimited computational capabilities and superhuman powers to understand the physical world from computations of unimaginable quantity and complexity, Eq. (1.63)—not to mention this book—would be superfluous. This equation is a compromise in which we settle for less than the maximum amount of information and cast (what is at heart) an equation of motion in terms of a relatively miniscule number of measurable macroscopic variables (e.g. temperature, pressure, density). We don't lose much from this compromise if systems of interest are composed of a great many molecules. Averaging removes complexity. But, given the origin of Eq. (1.63), why should we write it as $dU = đQ + đW$?

Although the time integral of dU/dt is trivial, that of W or of Q is not. Even pdV/dt cannot be integrated without knowing how p and V depend on time because they are *independent*, in general. For a few idealized processes (e.g. isothermal, isobaric, adiabatic), the integral of W depends only on initial and final states. There is often no tidy analytical expression for W, or it may not be expressible in terms of thermodynamic variables (see the end of Section 1.8). But we may be able to calculate integrated W from *measurements* of p and V in some processes. Q is even more difficult to pin down, but we may be able to measure its integral. Any smooth curve (path) in the pV-plane can be specified parametrically as $[p(t), V(t)]$, where the independent parameter t is time. The integral of pdV/dt depends on this curve. If t does not specify a unique (p, V), for example, if the curve is closed, we must decompose it into separate parts. Any two points $[p(t_1), V(t_1)]$ and $[p(t_2), V(t_2)]$

can be connected by indefinitely many curves. In mathematical jargon, Q and W are separately *curve-functionals* that assign a number to a curve, but their sum is *not*, depending only on the difference in U at the end points of the curve. Line, surface, and volume integrals are functionals in one, two, and three dimensions. According to the fundamental theorem of calculus, every continuous function $f(t)$ has an antiderivative $F(t)$; that is, $f = dF/dt$. Thus, if pdV/dt is continuous it has an antiderivative. but it may be elusive and not worth the effort to chase down. The integral of W is a curve-functional depending on the *entire* curve. Given the curve, we don't need to know the antiderivative of pdV/dt. This is more general, more useful, and leads to a geometrical interpretation of work done in a cycle. Analytical expressions for curves in the pV-plane exist mostly in textbooks.

How did differentials get into thermodynamics, given its origins in improving engines that do *finite* work in cycles lasting fractions of a second? We can only speculate. The temperature and pressure of any system are well defined only for *spatially uniform* equilibrium states. Processes occurring in the cylinders of engines (internal or external combustion) are a succession of nonequilibrium states driven by *spatial gradients*. The apparent function of differentials is to signal that equilibrium thermodynamics is strictly *inapplicable* to real processes. But there is a world of difference between inapplicable and "good enough." Writing work as pdV is a way of indicating that in no real process is work strictly reversible and infinitesimal. It is more honest to write pdV/dt accompanied by a caveat that this is strictly inapplicable to real processes. Equilibrium thermodynamics applied successfully to rapid compression and expansion of a gas is in Section 3.3. We have no qualms about applying the ideal gas law to real gases, even though no gas is strictly ideal. The trajectory of a projectile fired obliquely from the surface of the real Earth is a parabola only in physics textbooks. It is based on many idealizations: an infinitely flat, nonrotating Earth with uniform g and no atmosphere. This is "good enough" for many purposes. A few reviewers of the first edition excoriated us for our criticism of differentials. To be consistent, they would have to be just as critical of nonexistent parabolic trajectories in physics textbooks, not to mention all the other idealizations that lead to results that are usually "good enough."

In the following subsection we elaborate further on curves in the pV-plane, their historical importance in thermodynamics, and their continued importance in an unsuspected area.

We devote considerable space to criticizing the mathematics traditionally used in thermodynamics. If you have read this far, you may wonder why we have taken the trouble. Why don't we just eagerly embrace differentials? Our experience, like Bridgman's, is that an impediment to the teaching of thermodynamics is its confusing, illogical, if not ludicrous, notation. The mathematics of thermodynamics is inherently simple but made difficult, if not by conscious design, at least by failing to remove weeds from the garden of thermodynamics.

Now we can respond to Bridgman's lament at the beginning of this chapter about the difficulty of teaching thermodynamics because of its "unfamiliar mathematics." Thermodynamics can be done readily with *familiar* mathematics, and it should be the *easiest* subject to teach because, unlike, say, electromagnetic theory, fluid mechanics (i.e. the Navier–Stokes equation), and quantum mechanics, it is immediately applicable to everyday physical phenomena, which to understand does not require vectors, tensors, and partial differential equations (sometimes nonlinear). We give hundreds of examples, especially in the problems. The atmosphere and ordinary kitchens are thermodynamic laboratories that do not require expensive and complicated instruments. Anyone with an infrared thermometer and a thermometer can be an experimental thermodynamicist.

Pure and Impure Thermodynamics: The Indicator Diagram

In 1930 Gilbert N. Lewis, a giant of physical chemistry, published a paper (*Science*, Vol. 71, pp. 569–77) on the symmetry of time in physics, in which he averred that "Time is not one of the variables of pure thermodynamics." The important qualifier here is "pure." This leaves the door open for time in *impure* thermodynamics, which may be good enough for many purposes. If $U_2 = U_1$ in Fig. 3.1, the resulting closed curve exemplifies an *indicator diagram*: p and V versus *time*, in one cycle of a reciprocating steam engine, traced as a closed curve in the pV-plane by an instrument—the indicator—attached to a working cylinder. We now know that the area enclosed by an indicator diagram is the work done in a cycle. Although James Watt did not know this, he learned from experience rather than inadequate or incorrect theory that he could use these diagrams to improve steam engines. Fortunately, he did not know that indicator diagrams are forbidden in pure thermodynamics unless the cycle time is infinite so that p is uniform, and hence the power is zero. The demand for zero-power steam engines was not high. The cycle time for the earliest steam engines was of order 0.2 s.

He did not invent the (finite power) steam engine but he greatly improved it; his assistant, John Southern was the inventor. Watt considered steam to be a compound of material heat and water. First, "sensible" heat increases the temperature of water to 100 °C(at atmospheric pressure), then suddenly, almost magically, "latent heat" takes over the continued generation of steam (water vapor) despite the heating source being the same. But even before this the evaporation rate of water steadily increases as its temperature increases (see Chapter 5, especially Sections 5.3 and 5.10). Steam engines are *external* combustion engines in which liquid water is converted to vapor, its pressure increased, and a pressure difference drives a piston in a cylinder. In modern diesel and gasoline engines (*internal*) combustion occurs within cylinders.

Although the age of *reciprocating* steam engines has mostly passed (steam *turbines* are used for about 80% of world electric power generation), the industrialized world of today, for better or worse, has its origins in steam engines for powering locomotives, pumps for draining mines, steamships, water and sewage pumps, textile mills, and so on. The first stirrings of the Industrial Revolution coincided with Watt's improved steam engines toward the end of the eighteenth century. And indicator diagrams played no small part. For many years their construction was a closely guarded secret, and it was not until the 1820s that their details were widely known. Students today are unlikely to be told about indicator diagrams. But they survive in an unsuspected area: cardiology. A heart is a pump, *not* powered by a temperature difference, and hence not a thermodynamic engine. But a pulsating heart *must* do work on blood to cause it to flow. Pumps increase (mostly) the mass–motion kinetic energy of fluids, not their internal energy. Figure 3.2 shows a cardiac pressure–volume loop: pressure versus volume in human hearts beating at ≈ 1–2 Hz. Pressure–volume loops are used mostly for diagnosis, but they indicate the power output of a heart at different levels of exertion. We show only two loops, the average of measurements for seven healthy males 27–47 years old. The work done in each heartbeat is the area enclosed by a closed curve.

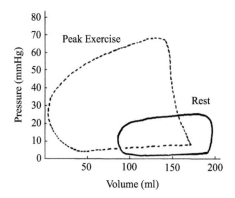

Fig. 3.2: Cardiac pressure–volume loop.

Adapted from Cornwell, W. J. *et al.* (2018) "New insights into right ventricular performance during exercise using high-fidelity conductance catheters to generate pressure volume loops," *Circulation*, 138, A13179.

If we soberly consider time and space variations of atmospheric pressure and temperature, we need not shrink from applying pure thermodynamics to impure processes. Over a vertical distance of ≈ 100 m in the atmosphere, the *relative* change of pressure is $\approx 2\%$, much less over the same horizontal distance. Over a vertical distance of ≈ 100 m absolute temperature changes by $\approx 0.5\%$, much less over horizontal distances. Near surfaces, temperature gradients can be large but only over distances of meters or less. Thus, at any time a 10^{6} m^{3} volume of air is approximately uniform. Air temperature at a particular point may vary by $\approx 1\,^{\circ}\mathrm{C}$ in an hour, pressure by ≈ 1 hPa. At each instant this volume of air is approximately in equilibrium. But there are smaller temperature fluctuations on smaller space and time scales, especially in the boundary layer, and are measured to investigate turbulence.

Impossible Processes

Various impossible processes underlie classical thermodynamic theory. For example, consider an ideal gas in a cylinder with an infinite ratio of thermal conductivity to wall thickness and fitted with a movable frictionless and massless piston.

The cylinder is completely immersed in an infinite, constant-temperature bath that is absolutely uniform and stable. An impossible process is isothermal, adiabatic ($Q = 0$) compression of the gas. Compression increases its internal energy, manifested by an increase in its temperature. To instantaneously compensate for this, Q must be > 0 for the process to be isothermal—but then it can't be adiabatic.

A further dose of reality: a 20-liter constant-temperature water bath costs a few thousand dollars, with a temperature uniform only to within $0.2\,°C$ and about the same stability. For this outlay of money, you get only a finite, *almost* constant-temperature bath. And a search of the Internet did not yield any vendors of frictionless pistons and cylinders with an infinite ratio of conductivity to wall thickness.

Labeling processes as "infinitesimal" or "quasistatic" are euphemisms for the more honest term "impossible." Nevertheless, they may be useful if they lead to measurable quantities and describe *possible* relations and processes with acceptable fidelity. As far as we know, there are no exact theories of macroscopic states and processes. The magnetic moment of the electron has been measured to about nine decimal places and agrees with theory to about seven. But a simpler system than a *single* electron is difficult to imagine. It supposedly doesn't even have a finite size, only a finite mass and charge. It is all downhill from the electron to the properties of macroscopic matter.

Physics is filled with impossibilities: absolutely smooth, infinite in lateral extent interfaces between absolutely homogeneous materials, plane harmonic waves, zero friction, an absolutely smooth, nonrotating spherical Earth without an atmosphere, ideal gases, inviscid fluids, point particles, rigid bodies, spherical cows—the list is long. All the cylinders, pistons, and containers in this book used to derive thermodynamic relations are fictitious.

If, in the following pages, you come across processes that seem impossible or logically inconsistent, it probably is because they are. The ultimate test of theories based on idealizations, assumptions, and approximations is not their purity but their acceptable agreement with experiment. In the first sentence of Section 1.1 we state that thermodynamics is a "science of measurable quantities." We have not changed our minds. Theory suggests what to measure, how it might be measured, how to correlate different measurements, and gives us a symbolic language for thinking and talking about them.

3.2 Specific Heats and Enthalpy

For simple systems as defined in Section 3.1, four essential variables are U and T, which are uniquely thermodynamic, accompanied by p and V. From them follow others that are *convenient*. They help us to think fruitfully and economically about physical phenomena loosely classified as thermal in ways that might not have occurred otherwise. This resembles pure mathematics, in which theorems are deduced from axioms. There is nothing in a theorem that is not in its axioms. Theorems are axioms repackaged according to rules of logic.

In preceding chapters we give *operational*, rather than abstract, definitions of U, T, and p. *Differences* in U may be measured by work done in adiabatic *processes* occurring in nonzero time and proportional to that time. The *state* variable T for a system in equilibrium is measured with thermometers; p is measured with manometers and barometers.

We now can combine the first law with the ideal gas law to obtain results that shed light on atmospheric processes. The first step is the tedious (but necessary) one of defining various physical quantities that occur frequently in thermodynamic applications.

With the chain rule Eq. (3.4), the first law can be put in the form

$$Q = \frac{\partial U}{\partial T}\frac{dT}{dt} + \left(\frac{\partial U}{\partial V} + p\right)\frac{dV}{dt}, \tag{3.36}$$

where $\partial U/\partial V$ has the dimensions of pressure and sometimes is called the *internal pressure*, an isothermal change in internal energy because of a change in intermolecular potential energy, which depends on the average intermolecular separation, and hence, volume V. $U(T,V)$ is sometimes called a *caloric equation of state* as distinguished from a *thermal equation of state* $p(T,V)$; p, V, and T are *state variables*. According to Eq. (3.36), Q can effect a *simultaneous* change of temperature and volume. It symbolizes interaction of the system with its surroundings.

Although U is defined only to within a constant, its partial derivatives are definite. The first partial derivative on the right side of Eq. (3.36) appears sufficiently often that it has acquired a name, *(total) heat capacity (at constant volume*, or *isochoric)*, and a special symbol:

$$C_V = \frac{\partial U}{\partial T}. \tag{3.37}$$

The reason for this name follows from Eq. (3.36) for a constant volume process $(dV/dt = 0)$:

$$Q = C_V \frac{dT}{dt}, \qquad V = \text{const.} \tag{3.38}$$

Heat capacity, a relic of the caloric theory, is used in the same way that we refer to a capacity for strong drink. Someone with a large capacity can drink heroic amounts without getting drunk. The rate of drinking is analogous to Q (process), the rate of getting drunk is analogous to dT/dt (change of state). The higher the drinking capacity, the lower the rate of getting drunk for a given rate of drinking. Similarly, the higher the heat capacity, the lower the rate of temperature change for a given heating rate. The term "cold capacity" makes as much sense as "heat capacity." Equation (3.37) is unchanged if $Q \to -Q$ and $dT/dt \to -dT/dt$ for $C_V > 0$. *Thermal capacity*, used by some authors, is more descriptive than the more familiar heat capacity. Joseph Black settled on the term heat capacity after abandoning the even worse terms *faculty for receiving heat* and *appetite for heat*.

In an adiabatic, constant volume process with $W \neq 0$, $C_V = W_{tot}/\Delta T$. This implies that *heat* capacity with equal reason could be called *work* capacity because the *same* ΔT could be obtained by working or heating (see Problem 80). Joule measured the *absolute* heat capacity of water solely by measuring work done and temperature increases (Section 1.8). Heat capacities *relative* to that of water can be determined in *adiabatic* mixing processes (see Problem 70).

Because C_V is *assumed* to depend on the mass M of the system, we often use the heat capacity per unit *mass* or *specific heat capacity* or simply *specific heat* (at constant volume):

$$c_v = \frac{C_V}{M}. \tag{3.39}$$

The specific heat capacity per unit *volume*, ρc_v, is often the relevant property in energy transfer (see Chapter 7). Another specific heat, much loved by chemists, and for good reasons, is the *molar specific heat*:

$$c_{v,m} = \frac{N_a}{N} C_V, \tag{3.40}$$

where N/N_a is the number of moles of the substance. Molar specific heat is an intrinsic property of molecules in a given environment. To convert a specific heat capacity per unit mass (kg) to molar in $J \, mol^{-1} K^{-1}$, multiply by the molecular weight and divide by 1000. Unless stated otherwise (Section 3.7), the system of interest is composed of a *single* component and its state variables are uniform. The specific heat capacity of a 1963 Volkswagen bus is meaningless, but its seating capacity is not.

One combination of thermodynamic variables occurs sufficiently often that it has acquired a special name and symbol, *enthalpy*, H:

$$H = U + pV. \tag{3.41}$$

Whether enthalpy should be pronounced with the emphasis on the first or second syllable has not yet been agreed upon at an international scientific congress held on some Caribbean island in the depths of the mid-latitude winter. We stress the first syllable because enthalpy then stands on equal phonetic grounds with energy and entropy. Like internal energy, enthalpy is defined only to within a constant, and hence physically relevant quantities are enthalpy differences. We encounter enthalpies often in thermodynamics. Seventy years ago Austin O'Leary argued that "we ought to rid ourselves of the ideas of heat, flow of heat, and flow of energy ... thermal effects can be accurately described in terms of *enthalpy*, *thermal transfer*, and *energy transfer* with no connotation in regard to a fluid." Enthalpy is not only shorthand for a particular combination of thermodynamic variables: it is also a means by which we can purge heat from our scientific vocabulary. For example, what are called latent heats (Section 5.3) are enthalpy *differences*. The term enthalpy skirts the problem of using heat as a noun. In some fields, enthalpy is called heat content, an archaic term harkening back to when heat was thought to have a material existence. Physical chemists seem to have recognized that it does not and have replaced various "heats" with enthalpies, for example, of reaction, mixing, hydration, combustion, formation, vaporization, sublimation, and fusion, which are enthalpy differences. The term *thermal enthalpy* is sometimes used to distinguish between enthalpies that are continuous functions of temperature and enthalpies that are discontinuous at the temperatures of phase transitions.

Could we do thermodynamics without enthalpy? Certainly—just as hydrodynamics and electrodynamics once were done in mind-numbing detail without vectors. But they definitely make equations more compact, easier to manipulate, and to grasp.

Given the definition Eq. (3.41) of enthalpy, the first law, Eq. (3.3), can be written

$$Q = \frac{dH}{dt} - V\frac{dp}{dt}. \tag{3.42}$$

In a constant-volume process, internal energy changes, whereas in a constant-pressure process enthalpy changes (if $Q \neq 0$).

We may consider H to be a function of the two independent variables pressure and temperature. The chain rule

$$\frac{dH}{dt} = \frac{\partial H}{\partial T}\frac{dT}{dt} + \frac{\partial H}{\partial p}\frac{dp}{dt} \tag{3.43}$$

enables us to write the first law as

$$Q = \frac{\partial H}{\partial T}\frac{dT}{dt} + \left(\frac{\partial H}{\partial p} - V\right)\frac{dp}{dt}. \tag{3.44}$$

The first partial derivative on the right side of this equation occurs sufficiently frequently that it has acquired a name—(total) *heat capacity* (at *constant pressure*, or *isobaric*)—and a special symbol:

$$C_p = \frac{\partial H}{\partial T} \qquad p = \text{const.} \tag{3.45}$$

For a constant-pressure process

$$Q = C_p \frac{dT}{dt}. \tag{3.46}$$

The corresponding specific heat per unit mass is

$$c_p = \frac{C_p}{M}, \tag{3.47}$$

and the molar specific heat is

$$c_{p,m} = \frac{N_a}{N}C_p. \tag{3.48}$$

The specific heat capacity per unit volume is ρc_p. Because intermolecular forces are short range, c_v, c_p are independent of M for sufficiently large systems.

Equations (3.37) and (3.45) are *definitions*, and you can't argue with a definition. Measuring C_V and C_p is another matter. To theoreticians and the authors of many textbooks, they come from God. In reality, they come from the labors of scientists spending long hours in laboratories taking great pains to ferret out experimental errors, but not becoming rich or famous from doing so.

C_p becomes meaningless if chemical reactions or phase changes occur. In an adiabatic, isobaric chemical reaction the temperature *could* change (Section 3.9), which implies that $C_p \to 0$. In a diabatic, isobaric phase change the temperature does *not* change (Section 5.3), which implies that $C_p \to \infty$.

That different materials have different "capacities for heat" was evident more than 200 years ago via simple experiments. If equal masses of water at different temperatures were mixed in an insulated container, the equilibrium temperature was their arithmetic average. But if equal masses of water and *different* liquids were mixed, the equilibrium temperature was *not* this average— sometimes not even close. Mixing equal volumes did not change this general result. Because of these observations, the logical conclusion was that different materials have different characteristic properties. As to the reason for this, Black, usually credited with the discovery of specific heats, acknowledged that "no general principle ... can yet be assigned." Nor can it be today within macroscopic thermodynamics.

Isobaric and isochoric processes are easy to define, visualize, and even create in the laboratory. But real processes can be partly isothermal, isobaric, and isochoric— or even more difficult to characterize.

There are at least *ten* "heat capacities": total, per unit mass, per unit volume, per mole, and per atom-mole, either for constant pressure or constant volume. Each one tells a different story. Yet heat capacity and specific heat capacity are often unqualified, leaving readers to guess which is meant—a source of confusion and misconceptions. The SI term for heat capacity per unit mass is specific heat capacity, its derived SI unit J kg^{-1} K^{-1}.

T, p, and V are state *variables* that do not characterize a material, whereas c_p, c_v, and ρ are *properties* of a material in a given state. The only property in the ideal gas law Eq. (2.13) that distinguishes one gas from another is R.

The two heat capacities C_p and C_V are not independent, which we can show as follows. From the definition of enthalpy and the chain rule, we have for a system of *fixed mass*

$$\frac{\partial H}{\partial T}\frac{dT}{dt} + \frac{\partial H}{\partial p}\frac{dp}{dt} = \frac{\partial U}{\partial T}\frac{dT}{dt} + \frac{\partial U}{\partial V}\frac{dV}{dt} + p\frac{dV}{dt} + V\frac{dp}{dt}. \tag{3.49}$$

Only two of the variables p, V, and T are independent. We may take V to be a function of p and T. From the chain rule for the time derivative of V in terms of the time derivatives of p and T, together with Eq. (3.49),

$$\left(\frac{\partial H}{\partial T} - \frac{\partial U}{\partial T} - \frac{\partial U}{\partial V}\frac{\partial V}{\partial T} - p\frac{\partial V}{\partial T}\right)\frac{dT}{dt} + \left(\frac{\partial H}{\partial p} - V - \frac{\partial U}{\partial V}\frac{\partial V}{\partial p} - p\frac{\partial V}{\partial p}\right)\frac{dp}{dt} = 0. \tag{3.50}$$

Because p and T are independent variables and we are considering an arbitrary process, Eq. (3.50) can be satisfied only if the coefficients of dp/dt and dT/dt separately vanish. In particular, vanishing of the coefficient of dT/dt yields a relation between the two heat capacities:

$$C_p = C_V + \left(\frac{\partial U}{\partial V} + p\right)\frac{\partial V}{\partial T}. \tag{3.51}$$

This is a general result, which is simpler for an *ideal* gas. Such a gas is one in which intermolecular forces are negligible, which means that the internal energy does not change as volume (intermolecular separation) changes for a fixed mass:

$$\frac{\partial U}{\partial V} = 0. \tag{3.52}$$

This equation, sometimes called *Joule's law* (although it was obtained earlier by Gay-Lussac), *defines* an ideal gas. But in the real world it is an approximation, not a law. From Eq. (3.52) and the ideal gas law in Eq. (3.51) it follows that

$$C_p = C_V + Nk. \tag{3.53}$$

The corresponding expressions per unit mass and per mole are

$$c_p = c_v + R \tag{3.54}$$

and

$$c_{p,m} = c_{v,m} + R^*. \tag{3.55}$$

Because N and k are positive, C_p is always greater than C_V (for a fixed mass).
A corollary of Eq. (3.52) is

$$\frac{\partial C_V}{\partial V} = \frac{\partial}{\partial V}\frac{\partial U}{\partial T} = \frac{\partial}{\partial T}\frac{\partial U}{\partial V} = 0 \tag{3.56}$$

provided that U is sufficiently well behaved that its cross partial derivatives are equal. Thus, the heat capacity at constant volume of an ideal gas may depend on temperature but not volume. The interpretation of $\partial U/\partial V \neq 0$ is that a volume change at constant temperature changes the average separation between molecules, and hence, the *internal* potential energy. U, in general, is not determined solely by temperature.

We obtained Eq. (3.54) by strictly mathematical reasoning. It is always wise to check mathematical results by physical reasoning. In so doing you acquire physical intuition and might even unearth errors. Mathematical manipulations are not always carried out infallibly. If your mathematical results don't square with physical reasoning, something may be wrong with your mathematics. To show that $c_p > c_v$ makes physical sense, consider an ideal gas confined to a cylinder fitted with the obligatory frictionless but tightly fitting piston. Fix the piston in place (with a pin, for example) and heat the gas for a certain amount of time. The temperature of the gas rises in this constant volume process. Remove the pin, the piston moves freely, and the pressure is constant. Heat the gas for the same amount of time as previously. Again, the gas temperature increases, but in this process the piston rises, work is done *by* the gas, and consequently its internal energy doesn't increase as much as it did in the constant-volume process. This implies that the temperature increase isn't as great, and hence, c_p is greater than c_v, consistent with what we derived.

With a physical understanding of the relation between the two specific heats, we can return to determine if Eq. (3.51) makes physical sense even for other than an ideal gas. According to this equation $C_p > C_V$ by an amount determined by the work done by the pressure p *plus* the internal pressure $\partial U/\partial V$ in an *assumed* isobaric thermal expansion $\partial V/\partial T = -(1/M\rho^2)(\partial\rho/\partial T)$. The physical interpretation is unchanged if we include this internal pressure, but now we face an apparent conundrum because what we call an expansion could be a compression. And we don't have to search any further than in a refrigerator for a fluid with this property. Between about $0\,°C$ and $4\,°C$ the density of liquid water *increases* with increasing temperature (see Fig. 3.9b). The way out of this conundrum is that $\partial U/\partial V$ can be negative, its absolute value greater than p, and hence, the (net) work is positive. Equation (4.72) is a form of Eq. (3.50) showing that $C_p > C_V$ even if $\partial V/\partial T < 0$.

Before proceeding we play more with derivatives to obtain some useful general results:

$$\frac{dp}{dt} = \frac{\partial p}{\partial V}\frac{dV}{dt} + \frac{\partial p}{\partial T}\frac{dT}{dt}, \tag{3.57}$$

and

$$\frac{dV}{dt} = \frac{\partial V}{\partial p}\frac{dp}{dt} + \frac{\partial V}{\partial T}\frac{dT}{dt}. \tag{3.58}$$

These two equations can be combined to yield

$$\left(\frac{\partial p}{\partial V}\frac{\partial V}{\partial p} - 1\right)\frac{dp}{dt} + \left(\frac{\partial p}{\partial V}\frac{\partial V}{\partial T} + \frac{\partial p}{\partial T}\right)\frac{dT}{dt} = 0. \tag{3.59}$$

For this equation to be satisfied in an arbitrary process

$$\frac{\partial p}{\partial V} = \frac{1}{\frac{\partial V}{\partial p}}, \qquad \frac{\partial p}{\partial V}\frac{\partial V}{\partial T} = -\frac{\partial p}{\partial T}, \tag{3.60}$$

which demonstrates that *sometimes* one can naively treat the numerator and denominator in the *notation* for a partial derivative as if they were numbers. Partial derivatives often are—but need not be—written as quotients to remind us that they are obtained as the *limit* of a quotient. From Eq. (3.60), it follows that

$$\frac{\partial V}{\partial T} = -\frac{\partial p}{\partial T}\frac{\partial V}{\partial p}. \tag{3.61}$$

But, by cancelling ∂V in Eq. (3.60) or ∂p in Eq. (3.61), we obtain contradictions—more evidence that notational tricks to aid the memory don't always work (see Section 3.1). Note the unexpected minus sign. Where did *that* come from?

If $\partial V/\partial T = 0$, it follows from Eq. (3.51) that $C_p = C_V$ provided that $\partial U/\partial V$ is finite. Near room temperature, these two heat capacities for water (but not for all or even most liquids) and for solids are approximately equal, and hence, often written without a subscript. But as evidenced by Eq. (4.72), which expresses their difference as a function of temperature, compressiblity, and coefficient of thermal expansion, approximate equality is not always valid.

Before proceeding to adiabatic processes, we milk one more result from Eq. (3.50):

$$\frac{\partial H}{\partial p} - V - \frac{\partial U}{\partial V}\frac{\partial V}{\partial T} - p\frac{\partial V}{\partial p} = 0. \tag{3.62}$$

From this equation, the ideal gas law, and Eq. (3.5) it follows that

$$\frac{\partial H}{\partial p} = 0. \tag{3.63}$$

This is the enthalpic equivalent of Eq. (3.52) for an ideal gas.

Although we cannot say how much heat a system has, we can say how much internal energy or enthalpy it has relative to a reference state. Moreover, in a constant-volume process (and only in such a process) the change in internal energy is equal to the time-integrated heating, just as in a constant-pressure process (and only in such a process) the change in enthalpy is equal to the time-integrated heating. A corollary of Eq. (3.63) is

$$\frac{\partial C_p}{\partial p} = \frac{\partial}{\partial p}\frac{\partial H}{\partial T} = \frac{\partial}{\partial T}\frac{\partial H}{\partial p} = 0, \tag{3.64}$$

provided H is sufficiently well behaved that its cross partial derivatives are equal. Thus, the heat capacity of an ideal gas at constant pressure does not depend on pressure.

Because of Eq. (3.52), U for an ideal gas depends only on temperature, and hence from Eq. (3.36)

$$U(T) - U(T_0) = \int_{T_0}^{T} C_V \, dT. \tag{3.65}$$

Because of Eq. (3.63), H depends only on temperature, and hence, from Eq. (3.45)

$$H(T) - H(T_0) = \int_{T_0}^{T} C_p \, dT. \tag{3.66}$$

Over the range $-40\,°C$ to $40\,°C$, the heat capacities of air vary by at most about 0.4%, so we can write

$$U(T) \approx C_V T, \qquad H(T) \approx C_p T, \tag{3.67}$$

where we set the constants equal to zero because we are often interested only in *changes* in internal energy and enthalpy. Because $C_p \approx C_V$ for solids and a *few* liquids (e.g. water), it follows from Eqs. (3.65) and (3.66) that enthalpy changes of these condensed phases are nearly the same as internal energy changes. At temperatures above $50\,^\circ$C, the two heat capacities of water begin to diverge, their ratio 1.12 at $100\,^\circ$C, which is of little consequence to meteorology. For gases, however, the difference between enthalpy and internal energy is *not* negligible and has profound consequences.

Specific heat capacity is often used to mean per unit mass, but when in doubt, check the units. We write specific internal energy, specific enthalpy, and specific volume, *per unit mass*, as u, h, and v, respectively. The corresponding quantities per unit volume are products with the mass density ρ, and usually not given separate symbols. We write molar quantities as u_m, h_m, and v_m. Except for the scale factor $1/N_a$, U/N is u_m. Joule's law is equivalent to $\partial u/\partial v = \partial u_m/\partial v_m = 0$.

Internal energy per unit mass obscures that molecular masses are largely irrelevant to the potential energy contribution to U of solids and liquids, which is of electrical origin. The often-privileged status of energies and enthalpies per unit mass is because mass is readily measured with an analytical balance and is invariant for a given sample. The number of moles must be calculated from the sample density and molecular weight.

The internal potential energy contribution to U is not necessarily proportional to the total number or mass of the molecules in a system (see Section 1.3). The potential energy first encountered by students is likely to be gravitational, and they may never learn about intermolecular potential energies. For ideal gases, they are negligible, although not for liquids and solids. But they are short range, whereas gravitational forces are long range. Consider a reference molecule in a uniform sea of n identical molecules per unit volume. The number in a shell of thickness Δr at a distance $r \gg \Delta r$ from this molecules is proportional to $nr^2\Delta r$. If the potential energy decreases as $1/r^p$, where r is the distance from the reference molecule, the contribution to its potential energy per unit thickness is proportional to $1/r^{p-2}$. Unless $p > 3$, the total potential energy of the reference molecule is dominated by the molecules *farthest* from it because there are more of them. Some theories of the *attractive* interaction between molecules yield a $1/r^6$ pairwise intermolecular potential energy. But the concept of molecules (or atoms) interacting pairwise is not valid for metals or covalent solids. Gravitational potential energy decreases as $1/r$, and hence, gravitational forces are long range. But because short-range intermolecular forces of electrical origin are so much greater, intermolecular gravitational forces are negligible. If this were not so, the thermodynamics of liquids and solids would be quite different from what we have come to consider normal or inevitable. Internal energy per unit mass or mole would depend on the size of a body. Although gravitational *intermolecular* forces are negligible, the *total* gravitational force exerted by the 10^{50} molecules of Earth is not. They make up for their individual feebleness by combining their always attractive forces. "Unity is strength" (Mattie Stepanek).

At very small intermolecular distances, weak attraction must give way to strong repulsion, which is not a force in the usual sense, but rather is a quantum-mechanical constraint on the overlap of atoms and molecules, sometimes called an *exchange interaction*. Although there is no simple expression for this interaction, it is often approximated as a potential energy that increases with decreasing distance as $1/r^{12}$. Don't be misled by the precision of Coulomb's law of electrostatic attraction and repulsion, the exponent in which is $2+q$, where q has been established by experiment to be no greater than about 10^{-16}. Although a molecule is a bound collection of positive protons and negative electrons individually obeying Coulomb's law, interaction between molecules, especially at very small distances is not just a simple superposition of Coulomb forces. Electrons crowded together in a small atomic volume behave differently from free electrons.

Is the Heat Capacity of Liquid Water Extraordinarily High?

We have read many times that liquid water has an extraordinarily high "heat capacity," usually unqualified. But even the qualifier *specific* is ambiguous because it could mean per unit *mass*, per unit *volume*, or per *mole* (i.e. per *molecule*). Only the last enables a molecule-for-molecule comparison. Although the specific heat capacity per unit mass of liquid water is about four times that of air (at constant pressure), the ratio of molar specific heat capacities decreases to 2.6 because the molecular weight of water (18) is less than that of air (29). The molar specific heat capacity of liquid water ($25\,^\circ\text{C}$) is about twice that of ice, but *less* than that of ethyl, methyl, propyl, and butyl alcohols, benzene, glycerin, butane, toluene, acetic acid, acetone, ammonia ... . That of paraffin wax is about ten times greater. Molecule for molecule, liquid water is an underperformer. But because it is a thousand times more dense, the specific heat capacity of water per unit *volume* is much greater than that of *air*. For this reason, the *total* heat capacity of a water layer about 2 m thick is equal to that of the *entire* atmosphere (per unit cross sectional area), *not* because of the extraordinary properties of the water molecule, but because of the thousand-fold greater density of liquid water.

According to the empirical law (rule) of Petit and Dulong, the molar specific heat capacity of *solid* elements at normal temperatures is about $3R^* = 25$ J mol^{-1}K^{-1}, a good approximation for many elements. The molar specific heat capacity of liquid water is about three times greater, of ice about 50% greater. Contrary to what has been passed down in countless books and papers for 200 years, this order of the names of the authors is *correct*, which can be verified by reading the original paper (see Section 4.2). In the same year (1819) it was published in French, it was translated into English and can be found on the Internet. If you desperately need a specific heat capacity (J kg^{-1} K^{-1}), divide 25 by the molecular weight and multiply by 1000. This is not a bad approximation even for air.

An Incompressibility Paradox: The Perils of Idealization

We once had a heated discussion with colleagues who insisted that if water were *absolutely* incompressible, pressure would still increase with depth in the ocean. We argued that it could not in such *idealized* water. At first glance it would seem that we were wrong because early in their education students are taught that pressure is "just" weight per unit area (see Section 2.3), and that of water at the bottom of the deepest ocean is about 1000 atm.

We offer the following in support of our argument. Consider a vertical column of identical molecules in a uniform gravitational field g. Each has a weight mg. We consider only nearest-neighbor forces between molecules. The column is in equilibrium. Molecules are labeled sequentially downward from 1 at the top. A force balance for each molecule is

$$mg + F_{2,1} = 0, \ mg + F_{1,2} + F_{3,2} = 0, \ mg + F_{2,3} + F_{4,3} = 0. \tag{3.68}$$

$F_{i,j}$ is the force on molecule j by molecule i. From Newton's third law, $F_{i,j} = -F_{j,i}$, the sum of all these equations up to molecule $N-1$ is

$$-(N-1)mg = F_{N,N-1}. \tag{3.69}$$

Assume that the force between molecules depends only on their separation. If this is everywhere the same, so is $F_{i,j}$. But this contradicts our result that $F_{N,N-1}$ increases with increasing N. Thus, the column of molecules cannot be absolutely incompressible. The force between neighboring molecules *must* increase because of a decrease in their separation. Interaction between real molecules that don't react chemically becomes strongly repulsive at sufficiently small separations (see Section 2.3). Even a miniscule decrease results in a large increase in this repulsion. As long as $F_{N,N-1}$ increases with increasing distance down the column, we obtain a force balance that makes sense: the weight of the entire column is balanced by the force between molecule N and molecule $N-1$. Molecule N, so to speak, is holding up all the molecules above it through a chain of interactions. This argument does not apply to air because the *attractive* force between two molecules at the average distance between them is negligible (Section 2.3).

The implication of Eq. (3.69) is that short-range *repulsion* between two molecules is at least as large as the *attractive* gravitational force exerted on a single molecule by Earth's approximately 10^{50} molecules. Repulsion is a better term than repulsive force. The stability of matter is a consequence of an interaction that is not a true force and cannot be understood by classical physics.

The total number of liquid water molecules with diameter 0.3 nm along a line 300 m long is about 10^{12}. A single water molecule at a depth of 300 m supports the weight of a trillion water molecules above it. At the greatest depth in the ocean the pressure is "only" 0.1 GPa. Much larger pressures are produced in laboratories.

We can apply a similar argument to a pillar of salt hundreds of times taller than Lot's wife. Salt is a crystalline solid with regular distances between its sodium and chlorine atoms on a lattice. If you could shrink to the size of an atom and wander through (hypothetical) incompressible salt, you would be hopelessly lost. You would have no way of determining that you had migrated from one height to another. The weight per unit area of the pillar above a given height would vary considerably from top to bottom but the pressure in the salt would be uniform because of no change in the lattice parameters. In a real salt pillar, pressure increases with distance from its top because the lattice changes ever so slightly to maintain equilibrium.

Beware of idealizations, especially ones invoking absolutes. If you push an idealization too far, it may turn around and bite you. By invoking a nonexistent entity in one context, we may obtain a valid approximation, whereas in a different context we may obtain nonsense. To estimate the increase in pressure with depth in water, we may *pretend*—but should not *believe*—that its density is constant.

Enthalpy of the (Hydrostatic) Atmosphere

To take some of the mystery out of enthalpy, let us bring it down to Earth or, better yet, put it into the atmosphere.

The total internal energy (per unit cross-sectional area) of the entire atmosphere is

$$U = \int_0^\infty \rho c_v T \, dz. \tag{3.70}$$

This integral and others like it embody the *assumption* of local thermodynamic equilibrium (Section 2.5). The total gravitational potential energy (relative to 0 at $z = 0$) is

$$P = \int_0^\infty \rho g z \, dz. \tag{3.71}$$

From the hydrostatic equation, Eq. (2.30), this integral can be written

$$P = -\int_0^\infty z \frac{dp}{dz} dz, \tag{3.72}$$

which can be integrated by parts

$$P = \int_0^\infty p \, dz - z p \big|_0^\infty. \tag{3.73}$$

At the lower limit of integration ($z = 0$) the pressure is finite. At the (symbolic) upper limit, the product zp becomes negligible because p decreases exponentially, and hence,

$$P = \int_0^\infty p \, dz. \tag{3.74}$$

With p from the ideal gas law Eq. (3.43) becomes

$$P = \int_0^\infty \rho R T \, dz, \tag{3.75}$$

and hence

$$U + P = \int_0^\infty \rho(c_v + R) T \, dz = \int_0^\infty \rho h \, dz = H. \tag{3.76}$$

The total enthalpy H (per unit area) of the atmosphere is the sum of its total internal energy and gravitational potential energy (within an arbitrary constant). If the atmosphere is heated, its enthalpy increases for two reasons: its internal energy increases, and its potential energy increases because the center of gravity of the atmosphere

$$\langle z \rangle = \frac{\int_0^\infty \rho g z \, dz}{\int_0^\infty \rho g \, dz} = \frac{P}{M_a g} \tag{3.77}$$

rises (see Section 3.10), where the total mass M_a of the atmosphere (per unit area) is fixed (at least over time scales of hundreds of thousands of years). This is the sense in which heated air does work against its environment, by which is meant Earth's gravitational field. *Changes* in H and P because of changes in the temperature structure of the atmosphere do not depend on any arbitrary constants.

From Eq. (3.76), the local enthalpy per unit volume is ρh, and hence, its time rate of change is the local energy flux density divergence or convergence (see Section 3.8). Pressure is not uniform in the atmosphere, but the enthalpy of an ideal gas does not depend on pressure [Eq. (3.63)]. Unlike for the kinds of systems considered in many thermodynamics books, gravitational potential energy changes of the atmosphere are not negligible, that is, g does not appear in Poisson's relations. The total mass per unit area of the atmosphere is about 9000 kg m^{-2}, whereas it is thousands of times smaller in a cylinder or room.

3.3   Adiabatic Processes: Poisson's Relations

An ideal *adiabatic* process is defined as one in which the system does not exchange energy with its surroundings because of a temperature difference between them. An adiabatic process need not be *isothermal* (constant temperature), and vice versa. An example of an adiabatic process that is not isothermal is a chemical reaction occurring within a container wrapped in a thick layer of insulation (Section 3.9). An example of an ideal isothermal process that is not adiabatic is compression (or expansion) of a gas in a thin-walled metallic cylinder with a uniform cross section and immersed in a constant-temperature bath. In an adiabatic, constant-volume (isochoric) process, internal energy is conserved; in an adiabatic constant-pressure (isobaric) process, enthalpy is conserved. Many processes in the atmosphere are nearly isobaric, because of which the physical quantity of relevance in our applications is often enthalpy, rather than internal energy. In a *diabatic* process $Q \neq 0$.

The first law of thermodynamics for an ideal gas undergoing a process in which reversible working is given by Eq. (1.65) is

$$Q = C_V \frac{dT}{dt} + p \frac{dV}{dt}. \tag{3.78}$$

In an adiabatic process this equation together with the ideal gas law $pV = NkT$ and the relation between heat capacities Eq. (3.53) yields

$$\frac{1}{\gamma - 1}\frac{1}{T}\frac{dT}{dt} + \frac{1}{V}\frac{dV}{dt} = 0, \tag{3.79}$$

where γ is the ratio of heat capacities:

$$\gamma = \frac{C_p}{C_v} \geq 1. \tag{3.80}$$

Under the assumption that γ is independent of temperature (Section 3.6), Eq. (3.79) can be integrated:

$$TV^{\gamma-1} = TV^{R/c_v} = \text{const.}, \tag{3.81}$$

where $R/c_v = \gamma - 1$. From the ideal gas law Eq. (2.11), Eq. (3.81) can be written in two equivalent ways:

$$pV^\gamma = \text{const.} \tag{3.82}$$

and

$$Tp^{(1-\gamma)/\gamma} = Tp^{-R/c_p} = \text{const.}, \tag{3.83}$$

where $R/c_p = (\gamma - 1)/\gamma$. Equations (3.81) and (3.82) are unchanged if volume is replaced by specific volume.

The preceding three equations, sometimes called *Poisson's relations*, are relations between all pairs of the three thermodynamic variables p, V, and T. In Section 2.1 we stressed that because the ideal gas law is a relation between three variables (for fixed N), nothing reliable can be said about the relation between any two of them unless this law can be supplemented by a constraint (e.g. V is constant). The constraint underlying the Poisson relations is that the process to which they apply must be adiabatic. Does such a process exist? A strictly adiabatic process would require perfect insulation of a system from its surroundings, and perfection is not of this world. Adiabatic processes are idealizations never realized in nature but nevertheless are sometimes good approximations. For example, as we shall see in Section 7.1, energy transfer between a system and its surroundings takes time. Thus, if a system undergoes a process so quickly that little energy is transferred, the process may be considered adiabatic to good approximation.

The Poisson relations are symmetric. Equation (3.81) written as $T_1 V_1^{\gamma-1} = T_2 V_2^{\gamma-1}$ is unchanged if $1 \to 2$ and $2 \to 1$. Thus, this idealized adiabatic process is said to be *reversible*, a mathematical consequence of the expression for reversible work in the first law (see Section 1.6). But no real processes are reversible. We return to this point in Section 4.1.

Frictionless piston–cylinder combinations exist only on paper. And in real compressions, say air in a bicycle pump, a piston with nonzero mass initially at rest must be accelerated to some speed, then decelerated to rest. Accelerations and decelerations are ignored in thermodynamics, which cannot cope with them, and so pretends that they don't exist and hopes for the best.

Again, the real and the ideal clash. Equation (3.81) is based on incompatible requirements: compression (or expansion) fast enough to be adiabatic but slow enough to be quasi-static. If a gas in a cylinder is compressed, its temperature increases. Diesel engines are based on this. The question is not *if* the temperature increases but by how much in finite times and how well Eq. (3.81) predicts increases for different compression ratios and values of γ. Measured temperatures also depend on cylinder diameter and composition, piston mass, and other factors missing from Eq. (3.81). We searched in vain for measurements of this dependence.

Among the limitations of Poisson's relations is restriction to temperatures and pressures such that the ideal gas assumption is valid. We note in Section 1.6 that the increase in the internal energy of a compressed gas is cumulative, that is, it increases with the travel distance d of the piston. From $\Delta U = C_V(T_f - T_i)$ and Eq. (3.81) it follows that $\Delta U = C_V T_i[(1 - d/d_i)^{1-\gamma} - 1]$, where i and f denote initial and final, and d_i is the initial distance from the piston to the bottom of the cylinder. If $d/d_i \ll 1$, then $\Delta U \approx RT_i d/d_i$. In the limit $d/d_i \to 1$, $\Delta U \to \infty$. This corresponds to compression to nothing, which requires an infinite amount of work. Before accepting equations, it is wise to probe their limits.

Measurements of γ extend over 200 years. From Eq. (3.82) $\gamma = -(V/p)(\partial p/\partial V) = p\kappa_s$, where κ_s is the *adiabatic compressibility*, and hence, γ is related to, in Robert Boyle's words, "the spring of the air," which literally underlies Rüchardt's method. Place a small piston in a tube attached to a larger volume of the gas of interest. If the piston is depressed and released, it will oscillate, as if it were attached to a spring, at a frequency determined by γ. Although γ is the ratio of two *equilibrium* properties of a gas, it can be measured in a *time-dependent* process because its speed (see Section 1.6) is much less than the speed of sound. Another method is to begin with the gas at a pressure p_i greater than ambient p_a, allow it to expand adiabatically to p_a, then returned to its original temperature isochorically: $\gamma = \ln(p_i/p_a)/\ln(p_i/p_f)$, where p_f is the final pressure. Both of these methods are based on theory for ideal gases and reversible adiabatic processes—yet another example of the difficulty of making measurements independent of theory. *If* one can obtain an acceptably accurate estimate of γ and *if* the relation Eq. (3.53) for an ideal gas is valid, then one can obtain the absolute heat capacities $C_p/N = k\gamma/(\gamma - 1)$ and $C_V/N = k/(\gamma - 1)$ for ideal gases.

According to Eq. (3.81), adiabatic compression or expansion results in temperature increases ("heating") or decreases ("cooling"), even though $Q = 0$ by definition. There seems to be no easy way out of this contradiction other than to remember that these temperature changes are the result of $W \neq 0$. The same contradiction is unavoidable in "frictional heating," which is a temperature increase because of *frictional working*.

The Poisson relations, based on an idealized adiabatic, quasistatic process, are in many textbooks unsupported by measurements. We found some in an automotive engineering journal. Peak temperatures in a compressed gas (mostly air) were measured as a function of compression ratio and engine speed (600, 1200, 1800 rpm). Agreement with Eq. (3.81) was good up to a ratio of about 10, beyond which measured temperature was lower than calculated. This is expected: the higher the temperature, the less likely that $|Q\tau/\Delta U| \ll 1$ is satisfied, where τ is the compression time. Peak temperature increases with increasing engine speed because τ is inversely proportional to this speed. Although τ is no greater than 0.1 s, and the temperature increase in this short time is hundreds of kelvins, measurements agree fairly well with a theory based on an indefinitely large τ.

3.4   (Dry) Adiabatic Processes in the Atmosphere

Underlying Poisson's relation Eq. (3.83) between temperature and pressure in an adiabatic process are implicit assumptions not always valid for air. The composition of the gas is assumed to be fixed, and hence so is γ, whereas air always contains variable amounts of water vapor. In Section 3.7 we show that for typical water vapor concentrations in air, γ is constant to within less than 1%. The first law in the form Eq. (3.78) does not include chemical reactions or phase changes. But if air is expanded adiabatically, *some* of its water vapor could condense, and the system would become liquid water droplets suspended in air, which is *not* an ideal gas. The volume fraction of liquid water is so small that this is of no concern. But the form of the first law is different if there are phase changes. As a consequence, a *decrease* of temperature because of expansion is partly opposed by an *increase* of temperature because of condensation. The easiest way to understand this is to recognize that the potential energy of water molecules *decreases* when they condense, which must be balanced by an *increase* in their kinetic energy manifested by a temperature increase (see Section 5.3).

Here we limit ourselves to the change of temperature as pressure changes for an ascending or descending parcel in "dry" processes, which does not mean that the air is dry, but rather that its water vapor component just goes along for the ride. A dry process is sometimes called an unsaturated process, in contrast to a moist or saturated process (see Section 5.1 for a definition of the misleading terms "unsaturated" and "saturated"). We consider saturated processes in Section 6.5. If we define the *potential temperature* Θ as the temperature a parcel of air would have if taken adiabatically, without condensation of water vapor, to a reference pressure p_0, usually 1000 hPa, the constant in Eq. (3.83) can be written as

$$\Theta p_0^{-R/c_p}, \tag{3.84}$$

so that

$$Tp^{-R/c_p} = \Theta p_0^{-R/c_p}, \tag{3.85}$$

and

$$\Theta = T\left(\frac{p_0}{p}\right)^{R/c_p}. \tag{3.86}$$

For a dry adiabatic process Θ is conserved because

$$\frac{d\Theta}{dt} = 0, \tag{3.87}$$

and the first law can be written as

$$c_p \frac{d\Theta}{dt} = \frac{\Theta}{T} q. \tag{3.88}$$

Θ of a parcel ascending or descending adiabatically in the atmosphere is constant and the parcel temperature can be calculated using Eq. (3.86). A parcel at $20\,^{\circ}\mathrm{C}$ ascending from 1000 hPa to 900 hPa will cool about $9\,^{\circ}\mathrm{C}$.

You are not likely to find potential temperature mentioned in physics textbooks. Meteorologists can claim this quantity as their own. Its roots can be traced to an 1888 paper by Herman von Helmholtz, one of the greatest scientists of his time (and, indeed, of all times). He used the symbol Θ to designate a quantity he called *Wärmegehalt*: the total heat contained in a mass of air. To measure this "heat" he considered the absolute temperature a mass of air would have if it were brought adiabatically to "normal" pressure. W. von Bezold suggested that *Wärmegehalt* be replaced by the more descriptive term "potential temperature," which met with Helmholtz's approval.

The adiabatic temperature variation of a parcel as it changes height can be obtained from the first law (per unit mass) by using the ideal gas law to rewrite Eq. (3.42) as

$$q = c_p \frac{dT}{dt} - \frac{1}{\rho} \frac{dp}{dt} = 0. \tag{3.89}$$

If we assume that the parcel is in hydrostatic equilibrium,

$$\frac{dp}{dt} = -\rho g \frac{dz}{dt}, \tag{3.90}$$

and Eq. (3.89) becomes

$$c_p \frac{dT}{dt} + g \frac{dz}{dt} = \frac{d}{dt} (c_p T + gz) = 0. \tag{3.91}$$

Dry static energy is defined as $c_p T + gz$ and is conserved for dry adiabatic processes except where nonhydrostatic effects are appreciable, for example, in intense convection, hurricanes, or severe storms. Because nonhydrostatic effects are ignored, this energy is called static. Although z has an arbitrary reference point, if $z = 0$ at 1000 hPa, $c_p T + gz \approx c_p \Theta$.

Conservation of dry static energy provides the basis for determining how the temperature of a parcel changes with height as the parcel ascends or descends.

The derivative with respect to time in Eq. (3.91) can be replaced with the total derivative with respect to height to give

$$c_p \frac{dT}{dz} + g = 0, \tag{3.92}$$

or

$$\frac{dT}{dz} = -\frac{g}{c_p}. \tag{3.93}$$

Since a *lapse rate* is a rate of *decrease* of something, the temperature decrease with height given by Eq. (3.93) is the dry adiabatic lapse rate

$$\Gamma_d = \frac{g}{c_p}. \tag{3.94}$$

During ascent, the parcel does not exchange energy with its surroundings because of a temperature difference between them or mixing (hence, the designation "adiabatic"), and any water vapor it contains is not transformed into liquid or ice (hence, the designation "dry").

If $g = 0$, $\Gamma_d = 0$, and hence, a parcel would ascend (or descend) isothermally. Moreover, from Eq. (2.28), the vertical pressure gradient also vanishes. Thus, two important (and simple) equations relevant to meteorology explicitly depend on gravity. On the laboratory scale (meters), the equilibrium temperature and pressure profiles of an isolated column of air are such that g can be set to zero without serious error. But the vertical scale of Earth's atmosphere is thousands of times greater.

Equation (3.91) can be expressed in a more physically revealing form:

$$\frac{dh}{dz} = -g, \tag{3.95}$$

where [see Eq. 3.67)] $h = c_p T$ is the specific enthalpy of the parcel. It decreases as the parcel rises because of negative *gravitational* work.

The dry adiabatic lapse rate is about $9.8\,°C\ km^{-1}$, which may be mistaken for what the lapse rate in the atmosphere *is* or *should be*. Actual lapse rates are not necessarily the dry adiabatic rate, because near surfaces (within a meter or so) lapse rates can be hundreds (or even thousands) of times larger. You can verify this for yourself. On a hot summer afternoon the temperature of an asphalt road may be $40\,°C$ or higher. If you are standing on such a road, your feet will be at $40\,°C$ but your nose may be sniffing air at $35\,°C$ (or lower). The distance from your feet to your nose is of order 1 meter, and hence, the lapse rate is about $5°C\ m^{-1}$, or 500 times the dry adiabatic lapse rate. We have measured lapse rates $\approx 1000\ °C\ m^{-1}$ a few millimeters above surfaces on a cloudy day in July. In Section 7.1 we discuss why lapse rates near the ground can be vastly greater than the dry adiabatic lapse rate.

For well-mixed layers of the atmosphere, far from surfaces, in which phase changes are not taking place, the dry adiabatic lapse rate is often approximately the lapse rate of air temperature (Section 4.4). Yet the significance of the dry adiabatic lapse rate is not that it is necessarily the actual lapse rate, but rather, as we show in Section 3.5, it demarcates the boundary between static stability and instability. Before showing this, we take a short side trip to discuss critically the preceding derivation.

What is the fundamental reason why temperature decreases with height in the atmosphere? The simple answer is conservation of energy. From a molecular point of view, if the potential energy of a molecule increases, its kinetic energy must decrease. If m is the mass of a molecule, g the acceleration due to gravity, and Δz the change in elevation, set the potential energy change equal to the change in the mean kinetic energy, $mg\Delta z = 3k\Delta T/2$. From this we obtain $\Delta T/\Delta z = 2mg/3k$; m follows from the mean molecular weight of air and Avogadro's constant. This is the microscopic approximation to the macroscopic Eq. (3.94) and corresponds to a lapse rate of 16 K km^{-1}, which is only about 63% higher than the dry adiabatic lapse rate. Either this is a happy coincidence, or we got the physics right. If we had gotten the exact value, you would have reason to believe that we cheated. By simple physical arguments we can make an intelligent guess that Γ_d must be proportional to the ratio of g to c_p—the only two relevant physical parameters.

Although we derived Eq. (3.95) by considering an ideal gas, it is more general and applies to liquids, for example, a lake. Take the z-axis downward. From the first law for an adiabatic process

$$\frac{dh}{dt} = \frac{1}{\rho}\frac{dp}{dt}.$$ (3.96)

Replace t by z, the distance downward into the water, and use the hydrostatic equation in the form $dp/dz = \rho g$ to obtain Eq. (3.95) with only a sign change. But the first derivation, although restricted, is more physically transparent.

Do Pistons and Cylinders Inhabit the Atmosphere?

Although the dry adiabatic lapse rate is a fundamental atmospheric stability parameter, its theoretical foundations may seem a bit shaky. We derived it by determining the rate of change of temperature of a rising parcel of air without defining exactly what is meant by a parcel. Even more disturbing, an essential ingredient in our derivation was the reversible working rate $-p\,dV/dt$, which we derived in Section 1.6 for a gas confined within an impermeable cylinder and acted upon by a moving, impermeable piston. Despite years of meteorological observations, such pistons and cylinders in the atmosphere have never been reported. Before we fully accept the dry adiabatic lapse rate, we should examine more carefully what went into its derivation.

Not so very long ago, the molecules now accepted as undeniably real were considered by many respectable scientists to be merely fictitious aids to theorizing without a real existence. That changed because of experimental evidence. Molecules could be localized, their number and masses measured. Although the particles and parcels of fluid mechanics bear a resemblance to their molecular counterparts, they continue to be the nebulous "molecules" of continuum theories. But there are short-lived, small-scale coherent structures in air. Call them parcels if you wish. They can be made visible by Schlieren techniques, videos of which will take your breath away—and make it visible.

Tidy thermodynamic relations are obtained for systems with well-defined boundaries, unlike air in the free atmosphere. Because parcels are fictitious, we can imagine a parcel to be air enclosed in a tiny impervious balloon. Assume that any work done in stretching it is negligible, as is its mass. We assume that the air pressure in the balloon is always that of its surroundings, but its volume and temperature can change. We also assume that there is no exchange of energy with the surroundings because of a temperature difference ($Q = 0$). The assumption of pressure equilibrium, but not temperature equilibrium, allows for the possibility of the density of the parcel to be different from that of its surroundings. Pascal, after whom the unit of pressure is named, observed the volume increase of a partially inflated bladder when carried up a hill. We can consider this bladder to be the first realization of a parcel. Imagine that we give the balloon a slight push upward or downward. The potential energy of the air in the balloon increases or decrease as it rises or falls in Earth's gravitational field, and pV work can be done on or by the air as the balloon volume decreases or increases. Underlying the assumption of pressure but not temperature equilibrium is that pressure equilibrates much faster than temperature.

For deriving the dry adiabatic lapse rate, the finite lifetime of a parcel is of no concern. Our derivation was based on the instantaneous rate of change of a parcel's temperature. Thus, the parcel need live for only a brief moment. Once its mathematical task has been completed, it is no longer of interest and may dissipate without any damage to our derivation. But the test of any theoretical result obtained by invoking fictitious parcels is agreement with observations, not the rigor of the derivation, and the derivation of the dry adiabatic lapse rate is not rigorous.

In a sense, a parcel is a fuzzy piston–cylinder with an irregular boundary, constantly changing, and of indefinite size. What happens to an idealized parcel looked upon as a small volume of an ideal gas containing a fixed number of molecules? Nowhere in sight is there a piston with a molecular number density thousands of times greater than that of the gas, and hence, the notion that the parcel pushes against or is pushed by surrounding molecules makes no sense. In Section 2.3 we give a molecular interpretation of pressure in an ideal gas as a momentum flux density and derive the vertical pressure gradient from this interpretation. Then, in Section 2.9 we argue that the pressure gradient force in a gas is not a fundamental force, but rather a rate of change of momentum. Thus, if the pressure within a parcel is different from that of its surroundings, there is a pressure gradient. This is the origin of what is called work done by the parcel on its surroundings (expansion) or by the surroundings on the parcel (compression). But because there is no solid interface between the parcel and its surroundings, this is just a way to describe macroscopically how an increase (decrease) of potential energy must be compensated for by an equal decrease (increase) of kinetic energy manifested by a decrease (increase) of temperature. *How* this is done in detail is beyond the reach of thermodynamics. But this is no different from the small increase in the internal energy, and hence, temperature, of a dropped ball (Section 1.4). If a small solid sphere with specific heat capacity c is dropped from rest a distance Δz before hitting the ground and stopping, and the mass–motion kinetic energy is converted entirely into internal energy of the sphere, its temperature change is $\Delta T = g\Delta z/c$, and hence $\Delta T/\Delta z = g/c$. Does this look familiar? The analogy here is almost perfect. We ignored drag on the sphere and its interaction with the ground. Similarly, a parcel neither experiences drag nor comes abruptly to a halt.

A major difference between atmospheric thermodynamics and that treated in physics and engineering textbooks is that they may not even mention g because it is irrelevant. On a laboratory scale the dry adiabatic lapse rate corresponds to temperature differences ($\approx 0.01\,^{\circ}\mathrm{C}$) difficult to measure, whereas in the atmosphere much larger differences are routinely measured because the scale is so much larger (see, e.g. Fig. 4.8 and figures in Chapter 6). Gases in the laboratory or in engines can be confined in cylinders fitted with pistons, whereas atmospheric air is free, but not absolutely because of gravity, which plays a similar role as pistons and cylinders in small-scale thermodynamics.

Without using the Poisson relation Eq. (3.83), we derive the lapse rate for a saturated parcel, Eq. (6.104), which reduces to the dry adiabatic lapse rate. The rate of working, Eq. (1.65) is nowhere to be seen in this derivation, but it does require first learning about entropy (Chapter 4).

3.5   Stability and Buoyancy

The concept of stability is fundamental in many branches of science and engineering. To determine the stability of a system we must first ascertain whether it is in equilibrium or not. Does the sum of all forces and torques on the system vanish? If so, it is in equilibrium. But it may be stable or unstable. For example, with great care we might be able to balance a bicycle in such a way that it would stay upright. But if a gnat were to land on a handlebar, the bicycle would topple to the ground. The bicycle was in equilibrium, but statically unstable. If we were to hop on the bicycle and put it in motion by cranking on the pedals, the bicycle would now be dynamically stable. While in motion, it would stay upright even though its rider might shift around and its wheels might be buffeted by encounters with ruts and bumps. An eagle could land on a moving bicycle without upsetting it, whereas a gnat was sufficient to topple the bicycle in static equilibrium.

To determine if a system in equilibrium is stable, we need to ask if a small perturbation to the system causes it to return to or depart from equilibrium? If it returns, the system is stable; if it departs, the system is unstable. The perturbation must be small. A moving bicycle hit by a truck will not return to the equilibrium position, but this is a catastrophe, not a perturbation.

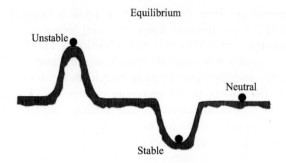

Fig. 3.3: A ball perched on top of a hill is in unstable equilibrium. Lying at the bottom of a depression, the ball is in stable equilibrium. Resting on a plane, the ball is in neutral equilibrium.

An example often given is that of a ball in a depression, on a mound, and on level ground (Fig. 3.3). Lying at the bottom of the depression the ball is in equilibrium. If it is given a slight nudge, it will return to the bottom (strictly speaking, it will roll back and forth with ever-decreasing amplitude until it finally settles down again at the bottom of the depression). The equilibrium is (statically) stable. When the ball is balanced on top of the mound, however, a slight nudge to the ball will cause it to roll off the mound. The equilibrium is unstable. When the ball is on level ground and nudged, its vertical position does not change. The equilibrium is neutral.

Buoyancy

Before making the connection between dry adiabatic lapse rate and static stability of the atmosphere, we must briefly discuss buoyancy.

Consider a fixed volume V of a fluid in a gravitational field; the volume is enclosed by a surface A (Fig. 3.4). The force on the volume exerted by its surroundings is a surface integral

$$-\int_A p\mathbf{n}\,dA, \tag{3.97}$$

where $\mathbf{n}$ is a unit vector normal to A and p is the pressure over this surface; the minus sign is required because the unit normal is directed outward from the surface. This surface integral is equivalent to a volume integral because of one of the several forms of the divergence theorem:

$$-\int_A p\mathbf{n}\,dA = -\int_V \nabla p\,dV. \tag{3.98}$$

Hydrostatic equilibrium is the assumption that this force is balanced by the weight of V:

$$-\int_V \nabla p\,dV + \int_V \rho\mathbf{g}\,dV = 0, \tag{3.99}$$

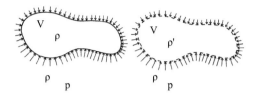

Fig. 3.4: A fluid (at rest) with density ρ in a volume V is acted on by two equal and opposite forces: its weight $(\rho g V)$ and the integral of the pressure p over the surface of V. If the fluid in V is replaced by a fluid with a different density ρ' in such a way that the pressure distribution is unchanged, a net force acts on the volume, the buoyancy force.

where $\mathbf{g}$ is the local gravitational field (per unit mass). Because V is arbitrary, Eq. (3.99) can be satisfied only if

$$\nabla p = \rho \mathbf{g}. \tag{3.100}$$

This is a more general form of Eq. (2.28).

Now assume that the fluid in V suddenly is replaced by a fluid with density $\rho' \neq \rho$, the surrounding fluid remaining undisturbed. According to Eq. (3.100), the pressure gradient is determined solely by density and gravity, and hence, the pressure outside V right up to its boundary A is what it was before replacement. Although the integrated pressure over A does not change, the weight of the fluid enclosed by it does. As a consequence, the total force

$$\mathbf{F} = -\int_A p\mathbf{n} \, dA + \int_V \rho' \mathbf{g} \, dV \tag{3.101}$$

acting on V does not vanish. If we use Eqs. (3.98) and (3.99), Eq. (3.101) becomes

$$\mathbf{F} = \int_V (\rho' - \rho)\mathbf{g} \, dV. \tag{3.102}$$

To be more specific, let $\mathbf{g}$ be $-g\mathbf{e}_z$, where g is constant:

$$\mathbf{F} = -g\mathbf{e}_z \int_V (\rho' - \rho) \, dV. \tag{3.103}$$

This is the buoyancy force on V. If $\rho' < \rho$, this force is upward and the volume (parcel of fluid) is said to be *positively* buoyant. If $\rho' > \rho$, this force is downward and the volume (parcel of fluid) is said to be *negatively* buoyant. The physical interpretation of buoyancy is straightforward: the *apparent* weight of an object in a fluid can be increased or decreased because of pressure differences over its upper and lower surfaces. Positive buoyancy is *negative* gravity, and hence hot air and helium balloons can be said to be anti-gravity machines. But the apparent weights in air of cloud droplets, raindrops, water-balloons, and humans (mostly water) are reduced by the *same* relative amount, about one part in a thousand, and hence, buoyancy cannot be why clouds "apparently" don't fall (see Section 7.2).

F is the difference between two weights: that of the body and that of its volume V of water, sometimes said to be "displaced by" the body. This is essentially Archimedes' principle, often invoked to "explain" buoyancy. But this is a description, not an explanation.

Buoyancy and drag are not even distantly related. Buoyancy depends on the difference between the density of an object and that of the surrounding fluid, and gravity. Drag depends on the speed of an object in a fluid, only its density, and gravity is irrelevant.

Buoyancy is also applicable to solids. Plate tectonics is driven by buoyancy, density differences because of temperature differences. However, the time scale is of order millions of years.

The buoyancy of an air parcel can be determined by applying Eq. (3.103) to an ideal parcel of volume V. The equation of vertical motion of the center of mass of this parcel is

$$F = -g(\rho_p - \rho_e)V = M\frac{d^2 z}{dt^2} = \rho_p V \frac{d^2 z}{dt^2}, \tag{3.104}$$

where the subscripts p and e denote the parcel and its environment, respectively. Because vertical air velocity w is defined as

$$w \equiv \frac{dz}{dt}, \tag{3.105}$$

$$\frac{dw_p}{dt} = g\left(\frac{\rho_e - \rho_p}{\rho_p}\right). \tag{3.106}$$

Thus, if the density of the parcel is less than that of its surroundings, the parcel will be accelerated upward. By supposition, the pressure p of the parcel is always that of its surroundings:

$$p = \rho_e T_e R = \rho_p T_p R. \tag{3.107}$$

Equations (3.106) and (3.107) combined give

$$\frac{dw_p}{dt} = g\left(\frac{T_p - T_e}{T_e}\right). \tag{3.108}$$

If we apply Eq. (3.108) to a parcel at a temperature higher than the surrounding temperature T_e, assume that the parcel pressure is approximately that of its surroundings, we obtain the useful result that its upward acceleration (force per unit mass) is approximately $g\Delta T/T_e$, where $\Delta T = T_p - T_e$. This displays the essential physics of free or buoyancy-driven convection in the atmosphere: a relative temperature difference and gravity (see Section 7.1).

Dry Adiabatic Lapse Rate and Stability

The stage now has been set to combine the previous two subsections. We consider the stability of a parcel initially in thermal equilibrium with its surroundings at

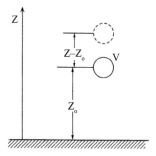

Fig. 3.5: A parcel of air with volume V is in stable equilibrium at height z_0. When the parcel is perturbed slightly, its displacement from equilibrium $\delta z = z - z_0$ results in a temperature change of $-\Gamma_d(z - z_0)$.

initial height z_0 and slightly displaced by $\delta z = z - z_0$ from this height (Fig. 3.5). Equation (3.108) can be written as

$$\frac{dw_p}{dt} = \frac{d^2}{dt^2}(\delta z) = g\left(\frac{T_p(z) - T(z)}{T(z)}\right), \tag{3.109}$$

where we omit the subscript e from the environmental temperature $T(z)$. If we slightly displace the parcel for the stability test, both the parcel and environment temperatures will change. For dry motions (no condensation) the parcel temperature change is the dry adiabatic lapse rate Γ_d, so that

$$T_p(z) = T_{p0} - \Gamma_d \delta z, \tag{3.110}$$

where T_{p0} is the temperature of the parcel at z_0. The temperature of the environment in Eq. (3.109) can be expanded in a Taylor series

$$T(z) = T_0 + \frac{\partial T}{\partial z}\delta z + \frac{1}{2}\frac{\partial^2 T}{\partial z^2}(\delta z)^2 + \dots \tag{3.111}$$

Because the lapse rate in the environment surrounding the parcel is

$$\Gamma = -\frac{\partial T}{\partial z} \tag{3.112}$$

and the higher order terms in the Taylor series expansion are small, Eq. (3.111) can be approximated as

$$T(z) = T_0 - \Gamma \delta z. \tag{3.113}$$

Equations (3.110) and (3.113) in Eq. (3.109) give

$$\frac{dw_p}{dt} = g\left(\frac{T_{p0} - T_0 - (\Gamma_d - \Gamma)\,\delta z}{T_0}\right),\qquad(3.114)$$

where we ignore the vertical temperature dependence of the denominator because it is much larger than the numerator and temperature variations are small compared with absolute temperature. We now can use Eq. (3.114) to make our stability tests. If the parcel is at rest at z_0 with the same temperature as its surroundings. $T_{p0} = T_0$, and hence,

$$\frac{dw_p}{dt} = \frac{-g\,(\Gamma_d - \Gamma)\,\delta z}{T_0}.\qquad(3.115)$$

Now consider a small vertical displacement of the parcel and examine the sign of the right side of Eq. (3.115) for different environmental lapse rates. If it is less than the dry adiabatic lapse rate ($\Gamma < \Gamma_d$) and displacement is upward ($\delta z > 0$) the buoyancy force in Eq. (3.115) will be downward. This will force the parcel back to its initial equilibrium level. Similarly, if the initial displacement were downward, the buoyancy force would be upward, and it would force the parcel back to its initial equilibrium level. The atmospheric is *locally* (at $z = z_0$) *stable*. If $\Gamma > \Gamma_d$, the buoyancy force would be upward for $\delta z > 0$ and the parcel would be accelerated upward away from the equilibrium level. For $\delta z < 0$, the parcel would be accelerated downward. The atmosphere is locally *unstable*. If $\Gamma = \Gamma_d$, the atmosphere is locally *neutral*. Thus, the criteria for static stability can be summarized as

$$\Gamma - \Gamma_d < 0,\quad \text{stable}\qquad(3.116)$$

$$\Gamma - \Gamma_d > 0,\quad \text{unstable}\qquad(3.117)$$

$$\Gamma - \Gamma_d = 0.\quad \text{neutral}.\qquad(3.118)$$

These criteria can be applied to any level in the atmosphere and are illustrated in Fig. 3.6 for an idealized temperature structure. At any level, the stability can be visualized by noting that the temperature of a parcel displaced from a given level will change at the dry adiabatic lapse rate. For example, for the stable layers shown in Fig. 3.6, a parcel displaced upward will be colder than its surroundings and negatively buoyant. A layer where the temperature increases with height has a negative lapse rate and is called an *inversion*. Inversions are stable and are important features in the atmosphere because they inhibit vertical mixing. Low-level inversions can trap pollution in the lower layers, and convection in the atmosphere is often capped by an inversion. The tops of deep convection associated with severe storms sometimes extend to the tropopause—a stable layer that is the boundary between the troposphere and the stratosphere. Unstable layers are *superadiabatic* layers and are not prevalent in the free atmosphere since these layers will quickly overturn and

(a) (b)

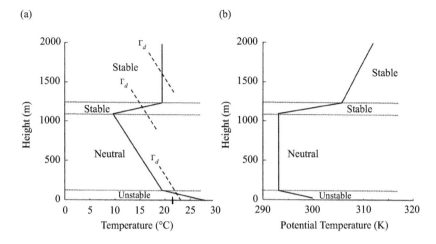

Fig. 3.6: a) Idealized temperature profiles in the lower atmosphere illustrating layers of different static stabilities. The dry adiabatic lapse rate shown can be compared with the atmospheric lapse rate in each layer. b) The equivalent potential temperature profiles for these layers.

become well mixed. Superadiabatic layers are observed above a strongly heated surface and generate convective enthalpy fluxes that heat the atmosphere above this layer.

Again, as in our derivation of the dry adiabatic lapse rate, the finite lifetime and fuzziness of a parcel do not cast doubts on our determination of atmospheric stability. We used the mathematical device of comparing the instantaneous rate of change of temperature of a parcel with that of its surroundings. Static stability at a point is a state of the atmosphere, not of hypothetical parcels with ill-defined boundaries and finite lifetimes.

Static stability criteria also can be represented in terms of potential temperature by using the ratio between temperature and potential temperature Eq. (3.86) in Eq. (3.109) to give

$$\frac{dw_p}{dt} = \frac{d^2}{dt^2}(\delta z) = g\left(\frac{\Theta_p - \Theta(z)}{\Theta(z)}\right), \tag{3.119}$$

where the parcel pressure is the same as that of the environment. For a vertical perturbation of a parcel in equilibrium, Θ_p will not vary with height because it is conserved for dry adiabatic motions. The atmospheric Θ distribution can be represented by the Taylor series expansion as

$$\Theta = \Theta_0 + \frac{\partial \Theta}{\partial z}\delta z + \ldots \tag{3.120}$$

If the higher-order terms in this expansion are neglected, and if the parcel has the same potential temperature as its environment, Eq. (3.119) can be written as

$$\frac{dw_p}{dt} = -\frac{g}{\Theta}\frac{\partial\Theta}{\partial z}\delta z. \tag{3.121}$$

Applying the stability test, as we did previously, using Eq. (3.121) gives the static stability in terms of the gradient of potential temperature:

$$\frac{d\Theta}{dz} > 0, \quad \text{stable} \tag{3.122}$$

$$\frac{d\Theta}{dz} = 0, \quad \text{neutral} \tag{3.123}$$

$$\frac{d\Theta}{dz} < 0. \quad \text{unstable}. \tag{3.124}$$

The relationship of the potential temperature stability criteria compared with those using temperature lapse rates can be formalized by taking the logarithm of Eq. (3.86) to give

$$\ln\Theta = \ln T - \frac{R}{c_p}\ln p + \frac{R}{c_p}\ln p_0. \tag{3.125}$$

The derivative of Eq. (3.125) with respect to z gives

$$\frac{1}{\Theta}\frac{d\Theta}{dz} = \frac{1}{T}\frac{dT}{dz} - \frac{R}{c_p p}\frac{dp}{dz}, \tag{3.126}$$

and applying the hydrostatic equation to the dp/dz term gives

$$\frac{T}{\Theta}\frac{d\Theta}{dz} = -\Gamma + \Gamma_d. \tag{3.127}$$

A knee-jerk response to questions about stability is that the atmosphere is stable if temperature increases with height because less-dense air overlies denser air. Conversely, the atmosphere is unstable if temperature decreases with height. This response is based on the notion that density is inversely proportional to temperature, which is not true unless pressure is constant, which it is *not* in the atmosphere. But, if by temperature we mean *potential* temperature, which incorporates both temperature *and* pressure, we obtain the correct stability criteria Eqs. (3.122)–(3.124). These are easier to apply than those with lapse rates because only the slope of Θ is needed and not the difference between the environmental lapse rate and the dry adiabatic lapse rate. This is illustrated in Fig. 3.6), where the Θ profiles corresponding to the temperature profiles for the different layers are shown.

If we naively take the criterion for atmospheric *instability* to be $d\rho/dz > 0$, then from the ideal gas law and the hydrostatic equation the temperature gradient must satisfy $dT/dz < -g/R$, whereas the correct criterion is $dT/dz < -g/c_p$, where $c_p > R$. Thus, the atmosphere can be unstable, even where density *decreases* with height.

Mineral and vegetable oils float on water because of their lower density. Confusion about stability in the atmosphere may arise from a false analogy with liquids. The criteria for stability of the two fluids are different. To the extent that incompressibility is a good approximation, the density of a liquid parcel does not change (much) in a small adiabatic displacement (see Section 4.3). Depending on the temperature profile in the liquid and its coefficient of thermal expansion, Eq. (3.150), the density of the parcel's surroundings can be greater (positive buoyancy) or smaller (negative buoyancy). Water is an atypical liquid in that its density can both increase and decrease with temperature (Fig. 3.9).

Parcel Oscillations

Whenever a mechanical system has a position of stable equilibrium, small-amplitude oscillations are possible. For a parcel in equilibrium, Eq. (3.121) can be written as

$$\frac{d^2}{dt^2}(\delta z) = -\frac{g}{\Theta}\frac{\partial \Theta}{\partial z}\delta z, \tag{3.128}$$

where a positive gradient of potential temperature corresponds to static stability. Equation (3.128) is the equation of motion of a simple harmonic oscillator with frequency ω_g given by

$$\omega_g^2 = g\left(\frac{1}{\Theta}\frac{\partial \Theta}{\partial z}\right)_o. \tag{3.129}$$

This is the *Brunt–Väisälä* frequency, a fundamental frequency of the atmosphere determined by the product of g and the vertical gradient of potential temperature. The greater this gradient, the more stable the atmosphere, the stronger the restoring force acting on a parcel, and the higher its oscillation frequency. For a typical tropospheric potential temperature gradient of $3\,°\mathrm{C\,km^{-1}}$, the period $(2\pi/\omega_g)$ of a buoyancy oscillation is about ten minutes; for an inversion layer in which the potential temperature gradient is $10\,°\mathrm{C\,km^{-1}}$, this period decreases to about three minutes. In a 1927 paper, Brunt derived the buoyancy oscillation frequency (Eq. 3.129) and noted that the frequencies he obtained for typical tropospheric potential temperature gradients were consistent with frequencies of pressure oscillations sometimes observed in microbarograph traces. He also noted that after submitting his original manuscript he learned that Väisälä had published an article in a Finnish journal two years previously, giving a similar derivation for the frequency of buoyancy oscillations. Hence the designation Brunt–Väisälä frequency, or, to be historically accurate, Väisälä–Brunt frequency.

The Brunt–Väisälä frequency is related to the frequency of internal gravity (or buoyancy) waves. As with simple one-dimensional buoyancy oscillations, the restoring force associated with these waves is buoyancy, although parcel motions are not confined to only vertical oscillations. These transverse waves can propagate both vertically and horizontally in the atmosphere. Internal gravity waves should not be confused with external gravity waves (such as those on the ocean), which propagate along the interface between two dissimilar fluids. In the last few decades there has been a move to call internal gravity waves buoyancy waves because buoyancy is the restoring force (buoyancy requires both gravity and density differences). Furthermore, the term *buoyancy waves* avoids confusion with the gravitational waves of general relativity.

Buoyancy waves are associated with mountain lee waves and can cause the clear-air turbulence sometimes experienced by aircraft. Air deflected by mountain barriers and convective elements impinging upon a stable layer are two mechanisms for generating buoyancy waves in the atmosphere. Lines of clouds parallel to mountain ridges often mark the upward-moving air associated with lee waves. The distance between these lines is the wavelength, and the corresponding frequency is proportional to the Brunt–Väisälä frequency. Deep convection impinging on a stable layer can generate buoyancy waves that propagate out from the center of the convection.

3.6   Specific Heats of Gases

Underlying our derivation of the dry adiabatic lapse rate was the assumption that γ, the ratio of specific heats, is independent of temperature. We now take a closer look at specific heats.

Although we often refer to molecules in preceding sections, we really considered point masses, which by definition have no structure. Molecules, however, do have an internal structure. They are composed of atoms, which themselves have an internal (electronic and nuclear) structure and (fuzzy) size. The closest thing to a point mass in nature is an electron, which (as far as is known) has neither size nor structure.

From Eq. (2.10) the internal *translational* kinetic energy of N molecules of *any* ideal gas is

$$U = N \left\langle \frac{mv^2}{2} \right\rangle = \frac{3}{2}NkT, \tag{3.130}$$

which is the *only* internal energy of an ideal gas of structureless atoms. Its heat capacities are therefore

$$C_V = \frac{3}{2}Nk, \qquad C_p = \frac{5}{2}Nk, \tag{3.131}$$

the ratio of which is

$$\gamma = \frac{\frac{5}{2}}{\frac{3}{2}} \frac{Nk}{Nk} = \frac{5}{3}. \tag{3.132}$$

Measured values of γ are about 1.67 for the monatomic gases helium, neon, argon, krypton, and xenon, which is comforting. But the value for nitrogen and oxygen, about 1.4, is discomforting. The source of this discrepancy is the internal structure of these diatomic molecules. The value for argon, when it was still a mysterious component of air, was evidence for its monatomicity. It was not at first welcomed into the family of elements, but as other noble elements were discovered, the Periodic Table had to be restructured (see References at the end of Chapter 2).

We showed in Chapter 2 that the kinetic energy of a system of point masses is the sum of the kinetic energy of the center of mass and the kinetic energy about the center of mass. We take as our diatomic molecule a system of two identical point masses (structureless atoms) separated by a variable distance (Fig. 3.7). For simplicity, the atoms are assumed to always lie in a plane. The origin of a coordinate system in this plane is the center of mass of the molecule. Two coordinates, x and y, specify the position of an atom relative to the center of mass of the molecule. To simplify further, we assume that the line from the origin to each atom makes the same angle θ with the x axis. We envision the atoms as lying at the ends of a rod that can vary in length but does not bend. Because the atoms are identical (in mass) and the origin of coordinates is the center of mass of the molecule, each atom is at the same distance r from the origin. The x and y coordinates of one of the atoms are related to its polar coordinates r and θ by:

$$x = r\cos\theta \text{ and } y = r\sin\theta. \tag{3.133}$$

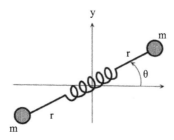

Fig. 3.7: Idealized diatomic molecule: two point masses of equal mass m attached by a spring, which can stretch but not bend. Here, the motion of the molecule is confined to a plane.

The kinetic energy (about the center of mass) of this atom is

$$K_a = \frac{1}{2}m(\dot{x}^2 + \dot{y}^2), \tag{3.134}$$

where a dot over a symbol indicates differentiation with respect to time. By differentiating Eq. (3.133) with respect to time and substituting the result in Eq. (3.134)

$$K_a = \frac{1}{2}mr^2\omega^2 + \frac{1}{2}m\dot{r}^2, \tag{3.135}$$

where $\omega = d\theta/dt$ (angular velocity). Because the x and y coordinates of one atom are the negative of those of the other, the total kinetic energy (about the center of mass) of the entire molecule is twice Eq. (3.135):

$$K_m = mr^2\omega^2 + m\dot{r}^2. \tag{3.136}$$

The first term is called the *rotational* kinetic energy (which strictly should be vibrational–rotational kinetic energy because r can vary). K_m is the total kinetic energy about the center of mass of a molecule considered as a (small) system composed of atoms; the second term is the *vibrational* kinetic energy.

A molecule is a collection of atoms more or less bound together. They travel as a group rather than independently. This togetherness is enforced by binding forces. We assume that the force binding our two atoms is derivable from a potential energy that depends only on the distance between them. Let r_0 be the distance of the atoms from the origin such that they exert no force on each other. We expand the potential energy P in a Taylor series about this equilibrium position:

$$P(r) = P(r_0) + \left(\frac{\partial P}{\partial r}\right)_{r_0} (r - r_0) + \frac{1}{2}\left(\frac{\partial^2 P}{\partial r^2}\right)_{r_0} (r - r_0)^2 + \dots . \qquad (3.137)$$

Potential energies are defined only to within an additive constant, so we can ignore the first term. By supposition, the atoms vibrate about an equilibrium position, and hence, the second term vanishes. If we retain only the third term and ignore all others, we obtain the potential energy of vibration in the *harmonic approximation*:

$$P \approx \frac{1}{2}\left(\frac{\partial^2 P}{\partial r^2}\right)_{r_0} (r - r_0)^2. \qquad (3.138)$$

Students familiar with the simple harmonic oscillator may recognize the form of P. The force on a simple spring is given by Hooke's law, $F = -kx$, where k is the spring constant (not Boltzmann's constant), and x is the displacement of the spring from its equilibrium position. The corresponding potential energy of the spring is $kx^2/2$, which has the same form as Eq. (3.138), where k is the second derivative of the potential at the equilibrium position. The two atoms vibrate *as if* connected together by a spring.

The total energy E of the molecule now can be written

$$E = K_{tran} + K_{rot} + K_{vib} + P_{vib}, \qquad (3.139)$$

where the subscript *tran* indicates translational (kinetic) energy, *rot* indicates rotational (kinetic) energy, and *vib* indicates vibrational energy (kinetic and potential).

It is plausible that when molecules collide (interact), translational kinetic energy can be partly transformed into rotational and vibrational energy, and vice versa. This expectation is borne out countless times every day on pool tables all over the world.

We showed in Section 2.1 that the mean translational kinetic energy of a point mass is $3kT/2$. The translational kinetic energy of a point mass is

$$\frac{1}{2}m(v_x^2 + v_y^2 + v_z^2). \qquad (3.140)$$

The average of each of the three terms in this kinetic energy is the same and equal to $kT/2$. According to the *equipartition theorem*, this is a general result: the average of each term in the total energy with the form of a square of coordinates or velocities is $kT/2$. If there are s such terms, the thermodynamic internal energy U of N molecules is

$$U = s\frac{1}{2}NkT. \tag{3.141}$$

Although P_{vib} is defined only to within an additive constant, and hence, so is U, we can ignore this because our goal is C_v, the derivative of U with respect to T. From Eqs. (3.53) and (3.141), the ratio of specific heats is

$$\gamma = \frac{s+2}{s}. \tag{3.142}$$

We emphasize that the equipartition theorem is true only to the extent that classical mechanics is true. According to this theorem, the average kinetic energy of a gas molecule, considered as a point mass is $3kT/2$, which brings us back to the fallacy discussed in Section 2.1, namely, that absolute zero is supposedly the temperature at which all motion comes grinding to a halt. Although we may apply the equipartition theorem with reasonable confidence to gases at normal terrestrial temperatures, at very low temperatures classical mechanics becomes an unreliable guide to the properties of matter. We cannot extrapolate to low temperatures a classical theorem approximately valid at high temperatures. Warning labels are required on packages of cigarettes and bottles of booze. Such labels should also be attached to all theories, physical laws, and suchlike: *Warning*: This theory has a restricted range of applicability. Use outside this range may be dangerous to your mental health.

In previous paragraphs we considered a gas of point masses. We now extend our treatment to molecules that cannot be considered as point masses. For such molecules, rotations and vibrations also can contribute to the number s of terms in Eqs. (3.141) and (3.142), and we must take some care in adding up these contributions. Before applying Eq. (3.142) to atmospheric diatomic gases, we acknowledge that molecules are not constrained to rotate only in a plane. Thus, the rotational term in Eq. (3.136) has two components. And because an atom is not really a point mass but has a finite size, we can imagine a third rotational contribution: a diatomic molecule—looked upon as a tiny dumbbell—also can rotate round the line connecting its two atoms. According to classical physics, this rotation also contributes to s. We therefore expect s for a diatomic molecule to be $3 + 3 + 2 = 8$ (three for translation, three for rotation, two for vibration), and hence, $\gamma = 10/8 = 1.25$. Alas, this does not square with measurements, which indicate a value closer to 1.4 for nitrogen and oxygen. But if we make the seemingly *ad hoc* assumption that a molecule is a rigid rod, and hence, its vibrational energy is, so to speak, frozen, and if we also neglect rotation about the line joining the two atoms, we obtain $\gamma = 7/5 = 1.4$. Measurements for polyatomic gases molecules support Eq. (3.142). As the number of atoms increases, γ tends toward 1.

We obtained agreement at the expense of ignoring the vibrational energy of a diatomic molecule and one of its rotational energies, which is a bit disturbing. And when we cogitate a bit more, we find reasons for deeper distress. Not only do molecules have structure, but also do atoms, made up of electrons surrounding nuclei, themselves composed of protons and neutrons in motion and bound together by nuclear forces. An atom of nitrogen has several electrons, as does an atom of oxygen, yet the intra-atomic kinetic and potential energies of these electrons (as well as the intranuclear kinetic and potential energies of neutrons and protons) seem not to contribute to the total molecular energies of nitrogen and oxygen, and hence, to their specific heats. To obtain agreement with measurements of the ratio of specific heats of diatomic molecules, we had to ignore vibrational, one rotational, and the internal electronic and nuclear contributions to the total energy of a molecule, which seems tantamount to cheating.

To get ourselves out of the embarrassing position of having obtained the correct answer at the expense of playing fast and loose with energies, we have no choice but to turn to quantum mechanics. According to classical mechanics, the translational kinetic energy of a molecule can have *any* value whatsoever, and so can the rotational and vibrational energies. According to quantum mechanics, however, although the translational kinetic energy can have any value, the rotational and vibrational energies are *quantized*. For a bound system they can have only a discrete set of values called *energy levels*. Only differences between these energies have a physical meaning. The average energy of a molecule is of order kT, and hence, this is the magnitude of the energy that can be exchanged in an average collision. At normal temperatures kT is appreciably less than the separation between vibrational energy levels but not between rotational energy levels (Fig. 3.8). As a consequence, except in rare collisions between molecules much more energetic than

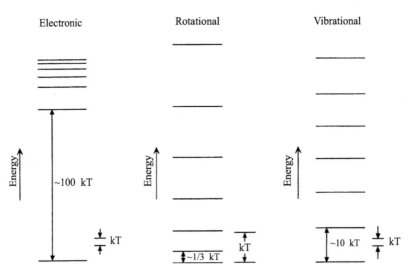

Fig. 3.8: Vibrational, rotational, and electronic energy levels (schematic) of a molecule, showing the spacing between levels relative to kT, where k is Boltzmann's constant and T is absolute temperature.

average, sufficient energy is not available to cause a molecule to increase its vibrational energy from one level to a higher level. For this reason, molecular vibrations do not contribute (much) to the specific heats of gases at ordinary temperatures. The spacing between electronic energy levels, in general, is even greater than that between vibrational energy levels (Fig. 3.8), which is why we can ignore electrons in determining the energies of gas molecules. The spacing between nuclear energy levels is even greater than that between electronic energy levels. The rotational energy levels shown in Fig. 3.8 are those associated with rotation of a diatomic molecule about an axis perpendicular to the line joining its two atoms. The spacing between the energy levels associated with rotation about this line is much greater than kT, which is why we could omit this rotational mode of motion in determining specific heats. For the same reason, rotational energies associated with the spinning of atoms of finite extent about three mutually perpendicular axes do not enter into the determination of specific heats of atomic gases at normal temperatures. According to classical mechanics, a rotating spherical object has rotational energy, which you can verify for yourself by trying to stop with your hand a rapidly spinning (not translating) bowling ball. But according to quantum mechanics, the quantized energy levels of similarly spinning atoms are widely spaced relative to kT. This is why we obtained good agreement with measurements of γ for monatomic gases by considering them to be point masses, even though they are not. It is not enough to be content with getting right answers by making a lucky guess. Without understanding why your guess was correct, your next guess might not be so lucky.

If all the internal modes of motion of molecules, including those of their electrons, neutrons, and protons, contributed to the specific heats of gases, the world would be different. For example, consider the temperature change in an adiabatic expansion or compression for a gas with $\gamma \approx 1$. According to Eq. (3.81), the temperature would hardly change. Diesel engines wouldn't work. The work done in compressing a gas with $\gamma \approx 1$ would go into increasing the translational kinetic energy *and* the energy of each of many internal modes of motion (vibrational, rotational, electronic, nuclear). According to Eq. (3.141), the greater the number of modes, the smaller the share each one gets of a given amount of work done (i.e. a given change in total internal energy ΔU), and each share is proportional to the temperature change ΔT:

$$\frac{\Delta U}{s} = \frac{1}{2} Nk \, \Delta T. \tag{3.143}$$

Many mouths to feed with a fixed amount of food means that everyone starves. As s increases, the specific heat capacity c_p in Eq. (3.94), increases, Γ_d decreases, and $\gamma \to 1$. Absent the constraints of quantum mechanics, the dry adiabatic lapse rate would be much smaller (closer to isothermal).

There is nothing inherent in the definition of an ideal gas that requires its specific heats to be independent of temperature. Ideality is determined by the extent to which interactions *between* molecules in a gas can be neglected, whereas interactions *within* a gas molecule determine the temperature dependence of specific heats. According to our discussion of quantized energy levels, we expect the specific heat of a gas to increase with increasing temperature. As kT increases, changes in vibrational energy states are more likely to be excited by intermolecular collisions. Our expectation is supported by measurements. But as noted in Section 3.2, over a wide range of temperatures in the troposphere, the temperature dependence of c_p and c_v for air is negligible. Because γ is a ratio of specific heats that differ by a constant factor, the fractional change in γ with temperature is even less, which is evident from differentiating γ with respect to temperature:

$$\frac{1}{\gamma}\frac{d\gamma}{dT} = -(\gamma - 1)\frac{1}{c_p}\frac{dc_p}{dT}. \tag{3.144}$$

We do not have to lose sleep worrying about the temperature dependence of c_p, c_v, and γ for air. We safely may take them to be constant over temperature ranges of meteorological interest.

The Ratio of Working to Heating at Constant Pressure

In a constant-pressure process

$$Q = C_p\frac{dT}{dt}, \tag{3.145}$$

and hence, the ratio of working to heating is

$$\frac{W}{Q} = -p\frac{dV}{dt}\bigg/C_p\frac{dT}{dt}. \tag{3.146}$$

In a such a process

$$\frac{dV}{dt} = \frac{\partial V}{\partial T}\frac{dT}{dt}, \tag{3.147}$$

which yields

$$\frac{W}{Q} = -\frac{p}{C_p}\frac{\partial V}{\partial T}. \tag{3.148}$$

This can be written as

$$\frac{W}{Q} = -\frac{p\alpha}{\rho c_p}, \tag{3.149}$$

where

$$\alpha = \frac{1}{V}\frac{\partial V}{\partial T} = -\frac{1}{\rho}\frac{\partial \rho}{\partial T} \tag{3.150}$$

is the (isobaric) *coefficient of thermal expansion.* Equation (3.149) is general, subject only to the condition that $W = -pdV/dt$. For an ideal gas

$$\alpha = \frac{1}{T}. \tag{3.151}$$

Equations (3.149) and (3.151) together with the ideal gas law can be combined to yield

$$\frac{W}{Q} = \frac{R}{c_p} = -\frac{\gamma - 1}{\gamma}. \tag{3.152}$$

This gives us another way of looking at γ. When an ideal gas is heated (or cooled) at constant pressure, work is also done by (on) the gas, the amount of work being determined solely by the ratio of the heat capacity at constant pressure to that at constant volume. According to our sign convention for work, $-W$ is work done *by* a system. Thus, we may interpret the quantity $-W/Q$, where Q is positive, as a kind of thermal efficiency of an ideal gas undergoing an isobaric process, about 29% for air. In the limit as γ approaches 1, the efficiency becomes zero. Heating causes no increase in temperature, and hence, no increase in translational kinetic energy—the kinetic energy that causes the expansion.

When liquids or solids are heated (or cooled) at constant pressure, work is done, but not nearly so much as for gases. If we express the coefficient of thermal expansion (Eq. 3.151) as

$$\frac{1}{V}\frac{\partial V}{\partial T} = -\frac{1}{\rho}\frac{\partial \rho}{\partial T}, \tag{3.153}$$

we may rewrite Eq. (3.149) as

$$\frac{W}{Q} = \frac{p}{\rho c_p}\frac{1}{\rho}\frac{\partial \rho}{\partial T}. \tag{3.154}$$

Near sea level, atmospheric pressure is about 10^5 Pa. The density of liquid water is about 10^3 kg m^{-3}, and its specific heat capacity is about 4.2×10^3 J kg^{-1}K^{-1}. With these values in Eq. (3.154)

$$\frac{W}{Q} \approx 0.025\frac{1}{\rho}\frac{\partial \rho}{\partial T}. \tag{3.155}$$

To proceed, we need to know how the density of liquid water varies with temperature (at a constant pressure of one atmosphere). This is shown in Fig. 3.9 for temperatures between $0\,^\circ$C and $100\,^\circ$C. Water has a broad density maximum at about $4\,^\circ$C, but at higher temperatures expands. The coefficient of thermal expansion of water is not constant, so we take its average

$$\frac{\int_{T_1}^{T_2} \frac{1}{\rho}\frac{\partial \rho}{\partial T}dT}{T_2 - T_1} = \frac{\ln\left(\rho_2/\rho_1\right)}{T_2 - T_1}, \tag{3.156}$$

which, over the temperature, range from $0\,^\circ$C to $100\,^\circ$C is $-0.00043\,^\circ$C^{-1}. With this value for the coefficient of thermal expansion in Eq. (3.155), the ratio of W to Q for liquid water is about -10^{-5}. Not much work is done when liquid water is heated at a pressure of one atmosphere.

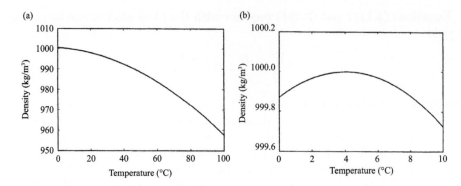

Fig. 3.9: Density of water versus temperature. Although (a) indicates a general decrease of density with temperature from $0\,^\circ$C to $100\,^\circ$C, a close look at the region between $0\,^\circ$C and $10\,^\circ$C (b) reveals a density maximum for water at about $4\,^\circ$C, one of the many peculiar properties of water.

Densities taken from N. E. Dorsey's Properties of Ordinary Water Substance.

3.7 Heat Capacities of Mixtures of Gases

A corollary of the equipartition theorem is that the internal energy of a mixture of (ideal) gases is the sum of the internal energies of each of its constituents. Not only do all the modes of motion of a given molecule get the same share of energy ($kT/2$) on average, so does each different species of molecule in a mixture. From Dalton's law, partial pressures of ideal gas mixtures are also additive, and hence so are enthalpies.

Because internal energies and enthalpies of mixtures of ideal gases are additive, so are heat capacities:

$$C_p = \frac{N}{N_a}\Sigma_j c_{p,m}^j f_j, \quad c_p = \frac{\langle c_{p,m}\rangle}{\langle M\rangle}, \tag{3.157}$$

and

$$C_V = \frac{N}{N_a}\Sigma_j c_{v,m}^j f_j, \quad c_v = \frac{\langle c_{v,m}\rangle}{\langle M\rangle}, \tag{3.158}$$

where $c_{p,m}^j$ and $c_{v,m}^j$ are molar specific heat capacities, $f_j = N_j/N$ is the mole fraction (or number fraction) of the jth species, and $\langle M\rangle$ is the mean molecular weight. The ratio of heat capacities for a mixture is a weighted average of the ratios of heat capacities of its constituents:

$$\gamma = \frac{\Sigma_j \gamma_j c_{v,m}^j f_j}{\Sigma_j c_{v,m}^j f_j}. \tag{3.159}$$

We also may express the specific heat capacities of a mixture in terms of the mass-weighted average of the specific heat capacities of its constituents:

$$c_p = \Sigma_j c_p^j w_j, \quad c_v = \Sigma_j c_v^j w_j, \tag{3.160}$$

where c_p^j and c_v^j are the specific heat capacities of the jth component and w_j is its mass relative to the total mass of the mixture. We also can express γ of a mixture as

$$\gamma = \frac{\Sigma_j \gamma_j c_v^j w_j}{\Sigma_j c_v^j w_j}. \tag{3.161}$$

Water Vapor Demystified

Water is anomalous, as we have been told many times. And what is anomalous slowly becomes magical. Water is invoked to explain everything under the sun. But the most salient property of water vapor (from a meteorological point of view) is that it alone among the constituents of air can condense at ordinary terrestrial temperatures (Section 5.4). In Section 3.2 we address the supposed magical properties of *liquid* water. As just another component of air along with nitrogen, oxygen, and argon, water *vapor* does not impart any special properties to air. The molar specific heat capacities of nitrogen, oxygen, and argon are 29.1 J mol^{-1}K^{-1}, 29.4 J mol^{-1}K^{-1}, and 20.8 J mol^{-1}K^{-1}, with $\gamma = 1.4$, 1.39, and 1.66, respectively. The molar specific heat capacity at constant pressure of dry air from Eq. (3.157) is 29.1 J mol^{-1}K^{-1}, and from Eq. (3.159) $\gamma \approx 1.4$. The molar specific capacity of water vapor is 33.5 J mol^{-1}K^{-1} with $\gamma = 1.33$. This order of the values of γ is consistent with the theory in Section 3.6. It is highest for the monatomic gas, lowest for the triatomic gas, and the intermediate values for the two diatomic gases are about equal. If we assume a 2% relative number concentration of water vapor and consider dry air to be a single gas, γ for this moist air from Eq. (3.159) is the same as that for dry air to within less than 1%.

Although $c_{p,m}$ of water vapor is greater (although not dramatically) than that of dry air, water vapor is *at most* about 4% of moist air, especially since the number density of water vapor falls off with height more rapidly than that of the other molecules in air. A more typical value is 2%, or even less. If we assume that moist air is a mixture of 2% water vapor and 98% dry air with average molecular weight 28.94 and molar specific heat capacity 29.1 J mol^{-1}K^{-1}, the specific heat capacity at constant pressure per unit mass of this moist air is 1027 J kg^{-1}K^{-1}, which is only 2% greater than that of dry air. The same is true for other thermodynamic properties of air and also for its transport properties, for example, thermal conductivity and coefficient of viscosity (Chapter 7). If you are tempted to invoke water vapor to explain some atmospheric phenomenon, pause long enough to do a back-of-the-envelope calculation. Ask yourself if a few percent water vapor in air can markedly affect its thermodynamic properties. More often than not, the answer is "No," as in this example. A colleague wrote to us about a TV meteorologist who asserted that "dry air heats up faster than moist air," a supposed explanation of why temperatures in the middle of the United States were so high one summer. Explanations aimed at general audiences or even students, no matter how obviously ridiculous, are unlikely to be completely rejected. Here's one to try. Question: Why are days longer in the summer? Answer: Thermal expansion.

Isobaric, Adiabatic Mixing of Moist Parcels

As an application of enthalpy conservation and additivity of enthalpies, consider what happens when two parcels of moist air with different masses mix isobarically and adiabatically. The two parcels are initially at different temperatures and partial pressures of water vapor. Let e denote the partial pressure of the water vapor in the mixture and T its temperature after equilibrium is attained. If we plot e versus T, we obtain a curve, the points on which represent all possible moisture states of the resulting mixed parcel. All these states are for different relative masses of the two parcels. What is the shape of this curve, assuming no condensation of water vapor?

To help us visualize this mixing process, imagine the two parcels to be enclosed within an insulated cylinder fitted with a weighted piston that may move freely (Fig. 3.10). Initially, the two parcels are separated by an insulated partition. The partition is removed, and the parcels mix at a constant total pressure determined by the weight per unit area of the piston. Given the initial equilibrium states of the two parcels, what is the final equilibrium state of the mixture?

In an isobaric adiabatic process enthalpy is conserved. Let subscripts 1 and 2 denote the two parcels and i and f their initial and final states. Conservation of enthalpy requires that

$$H_{1i} + H_{2i} = H_{1f} + H_{2f}, \tag{3.162}$$

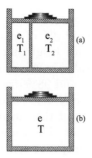

Fig. 3.10: Isobaric mixing of two parcels of air with different initial temperatures T_1 and T_2 and vapor pressures e_1 and e_2. When the insulated partition separating the two parcels (a) is removed, the parcels mix to form a parcel with temperature T and vapor pressure e (b).

where

$$H_{1i} = C_{p1}T_1, \quad H_{2i} = C_{p2}T_2, \quad H_{1f} = C_{p1}T, \quad H_{2f} = C_{p2}T. \tag{3.163}$$

If these equations are combined,

$$T = \frac{C_{p1}}{C_{p1} + C_{p2}}T_1 + \frac{C_{p2}}{C_{p1} + C_{p2}}T_2. \tag{3.164}$$

Not surprisingly, the final temperature lies between the temperatures of the two parcels that went into the mixture.

The partial pressures of water vapor in the parcels are

$$e_1 = N_{w1}\frac{kT_1}{V_1}, \; e_2 = N_{w2}\frac{kT_2}{V_2}. \tag{3.165}$$

The total pressure p (which is the same in both parcels) is

$$p = (N_{a1} + N_{w1})k\frac{T_1}{V_1} = (N_{a2} + N_{w2})k\frac{T_2}{V_2}. \tag{3.166}$$

If these two sets of equations are combined, we obtain

$$e_1 = \frac{N_{w1}}{N_{w1} + N_{a1}}p, \quad e_2 = \frac{N_{w2}}{N_{w2} + N_{a2}}p. \tag{3.167}$$

Again, this weighted average is not surprising. The subscripts a and w denote dry air and water vapor, respectively.

After mixing, we have

$$e = \frac{N_{w1} + N_{w2}}{N_{w1} + N_{w2} + N_{a1} + N_{a2}}p. \tag{3.168}$$

From Eqs. (3.167) and (3.168)

$$N_{w1} = \frac{e_1}{p}(N_{w1} + N_{a1}), \quad N_{w2} = \frac{e_2}{p}(N_{w2} + N_{a2}). \tag{3.169}$$

If we combine these two sets of equations,

$$e = e_1\frac{N_{w1} + N_{a1}}{N_{w1} + N_{w2} + N_{a1} + N_{a2}} + e_2\frac{N_{w2} + N_{a2}}{N_{w1} + N_{w2} + N_{a1} + N_{a2}}. \tag{3.170}$$

Again, no surprises. The final vapor pressure of the mixture is a weighted average of the vapor pressures of its two components.

Now, we assume that the relative amount of water vapor in both parcels is small ($N_w \ll N_a$). With this assumption (good for ordinary air), the heat capacities are proportional to the number of dry air molecules, and we have (approximately)

$$T = \frac{N_{a1}}{N_{a1} + N_{a2}}T_1 + \frac{N_{a2}}{N_{a1} + N_{a2}}T_2, \tag{3.171}$$

and

$$e = \frac{N_{a1}}{N_{a1} + N_{a2}}e_1 + \frac{N_{a2}}{N_{a1} + N_{a2}}e_2. \tag{3.172}$$

These equations have the form

$$e = xe_1 + (1 - x)e_2, \quad T = xT_1 + (1 - x)T_2, \tag{3.173}$$

where

$$x = \frac{N_{a1}}{N_{a1} + N_{a2}}. \tag{3.174}$$

We can eliminate x in Eq. (3.173) to obtain e versus T:

$$e = \frac{e_1 - e_2}{T_1 - T_2}T + \frac{e_2T_1 - e_1T_2}{T_1 - T_2}. \tag{3.175}$$

Subject to approximations, e is a linear function of T (Fig. 3.11). This linearity and the *nonlinearity* of the dependence of saturation vapor pressure on temperature (Section 5.3) are two of the necessary ingredients of *mixing clouds* (such as the "steam" formed on your breath in winter). Cloud formation requires a degree of supersaturation and condensation nuclei (Sections 5.3 and 5.10), of which there usually is no shortage in the atmosphere. For more about mixing clouds see Section 6.8. To understand them requires Eq. (3.175).

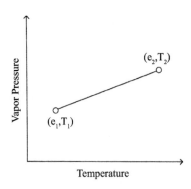

Fig. 3.11: All final states of the system depicted in Fig. 3.10 lie on a straight line connecting the initial states of the two mixed parcels.

3.8   Atmospheric Applications of the First Law

The first law underlies one of the basic equations used to determine the time evolution of the thermodynamic state of the atmosphere. To put this law in a form useful for atmospheric applications, divide Eq. (3.78) by the mass M to give

$$q = c_v \frac{dT}{dt} + p \frac{dv}{dt}, \tag{3.176}$$

where $q = Q/M$ is the heating rate per unit mass and $v = 1/\rho$. From the ideal gas law $v = RT/p$, Eq. (3.176) can be written as

$$q = (c_v + R) \frac{dT}{dt} - \frac{1}{\rho} \frac{dp}{dt} = c_p \frac{dT}{dt} - \frac{1}{\rho} \frac{dp}{dt}. \tag{3.177}$$

The first law in this form can be used to write the time evolution of temperature as

$$\frac{dT}{dt} = \frac{1}{c_p \rho} \frac{dp}{dt} + \frac{q}{c_p}. \tag{3.179}$$

The total derivative (advective derivative, material derivative, hydrodynamic derivative, etc.) of temperature describes the change of temperature following a parcel. But for many applications we need to consider the temporal variation of temperature at a *fixed* point in space. Weather forecasts, for example, are made for specific locations and not following a parcel. Numerical models of the atmosphere do not consider the time evolution of swarms of parcels, but rather represent the time evolution of variables in small volumes at fixed points in space. The total derivative in Eq. (3.179) can be expanded in Cartesian coordinates using the chain rule to give

$$\frac{dT}{dt} = \frac{\partial T}{\partial t} + \frac{\partial T}{\partial x} \frac{dx}{dt} + \frac{\partial T}{\partial y} \frac{dy}{dt} + \frac{\partial T}{\partial z} \frac{dz}{dt}, \tag{3.180}$$

where the partial derivatives are defined at (x, y, z). Following the notation used in meteorology, we define

$$u \equiv \frac{dx}{dt}, \quad v \equiv \frac{dy}{dt}, \quad \text{and } w \equiv \frac{dz}{dt}, \tag{3.181}$$

where u is the zonal wind component (positive from west to east), v is the meridional wind component (positive from south to north), and w is the vertical wind component. Using these definitions in Eq. (3.180) gives the local time derivative of temperature as

$$\frac{\partial T}{\partial t} = -\left(u \frac{\partial T}{\partial x} + v \frac{\partial T}{\partial y} \right) - w \frac{\partial T}{\partial z} + \frac{1}{c_p \rho} \frac{dp}{dt} + \frac{q}{c_p}. \tag{3.182}$$

The total derivative dp/dt also can be expanded in terms of partial derivatives to give

$$\frac{dp}{dt} = \left(\frac{\partial p}{\partial t} + u\frac{\partial p}{\partial x} + v\frac{\partial p}{\partial y} + w\frac{\partial p}{\partial z} \right).$$ (3.183)

In this expansion, the vertical gradient of pressure term will be much larger than any of the other terms so that

$$\frac{1}{c_p\rho}\frac{dp}{dt} = \frac{w}{c_p\rho}\frac{\partial p}{\partial z}.$$ (3.184)

Applying the hydrostatic approximation to this equation gives

$$\frac{1}{c_p\rho}\frac{dp}{dt} = -\frac{g}{c_p}\frac{dz}{dt}.$$ (3.185)

Because the lapse rate $\Gamma = -\partial T/\partial z$ and the dry adiabatic lapse rate $\Gamma_d = g/c_p$, Eq. (3.182) may be written as

$$\frac{\partial T}{\partial t} = -\left(u\frac{\partial T}{\partial x} + v\frac{\partial T}{\partial y} \right) - w\left(\Gamma_d - \Gamma \right) + \frac{q}{c_p}.$$ (3.186)

The first term in brackets is horizontal temperature advection. Thus, the local temperature change (*temperature tendency*) depends on the horizontal temperature advection, the vertical velocity times a stability factor $(\Gamma_d - \Gamma)$, and local heating or cooling by diabatic processes. Eq. (3.186) is sometimes referred to as the temperature tendency equation, although it is just a form of the first law that accounts for mass–motion (wind) affecting the temperature of an *open* system (see Section 1.7), the neighborhood around a point of interest in air. The change in temperature in that neighborhood is determined not only by local processes, but also by surrounding air flowing into and out of it. Examples of local processes are (net) radiative heating of ground during the day, and hence, air in contact with it, and its converse, (net) radiative cooling at night. Temperature changes are both domestic and imported products.

A general sense of the magnitude of the terms in the temperature tendency equation can be made by considering typical values for large-scale (synoptic-scale) weather systems. For example, if the temperature decreases northward at 10 °C per 1000 km $\partial T/\partial y = -10^{-5}\,°C m^{-1}$, and the wind is from the north at 5 m s^{-1} $(v = -5\,\text{m s}^{-1})$, the local temperature tendency would be about $-4\,°C$ per day solely by meridional temperature advection. Meteorologists call this *cold advection* because colder air is being advected to a local point. When a strong cold front at mid-latitude passes through a location, the temperature tendency can be a few degrees cooling per hour, although the air temperature following a parcel in the air mass will change much more slowly.

The magnitude of the temperature tendency due to vertical motion can be estimated for synoptic-scale atmospheric flows by considering a vertical velocity w of

1 cm s^{-1} (upward motion) and a lapse rate Γ of 6.5 °C km^{-1} (the lapse rate of the U.S. Standard Atmosphere from the surface to 11 km) at a fixed point above the surface. At this point, the upward motion would result in a negative temperature tendency of about -3 °C per day. A downward motion of the same magnitude would result in a warming of about 3 °C per day. For a layer where the lapse rate is dry adiabatic, the vertical velocity will have no effect on the local temperature. In a layer where temperature increases with height (an inversion layer; $\Gamma < 0$), a negative w (subsidence) can warm the layer substantially; q/c_p is an important component of the temperature tendency equation for many atmospheric applications. But some effort is needed to demystify this term and quantify it for practical applications. If we apply this term to a volume of air with a horizontal area A and a thickness Δz with a volume-averaged density $\bar{\rho}$, we can represent the mass of the volume as $\Delta m = \bar{\rho}\Delta z A$ and write q/c_p in Eq. (3.186) as

$$\frac{q}{c_p} = \frac{1}{c_p}\frac{Q}{\bar{\rho}\Delta z A}. \tag{3.187}$$

Because Q/A is the *net* energy flux density (Wm^{-2}) into the volume, Eq. (3.187) can be written as

$$\frac{q}{c_p} = -\frac{1}{c_p}\frac{(F_{out} - F_{in})}{\bar{\rho}\Delta z}, \tag{3.188}$$

where the net energy flux density is the difference between the flux density into the volume and the flux density out $(F_{in} - F_{out})$. If F_{out} is greater (less) than F_{in}, the air will cool (warm). The quantity q/c_p is the temperature tendency due to energy gains or losses. An estimate of its magnitude can be made by assuming a net energy flux density of 100 Wm^{-2} applied to a 1-km-thick atmospheric layer with a density of 1 kg m^{-3}. This flux density would result in a temperature tendency of about 10^{-4} °Cs^{-1}, about 10 °C per day.

If we take the limit as $\Delta z \to 0$ of Eq. (3.188), we can write

$$\frac{q}{c_p} = -\frac{1}{c_p\rho}\frac{\partial F}{\partial z}, \tag{3.189}$$

where ρ is the density at a point. If this form is used in the tendency equation, the temperature tendency will be defined at a point. In Eq. (3.189), $\partial F/\partial z$ is an energy flux density divergence. If the divergence is positive, q will be negative (cooling) since more energy is leaving the system than is going into it. If the divergence is negative, which is convergence, q will be positive (warming). In Eq. (3.189), the flux density divergence is the vertical component of the vector flux density $\mathbf{F}$. The flux density divergence in Cartesian coordinates is

$$\nabla \cdot \mathbf{F} = \frac{\partial F_x}{\partial x} + \frac{\partial F_y}{\partial y} + \frac{\partial F_z}{\partial z}. \tag{3.190}$$

Here, the divergence is defined at a point in space, which is the local divergence. In Eq. (3.189) the flux is the vertical component F_z. The divergence for a finite volume is given by the divergence theorem

$$\int_V \nabla \cdot \mathbf{F} dV = \int_A \mathbf{F} \cdot \mathbf{n} dA, \qquad (3.191)$$

where the unit vector $\mathbf{n}$ at any point of the surface A enclosing the volume V is directed *outward*. Thus, if the surface integral is *positive* there is a net *outward* transport of the quantity represented by $\mathbf{F}$. This corresponds to a divergence in the sense of something spreading outward from a point. If it is negative, this corresponds to convergence—the opposite of divergence. The average divergence of $\mathbf{F}$, defined by

$$\overline{\nabla \cdot \mathbf{F}} = \frac{\int_A \mathbf{F} \cdot \mathbf{n} dA}{V}, \qquad (3.192)$$

is local. It tells us how much is going into or out of a unit volume at a point. It can be positive or negative at different points. If it is positive, this corresponds to divergence in the physical sense. If it is negative, it corresponds to convergence. In nearly all atmospheric applications, the volume-averaged flux density divergence is used, although the scale of the volume used can vary substantially. Further, the divergence of the horizontal fluxes is much smaller than the vertical flux divergence. Thus, for most practical applications, the finite difference form given by Eq. (3.188) is used, where the temperature tendency will be for the mean temperature of the layer and the vertical divergence of the flux density is the mean divergence over the volume of interest. When the thickness of the layer in question is of order 10 m, the corresponding flux divergence can be considered a local value in most applications.

Pressure is frequently a vertical coordinate in the equations used for atmospheric applications. This coordinate has the advantage that density in many of the terms in these equations is eliminated. For example, applying the hydrostatic approximation to Eq. (3.189) gives

$$\frac{q}{c_p} = \frac{g}{c_p}\frac{\partial F}{\partial p} = \Gamma_d \frac{\partial F}{\partial p}, \qquad (3.193)$$

from which density is absent. To get a sense of the magnitude of the warming (cooling) rate given by Eq. (3.193), we can write it in finite difference form following the format of Eq. (3.188) as

$$\frac{q}{c_p} = -\Gamma_d \frac{(F_{out} - F_{in})}{|\Delta p|}, \qquad (3.194)$$

where $|\Delta p|$ is the absolute value of the vertical pressure difference across the layer and is proportional to the mass (per unit area) in the layer. The magnitude of this term can be illustrated by considering a layer with a pressure difference of 100 hPa subject to a net flux density gain (loss) of 100 Wm^{-2}. Because $\Gamma_d \approx 10\,°C$ km^{-1} and one day is about 10^5 s, such a flux convergence (divergence) would give a heating (cooling) of about 10 °C per day. Using more exact values in Eq. (3.194) gives a temperature tendency of 8.5 °C per day. This tendency (for a 100 Wm^{-2} net flux and a 100 hPa layer) is a useful number to remember and can be easily scaled to any net flux density and pressure interval. When applying Eq. (3.186) to estimate a temperature tendency, always check the sign of your result. An energy decrease (increase) results in a cooling (warming). As noted previously, when this equation is applied to a finite layer, the temperature tendency is for the average temperature of the layer.

The ratio of net flux density to a pressure difference has the units of a velocity. For a net flux density 100 W m^{-2} and a pressure interval 100 hPa, this velocity is about 1 cm s^{-1}, which is of the same magnitude as synoptic-scale vertical velocities in the atmosphere; the temperature tendency is the product of this velocity and the dry adiabatic lapse rate. If more energy is transported out of the layer than in, then $F_{out} > F_{in}$ (flux density divergence) and the velocity is positive (upward), the result is cooling. If there is an energy flux density convergence and the velocity is negative (downward), the result is warming.

In the atmosphere, radiative flux densities (irradiances) are important for determining q/c_p in Eq. (3.189) in many applications. For example, the clear-sky atmosphere is cooled by a longwave radiative flux density divergence. The globally averaged clear-sky longwave flux density divergence through the atmosphere (1000 hPa) is about 200 Wm^{-2}. If this energy decrease is from the entire troposphere, where the pressure at the tropopause (the boundary between troposphere and the stratosphere) is about 200 hPa, the mean cooling rate would be about 2 °C per day. When clouds are present, the vertical profile of longwave cooling can have lots of structure. For example, low-level clouds in the atmosphere can experience substantial cooling at cloud top by a strong net flux density divergence. Cloud tops of low-level stratus clouds (tops 0.5 to 1 km) in the subtropics, for example, can experience a longwave flux density divergence on the order of 100 Wm^{-2} in a layer less than 50 m deep to give a cooling rate of about 1.7 °C per day ($\approx 0.07\,°C$ h^{-1}). This may appear to be substantial cooling, but other terms in the tendency equation tend to balance this strong cooling at cloud top (see Problem 73). In atmospheric layers near the Earth's surface (boundary layers), there can be large energy flux densities from the surface that can warm these layers.

The q/c_p term is often called a heating (or cooling) rate because its dimensions are temperature per unit time. Temperature, not enthalpy. remains our go-to variable when we need to sense the magnitude of thermodynamic quantities and processes.

The temperature tendency equation can be used to examine how processes affect the time evolution of the lapse rate (static stability) of a layer. Taking the derivative of Eq. (3.186) with respect to height for a layer where the lapse rate does not vary with height gives

$$\frac{\partial}{\partial z}\frac{\partial T}{\partial t} = -\frac{\partial}{\partial z}\left(u\frac{\partial T}{\partial x} + v\frac{\partial T}{\partial y}\right) - (\Gamma_d - \Gamma)\frac{\partial w}{\partial z} + \frac{\partial}{\partial z}\left(\frac{q}{c_p}\right). \qquad (3.195)$$

If we interchange the time and height derivatives and because $\Gamma = -\partial T/\partial z$, Eq. (3.195) may be written as

$$\frac{\partial \Gamma}{\partial t} = \frac{\partial}{\partial z}\left(u\frac{\partial T}{\partial x} + v\frac{\partial T}{\partial y}\right) + (\Gamma_d - \Gamma)\frac{\partial w}{\partial z} - \frac{1}{c_p}\frac{\partial q}{\partial z}. \qquad (3.196)$$

This equation shows how various processes affect the stability of a layer. Common to all three terms is that if the lower part of a layer is heated (cooled) relative to the top, the layer will be stabilized (destabilized) with the lapse rate decreasing (increasing). The first term is the vertical derivative of the temperature advection. Thus, warm advection decreasing with height will destabilize the layer. If there is stronger cold air advection at the base of the layer than at the top, the lapse rate will decrease with time, which is a stabilization. Upward motion increasing with height will destabilize the layer. Downward motion increasing in magnitude with height will stabilize the layer. Thus, clouds tend to form in areas of upward motion and are suppressed in areas of subsidence (downward motion). Finally, if the bottom of the layer is heated (cooled) by diabatic processes, the layer will be destabilized. (stabilized).

The first law can be applied to lakes and oceans to estimate the temperature tendency of layers of water at Earth's surface. Generally, the advective terms are small and the pressure changes with time are irrelevant. Under these conditions the time tendency of the temperature of a layer of water of finite thickness may be estimated using Eq. (3.186) with the density of water $\rho_w = 1000\ \mathrm{kg\,m^{-3}}$ and specific heat $c_w = 4200\ \mathrm{J\,kg^{-1}\,K^{-1}}$. Thus, for a given flux density convergence, the temperature tendency for a layer of water of given depth will have a temperature tendency that is about 0.02% of an equivalent depth in the lower atmosphere. For a flux density convergence of 100 $\mathrm{Wm^{-2}}$ in a 1-m layer of water, the temperature tendency would be about 2 °C per day.

3.9 Chemical Reactions and Temperature Changes

To this point we have considered processes only in systems of constant composition. We ignored the possibility of chemical reactions. But without their ceaseless humming in our bodies (metabolism) we soon would perish. Chemical reactions also occur in planetary atmospheres and in the combustion chambers of engines. The following is a simple approach to chemical thermodynamics.

Consider a *closed* system in which chemical reactions take place adiabatically and isobarically. The possibility of reactions is implicit in the first law by way of the dependence of the total enthalpy H on the number of each distinguishable molecular species, which could *change* because of dissociation and combination.

We assume a system with two molecular components, ideal gases that can combine to form a third ideal gas; N_1, N_2, N_3 are the numbers of moles of each. One molecule of 1 and one molecule of 2 combine to form one molecule of 3, and hence,

$$\frac{d}{dt}(N_1 + N_2 + N_3) = 0. \tag{3.197}$$

The *total* pressure, the sum of all partial pressures, is constant. Take the molar specific enthalpy of each species to be $h_j(T)$ and assume that the partial enthalpies are additive:

$$H = N_1 h_1 + N_2 h_2 + N_3 h_3. \tag{3.198}$$

In an adiabatic, isobaric process, enthalpy is conserved:

$$\frac{dH}{dt} = N_1 \frac{dh_1}{dt} + h_1 \frac{dN_1}{dt} + N_2 \frac{dh_2}{dt} + h_2 \frac{dN_2}{dt} + N_3 \frac{dh_3}{dt} + h_3 \frac{dN_3}{dt} = 0$$

$$= \left(N_1 \frac{\partial h_1}{\partial T} + N_2 \frac{\partial h_2}{\partial T} + N_3 \frac{\partial h_3}{\partial T} \right) \frac{dT}{dt} + h_1 \frac{dN_1}{dt} + h_2 \frac{dN_2}{dt} + h_3 \frac{dN_3}{dt}, \tag{3.199}$$

where

$$\frac{dN_1}{dt} = \frac{dN_2}{dt} = \frac{dN}{dt}, \tag{3.200}$$

and hence,

$$\frac{dN_3}{dt} = -2 \frac{dN}{dt} > 0. \tag{3.201}$$

From the previous five equations, it follows that

$$\left(N_1 \frac{\partial h_1}{\partial T} + N_2 \frac{\partial h_2}{\partial T} + N_3 \frac{\partial h_3}{\partial T} \right) \frac{dT}{dt} + (h_1 + h_2 - 2h_3) \frac{dN}{dt} = 0. \tag{3.202}$$

The first term can be interpreted as a rate of change of kinetic energy, the second as a rate of change of potential energy—their sum is constant. The reaction ends when $dN/dt = 0$, and hence, $dT/dt = 0$. Because we assumed an isobaric process, some work (positive or negative) may be done.

Rewrite Eq. (3.202) as

$$\left(N_1 \frac{\partial h_1}{\partial T} + N_2 \frac{\partial h_2}{\partial T} + N_3 \frac{\partial h_3}{\partial T} \right) \frac{dT}{dt} = -(h_1 + h_2 - 2h_3) \frac{dN}{dt}. \tag{3.203}$$

The dimensions of this equation are power. The factor multiplying dT/dt is a total heat capacity, which is not constant, but positive because $\partial h_j / \partial T > 0$. The term on the right is carelessly called the rate of "heat generation," but can be interpreted as pseudo-heating, a rate of enthalpy change, and hence, temperature change *solely* because of an adiabatic *transformation* of molecules 1 and 2 into molecule 3. Because $dN/dt \leq 0$, $dT/dt \geq 0$ *provided that* $h_1 + h_2 - 2h_3 = \Delta h > 0$; Δh is a result of molecular rearrangements, the essence of chemical reactions. A decrease of molecular potential energy is accompanied by a corresponding increase in molecular kinetic energy, manifested macroscopically by a temperature increase.

But there is no physical reason why Δh cannot be *negative*. Fill an insulated container with tap water and let it come to room temperature. Add enough salt approximately at room temperature to make the resulting brine undrinkable. Measure the initial temperature of the salt and water and the final temperature of the brine. We did this with water initially at $21.5\,°C$ and salt initially at $20.8\,°C$. The final brine temperature was $20.1\,°C$, *lower* than both temperatures. Was this "cold generation" or was a temperature *decrease* the result of molecules interacting in such a way that potential energy increased at the expense of kinetic energy? We can measure only temperature changes. No "heat" is "absorbed" or "released" in chemical reactions if $Q = 0$. The terms "endothermic reaction" and "exothermic reaction" are jargon, not explanations. And it is misleading jargon; exo means *outside*, endo means *inside*, and yet both kinds of reactions occur within a *closed*, insulated container. Either the potential energy increases and the kinetic energy decreases, or vice versa, measured by a temperature decrease or increase.

We can approach this from a slightly different direction. For an adiabatic, isobaric process in which $dH(T, N_1, N_2 \ldots)/dt = 0$

$$\frac{dT}{dt} = -\left[\frac{\left(\frac{\partial H}{\partial N_1}\frac{dN_1}{dt} + \frac{\partial H}{\partial N_2}\frac{dN_2}{dt} + \cdots\right)}{\frac{\partial H}{\partial T}}\right]. \tag{3.204}$$

$\partial H/\partial T$ is positive because it is the heat capacity at constant pressure and composition. According to the numerator some species may be decreasing because of combining with other species to form new species. Mass is conserved (to very good approximation) but rearranged. The numerator could be negative or positive, and hence, the temperature could increase or decrease. No heat or cold is "generated" because this is an adiabatic (but *not* an isothermal) process. To call this "heat generation" is to conflate heat and temperature—something still being done long after it should have been buried in the same grave with caloric.

Because this reaction is assumed to be isobaric and adiabatic, enthalpy is constant. To *restore* the products of reaction to the initial temperature requires a diabatic process: $Q < 0$ if the temperature increases, $Q > 0$ if the temperature decreases. Thus, the difference between the final and initial enthalpies is negative if the temperature increases (in the reaction), and positive if it decreases.

This analysis cannot shed any light on exactly *how* molecules interact or how fast. Chemical *kinetics* is a large field unto itself, and is very complicated, especially because many chemical reactions are not simply $A + B \rightarrow C + D$, but often proceed through a series of intermediate reactions with different reaction rates. We were careful to note in Section 1.8 that a catalyst is required for combustion of hydrogen in oxygen, without which the reaction of two moles of hydrogen with one mole of oxygen to produce two moles of water vapor would be half complete in about 10^{25} years, whereas a spark or flame shortens this to 10^{-6} seconds.

If $Q \neq 0$ and $dT/dt = 0$,

$$-(h_1 + h_2 - 2h_3)\frac{dN}{dt} + Q = 0. \tag{3.205}$$

The core temperature of humans is usually higher than ambient temperature $(Q < 0)$, so this equation tells us that at least some internal chemical reactions (metabolism) can keep core temperature more or less constant. The first (metabolic) term in this equation is a rate of change of enthalpy equal and opposite to the cooling rate Q. Because humans walk and run, we must add mass–motion kinetic energy changes to the first law. Even couch potatoes lift their legs to walk to the refrigerator, so we must add gravitational potential energy changes. From Eq. (1.30) the energy budget equation therefore becomes

$$-(h_1 + h_2 - 2h_3)\frac{dN}{dt} + Q - \frac{dK_{cm}}{dt} - \frac{dP^e}{dt} = 0. \qquad (3.206)$$

This is a highly simplified energy budget for humans and animals if we ignore that they are *not* closed systems (Section 1.7). The food we eat is devoted in part to maintaining a nearly constant core temperature, not to "heat generation." Food is a source of potential energy that can be transformed partly and quickly into energy forms necessary to maintain our existence and partly stored as fat to be used as the need arises. This is like a paycheck partly directed at immediate needs and partly salted away for future needs.

Chemical reactions in our bodies are not entirely devoted to keeping our core temperature approximately constant. Because thermodynamics emphasizes engines that do work driven by appreciable temperature differences (Section 4.4), it is easy to ignore that there must be engines (or motors) at the molecular and cellular level that do work by other means and accompanied by negligible temperature differences. The proof is that we can lift our arms and that some part of a human body (e.g. lungs, heart, blood), even in slumber, is always in motion. Small temperature increases are byproducts. The notion that food provides "fuel" (e.g. carbohydrates) that we "burn" is a misleading analogy in that our direct experience of burning fuel—wood, coal, oil, natural gas—is rapid oxidation of hydrocarbons resulting in temperature increases of hundreds of degrees. Oxidation of carbohydrates in our bodies is more akin to the slow rusting of nails than to rapid, high-temperature combustion in a wood stove.

We can go a bit further. Consider a system with a fixed volume V. The initial enthalpy of the system is $H_{si}(T_i, p_i)$ and it undergoes an isochoric, adiabatic change of state to $H_{sf}(T_f, p_f)$ because of chemical reactions. Mass is conserved but the composition of the system changes, and hence, the initial (i) and final (f) systems are *not* the same. From the first law

$$H_{sf}(T_f, p_f) - H_{si}(T_i, p_i) - V(p_f - p_i) = 0. \qquad (3.207)$$

If the system in this final state is restored in a *diabatic* process to its initial temperature and pressure, but with no further change in composition,

$$Q_{tot} = H_{sf}(T_i, p_i) - H_{sf}(T_f, p_f) - V(p_i - p_f). \qquad (3.208)$$

Add these two equations to obtain

$$Q_{tot} = H_{sf}(T_i, p_i) - H_{si}(T_i, p_i). \qquad (3.209)$$

This is an enthalpy difference between a system at a given temperature and pressure *solely* because of chemical reactions, and (in principle) it is measurable.

The vague term energy content of food needs clarification. According to the first law, there is *no* change in the internal energy of (dried) food that undergoes adiabatic combustion in a rigid container (calorimeter). But the molecular *potential* energy of the food decreases accompanied by an equal increase in its *kinetic* energy manifested by a temperature increase that can be directed mostly to increasing the temperature of a mass M of water. $Mc\Delta T$, where c is the specific heat capacity per unit mass of water and ΔT is its temperature increase, is a measure of the *potential energy change* (oxidation) of a given mass of food. In a human body, this (slow) change keeps our core temperature almost constant and powers our internal motors. Energy content of food is shorthand for its possible potential energy decrease.

3.10   Residence Time of the Internal Translational Kinetic Energy of Earth's Atmosphere

We note in Section 3.2 that the internal energy of a gas (or anything) is defined only to within an arbitrary constant. But we can assign an absolute value to the internal translational kinetic energy of an *ideal* gas. Kinetic energy of the winds is not included. From Eq. (2.9) it follows that the kinetic energy density is $3RT\rho/2$. If the Sun were turned off and the atmosphere completely insulated from Earth, how long would it take for the total kinetic energy of the atmosphere (per unit area) to decrease by a factor of 2?

Assume an isothermal atmosphere. The absolute temperature of the troposphere, where most of the mass of the atmosphere resides, is almost uniform. Assume an exponential decrease of mass density with height, where H_ρ is the scale height. The total internal kinetic energy per unit area is $3\rho_o RTH_\rho/2$, where ρ_o is the mass density at the surface, and T is the assumed uniform temperature of the troposphere. For $\rho_o = 1\text{kg m}^{-3}$, $T = 300$ K, and $H_\rho = 8000$ m, the initial total kinetic energy is about 109 J m^{-2}. Take the average emission at the top of the atmosphere to be 200 Wm^{-2}. Assume that this does not change appreciably during the time interval of interest. The residence time τ (in seconds) is determined by $200\tau = 10^9/2$, about thirty days. It is pointless to aim for more precision because the residence time is not a precise number. The atmosphere is nonuniform, emission by it varies in space and time, and we made several rough approximations.

If the internal kinetic energy decreases, it is plausible that the mass distribution (density profile) will change. If mass is shifted to lower altitudes the potential energy will decrease, and vice versa. We can write the center of gravity of the atmosphere (flat Earth, uniform g), Eq. (3.77) as

$$\frac{\int_0^\infty z\rho_o \exp(-z/H_\rho)dz}{\int_0^\infty \rho_o \exp(-z/H_\rho)dz}. \tag{3.210}$$

The mass of the atmosphere is constant. It loses energy, not mass (Section 2.4). The center of gravity must be proportional to the scale height because it is the only physical quantity with the dimensions of length. The proportionality factor happens to be 1 and $H_\rho \approx RT/g$ (Section 2.2). As the average tropospheric temperature T decreases because of emission of radiation, we expect the center of gravity to descend, and hence, the potential energy to decrease. Potential energy and kinetic energy decrease, both depending on the decrease of the average temperature and in the same way. It is reasonable to assume as a first cut that their decreases are about the same. Even if the Sun were to be magically turned off, Earth's atmosphere still would be in contact with a slowly changing terraqueous globe.

Annotated References and Suggestions for Further Reading

Percy W. Bridgman's comments about the mathematics of thermodynamics are in his 1961 *The nature of thermodynamics*. New York: Harper, p. 4.

For a critical discussion of the mathematics of thermodynamics, see Karl Menger (1950) "The mathematics of elementary thermodynamics," *American Journal of Physics*, 18, pp. 89–103.

For mathematicians' views about differentials, see C. B. Allendoerfer (1952) "Editorial [Differentials]," *American Mathematical Monthly*, 59, pp. 403–6; Mark Kac and J. F. Randolph (1942) "Differentials," *American Mathematical Monthly*, 49, pp. 110–12. For these papers, others about differentials, and many more, see Tom M. Apostol *et al.* (eds.) (1969) *Selected papers on calculus*. Providence: American Mathematical Association. To fully understand present-day science and mathematics requires knowing their real history rather than mythology. For the evolution of the differential and derivative concepts see Henk J. M. Bos (1974) "Differentials, higher-order differentials, and the derivative in the Leibnizian calculus," *Archive for History of the Exact Sciences*, 14, pp. 1–90. Even a quick reading is worth the time. The controversial development of calculus over centuries by mathematicians of the caliber of Newton, Leibniz, Euler, Gauss, Cauchy ... may dampen feelings of thick-headedness because of failure to fully grasp calculus after a one-semester course. If you were confused, take heart: so were your illustrious predecessors.

For a history of the many different symbols used for derivatives and integrals, see Florian Cajori (1952) *A history of mathematical notations*, vol. II. Chicago: Open Court.

Clifford Truesdell's ridicule of the mathematical notation in thermodynamics is in his 1969 *Rational thermodynamics* (New York: McGraw-Hill, p. 4.

The subsection title *"Those Accursed Differentials"* in Section 3.1 was inspired by Clifford Truesdell's condemnation of "those accursed differentials famous as vehicles of thermodynamic obscurity" in Clifford Truesdell (1980) *The tragicomical history of thermodynamics 1822–1854*. New York: Springer, p. 21. He further notes (p. 22) that "when apparently timeless variables [such as

V and T] change, they do so in the course of time; time is the basic descriptor of natural changes." He wrote that from the first thermodynamics was "a dismal swamp of obscurity" and today [1969] "in common instruction it [still] is" (p. 6).

The suggestion that enthalpy replace heat was made by Austin J. O'Leary (1950) "Enthalpy and thermal transfer," *American Journal of Physics*, 18, pp. 213–21.

For the history of the term enthalpy and why its symbol is H, see Imgard K. Howard (2002) "H is for enthalpy, thanks to Heike Kammerlingh Onnes and Alfred W. Porter," *Journal of Chemical Education*, 79, pp. 697–8.

Density versus temperature data for water were taken from N. Ernest Dorsey (1940) *Properties of ordinary water-substance.* New York: Reinhold. This is an invaluable compendium of the properties of water.

Textbook treatments of specific heats of gases are discussed critically and thoroughly by Clayton A. Gearhart (1996) "Specific heats and the equipartition law in introductory textbooks," *American Journal of Physics*, 64, pp. 995–1000. Gearhart concludes that very few physics textbooks ($\approx$ 20%) treat specific heats correctly—yet another sad example of error propagation.

An extensive search unearthed only one book explicitly stating that γ is not necessarily $\approx$ 1 for all fluids. See Dudley Towne (1967) *Wave phenomena.* London: Addison-Wesley, p. 27: "the value of γ for most fluids is very close to unity, though there are notable exceptions, such as turpentine, for which $\gamma = 1.27$." But there are *many* exceptions: γ for 25 liquids, from water (1.007) to carbon disulfide (1.566), are given by S. Parthasarathy and D. Guruswamy (1955) "Sound absorption in liquids in relation to their specific heats," *Annalen der Physik*, 451, pp. 31–42. Water is the most flagrant nonconformist (Table 1).

For textbooks in which dx is sometimes daringly expunged from integrals, see R. Creighton Buck (1956) *Advanced calculus.* New York: McGraw Hill, and Karl Menger (1955) *Calculus: a modern approach.* 3rd edn. Oxford: Ginn & Company. Buck is even so rash as to write double integrals without $dxdy$. Both books are rigorous without being oppressive.

For observations of superadiabatic lapse rates see E. S. Takle (1983) "Climatology of superadiabatic conditions for a rural area," *Journal of Climate and Applied Meteorology*, 22, pp. 1129–32.

The consequences of buoyancy are much easier to observe in water—floating corks, buoys, ships—because of its well-defined upper boundary than in boundless air. A striking example is icebergs. Although the lower density of ice than of water is necessary for them to be stable, most of them are not because of their shape. For an interesting short article on how the stability of floating objects depends on their shape and density contrast see Henry Pollack (2019) "Tip of the iceberg," *Physics Today*, 72(12), pp. 70–1.

For the latest paper on measuring γ, with a long list of references, see M. T. Caccomo, G. Castorina, F. Catalano, and S. Magasù (2019) "Rüchardt's experiment treated by Fourier transforms," *European Journal of Physics*, 40, 025703. For the method based on an adiabatic followed by an isochoric process, see William M. Moore (1984) "The adiabatic expansion of gases and the determination of heat capacity ratios," *Journal of Chemical Education*, 61, pp. 1119–20.

For measured peak temperature as a function of compression ratio and engine speed, see K. C. Tsao, P. S. Myers, and O. A. Uyehara (1962) "Gas temperatures during compression in motored and fired diesel engines," *SAE Transactions*, 70, pp. 136–45, figure 3.10.

Except for the largest raindrops, a constant drag coefficient of 0.4 is not valid. At terminal speed it depends on the Reynolds number, which itself depends on the terminal speed. And the dependence on Reynolds number is not a simple analytical expression. For measured terminal velocities see Ross Gunn and Gilbert D. Kinzer (1948) "The terminal velocity of fall for water droplets in stagnant air," *Journal of Meteorology*, 6, pp. 243–48. The terminal velocity for a wide range of sizes varies from 0.4 to 9 m s^{-1}, a factor of 22, whereas if the drag coefficient were constant this factor would be about 7.

For a history of the indicator diagram see Davis Baird (2004) *Thing knowledge: a philosophy of scientific instruments*. Berkeley: University of California Press, ch. 8. The "central message" of his book is that "our material creations bear knowledge ... independently of theory or in spite of bad theory." For unraveling the mystery of when and by whom the indicator diagram was invented see David Philip Miller (2011) "The mysterious case of James Watt's '1785' steam indicator: forgery or folklore in the history of an instrument?" *International Journal for the History of Engineering and Technology*, 81, pp. 129–50.

For the history of specific heats see Douglas McKie and Niels H. De V. Heathcote (1935) *The discovery of specific and latent heats*. London: Edward Arnold.

Observations and assimilated data sets are used to estimate global clear-sky longwave radiative cooling by R. P. Allan (2006) "Variability in clear-sky longwave radiative cooling of the atmosphere," *Journal of Geophysical Research* [online], 111(D22). Researchers have used the global clear-sky cooling rate as a measure of the strength of the global hydrological cycle. The relationships between the global clear-sky cooling and the hydrological cycle are presented in R. P. Allan (2009) "Examination of relationships between clear-sky longwave radiation and aspects of the atmospheric hydrological cycle in climate models: re-analyses, and observations," *Journal of Climate*, 22, pp. 3127–45.

For a simple experiment with salt and ice, see Craig F. Bohren (1993) "Melting with salt and heating with ice," *Weatherwise*, 46(6), pp. 46–8.

One of the two papers in which what is now called the Brunt–Väisälä frequency is derived is by David Brunt (1927) "The period of simple vertical oscillations in the atmosphere," *Quarterly Journal of the Royal Meteorological Society*, 53, pp. 30–32.

The statement in Section 3.9 about combustion of hydrogen in oxygen is from Keith J. Laidler (1987) *Chemical kinetics*. 3rd edn. New York: Harper & Row, p. 3.

Problems

1. In 1845 James Prescott Joule suggested that the water at the bottom of a water-fall should be warmer than at the top, in particular, for Niagara Falls (about 160 feet) one-fifth of a Fahrenheit degree. How did he obtain this number? The specific heat capacity of liquid water is about $4200\,\mathrm{J\,kg^{-1}\,K^{-1}}$. (*Hint*: If you have trouble imagining a waterfall, consider a single water droplet falling the same vertical distance).

   There is world of difference between what "should be" and what "is." We scoured the works of Joule without finding any evidence that he measured the calculated temperature increase, and for good reason (see Problem 5). Text-books and popularizations assert that a falling object upon impact is warmed a certain (unstated) amount without adducing confirming measurements.

2. The fuel–air mixture in gasoline engines is ignited by a spark plug. But diesel engines do not have spark plugs. Air enters the cylinders of diesel engines at normal atmospheric pressure and is compressed until the temperature is high enough to ignite fuel sprayed into the cylinders. The ratio of the maximum to minimum cylinder volume (determined by the travel of the piston) is called the *compression ratio*. Suppose that the ignition temperature of diesel fuel in air is 800 K. From this, determine the minimum compression ratio.

3. A diesel engine is not all give and no take. If you believe the first law, it takes work to compress air. Determine how much power this takes relative to the output power for a vehicle of your choice with a given engine displacement, engine speed (rpm), and power at this speed. We chose a Mercedes Benz diesel, S-class W 220, with a 3.2-liter displacement and rated power 145 kW at 4200 rpm. You may be surprised by the result. We were.

4. Show that the total work done in an adiabatic process for an ideal gas can be obtained by integrating Eq. (1.65) using the Poisson relation Eq. (3.82), the gas law Eq. (2.11), and Eq. (3.53). If you do this problem correctly you should immediately see that you could have done it more easily.

5. Problem 1 was to determine the physical basis of Joule's suggestion that the water at the bottom of a waterfall should be warmer than at the top, in par-ticular, for Niagara Falls (160 feet) one-fifth of a Fahrenheit degree. On the basis of what you know about the atmosphere, discuss why it would be diffi-cult to verify Joule's hypothesis by making temperature measurements at, for example, the top and bottom of Niagara Falls.

6. We considered four different kinds of processes: isobaric (constant pressure); isochoric (constant volume); isothermal (constant temperature); and adiabatic ($Q = 0$). We defined heat capacities for only two of these processes: constant

pressure and constant volume. Why did we not consider heat capacities for the other two processes?

7. Problem 9 in Chapter 1 was to determine the muzzle velocity of a bullet as a function of barrel length under the assumption that the explosion is a constant-pressure process. Do the same problem but assume that the process is adiabatic. What do you conclude from your result?

8. Commercial jet aircraft fly at cruising altitudes between about 30,000 and 40,000 feet. Yet such aircraft carry heat exchangers to cool cabin air while flying at these altitudes.
   (a) Estimate the air temperature at which jet aircraft fly.
   (b) Explain why it is necessary to cool cabin air at cruising altitudes. *Hint*: The ratio of the heat capacities of air at constant pressure and volume (γ) is about 1.4.

9. In 1936 Robert Emden published a note in which he posed the following question: "Why do we have Winter Heating? The layman will answer: 'To make the room warmer.' The student of thermodynamics will perhaps so express it: 'To impart the lacking (inner, thermal) energy.' If so, then the layman's answer is right, the scientist's wrong." Discuss. Consider a room in an ordinary house heated by a furnace. First you must show that the total internal energy of the air in the room is constant. But this air is being heated. What happened to the energy?

10. Students are taught that when air is heated it expands. Then they are taught (but not in the same lesson) that when air expands it cools. Some students find these two explanations to be contradictory. Please resolve the apparent contradiction.

11. Consider an insulated, rigid container divided into two regions: one containing a gas, the other empty, separated by a partition. When the partition is punctured, the gas freely expands to fill the entire volume of the container. In equilibrium, the temperature of the gas before and after expansion is the same *if* the gas is ideal. When *real* gases expand under these conditions, however, their temperature decreases slightly. What does this result tell you about the nature of the average forces between gas molecules?

12. Equation (3.173) is a good approximation if the number fraction of water molecules in air is small. Show that this equation would be exact (subject to the other assumptions made) for *any* fraction of water vapor if its specific heat capacity were equal to that of dry air. A corollary of this result is that Eq. (3.173) is a better approximation than we might think on the basis of water vapor fractions because the ratio of these specific heat capacities is less than two.

13. Estimate the rise in sea level if the average temperature of the world's oceans were to increase by one degree Celsius. This is an open-ended question. You can make a quick-and-dirty guess by using the thermal expansion coefficient of pure water given in this chapter. But you can do better by using the thermal expansion coefficient of salt water (which, moreover, depends on salinity

and pressure). You need to obtain (or guess) such quantities as the average temperature and depth (volume divided by area) of the oceans.

14. In Section 3.7 we considered isobaric, adiabatic mixing of two parcels with different temperatures and water vapor partial pressures. Under what conditions (if any) is this process also isovolumetric (excluding the trivial example of parcels with identical initial temperatures)? *Hint*: Keep in mind that air is a mixture of dry air and water vapor.

15. A TV meteorologist advanced the following explanation for two appreciably different maximum temperatures on a clear day in two American cities at nearly the same latitude, Phoenix and Atlanta. The forecast for Atlanta was 91 °F, whereas that for Phoenix was 105 °F. He argued that Atlanta is more humid and made the following analogy: it takes less time to bring water to a boil when there is not much water in the pan than when the pan is full. Discuss this explanation.

16. On an early spring day, the lapse rate above Denver and its surroundings is about the average value, two-thirds the dry adiabatic lapse rate. For reasons we need not dwell on, air over the 14,000 foot mountains to the west begins to flow down-slope into Denver. Do you expect the air temperature in Denver to increase, decrease, or remain the same?

17. We are often told that water is incompressible. If this were literally true, the world would be a different place. In particular, sea level would be higher (for the same mass of water). Estimate the change in sea level if water were truly incompressible. The isothermal compressibility is defined as

$$-\frac{1}{V}\frac{\partial V}{\partial p}.$$

You may approximate the isothermal compressibility of water (pure or salt) as $5 \times 10^{-5} \text{atm}^{-1}$.

Problem 13 was to determine the rise in sea level because of *thermal expansion* for a one degree Celsius increase in temperature, all else being equal. But all else is not equal. The isothermal compressibility of water depends on temperature (and pressure). Water becomes more incompressible as its temperature rises. Table 110 in N. E. Dorsey's compilation *Properties of Ordinary Water Substance* gives the following values for the isothermal compressibility of sea water (for a pressure of 150 atm): $46.2 \times 10^{-6}\text{atm}^{-1} (0\,°\text{C})$; 10^{-6} atm^{-1} $(0\,°\text{C})$; $43.4 \times 10^{-6}\text{atm}^{-1} (10\,°\text{C})$; $41.4 \times 10^{-6}\text{atm}^{-1} (20\,°\text{C})$. From these data estimate the additional rise in sea level (above that accounted for solely by thermal expansion) because of the temperature dependence of the compressibility of sea water (for a one degree Celsius temperature increase).

18. In Section 2.1, in the subsection on atmospheric temperature measurements, we stated that aerodynamic heating of temperature probes mounted on aircraft can be as much as 5 °C at an airspeed of 100 ms^{-1}. By simple arguments, based solely on what you have learned in this and the first chapter, you should be able to obtain this temperature increase. Please do so. *Hint*: Only one

short equation is necessary. Take the specific heat capacity of air (at constant pressure) to be $1000 \, \mathrm{Jkg^{-1} \, K^{-1}}$.

19. A man once called one of the authors to ask him about clouds. He began by saying that he knew clouds didn't fall out of the sky because when water condenses, it weighs less than air, and hence, clouds are buoyant. Critically discuss this explanation.

20. If you ever have inflated the tire of a bicycle with a hand pump, you know that it gets warmer. You may encounter people who claim that this is a consequence of "friction" in the pump. Devise a simple but convincing demonstration to show that friction has essentially nothing to do with why the pump gets warmer. Keep in mind that the people you are trying to convince are likely to not know any physics, so you won't get anywhere by invoking mathematical expressions or physical laws.

21. What is the depth of the ocean layer (per unit area) with the same total heat capacity as the entire atmosphere?

22. The year 1824 marks the publication of one of the gems of thermodynamics, Sadi Carnot's *Reflections on the Motive Power of Fire*, in which he offers the following proposition: "When a gas passes without change of temperature from one definite volume and pressure to another volume and another pressure equally definite, the quantity of caloric absorbed or relinquished is always the same, whatever may be the nature of the gas chosen as the subject of the experiment." By "caloric" Carnot meant what in modem terms is integrated Q. You should be able to prove Carnot's proposition in a few lines. Please do so.

23. Show that it is possible to determine γ, the ratio of specific heats at constant pressure and at constant volume, of a gas by measuring only three volumes in two processes. First, a fixed amount of the gas is adiabatically compressed (or expanded) beginning at some initial temperature, then cools (or warms) at constant pressure to its initial temperature. Devise a simple experiment for measuring γ based on these two processes.

24. A contributing factor to air pollution in the Los Angeles Basin is the prevalence of temperature inversions. When Morris Neiburger was a meteorologist with the Air Pollution Foundation in Los Angeles, he had to evaluate schemes (other than the obvious one of reducing the automobile population) for reducing air pollution. One such scheme entailed eliminating inversions. To evaluate this scheme Neiburger considered an average early morning temperature profile over Long Beach in September. The surface temperature was about $18 \, °\mathrm{C}$ and decreased at approximately the dry adiabatic rate to a height of 475 m, at which height the temperature increased at about $14.6 \, °\mathrm{Ckm^{-1}}$ to 1055 m. Above this height the temperature again decreased at approximately the dry adiabatic rate. Calculate how much energy (per unit area) it would take to eliminate the inversion. On the basis of your calculations, what do you think of this scheme? *A few hints*: Don't waste time trying to obtain an exact solution to this problem. All that is wanted is a crude estimate, which you can obtain in one line. Your result will be easier to interpret if you assume that the inversion

would have to be eliminated anew each morning, and hence, you can express your result as an (average) energy flux (energy per unit area and time).

Neiburger discussed his evaluations of the inversion–elimination scheme and several others in a fascinating paper in *Science*: Weather modification and smog, Vol. 126, 1957, pp. 6637–45.

25. Show that if γ (the ratio of specific heats) of an ideal gas is independent of temperature, it necessarily follows that c_p (and c_v) is independent of temperature and conversely.

26. Sound waves in a fluid are waves of compression and expansion. Thus, the velocity of sound in a fluid should depend on its compressibility, which is a measure of how easily the fluid is compressed, and its density, which is a measure of the inertia of the fluid. And indeed the velocity of sound c in a fluid is

$$c = \frac{1}{\sqrt{\kappa \rho}},$$

where

$$\kappa = -\frac{1}{V}\frac{\partial V}{\partial p}$$

is its compressibility, but which one? Newton took it to be the isothermal compressibility. The velocity of sound in air so obtained is about 20% lower than that measured. Laplace argued that compressions and expansions of sound waves in air are adiabatic rather than isothermal. Calculate the velocity of sound in air using both the isothermal and adiabatic compressibility. Which agrees better with measurement? The velocity of sound in dry air at 15 °C is about 341 m s^{-1}. *Hint*: Use Eq. (3.82).

27. This problem is related to the previous one. Water roiled by the propellers of ships or submarines or by breaking waves contains air bubbles. One might think that the velocity of sound in bubbly water lies between that of air and of water, although much closer to that of water if the bubble volume fraction (fraction of total volume occupied by air) is small, say 0.001 or less. Here is an example in which one's intuitive first guess may be wrong. Show that the velocity of sound in bubbly water not only can be appreciably less than that in bubble-free water, but even less than the velocity of sound in air. The density of air is about 1000 times less than that of water; the velocity of sound in water is about four times that in air; the compressibility of air is about 10^5 times greater than that of water. *Hint*: Use the previous expression for the speed of sound of a fluid, the definition of compressibility, and make judicious approximations wherever possible.

28. Ultimately, all winds are a consequence of density differences, which themselves depend on temperature and humidity differences. Estimate the maximum speed (but still realistic) that can be attained by a parcel accelerated upward from rest at the surface by buoyancy forces. Compare your estimate with the highest recorded wind speed. What do you conclude?

29. The high specific heat capacity of liquid water (about four times that of air) is said to be responsible for ironing out atmospheric temperature fluctuations, the

oceans acting as a huge buffer for the atmosphere. The implication of this is that if the specific heat capacity of ocean water were, say, that of air, we would be in a pickle, subject to more violent changes of weather. Perhaps the folks who make this assertion are correct, although we never have seen an irrefutable proof of it. It just seems to be one of those immutable truths that everyone knows but no one has proven. Critically examine the allegedly beneficial consequences of the high specific heat capacity of water, for which we all are supposed to be grateful. The ocean mixed layer is of order 100 m deep (although it varies seasonally and geographically). This layer is the surface layer that is nearly isothermal because of mixing by wave motion and by motion driven by density gradients resulting from temperature and salinity gradients. *Hint*: One way to address this problem is by considering a simple mathematical model of an atmosphere–ocean system, where each is considered to be isothermal with a fixed total heat capacity. The model ocean is heated (and cooled) periodically by radiation, and in turn the ocean (i.e. the mixed layer) heats (and cools) the atmosphere at a rate proportional to the temperature difference between them. By setting up the two coupled energy-balance equations, you'll acquire some insights. And if you can solve these equations (for a periodic ocean heating function), you'll understand even more about atmosphere–ocean coupling. Whatever conclusions you reach about the benefits of the high specific heat capacity of water, be sure to qualify them.

30. Dirigibles (blimps) and balloons are sometimes referred to as lighter-than-air aircraft. Why?

31. The annual Darwin Award is given to the person who did the gene pool the biggest service by killing himself in the most extraordinarily stupid way. The 1997 Award went to Larry Waters of Los Angeles (who survived his award-winning accomplishment). He purchased 45 weather balloons and several tanks of helium, inflated the balloons, attached them to a lawn chair, and tied himself to the chair. When he cut the cord attaching the chair to his Jeep he shot into the air and didn't stop climbing until about 11,000 feet. He was rescued by a helicopter but immediately arrested for violating Los Angeles International Airport's airspace. Assume that the total mass of Larry, his lawn chair, and the six-pack of beer he took with him on his flight was 70 kg. Assume that the surface temperature was 25 °C. The balloons were described as being more than four feet in diameter, so you may assume that their initial diameter is 4.5 feet. From these data and the information in the previous paragraph you should be able to estimate the diameter of a balloon at Larry's cruising altitude of 11,000 feet. The molecular weight of helium is four, whereas that of air is about 28.8 and its gas constant is about $287\,\mathrm{J\,kg^{-1}K^{-1}}$. The initial pressure in the balloon is unknown. This pressure must be greater than atmospheric pressure, but on physical grounds (the balloons do not burst) it cannot be much greater. Also, the initial diameter of a balloon is not known exactly. How do uncertainties in the initial helium pressure and the initial balloon diameter affect your estimate of the balloon diameter at 11,000 feet?

32. For Problem 2 you need the ratio of specific heats for air–fuel mixtures. Is assuming γ for air (≈ 1.4) a good approximation? To answer this requires knowing realistic air–fuel ratios, the mass of air per unit mass of fuel in the combustion chamber. The gas in the chamber of a diesel engine is a mixture of air and various complex polyatomic molecules. Based solely on what is in this chapter you should be able to predict whether γ will increase or decrease if vaporized fuel is mixed with air and even make an educated guess at by how much.

33. In a discussion of buoyancy, we found the assertion that "The density of water in a deep pool increases with depth, which means that the weight of a fixed volume of water is greater the deeper one goes." This implies that buoyancy is a consequence of an increased density with depth, rather than increased pressure. Criticize this. HINT: Assume constant temperature and take the isothermal compressibility of water to be 46×10^{-11} (SI units). The compressibility of water does not depend much on pressure or temperature (see Rana A. Fine and Frank J. Millero, 1973: Compressibility of water as a function of temperature and pressure. Journal of Chemical Physics, Vol. 59, pp. 5529-36).

34. In the previous problem, we ignored changes in the gravitational potential energy of the atmosphere. First, convince yourself that as the thermodynamic internal energy of the isolated atmosphere decreases its potential energy could also change. Explain by physical arguments. To determine how it changes (increase or decrease) requires just a bit of analysis. HINT: Think center of gravity. Ask yourself if your result makes physical sense. With a bit more thought you even can estimate by how much they both change.

35. This problem is related to the first problem. In John Tyndall's *Fragments of Science* (p. 6) he wrote that "Under the operation of this force [gravity] a stone falls to the ground and is warmed by the shock." What material would you choose for a "stone" to obtain its greatest possible temperature increase? This requires searching for thermophysical properties of materials. Using this stone can you at least conceive of an experiment in which the temperature increase might be measurable? It takes much more thought to measure something than to blithely assert what should be measured.

36. To estimate the temperature increase in the previous problem, we (and many others) have snuck in assumptions that are certainly not correct. What are they? This is the most important problem about this subject because you have to think about why a solution is *wrong* rather than right. Textbook problems usually have nice tidy solutions often bearing little resemblance to reality. HINT: Falling objects have a finite size. How does this change the solution? Don't try to solve exactly the problem of the temperature increase upon impact. Explain physically why this is much more difficult than is usually acknowledged. You will need to know some of the results in Chapter 7.

37. The rather small temperature increase of a falling object upon hitting the ground (see previous two problems) is because its terminal velocity is not especially high. The velocity of a bullet fired from a rifle is much higher. Could the temperature increase of a rifle bullet immediately after impact with a

hard target (e.g. steel plate) increase sufficiently to cause a forest fire? This is not a make-believe problem. See Finney *et al.* cited in the Bibliography for calculations and measurements well worth reading about. These should be presented in textbooks instead of tiny temperature increases of falling objects upon impact that to our knowledge have not been measured.

38. Where on Earth would you expect to find the driest air, that is, with the smallest fraction of water vapor molecules? Before giving a knee-jerk answer, stop to think. When you have thought of a few candidates try to determine which one is the winner.

39. Can the change in density of water with depth because of temperature gradients be comparable to or exceed (in magnitude) the increase in density because of the pressure gradient? HINT: You will need the coefficient of thermal expansion of water. Choose a temperature not close to the temperature at which the density is a maximum (see Fig. 3.8).

40. Measurements show that the isobaric coefficient of thermal expansion of gases is usually much greater than that of liquids. But what is the physical reason? *HINT*: It may help to think about how one would measure this coefficient.

41. We have seen the statement that, after a storm, ocean water temperature increases because motion of the water whipped up by winds is "converted into heat." Do you believe this? Assume that the change in mass motion kinetic energy of the water is converted entirely into internal energy. The water speed increases to V because of the wind, and when it subsides, the water returns to its quiescent state. Support your answer with simple calculations. V is *not* the wind speed.

42. We usually think of the density of Earth's atmosphere as decreasing with elevation. For what lapse rate would the density be constant with elevation (over a limited distance)? Try to give an explanation in words first. Does this lapse rate correspond to stability? Under what conditions might it be observed? Can you find any observations?

43. What can you infer about your body because you can almost float without effort in fresh water?

44. What is the minimum volume for a hot air balloon at sea level to lift one kilogram? Neglect the mass of the balloon. You will have to assume a reasonable temperature (i.e. one that doesn't melt or weaken the balloon) for the air in the balloon. With this result determine the minimum balloon volume necessary to lift four people, not including the gondola and the burners and fuel. Make any reasonable assumptions about the weight of the passengers. Convert the volume to a linear dimension. If you happen to see a hot air balloon floating overhead, estimate its size by the size of the passengers and compare this with what you calculated. What is the volume of a balloon that can just lift itself (with gondola)?

45. What is the rate of decrease of relative lift with height for a hot air balloon? Make any reasonable assumptions. Don't agonize over the difference between

scale heights for pressure and density, which is equivalent to ignoring the decrease of temperature with height.

46. Determine the change in relative lift taking account of the difference between the scale heights for density and pressure. Make any reasonable assumptions for the lapse rate. The solution is simpler if you assume that z is appreciably less than the scale height.

47. How high could a helium balloon rise in Earth's atmosphere? Assume that the balloon is inelastic (e.g. made of mylar) and is strong enough not to burst. Ignore the mass of the balloon.

48. Make the previous problem a bit more difficult by including the weight of the balloon. Before doing any analysis ask yourself if the 17 km height in the previous problem will increase or decrease. Make use of any information about mylar balloons you can find or know from experience.

49. We implicitly assumed a balloon that will not burst regardless of the pressure difference between inside and outside. Again, ask yourself why the balloon will burst. Why does the pressure difference change and how? For this problem we cannot determine the height at which the balloon will burst. And the best we can do is estimate the maximum pressure difference in the troposphere.

50. This is related to the previous problem, but now we have a better hope of obtaining a more definite answer. Consider an ordinary elastic rubber (latex) balloon. Fill it with helium and release it at the surface. What happens? Ignore the (small) pressure difference between inside and outside. Ignore the change of air temperature with height. You can justify this afterwards. Get some toy balloons and inflate them until they break. Note the volume change. Why do they break?

51. This problem is related to the previous problem. Blaise Pascal wrote that "if a balloon, only half inflated, not fully so, as they usually are—were carried up a mountain, it would necessarily be more inflated at the mountain top, and would expand to the degree to which it was less burdened." We could find no evidence that he did this, and the words "if," "were," and "should" suggests otherwise. How high would a mountain have to be before one could accurately measure the increased volume of a balloon? Keep in mind that one would have to measure linear dimensions (cube root of volume) and balloons are rarely perfect spheres. If you have access to a sufficiently high mountain, take some balloons up it and observe any volume increases. Think of a clever way of measuring the size change. If the mountain is sufficiently high, you should measure the temperature change if any.

52. Specific heat capacity per unit mass can be measured with a thermometer and items in a kitchen. If you can figure out how, measure the specific heat capacity of a metal object of known composition. The original definition of "amount of heat" (Chapter 1) was based on a one-degree temperature rise of one gram of water. Thus, specific heat capacities were relative to that for water. We did a quick experiment and obtained a value for a lead object about two times too

large. You'll probably have better luck using a liquid that mixes with water and doesn't react with it. We used isopropyl alcohol (70%) and got a better result.

53. We give the isobaric coefficient of thermal expansion and the isothermal compressibility for which measurements are readily found. What about the (unnamed) isochoric coefficient $(1/p)(\partial p/\partial T)$? It is conspicuous by its absence. What is $\partial p/\partial T$ for an ideal gas? What is it for a liquid such as water? Obtain a general expression for $\partial p/\partial T$ using Eq. (3.61) and the definition of compressibility. Calculate $\partial p/\partial T$ for water (at, say, 20–25 °C) and compare it with that for standard air. What do you conclude? Can you think of a simple experiment with water? Try to find bursting strengths of plastic and glass bottles.

54. Suppose that U is a function of V and T. Give a physical interpretation of $\partial U/\partial V$ on the basis of which give its expected sign. Can you find an example in which your expectation is not met? HINT: Use Eqs. (3.51) and (4.72). Give a physical interpretation of $\partial U/\partial T$.

55. Why might it be more difficult to measure the coefficient of thermal expansion of liquids than of solids or gases?

56. The maximum possible lift of a balloon (or airship) would be obtained with a (hypothetical) vacuum balloon, the idea for which goes back centuries. How much greater lift (per unit volume and at the same temperature and pressure) could be obtained by this balloon compared with one filled with molecular hydrogen. What would be the advantages of such a balloon (if it could be built)? What would be the disadvantages?

57. If a vacuum airship is not (yet) possible on Earth, there are other planets in our solar system. Can you find one with an atmosphere where vacuum airships might be feasible? Find the volume necessary to lift one kilogram.

58. It is difficult to measure thermal expansion of liquids, especially ones like mercury with small coefficients of thermal expansion, because liquids must be contained in glass tubes, which also expand. Using only one tube the combined expansion of glass and liquid is measured. But in the early 1600s Dulong and Petit came up with an ingenious way in which only expansion of the liquid was measured by using *two* vertical tubes. How? *Hint*: To answer this question it helps if you understand how an ordinary mercury-in-glass barometer works. Sketch possible arrangements of the two tubes.

59. Pistons that inhabit textbooks are magical: they can be accelerated instantaneously from rest to a constant speed at no cost in energy. Suppose that a typical real automobile piston has a mass of 0.5 kg.
   (a) How much energy is required to accelerate this piston to a constant speed of 10 ms^{-1}?
   (b) Assume that this real piston travels only a small fraction of the total stroke length to reach this speed from its highest position in the cylinder.
   Assume any reasonable volume for a real cylinder. Assume that the gas in the cylinder is mostly air (a good approximation). Assume any reasonable increase in temperature of this air because of compression. How much energy

is required? What do you conclude? Don't fret over tiny details. All that is wanted is an approximation.

60. The concept of heat capacity arose from the following observation made more than 200 years ago. When equal volumes of water with different temperatures were mixed (adiabatically), the final temperature was the arithmetic average of the temperatures. But when a volume of water was mixed with an equal volume of much denser mercury, the result was unexpected: the final temperature was close to the arithmetic average. To account for this, a different "fudge factor," called the "capacity for heat," without knowing what it meant, had to be applied to each volume (equivalent to density). Determine how to measure c (per unit mass) of a material relative to that of water. State all assumptions. Try your result on mercury. What do you think was the expected weighting of the two temperatures? Suppose that instead of mixing equal volumes, equal masses had been used. What would have been the result?

61. From the cardiac pressure–volume loop (Fig. 3.2) estimate the power output of a heart in watts at peak output. You will have to assume a peak pulse rate.

62. It is tempting to jump to the reasonable conclusion that adiabatic compressions and expansions are symmetric. That is, beginning with the same initial temperature and volume, in a compression the relative increase in temperature for a given relative decrease in volume (in magnitude) is equal and opposite to the relative decrease of temperature for the same relative increase in volume. Show that this is not true except in the limit of zero relative volume change.

63. In Section 1.8 we described shaking water in an insulated container and determining its rise in temperature. This is an (approximately) adiabatic process in which work was done. Determining the details of this would be difficult if not impossible. But we can measure the work. How? Do the experiment. Express your result in food calories.

64. If cloud droplets (or anything) did not experience drag (which does not exist in many physics textbooks), how long would it take a cloud droplet to fall 1000 m from rest (assuming no updrafts)? State all your assumptions. What do you conclude from this?

65. How long does it take a rain drop to fall 1000 m if drag is accounted for? You make take the drag coefficient to be 0.4 and constant. You may assume that the drop reaches its terminal velocity in a few tens of meters. First, determine a general expression for the terminal velocity as a function of droplet diameter. Then assume a diameter of 1 mm. State all your assumptions. If you are observant in the first few minutes when rain drops begin to fall on initially dry ground, you will see evidence to support your analysis for the first part of this problem. Can you think of a simple experiment that will obviate the problem of the ground becoming uniformly wet?

66. What implicit simplifying assumption did you probably make to solve the previous three problems? Suppose you remove this assumption. How does it affect your results? HINT: See Section 5.1.

67. A sealed room 3 m × 5 m × 5 m contains air at 10 °C and one atmosphere. What is its mass and weight per unit area? If a 1000 W heater in the room runs for 30 minutes, what would be the final air temperature and pressure? Assume that the walls, ceiling, and floor of this idealized room are not affected. These dimensions are roughly those of a large room in a residence. Your calculated temperature increase may be at odds with your experience of how fast air temperature changes when you turn on a small heater or turn up a thermostat. Why? There is no single reason. This is a multi-faceted problem with no tidy solution. Welcome to reality! Chapter 7 may help.

68. A blunt-nosed object (e.g. airplane) moving at constant ground speed v_∞ through air is aerodynamically equivalent to air at this speed flowing past the object on the ground (which is why aircraft designs can be tested in wind tunnels). Because air cannot penetrate the nose, the speed there is zero. Thus, air is compressed as it piles up against it. If this occurs adiabatically, the temperature of the air will increase (called compressional aerodynamic "heating"). Determine how this temperature increase depends on v_∞ and air properties. The easiest (and best) way to do this is by physical reasoning. HINT: Use the first law of thermodynamics including mass–motion kinetic energy. The more time-consuming way, and not as good for developing physical intuition, is to use Bernoulli's law, a Poisson relation, the ideal gas law, and make an approximation. A typical cruising speed for commercial jet aircraft is about 900 kph. Determine the aerodynamic heating. If you use your expression for rocket reentry in Earth's atmosphere, what do you conclude?

69. To convince yourself that the minus sign in Eq. (3.61) is not a typographic error, verify that it is correct for an ideal gas. A good habit to get into is checking results, especially if they don't look right, by giving them a test they must pass. If they do, this does not prove that they are correct, but if they don't it does prove that they are wrong.

70. Joseph Black (c. 1760) mixed equal masses of gold and water, the gold initially at $T_g = 150\,°\mathrm{F}$, the water at $T_w = 50\,°\mathrm{F}$. Their common equilibrium temperature was $T_f = 55\,°\mathrm{F}$. From these temperatures and an expression for the final temperature T_f, the ratio of the specific heats per unit mass of gold and water follows. Interpret T_f so that you will remember a general result (subject to assumptions). Check your results against modern data for gold and water.

71. An iron cylinder has been lying on its side long enough that its density is more or less uniform. We suddenly tilt it to vertical. Estimate the change in length because of compression of the cylinder by its own weight (gravity). Take the isothermal compressibility of iron to be $10^{-11}\mathrm{Pa}^{-1}$ and its density to be 7900 kg m^{-3}. Assume a large but reasonable length for the cylinder. State all assumptions and approximations. The *time* over which this length change occurs is not readily obtained.

72. Hang a cold object, for example, a bag of frozen peas, in still air at a higher temperature. What is the expected pattern of air near the object? Verify

your expectation with an experiment that is dead simple. Baby powder and a flashlight won't help. Why?

73. We note in Section 3.8 that the difference between the longwave radiation emitted *out* of the top of the atmosphere and the net longwave radiation emitted from the surface is about 200 W m^{-2}. Assume that this energy loss is from the entire troposphere where the tropopause (the boundary between troposphere and the stratosphere) is at about 200 hPa. Verify that this energy loss would give an average cooling rate for the troposphere of about 2 °C day^{-1}. Hint: This problem is most easily done by noting that the mass (per unit area) of the troposphere is proportional to the difference between the surface pressure and the pressure at the tropopause.

   Estimate what vertical velocity would be needed to balance this cooling assuming a lapse rate of 6.5 °C km^{-1} in the troposphere. Due to mass conservation, the descending air in clear-sky areas will be balanced by air ascending in cloud updrafts. If these active clouds were to occupy 1% of the clear-sky area, give a rough estimate of the magnitude of the average vertical velocity in the active cloud areas. For this estimate, assume that the mean density is the same in the ascending and descending areas. What does this result tell you about the vertical velocities in clouds compared with the subsidence (negative vertical velocity) in larger clear areas around these clouds? Your experience flying through cloudy and clear areas in an aircraft might confirm this point. Researchers have used the global clear-sky cooling rate as a measure of the strength of the global hydrological cycle. An increase in the clear-sky cooling indicates an increase in the strength of the hydrological cycle with more precipitation. Use the results from this problem to discuss why there may be such a correlation.

74. Ice cream can be made at home by mixing rock salt with ice to obtain a low-temperature salty slush in which the container of ingredients is placed to freeze. Can you give a simple physical explanation of this lowering of temperature? You can assume that the salt and ice are initially at 0 °C and that mixing is adiabatic. It seems contradictory that salt can melt ice and at the same time lower the freezing point of the salty meltwater. All you need to do is appeal to basic principles rather than throw mantras at this problem. Ask yourself if mixing salt and ice (adiabatically) and ending up with a salt solution at a lower temperature violates any physical laws.

   Why doesn't this process work if sand is used in place of salt? *Hint:* Section 3.9.

   Can you explain by simple arguments why adding salt to water lowers its freezing point? Think about what must happen when liquid water makes the abrupt transition to ice and how dissolved salt in the water might affect this. All is wanted is a rough plausible explanation, without which this observation is just a mystery.

75. The specific gravity of 99 healthy men ranged from 1.02 to 1.099. What is the range of their weights when immersed in water, relative to their weights

in air? Expressed as fractions, the results are more dramatic and easier to remember.

76. Many finely divided metals are *pyrophoric*. They can ignite spontaneously even at room temperature. Explain. Consider metallic particles in air in a large, insulated container. *Hint*: See Section 3.9

77. So many "thought experiments" in textbooks are idealized that it is easy to lose sight of reality. Consider a real vessel, although perfectly insulated, containing water at the same temperature T_h. To this water add an equal mass at $T_c < T_h$. Assume that the specific heat capacity of water is independent of temperature What is the final temperature of the water? The textbook answer is the arithmetic average $(T_h + T_c)/2$. In the real world, will it be higher or lower? Change the order. The water and container are in equilibrium at T_c and water at T_h is added. Is the final temperature higher or lower? You can answer these questions and obtain an expression for the difference in the two final temperatures by physical reasoning.

78. We note in Section 1.8 that electric power generation is a sequence of energy transformations. Consider the motion of an ordinary car. Describe the entire sequence of energy transformations, classified only as kinetic or potential, mass–motion, and internal, that result in and maintain this motion. Begin with the ultimate energy transformation that underlies all the others. Distinguish between startup and constant speed. Assume a flat road.

79. Chemists have a rule that to dilute an acid with water, add the acid to the water, not conversely to avoid bad luck. Is this just a superstitious relic of alchemy or is there a sound physical reason for it? Suppose that you want to break this rule. How could you do so safely? *Hints*: Everything you need is in this chapter. The specific gravity of water is appreciably less than that of acids, especially ones that react strongly with water.

80. In Section 1.8 we describe an experiment in which we spun 200 ml of water in a food processor for 50 seconds resulting in a temperature increase of 0.2 °C. The rated power of the processor is 350 W. What is the expected temperature increase of the water? What do you conclude? As everyone knows, "more is better." For example, the temperature of water in a pan on a stove increases with time. But when we spun the water for only ten seconds the temperature increase was the *same*. Why? Can you think of a simple experiment that will resolve this puzzle? You may assume an adiabatic process. Can you think of a fairly common observation related to this problem? *Hint*: See Problem 1.19.

81. This problem is related to the experiment described in Section 1.8 in which ten minutes of shaking 150 ml of water resulted in a temperature increase of 4.4 °C. After shaking the water we continued to measure its temperature. The rate of decrease was greatest in the first five minutes, but the average rate over 25 minutes was 0.032 °C min^{-1}, about half the initial rate. What important lesson follows from the entire experiment? What was the minimum number of kilocalories required to increase the temperature of the water by shaking? Could you heat water for coffee by shaking?

82. Estimate the speed of buoyancy-driven updrafts. All that you need is what is in this chapter and what you learned in freshman physics.

83. Einstein's equation connecting energy and mass, $E = Mc^2$, has become a popular icon familiar to the many, understood by the few. We *could* use it to measure and express internal energy changes in mass units. If the temperature of one liter of water is increased from the freezing point to the boiling point, by how much does its mass increase? What do you conclude?

84. To get a feeling for buoyancy, estimate the time it would take a ping pong ball released from the bottom of a swimming pool 2 m deep to reach the surface *solely* because of buoyancy? If you have access to a swimming pool, try to measure this time. If you do, what do you conclude?

85. If gas molecules react to form different molecules, is the specific heat capacity of the reaction products the same as the average for the mixture of the reactants (both at the same temperature)? As a simple example, consider the reaction in which two molecules of hydrogen react with one molecule of oxygen to form one molecule of water vapor. *Hints*: See Section 3.7. You will have to search for data for the three molecules at about the same temperature and for the same units. We calculated the isobaric specific heat capacity in kJ kg^{-1}K^{-1}.

86. Does a closed, thin-walled flask filled with air, say, weigh more than when empty? Does your answer depend on how its weight is measured? With a spring scale the flask is suspended; with a balance scale, the flask is placed on a surface. State all assumptions. When you consider the flask on a balance, take it to be a cube, the system its solid parts, its surroundings the air inside and outside. Which method would you use to determine the mass of the air in the flask as precisely as possible?

87. In Section 3.2 we note that if intermolecular forces had the same form as gravitational forces, thermodynamics of liquids and solids would be difficult. Show that the total gravitational potential energy *per unit mass* of a body is proportional to its mass. There is an easy way and a hard way to do this. If you know the form of the gravitational potential energy of two point masses, you can guess the form of this energy for, say, a uniform spherical body (to within a dimensionless constant). This takes one line. With much more effort you can evaluate a double integral to obtain the dimensionless constant (which is useless for this problem). And you can evaluate this integral analytically only for uniform bodies of simple shape.

88. We note in Section 1.6 that the contribution of gravity to the work done in compression of a gas, say air, is negligible relative to the pV work. Show this. *Hints*: Equations (3.65) and (3.81) are useful.

89. We have encountered statements implying that if Earth were to suddenly stop by some (magical) process, the oceans would boil. By simple analysis you can check this. Assume for simplicity that Earth is entirely water. You know the angular speed of Earth about its axis of rotation. In textbooks you can find expressions for the rotational kinetic energy of a sphere. But you don't need this for a first cut. You need only three readily obtainable quantities.

4

Entropy

Entropy is a virus that has escaped from the laboratory and infected people, especially nonscientists with a literary bent. R. Z. Sheppard reviewing in *Time* (May 1, 1995) *The Information* by Martin Amis asserts that "By now the literary uses of entropy are threadbare even for Amis." And in an article by Ivan Hannaford, "The Idiocy of Race" (*Wilson Quarterly*, Spring 1994), we find "All that remained ... was the shell of an orthodox race-relations policy that only exacerbates the state of civic entropy." These examples were stumbled upon in everyday reading. Countless other examples can be found. You can amuse yourself by noting the ways in which entropy is used outside thermodynamics and puzzle over the connection, if any, between scientific and literary entropy.

In a way, increased use of *entropy* by people who haven't the foggiest idea what it means is good. As a term becomes part of everyday speech and writing, it loses its mystery, and people think they understand it. Familiarity breeds contentment. We usually don't agonize over the meaning of everyday words. Energy, for example, is not so enigmatic as entropy because energy is more familiar. It is used in so many different ways by so many different people, often in ways unrelated to or even incompatible with its use in physics, that students rarely find energy so formidable as entropy. Yet, in its scientific sense it is no more mysterious than energy. The defining characteristic of energy in the version of the first law in Section 1.1 is that it remains constant (for an isolated system), whereas the defining characteristic of entropy in the second law is that it can only increase (for an isolated system). Why is it more puzzling that one quantity can only increase whereas another quantity must remain constant?

Entropy used by nonscientists who want to appear up to date in a scientific world vaguely conveys undesirability, often associated with disorder or dysfunction. Civic entropy, in the example cited in the opening paragraph, presumably is meant to convey the breakdown of law and order and social cohesiveness. We critically examine the association of entropy with disorder, but first we determine the mathematical form of the entropy of an ideal gas and examine how it changes in various

Atmospheric Thermodynamics. Second Edition. Craig F. Bohren and Bruce A. Albrecht, Oxford University Press.
© Craig Bohren and Bruce Albrecht (2023). DOI: 10.1093/oso/9780198872702.003.0004

processes. This is more satisfactory than defining entropy as an abstract quantity. If we understand a concrete example, we are better prepared to grasp it in the abstract.

Although we state without qualification that entropy, like energy, is measurable, this is wishful thinking. Measurements of energy and entropy are almost invariably made for pure, homogeneous, uniform, single-component systems. That the energy and entropy of a rhinoceros exist must be taken on faith.

Entropy has many definitions and interpretations about which there is no consensus. Unlike energy, entropy is still the source of disputes that have been roiling the waters of science for more than a century. In a paper about the "many faces of entropy," Harold Grad (1961) wrote that bewilderment about entropy is a consequence of the many quantities bearing this name: macroscopic functions of state variables and microscopic quantities such as the logarithm of a probability. He opined that much

> of the confusion ... is traceable to the ostensibly unifying belief (possibly theological in origin!) that there is only one entropy ... It does not seem possible to give a precise mathematical definition of entropy or to create an abstract mathematical structure which is general enough to include all of the interesting applications.

If you are confused about entropy, so were some of the founding fathers of thermodynamics. "Lord Kelvin ... a pioneer in establishing the second law, never understood the idea of entropy at all." He was preceded by "Clausius, who did as much as anyone to establish the second law" and yet "believed all his life that it was a purely mechanical law."

Confusion and controversy over the second law and entropy appear to be mostly how it is defined, calculated, and interpreted at the microscopic level. Entropy is much too rich in nuances to be encapsulated in a mantra. The expressions with a limited range of applicability that we give are mostly beyond controversy and serve our modest needs.

We eased into the first law of thermodynamics, Eq. (1.63), by way of a mechanical model of matter as a system of interacting point molecules, then made the transition to a thermodynamic model in which molecules were off-stage actors. In this chapter, we reverse this order by deriving entropy from macroscopic thermodynamics and then giving it a microscopic interpretation.

4.1 Entropy of an Ideal Gas

The first law of thermodynamics applied to a closed system with reversible working rate $-p\, dV/dt$ is

$$Q = \frac{dU}{dt} + p\frac{dV}{dt}. \tag{4.1}$$

Recall that this applies to a sufficiently small system and sufficiently slow process that at any instant pressure and temperature are *approximately* uniform. This implies constraints on Q. For example, the system is completely immersed in a large constant and uniform temperature bath. From Eq. (4.1), we can obtain a quantity called *entropy*. We do so for an ideal gas, although the concept of entropy has a much wider applicability.

The rate of change of internal energy of an ideal gas, defined by $\partial U / \partial V = 0$, can be written

$$\frac{dU}{dt} = C_V \frac{dT}{dt}. \tag{4.2}$$

Divide Eq. (4.1) by T and use Eq. (4.2) and the ideal gas law to obtain

$$\frac{Q}{T} = \frac{C_V}{T} \frac{dT}{dt} + \frac{Nk}{V} \frac{dV}{dt}. \tag{4.3}$$

From Eqs. (3.53) and (3.80) we can write Eq. (4.3) as

$$\frac{Q}{T} = C_V \left(\frac{1}{T} \frac{dT}{dt} + \frac{\gamma - 1}{V} \frac{dV}{dt} \right). \tag{4.4}$$

If C_V is independent of temperature, the equation

$$\frac{dS}{dt} = \frac{Q}{T} \tag{4.5}$$

has the solution

$$S = S_0 + C_V \ln \left(\frac{T V^{\gamma - 1}}{T_0 V_0^{\gamma - 1}} \right), \tag{4.6}$$

where 0 indicates a reference state and S_0 is the value of S for this state. S is a thermodynamic variable, called *entropy*, with the property that its time derivative is Q/T. Entropy owes its name to Rudolf Clausius (b. Köslin, Prussia, 1822; d. Bonn, Germany, 1888), a German physicist who in 1850 wrote that

> I propose to name the magnitude S the *entropy* of the body from the Greek word ἡτροπή, a transformation. I have intentionally formed the word *entropy* so as to be as similar as possible to the word *energy*, since both these quantities ... are so nearly related to each other in their physical significance that a certain similarity in their names seemed to me advantageous.

As with energy, we know the dimensions of entropy: energy divided by absolute temperature, the same as those of heat capacity.

The entropy of an ideal gas with heat capacity independent of temperature sometimes is written carelessly as

$$S = C_V \ln(T V^{\gamma - 1}) = C_V \ln T + Nk \ln V. \tag{4.7}$$

As it stands, this is meaningless because T and V have dimensions, and hence, the values of the logarithms depend on the choice of units. Mathematicians deal mostly with pure numbers and are usually oblivious to the dimensions of physical quantities. Although we sometimes write Eq. (4.7) because of its simpler form, it can be used only to determine entropy *differences* because $\ln(a/b) = \ln a - \ln b$, and a/b is dimensionless if a and b have the same dimensions. Entropy is the sum of two logarithmic functions. Appearance of the thermophysical property $\gamma - 1$ in Eq. (4.7) may mislead. Nk is not such a property whereas C_V is. The volume contribution to the entropy of an ideal gas should not depend on its properties, but the temperature contribution should. The second way of expressing entropy displays this.

Because unconfined air has no definite volume, it is often more convenient to express the entropy of an ideal gas as a function of T and p. From Eq. (4.6) and the ideal gas law

$$S - S_o = C_p \ln \left(\frac{T}{T_o} \right) - Nk \ln \left(\frac{p}{p_0} \right). \tag{4.8}$$

The entropy of an ideal gas does not depend on its mass M, only on the *number* of molecules N. Substituting $N = M/m$ and $R = k/m$ in Eqs. (4.7) and (4.8), where m is the mass per molecule, so that $Nk = MR$, which may be convenient, but also misleading.

Having derived the entropy of an ideal gas by way of a particular idealized process in which $W = -p\,dV/dt$, we may forget about how we obtained entropy and consider it to be yet another thermodynamic (equilibrium) state variable. That is, the entropy of an ideal gas is *defined* by Eq. (4.6) or (4.8), which also happen to give the entropy change of an ideal gas in a particular idealized process. The best way to understand entropy is to examine how it behaves. As we show, it behaves differently from energy.

The entropy of an ideal gas defined as a logarithmic function of macroscopic thermodynamic variables is consistent with the microscopic definition, $S = k \log W$, where W is sometimes called the *multiplicity*, the number of microstates for a given macrostate, *not* the rate at which work is done. Although this equation is inscribed on Boltzmann's tombstone, he never wrote it. Planck did and called k Boltzmann's constant.

Entropy Change in a Free Expansion

Idealization Alert! In thermodynamics books—this one no exception—insulated containers in which processes are imagined to take place are usually, but implicitly, *perfectly* insulated. If you have a source of perfect insulation, please let us know. Real containers always take part in processes occurring within them.

Consider an ideal gas in a closed, insulated container divided into two chambers by an impermeable partition (Fig. 4.1). The volume of one chamber is V_1, that of the two chambers taken together is V_2. The temperature of the gas initially confined to V_1 is T, and hence, its entropy is

$$S_i = C_V \ln(TV_1^{\gamma-1}), \tag{4.9}$$

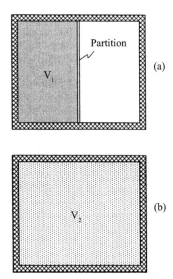

Fig. 4.1: Free expansion of an isolated ideal gas. If the partition separating the gas in volume V_1 (a) from a volume with a much smaller number of molecules is removed, gas expands to fill the entire volume.

where the subscript i denotes the initial state. If we remove the partition the gas is free to occupy both chambers. This is a spontaneous process requiring no assistance from an external agent. Because the gas is ideal, its temperature does not change. The path to the final equilibrium state and the time it takes are outside the scope of thermodynamics, but spontaneous does *not* imply instantaneous. Because the gas is ideal, its final *equilibrium* temperature does not change even though its volume increases, and hence, its final entropy is

$$S_f = C_V \ln(TV_2^{\gamma-1}).\tag{4.10}$$

Therefore, the entropy change (of the universe) is

$$\Delta S = S_f - S_i = C_V(\gamma - 1)\ln\left(\frac{V_2}{V_1}\right) = Nk\ln\left(\frac{V_2}{V_1}\right).\tag{4.11}$$

Because $V_2 > V_1$, $\Delta S > 0$. In this process, the entropy of the universe increased even though there was no heating and no working, and hence, no change in internal energy. This is the first indication that entropy behaves differently from energy. Always keep in mind the difference between make-believe and reality. The entropy difference, Eq. (4.11), is for an ideal gas and constant internal energy. More importantly, it is much easier to create an adiabatic process ($Q = 0$) on paper than in the laboratory. But this idealized process shows that *all* adiabatic processes are not isentropic. An isentropic process is one in which the entropy at the end equals that at the beginning. What happens between these two equilibrium states does not affect this definition.

We could *not* have obtained the entropy change in this free expansion by integrating

$$\frac{dS}{dt} = \frac{Q}{T}, \tag{4.12}$$

which would have given $\Delta S = 0$ because $Q = 0$. This is because underlying the *correct* use of Eq. (4.12) is the requirement that the working rate be $-p\,dV/dt$, whereas in an adiabatic free expansion $W = 0$. Although the volume of the gas increases in this process, no work is done on or by it. $W \neq 0$ requires *interaction* between a system and its surroundings.

Left to itself, the system will not return spontaneously to its state before the free expansion, when all the molecules were in V_1. This is equivalent to scrambling eggs. Once they have been scrambled, they cannot be unscrambled without effort (see Problem 4.9). And so it is with the free expansion of a gas. To restore it to its initial state requires doing work

$$W_{tot} = -\int_{V_2}^{V_1} p\frac{dV}{dt}\,dt \tag{4.13}$$

at constant temperature T. From the ideal gas law in Eq. (4.13)

$$W_{tot} = NkT \ \ln\left(\frac{V_2}{V_1}\right) = T\,\Delta S, \tag{4.14}$$

where ΔS is the entropy change in the free expansion (Eq. 4.11). Such spontaneous processes are in the direction of increasing entropy, but thermodynamics cannot tell us how fast. Rusting of iron in air is a very slow spontaneous process.

Dissolving ordinary salt in water (adiabatically) is another spontaneous process in that the entropy of the solution is greater than the sum of the entropies of the salt and the water. Sea water can be and is desalinated—but not at no cost.

The free expansion of a gas gives us insights into the limitations of thermodynamics. If $Q = 0$ and $W = 0$, from Eq. (1.63) $\Delta U = 0$, and $\Delta T = 0$ because the internal energy of an ideal gas depends only on its temperature. But wait! What is meant by $\Delta T = 0$? It can mean only a difference for two equilibrium states in which the temperature of the gas is approximately uniform. From the first law in the form Eq. (1.62), we have to reckon with the mass–motion kinetic energy streaming from the hole for a moment after it is punched in the partition. This energy doesn't come for free. It has to be paid for, so to speak, by a decrease in the random kinetic energy, and hence, temperature. During a brief moment the volume of the gas is not well defined, nor are the pressure and temperature uniform. Only after the dust has settled can we say that $\Delta T = 0$. What happens during the transition between the two macroscopically quiescent states is beyond the scope of thermodynamics. Indeed, it is probably unknowable and its molecular interpretation obscure or fanciful. The first law implies only that hidden mechanisms exist to balance the energy budget.

To get a better handle on this we did a simple experiment. On an overcast morning we inflated the tire on a wheelbarrow. After waiting for about an hour we depressed the tire valve while holding a thermocouple near its end during the few seconds of rapid outflow of air. Its temperature was between 3 °C and 10 °C *lower* than ambient air temperature, which also was the temperature of the tire to within a tenth or so of a degree. The outflowing air had directed kinetic energy *extracted* from random kinetic energy. This transformation must be mediated by molecular collisions (interactions) we can ignore for *some* purposes (e.g. the ideal gas law) but not all. For example, a strictly collisionless gas ($\sigma = 0$) does not have a finite thermal conductivity, viscosity, and diffusion coefficient (see Sections 7.1, 7.2, and 7.3).

A dramatic demonstration (more suitable for a classroom) we have only read about is to allow compressed carbon dioxide gas in a cylinder at room temperature to flow out of a hole. The resulting cloud of dry ice (solid carbon dioxide) particles is evidence that the temperature dropped to -78.5 °C. This does not require a thermometer but does require understanding the phase diagram (Section 5.5) for carbon dioxide.

Entropy Changes upon Heating and Cooling

As our next example of an entropy change, consider the heating or cooling of an ideal gas. For simplicity, we take a fixed volume of gas in contact with a much larger volume of the same gas, called a *reservoir*. You can imagine the gases to be enclosed in rigid metal tanks (Fig. 4.2). The two volumes are insulated from their surroundings. Together, they are our imaginary universe. Initially, the temperature of the smaller volume is T_1, and that of the larger volume is T_2, where $T_2 > T_1$. With the passage of time, both volumes are *assumed* to come to the same temperature T (see Section 4.4). In this process, total internal energy is conserved:

$$C_{V1}T_1 + C_{V2}T_2 = C_{V1}T + C_{V2}T. \tag{4.15}$$

This equation can be solved for the final temperature:

$$T = \frac{C_{V1}}{C_{V1} + C_{V2}}T_1 + \frac{C_{V2}}{C_{V1} + C_{V2}}T_2. \tag{4.16}$$

Volume does not change in this process, and hence, the entropy changes from Eq. (4.7) are

$$\Delta S_1 = C_{V1}\ln\left(\frac{T}{T_1}\right), \quad \Delta S_2 = C_{V2}\ln\left(\frac{T}{T_2}\right). \tag{4.17}$$

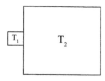

Fig. 4.2: Two rigid containers filled with an ideal gas at different initial temperatures T_1 and T_2 are placed in contact and isolated from their surroundings.

By using Eq. (4.16), the total entropy change can be written

$$\Delta S = \Delta S_1 + \Delta S_2 = C_{V1} \left[\ln \left(\frac{\mu + x}{1 + \mu} \right) + \frac{1}{\mu} \ln \left(\frac{1 + \mu/x}{1 + \mu} \right) \right], \tag{4.18}$$

where

$$\mu = \frac{C_{V1}}{C_{V2}}, \quad x = \frac{T_2}{T_1}. \tag{4.19}$$

Our task is simpler if we take the limiting case of an infinite reservoir ($\mu \to 0$). The first term in brackets in Eq. (4.18) poses no problems:

$$\lim_{\mu \to 0} \ln \left(\frac{\mu + x}{1 + \mu} \right) = \ln x. \tag{4.20}$$

The second term requires more effort because $1/\mu$ is infinite in the limit of zero μ, whereas the limit of the logarithmic term is zero. To evaluate the second limit, we use the expansion

$$ln(1 + z) \approx z \text{ for } |z| << 1. \tag{4.21}$$

For sufficiently small values of μ we have

$$ln \left(\frac{1 + \mu/x}{1 + \mu} \right) = \ln(1 + \mu/x) - \ln(1 + \mu) \approx \frac{\mu}{x} - \mu, \tag{4.22}$$

which yields the limit

$$\lim_{\mu \to 0} \frac{1}{\mu} \ln \left(\frac{1 + \mu/x}{1 + \mu} \right) = \frac{1}{x} - 1. \tag{4.23}$$

With this result, the entropy change of the universe is

$$\Delta S = C_{V1} \left(\ln x + \frac{1}{x} - 1 \right). \tag{4.24}$$

By supposition the reservoir is hotter than the volume V_1 of gas. If the reservoir is infinite, it follows from Eq. (4.16) that $T = T_2$ ($x \to 1$). Thus, we can describe this process as heating: the temperature of the smaller volume of gas increases, whereas that of the (infinite) reservoir does not change (measurably). What happens to the total entropy in this process? To answer this requires examining the sign of the function

$$f(x) = \ln x + \frac{1}{x} - 1 \tag{4.25}$$

in Eq. (4.24) for $x > 1$. Note that $f(1) = 0$ and

$$\frac{df}{dx} = \frac{1}{x} - \frac{1}{x^2} \tag{4.26}$$

is positive for $x > 1$. In this heating process, the entropy of the universe increases.

We now might be tempted to surmise that if entropy increases in a heating process, it decreases in a cooling process. This is a reasonable expectation, but untrue, which attests to another peculiarity of entropy. From Eq. (4.26) it follows that df/dx is zero only at $x = 1$, and hence, f has an extremum at this value. This extremum is a minimum because the second derivative of f is positive at $x = 1$. Because the minimum of f is zero at $x = 1$, $f > 0$ for $x > 1$ (heating) *and* for $x < 1$ (cooling). For both heating and cooling, the entropy of the universe increases.

Unlike for the free expansion, we could have obtained the entropy change for heating and cooling by integrating Eq. (4.12). This is because $W = 0$ and is equal to $-pdV/dt$ with V constant. Let Q be the rate of heating (cooling) of the smaller volume of gas. The rate of cooling (heating) of the reservoir is therefore $-Q$. By supposition, the reservoir is infinite, and hence, its temperature does not change during the process. The entropy change of the reservoir is therefore

$$\Delta S_2 = \int -\frac{Q}{T_2} dt = -\frac{1}{T_2} \int Q\, dt = -\frac{1}{T_2} C_{V1}(T_2 - T_1). \qquad (4.27)$$

The entropy change of the smaller volume is

$$\Delta S_1 = \int \frac{Q}{T} dt = \int \frac{1}{T} \frac{dU}{dt} dt = C_{V1} \int \frac{1}{T} \frac{dT}{dt} dt = C_{V1} \ln\left(\frac{T_2}{T_1}\right). \qquad (4.28)$$

Thus, we obtain the same total entropy change as in Eq. (4.24).

Before we try to make sense out of this result, we consider a process in which entropy does *not* increase. Such a process is adiabatic ($Q = 0$) with working rate $-p\, dV/dt$. We showed in Section 3.3 that in this reversible process

$$TV^{\gamma-1} = \text{const}. \qquad (4.29)$$

This result in Eq. (4.6) yields *no* entropy change—another example in which we get the same result using Eq. (4.12). The best we have been able to do is find a process in which entropy is constant. In the other processes considered, entropy increased. A process in which entropy is constant is said to be *isentropic*. For the first adiabatic process we considered (free expansion), entropy increased, whereas for the second adiabatic process, entropy was constant. Although some adiabatic processes are isentropic, not all are.

We obtained what, on reflection, is a disturbing result: the entropy of a fixed-volume (ideal gas) reservoir changed, yet its temperature remained constant. Suppose that before the smaller volume was put into contact with the reservoir, we measured its temperature. When our backs were turned, someone put the smaller volume in contact with the reservoir, then removed this volume when its temperature was that of the reservoir. We turned around, oblivious to what happened when we weren't looking, and measured the temperature of the reservoir. We recorded no change and would have had to conclude that nothing happened. But the entropy of the reservoir changed even though we could not make any macroscopic measurement that indicated a change in its state. The only way out of this embarrassing fix is to assert that infinite reservoirs are idealizations not found in nature. In any energy transfer process between a reservoir and a system, the temperature of the reservoir *must* change even though immeasurably small. Another way of looking at this problem is that according to Eq. (4.16), $T = T_2$ in the limit of an infinite reservoir (C_{V2} becomes infinite). But if we set C_{V2} to infinity and $T = T_2$ in Eq. (4.17), we obtain an indeterminate result for the entropy change of the reservoir. So we sneak up on the infinite reservoir by taking the limit of the entropy change as C_{V2} approaches (but never reaches) infinity. This *limit* exists:

$$\lim_{C_{V2} \to \infty} \Delta S_2 = C_{V1} \left(\frac{1}{x} - 1 \right) \tag{4.30}$$

even though the infinite reservoir does not.

We already have seen that if a reversible compression or expansion of an ideal gas is carried out adiabatically, the entropy of the universe does not change. What happens if the same process is carried out isothermally? To achieve this, we would have to put the cylinder in contact with a large (strictly, infinite) constant-temperature reservoir. The change in entropy of the gas in an isothermal expansion from volume V_1 to V_2 is exactly the same as that in the free expansion (Eq. 4.11). But now because the gas interacts with its surroundings, the reservoir, we have to compute its entropy change to obtain the total entropy change of the universe. We showed that the entropy change of a reservoir can be obtained by integrating Q_{res}/T:

$$\Delta S_{res} = \int_1^2 \frac{Q_{res}}{T} dt = \frac{1}{T} \int_1^2 Q_{res} \, dt . \tag{4.31}$$

Here, Q_{res} is the negative of that appropriate to the gas (cooling of the gas means heating of the reservoir, and vice versa), which from the first law for an ideal gas is $p \, dV/dt$ in an isothermal process. Thus, the entropy change of the reservoir is

$$\Delta S_{res} = \frac{1}{T} \int_1^2 -p \frac{dV}{dt} dt, \tag{4.32}$$

which is (using the ideal gas law)

$$\Delta S_{res} = -Nk \int_1^2 \frac{1}{V} \frac{dV}{dt} dt = -Nk \ln \left(\frac{V_2}{V_1} \right) . \tag{4.33}$$

When this entropy change of the reservoir is added to that of the gas in the cylinder (Eq. 4.11), the result is zero net entropy change. On the basis of what we have done to this point, we conclude that a reversible process is one in which the entropy of the universe does not change. A process that is not reversible is irreversible. All real processes are irreversible.

The entropy of the universe does change in heating or cooling of a volume of a gas in contact with an (infinite) reservoir (see Eq. 4.24). But in the limit as $x \to 1$ (the small gas volume and reservoir approach the same initial temperature), the entropy of the universe does not change. Could we carry out the heating or cooling of the small gas volume in such a way that the entropy of the universe does not change?

To answer this, consider the following idealized process. The small volume, which henceforth we call our system, is brought into successive contact with N infinite reservoirs with temperatures $T_1 < T_2 < T_3 < \ldots < T_N$, separated in temperature by $\Delta T/(N-1)$, where $\Delta T = T_N - T_1$ (Fig. 4.3). Initially, the system is in equilibrium with reservoir 1, then comes into equilibrium with each reservoir in turn. The succession of N equilibrium states of the system is an approximation to a quasistatic process; in the limit as N approaches infinity this discontinuous process approaches a continuous process (i.e. an ideal quasistatic process). The jth entropy change of the universe is

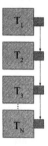

Fig. 4.3: A set of N ideal gas reservoirs with temperatures T_1, T_2, ... T_N, where the temperature difference between successive reservoirs is arbitrarily small. A rigid container filled with an ideal gas is successively placed in contact with each reservoir and allowed to come into temperature equilibrium with it. By this process the temperature of the gas in the container evolves in small steps from T_1 to T_N.

$$\Delta S_j = C_V \left[\ln \left(\frac{T_{j+1}}{T_j} \right) + \frac{T_j}{T_{j+1}} - 1 \right], \tag{4.34}$$

where C_V is the heat capacity of the system. After a bit of algebra, the total entropy change can be written

$$\Delta S = \Sigma_j \Delta S_j = C_V \left(\ln \frac{1}{y} - \frac{1}{N} \sum_{j=2}^{N} \frac{1}{\frac{y}{1-y} + \frac{j-1}{N-1}} \right), \tag{4.35}$$

where $y = T_N/T_1$. What is the limit of the sum in this equation as N approaches infinity? To answer, first change the index in the sum in Eq. (4.35) from j to $k+1$ and let $N = M+1$. With these transformations of indices the sum becomes

$$\frac{1}{M+1} \Sigma_{k=1}^{M} \frac{1}{\mu + z_k}, \tag{4.36}$$

where $z_k = k/M$ and μ is $y/(1-y)$. Because $M \gg 1$, Eq. (4.36) is to good approximation replaced by

$$\Sigma_{k=1}^{M} \frac{\Delta z}{\mu + z_k}, \quad 0 \le z_k \le 1, \tag{4.37}$$

where $\Delta z = z_{k+1} - z_k = 1/M$. In the limit as M becomes infinite, this sum becomes an integral:

$$\lim_{M \to \infty} \Sigma_{k=1}^{M} \frac{\Delta z}{\mu + z_k} = \int_0^1 \frac{dz}{\mu + z} = -\ln y, \tag{4.38}$$

which when substituted into Eq. (4.35) yields $\Delta S = 0$. To check this result we chose y at random, then computed the total entropy change (relative to C_V) from Eq. (4.35) for various values of N. For $N = 50,000$, the entropy change (relative to the heat capacity) was 0.0000221. This process can be said to be reversible. We have a huge set of reservoirs, each differing in temperature from its nearest neighbors by a small amount. We can successively put the system in contact with these reservoirs, one after the other, arranged in order of increasing temperature, and then reverse the process to restore the system to its initial temperature. And all without any permanent change in the universe. But to conjure up this reversible heating and cooling process we had to invoke a considerable degree of unreality: an infinite number of infinite reservoirs. But there is another element of unreality in this thought experiment. We ignored the entropy change of the unnamed agents (elves, robots, ghosts ... ?) who carried out the infinite shuffling of the system from one reservoir to the next.

What about the free expansion? We can imagine this process to be carried out as quasi-statically as we wish. The space into which the gas expands is divided by partitions into a set of thin chambers (Fig. 4.4). We remove each partition in turn, allowing the expanded gas to come into equilibrium each time. The result is a set of equilibrium states, and in the limit of an infinite number of partitions this set of states is continuous. Yet, even in this quasistatic process, the entropy of the universe increases according to Eq. (4.11). Moreover, this process would not be described as

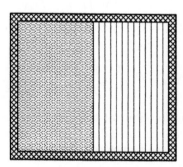

Fig. 4.4: The volume on the left encloses an ideal gas. The volume on the right initially encloses empty space, subdivided into compartments by a set of very many exceedingly thin partitions. Each partition is removed in turn and the gas is allowed to reach equilibrium before the next partition is removed. The limit of an infinite number of partitions corresponds to a continuous sequence of equilibrium states (i.e. a quasistatic process).

reversible. Thus, we conclude that for a process to be reversible it is necessary for
it to be quasistatic but not sufficient. Now we can *define* a reversible process as one
in which the entropy of the universe does not change. No real process is reversible.
Reversibility is an idealization. In all real (irreversible) processes the entropy of the
universe increases.

We can *define* the entropy change of any system between two states as

$$S_2 - S_1 = \int_1^2 \frac{Q}{T} dt, \tag{4.39}$$

where the only restriction on the process connecting states 1 and 2 is that it be
reversible (provided such a process exists). This definition is consistent with the
entropy change derivations for an ideal gas undergoing reversible processes. Note
that Eq. (4.39) is not in conflict with our assertion that a reversible process is
one in which the entropy change of the *universe* is zero. In such a process the
entropy of a system changes according to Eq. (4.39), and the change in entropy of
its surroundings is equal and opposite in sign so that the total entropy change of
the universe, the system together with its surroundings, is zero.

Alert readers may have noticed an apparent contradiction. Equation (4.39)
applied to a fixed volume of ideal gas heated or cooled by contact with a single reservoir yields the correct value for the entropy change of this volume, even
though the heating or cooling process is not reversible (the entropy of the universe
increases). This is because the volume could have evolved by a reversible process
between the same initial and final equilibrium states as in the irreversible process.
For example, instead of the volume interacting with a single reservoir, it could
have evolved from a given initial to a given final temperature by interacting with
an infinite sequence of reservoirs, and this process (as we showed) is reversible.
Entropy is a state variable: it depends only on the state of the system, not on its
history. Thus, the entropy change of the volume of gas when it is irreversibly heated
or cooled (entropy of the universe increases) is the same as when it is reversibly
heated or cooled (entropy of the universe is constant).

For a process to be reversible it is necessary that the working be reversible
($W = -p \, dV/dt$) but not sufficient. Again, consider the example of a fixed volume
of ideal gas interacting with a single reservoir at a temperature different from the
initial temperature of the volume. Here, the working rate is quasistatic (indeed, it
is static because the volume is fixed) and reversible, and yet the entropy of the
universe increases.

All the specific examples we have given for an ideal gas are special instances of
the second law of thermodynamics in the form

$$\Delta S_{universe} \geq 0. \tag{4.40}$$

It is with some trepidation that we present the law in the form of Eq. (4.40), for we
are well aware of Clifford Truesdell's assertion that "Every physicist knows exactly
what the first and second laws [of thermodynamics] mean, but it is my experience

that no two physicists agree about them." To forestall critics, we merely note that Eq. (4.40) is only one of the many forms of the second law of thermodynamics.

Another statement of the second law is that the entropy of an *isolated* system can never decrease. Such a system, by definition, does not interact with its surroundings, and hence, is a universe unto itself, to which the second law in the form Eq. (4.40) is applicable.

The Second Law and Stability

The second law gives us a criterion for the stability of an isolated system. One meaning of stability is resistance to change. A stable system will not change its state spontaneously or when slightly perturbed. Thus, an isolated system in a state of maximum entropy is stable. According to the second law, a change in the state of the system would entail an increase in its entropy (or, at best, no change). But if the entropy of the system is a maximum, there are no states of higher entropy accessible to it. When you are on the summit of a mountain, you cannot climb higher. Maximum here means a *local* maximum, not an *absolute* maximum. At the summit you can go higher, but to do so you have to undergo a transformation from the mortal to the immortal state. Similarly, a state of maximum entropy is one for which the entropy is larger than that of all states *accessible* to the system.

As an example, we apply this stability criterion to the free expansion considered previously. When the partition separating the gas from a region of very low pressure is removed or punctured, the gas evolves to a state of maximum entropy by filling the entire volume accessible to it. Once the gas has done so, it does not retreat into a smaller volume. To do so would entail a decrease in entropy, which is not allowed by the second law. Departures from the second law are not prohibited absolutely but are exceedingly unlikely. Thermodynamics applies to macroscopic systems, and thermodynamic variables are averages for such systems. Pressure and temperature, and hence, entropy, are averages and have meaning only for systems composed of many molecules. Wherever there are averages there are deviations from these averages (fluctuations) lurking in the background. On average, the entropy of a confined gas cannot increase, but there always will be fluctuations giving rise to transitory (and small) violations of the second law. A fluctuation so large that all the gas molecules in a room spontaneously migrate into one corner is possible, but don't lie awake at night in fear of asphyxiation.

Entropy of Mixtures: Entropy of Mixing and the "Gibbs Paradox"

Given that the atmosphere is a mixture of gases, we can hardly proceed without first determining the entropy of a mixture of noninteracting gases. Although we could make a judicious guess at the entropy of a mixture as a function of the entropies of its components, the safest approach is to begin anew. We showed in Section 3.7

that internal energies of ideal gases in a mixture are additive, and hence, the total rate of change of internal energy of a mixture is

$$\frac{dU}{dt} = \sum C_{Vj} \frac{dT}{dt}, \qquad (4.41)$$

where C_{Vj} is the heat capacity (at constant volume) of the jth component. From Dalton's law of partial pressures

$$p = \Sigma p_j, \qquad (4.42)$$

where p_j is the partial pressure of the jth component. Equations (4.41) and (4.42) together with the first law Eq. (4.1), yield

$$Q = \sum C_{Vj} \frac{dT}{dt} + \sum p_j \frac{dV}{dt}. \qquad (4.43)$$

Each component of the mixture satisfies the ideal gas law

$$p_j = \frac{N_j kT}{V}, \qquad (4.44)$$

where V and T are their common volume and temperature. If we combine the preceding equations, we obtain

$$\frac{Q}{T} = \sum C_{pj} \frac{1}{T} \frac{dT}{dt} + \sum \frac{N_j k}{p_j} \frac{dp_j}{dt} = \frac{dS}{dt} = \frac{d}{dt} \sum_j S_j, \qquad (4.45)$$

where the entropy of the jth component is

$$S_j = C_{pj} \ln T - N_j k \ln p_j. \qquad (4.46)$$

For simplicity we omit the constants of integration. From inspection of Eq. (4.8), the entropy of the mixture is the sum of the partial entropies of each of its components acting independently, which comes as no surprise.

Given that entropies of noninteracting gases are additive, we can determine entropy changes upon mixing of different ideal gases at constant pressure and temperature. Suppose that the gases are initially in chambers with volumes V_j separated by impermeable partitions. Each chamber contains an ideal gas at the same pressure p and temperature T. The initial entropy S_i of this system is the sum of entropies

$$S_i = \Sigma C_{pj} \ln T - \Sigma k N_j \ln p. \qquad (4.47)$$

When the partitions are removed, the gases mix freely until they occupy the same volume $V = \Sigma V_j$. The final entropy S_f is

$$S_f = \Sigma C_{pj} \ln T - \Sigma k N_j \ln p_j, \qquad (4.48)$$

where

$$p_j = \frac{N_j kT}{V}. \qquad (4.49)$$

We also have from the ideal gas law

$$p_j = \frac{N_j}{N} p, \tag{4.50}$$

where N is the total number of molecules of all kinds. From Eqs. (4.49) and (4.50)

$$S_f = \Sigma C_{pj} \ln T - \Sigma k N_j \ln p - \Sigma k N_j \ln \left(\frac{N_j}{N} \right). \tag{4.51}$$

From Eq. (4.47) it follows that

$$S_f = S_i - \sum k N_j \ln \left(\frac{N_j}{N} \right). \tag{4.52}$$

The quantity

$$- \sum k N_j \ln \left(\frac{N_j}{N} \right) \tag{4.53}$$

is called the *entropy of mixing* and is necessarily positive because $N_j/N \leq 1$. The origins of this entropy are not a deep mystery. After mixing, each component occupies a greater volume than it did initially. Yet, Eq. (4.53) leads to a paradox. Nothing in this expression for the entropy of mixing depends on any specific property (e.g. molecular weight, specific heat capacity) of the molecules in the mixture. Thus, if we had taken all the molecules in the separate chambers to be identical, we would have obtained the same entropy of mixing even though no apparent macroscopic mixing had taken place. This paradox, called the Gibbs paradox (of the first kind), can perhaps be made clearer by considering a simpler example. A fixed volume $2V$ is divided into two volumes V separated by an impermeable partition. Each subvolume contains N molecules of the same kind of ideal gas at the temperature T. The partition is removed. What is the entropy change? There are two possible answers. The first is the obvious, common sense one: there is no entropy change because, in effect, nothing happened. The second comes from applying Eq. (4.53), according to which the entropy change is

$$\Delta S = 2Nk \ln 2. \tag{4.54}$$

Both entropy changes cannot be correct. We arrived at this paradox because we applied Eq. (4.53) in the limit as the different kinds of molecules become the same. It was derived under the implicit assumption that the molecules are different. Once we decide they are, they are forever, and if we decide they are not, no matter how small the difference, we cannot use Eq. (4.53). Gibbs did not consider his eponymous paradox to be paradoxical.

The lack of consensus about the second law and what is meant by entropy is noted elsewhere in this chapter. Because there is a similar lack about the Gibbs paradox, we would be crazy to try to resolve it. But we can give an inkling of what is at issue. We faced something similar in Section 3.7. The ratio of specific heats of monatomic gases γ determined from classical theory was in accord with measurements. Not so for polyatomic gases. To resolve this we had to invoke quantum mechanics, according to which (internal) energy levels of molecules are quantized. Similarly, resolution of the Gibbs paradox seems to require invoking quantum mechanics (although some authors think otherwise). At the heart of this paradox is the meaning of *identical, distinguishable,* and *indistinguishable.* Particles (including atoms and molecules) are identical if they have the same *permanent* properties (mass, number of electrons, etc.). Classically, all identical particles are distinguishable in principle in the sense that the trajectory of every particle can be followed forever. The Gibbs paradox is (or appears to be) a consequence of the assumption of identical, distinguishable particles. But according to quantum theory they are indistinguishable. If two identical particles are distinguishable before interaction, they are not after. Imagine that the interaction occurs in a black box. Two particles went in, two came out, but what happened inside is neither known nor knowable. A further complication is that all properties of molecules are not permanent because their internal states change.

Entropy Changes upon Mixing Two Gases with Different Temperatures and Pressures

Now consider the entropy change when two gases, initially with different temperatures and pressures, mix to form a gas with uniform temperature and pressure. This problem is tackled most easily if we take the two gases to be in adjacent volumes separated by a partition. Subscripts 1 and 2 are appended to the thermodynamic variables of these gases before mixing. When the partition is removed, the gases mix to occupy the same volume $V = V_1 + V_2$ and come to the same temperature T. Because this process occurs adiabatically and without any work being done, total internal energy is conserved. It follows from Eq. (4.6) and additivity of entropy that the total entropy change is the sum of two entropy changes, one resulting from a temperature change, the other from a volume change:

$$\Delta S = \Delta S_{temp} + \Delta S_{vol}, \tag{4.55}$$

where

$$\Delta S_{temp} = C_{V1} \ln\left(\frac{T}{T_1}\right) + C_{V2} \ln\left(\frac{T}{T_2}\right) \tag{4.56}$$

and

$$\Delta S_{vol} = N_1 k \ln\left(\frac{V_1 + V_2}{V_1}\right) + N_2 k \ln\left(\frac{V_1 + V_2}{V_2}\right). \tag{4.57}$$

Because $V_1 + V_2$ is greater than both V_1 and V_2, the entropy change, Eq. (4.57), associated with the volume change is necessarily positive. The entropy change, Eq.

(4.56), associated with the temperature change is identical to what we obtained previously (Eq. 4.18):

$$\Delta S_{temp} = C_{V1} \left[\ln \left(\frac{\mu + x}{1 + \mu} \right) + \frac{1}{\mu} \ln \left(\frac{1 + \mu/x}{1 + \mu} \right) \right], \tag{4.58}$$

where μ and x are defined in Eq. (4.19). The quantity μ, the ratio of heat capacities of the two gases, is arbitrary. To determine the sign of the entropy change ΔS_{temp}, we examine the extrema of the function $f(x) = \Delta S_{temp}/C_{V1}$. The first derivative of this function

$$\frac{df}{dx} = \frac{1}{\mu + x} - \frac{1}{x(\mu + x)} \tag{4.59}$$

vanishes only at $x = 1$, and $f(1) = 0$. The second derivative of f

$$\frac{d^2 f}{dx^2} = \frac{1}{(x + \mu)^2} \left(\frac{2x + \mu}{x^2} - 1 \right) \tag{4.60}$$

is positive at $x = 1$. Thus, the minimum value of f is zero only when the initial temperatures of the two gases are the same, but otherwise, f is positive. In this mixing process, therefore, the entropy of the universe always increases, and hence the process is irreversible.

An Entropic Derivation of Joule's Law

For a reversible process

$$\frac{dS}{dt} = \frac{1}{T} \frac{dU}{dt} + \frac{p}{T} \frac{dV}{dt} = \frac{\partial S}{\partial T} \frac{dT}{dt} + \frac{\partial S}{\partial V} \frac{dV}{dt}, \tag{4.61}$$

from which follow

$$\frac{\partial S}{\partial T} = \frac{1}{T} \frac{\partial U}{\partial T} \tag{4.62}$$

and

$$\frac{\partial S}{\partial V} = \frac{1}{T} \left(\frac{\partial U}{\partial V} + p \right) \tag{4.63}$$

because V and T are independent variables. If we differentiate Eq. (4.62) with respect to V and Eq. (4.63) with respect to T, we obtain

$$\frac{\partial^2 S}{\partial V \, \partial T} = \frac{1}{T} \frac{\partial^2 U}{\partial V \, \partial T} \tag{4.64}$$

and

$$\frac{\partial^2 S}{\partial T \, \partial V} = \frac{1}{T} \left(\frac{\partial^2 U}{\partial T \, \partial V} + \frac{\partial p}{\partial T} \right) - \frac{1}{T^2} \left(\frac{\partial U}{\partial V} + p \right). \tag{4.65}$$

The cross partial derivatives of a function $f(x, y)$ are equal,

$$\frac{\partial^2 f}{\partial x \, \partial y} = \frac{\partial^2 f}{\partial y \, \partial x}, \tag{4.66}$$

if f is a continuous function of x and y. With the assumption that S is a continuous function of T and V, Eqs. (4.64) and (4.65) can be set equal to each other. And if U is a continuous function of T and V

$$\frac{\partial U}{\partial V} = T \frac{\partial p}{\partial T} - p. \tag{4.67}$$

This is a general result, which we can apply to liquids as well as gases (see the following paragraph). For an ideal gas

$$\frac{\partial p}{\partial T} = \frac{p}{T}, \tag{4.68}$$

from which follows

$$\frac{\partial U}{\partial V} = 0. \tag{4.69}$$

This is the defining equation for an ideal gas. What we have shown by way of entropy is that Eq. (4.69) is consistent with the ideal gas law, which is comforting, even if expected.

From Eq. (4.67) we can write heating for any fluid in terms of thermodynamic parameters:

$$Q = C_V \frac{dT}{dt} + \left(T \frac{\partial p}{\partial T} \right) \frac{dV}{dt}. \tag{4.70}$$

We now have enough tools to go beyond our discussion of specific heats in Section 3.2, in which we obtained the general result

$$C_p = C_V + \left(\frac{\partial U}{\partial V} + p \right) \frac{\partial V}{\partial T}. \tag{4.71}$$

By using Eq. (4.67), Eq. (4.71) can be written

$$C_p = C_V + T \frac{\partial p}{\partial T} \frac{\partial V}{\partial T}. \tag{4.72}$$

From Eq. (3.60) it follows that

$$C_p = C_V + \frac{TV\alpha^2}{\kappa_T}, \tag{4.73}$$

where the isothermal compressibility κ_T is the relative change in volume with change in pressure

$$\kappa_T = -\frac{1}{V} \frac{\partial V}{\partial p} = \frac{1}{\rho} \frac{\partial \rho}{\partial p}, \tag{4.74}$$

and the isobaric coefficient of thermal expansion α is the relative change in volume with change in temperature

$$\alpha = \frac{1}{V}\frac{\partial V}{\partial T} = -\frac{1}{\rho}\frac{\partial \rho}{\partial T}. \tag{4.75}$$

The compressibility ($1/\kappa_T$ is called the *bulk modulus*) is positive for materials that shrink when subjected to increased pressure. This is the norm, but there are exceptions. For materials with $\kappa_T > 0$, the second term on the right side of Eq. (4.73) is inherently non-negative ($\alpha^2 \geq 0$), and hence, $C_p \geq C_V$. It also follows from Eq. (4.73) that C_p for liquid water is identically equal to C_V at $\approx 4\,°\text{C}$, where the density of water is a maximum, and hence, $\alpha = 0$ (see Fig. 3.9).

The temperature dependence of the coefficient of thermal expansion of water at normal temperatures is why its isobaric and isochoric specific heats are almost equal and why they begin to diverge (Section 3.2) as α increases with temperature (see Fig. 3.8); the compressibility and density don't change much.

The internal energy U of a simple system can be considered a function of entropy S and volume V, and from the first law $\partial U/\partial S = T$ and $\partial U/\partial V = -p$. Some authors *define* T and p by these derivatives, thereby transforming readily measurable and intuitively understandable parameters into abstractions. Pedagogically, this is a step backwards and perpetuates the notion that thermodynamics is a parlor game in which the winner is whoever constructs the largest number of partial derivatives. Extra points are given for making them look as strange as possible.

Entropy and Disorder: A Persistent Swindle

Generations of desperate teachers, faced with the cries of anguished students for an immediate, concise, and readily digestible explanation of what this mysterious quantity entropy really is, have seized upon a pacifier—Don't fret, entropy is just a measure of disorder! But in fact, entropy *is* disorder. The greater the entropy, the greater the disorder. Indeed, entropy used by people who are ignorant of science, as in the examples at the beginning of this chapter, is merely a fancy scientific synonym for disorder, and entropy increases are taken to be indications of increased disorder in all its forms, ranging from a messy room to civil disturbances.

A standard bogus example of entropy increase is the shuffling of a deck of playing cards initially arranged in numerical order for each suit. Even if a deck of identical cards is shuffled it has the same entropy change as a shuffled ordinary deck: zero. But the entropy of the *universe* increases because of the efforts of a shuffler. For another absurd example, take a random array of identical *macroscopic* objects and rearrange them so that they are on a regular lattice. The entropy change of the array is zero whereas if you uncritically swallow the entropy = disorder mantra it should have decreased. What *increases* is the entropy of the universe because a human agent or a machine had to put the objects in order, which required energy transformations (e.g. metabolism, combustion of fuel). The same entropy increase could have been obtained by pushing a peanut with your nose along the ground a certain distance.

Good explanations are often characterized by the reduction of the unfamiliar to the familiar. Those who propose to explain entropy (unfamiliar) by saying it is merely disorder (familiar) are assuming that the consumers of this putative explanation have an intuitive and unambiguous sense of what is meant by disorder. If only this were true!

We began our discussion of entropy by showing that, in an adiabatic free expansion, the entropy of an ideal gas increases. When the volume of such a gas increases, does this unambiguously signal an increase in its disorder? Perhaps it does—to some people. Yet, it is not likely to be obvious to everyone that increased volume (and hence, increased entropy) means increased disorder unless we *define* this to be so. But if we do, we defeat our purpose of supposedly explaining entropy as disorder. That is, we define disorder by means of entropy, not the other way around.

The entropy of an ideal gas increases with temperature. Does this correspond to increased disorder? Again, it depends on what is meant by disorder. As the temperature of the gas increases, so does the mean kinetic energy of its molecules. If we agree that when gas molecules move faster, they are more disordered, we have another correlation between increased entropy and increased disorder. But again, we achieved this correlation by defining disorder in a particular way.

Section 5.3 shows how the entropy per unit mass of a solid, such as ice, is less than the entropy of its liquid at the same temperature, which in turn is less than the entropy of the vapor phase. This is an example in which greater entropy is associated with an increase in what many people might intuitively accept as increased disorder: A (crystalline) solid is more ordered than its liquid, which is more ordered than its vapor.

There is not always a clear connection between entropy increases and intuitively obvious increases in disorder. For example, suppose we have a subcooled liquid isolated from its surroundings. A subcooled liquid exists as a liquid at a temperature below the nominal freezing point (see Chapter 5). Suppose that some of this liquid freezes because of the introduction of a freezing nucleus. According to the second law, the entropy of this isolated system cannot decrease. Yet, we end up with a system that is more ordered, at least in outward appearance. It is indeed true that in the process of partial freezing of the subcooled water, the temperature of the system increases, which corresponds to an entropy increase. But now we have to argue that the entropy increase because of the temperature increase just happens to be greater than the entropy decrease because of the phase transition from liquid water to ice, which is by no means obvious. For this example, we could retain the entropy equals disorder equation, but only by stretching. Now turn to an example that really knocks the correlation between entropy and disorder into a cocked hat. Consider an isolated supersaturated solution—a liquid in which a solid has been dissolved to a concentration greater than it would be for equilibrium. Such a supersaturated solution is unstable (or, as is sometimes said, metastable), but can exist that way more or less indefinitely. A crystal suddenly and spontaneously forms in the solution. Again, the entropy of the system cannot decrease. Yet the appearance of the crystal certainly would be regarded as an increase in order. But in this example the temperature of the system could *decrease*. How can we retain the disorder interpretation of entropy when the system has undergone a partial transition from liquid to solid and its temperature also has decreased? Only by heroic efforts might it be possible to do so.

Our criticisms of the disorder interpretation of entropy are neither new, nor original. As long ago as 1944, K. K. Darrow, in a superb expository article on the concept of entropy noted that "We cannot … always say that *entropy is a measure of disorder* without at times so broadening the definition of 'disorder' as to make the statement true by definition only."

Many years later P. G. Wright published a critical essay on the association of entropy with disorder. He demonstrated by examples that "no exact correlation should be expected between precise quantitative concepts and imprecise qualitative concepts." Entropy is a thermodynamic variable that in principle is measurable. Disorder is not such a variable. Wright draws one of his examples (crystallization in a supersaturated solution) from an earlier paper by M. L. McGlashan, whose criticism of thermodynamics taught as "muddled metaphysics" and "bogus history" makes for great reading. Thus, there have been a few lone voices crying out against the dubious notion that entropy is disorder. Entropy is entropy, just as energy is energy, unpalatable though this may be. To understand entropy, use it in as many ways as possible, until in your mind it is an independent entity, complete and whole in itself, not a mere synonym for the vague concept of disorder. (For another example undermining the notion that entropy is synonymous with disorder, see Problem 31 in Chapter 5.)

Because our heretical remarks about equating entropy with disorder are likely to stir up violent emotions, especially in the countless pedagogues who have resorted to this shaky crutch, we briefly play devil's advocate and criticize the critics. The examples adduced by some of them entailed matter in metastable equilibrium (e.g. sub-cooled liquid, supersaturated solution), about which we have more to say in Section 5.4. It could be argued that entropy is not defined for matter in metastable states, that it is strictly the property of matter in stable equilibrium, and hence, that throwing cold water on the equivalence of entropy to disorder by invoking metastable states is fallacious. However, entropy is just one among several thermodynamic variables. If we argue that entropy is not defined for metastable states, we cannot stop there. We also would have to argue that temperature, pressure, internal energy, and even volume, are not defined. But we would be highly reluctant to abandon the view that the temperature and pressure of, say, subcooled water, are well defined—these variables are measured.

Frank L. Lambert, a retired chemistry professor, devoted the last two decades of his long life to severely criticizing the mantra that entropy is just disorder. And he seems to have convinced many teachers of chemistry to mend their ways. But mantras in circulation for more than a century are harder to kill than cockroaches.

If entropy is a physical *quantity* with dimensions energy divided by absolute temperature, it is in principle measurable. But there are no entropy meters. Entropy changes can be calculated from temperature and pressure measurements given a suitable theory. For example, Eq. (4.7) is a theoretical expression for the entropy change of a uniform ideal gas with a specific heat capacity independent of temperature; Eq. (4.98) is a theoretical expression for the entropy change of a uniform solid with a specific heat capacity independent of temperature. To go beyond these two idealized systems requires much more work, but even then, the possibilities are highly restricted. Although we may believe that the hopping of a kangaroo causes the entropy of the universe to increase, we can neither measure nor calculate it. And sweeping statements about entropy increases must be tempered by the realization that they are almost always unobservable and can only be imagined.

Microscopic Interpretation of Entropy

We now briefly apply the interpretation of entropy in Section 4.1 as the number of microstates per macrostate to an example sometimes tossed off as "disorder." Put a drop of milk at a certain temperature into an insulated cup of coffee at the same temperature. This is an isolated system, and hence, its entropy cannot decrease. In fact, it greatly increases. Before the drop disperses, it had a certain (large) number of microstates limited by its temperature and size (macrostate). And similarly for the coffee. The total number of microstates corresponding to the combined milk drop and coffee macrostates is the product of the two, and hence, their logarithms add (entropy is additive). But after the milk has uniformly dispersed in the coffee, the total number of microstates is much greater. Milk and coffee can occupy regions from which they had been excluded. Based solely on the increase in the number of microstates per macrostate, the entropy of the system has increased. Moreover, although it is not impossible for all the dispersed milk to spontaneously evolve (briefly) to the initial drop, the probability of this happening over the age of the universe is nil. Invoking ill-defined increased disorder is sound-bite science lacking explanatory power. Now we can address the messy desk argument. Consider two sheets of paper. The total entropy of the two taken together is the sum of their entropies. We can move then around and if their temperature is unchanged, the total entropy of the sheets does not change. This is true for any number of sheets, stacked neatly on or strewn haphazardly over a desk.

In his 1930 paper on the symmetry of time in physics (Section 3.1), G. N. Lewis distinguished between laws of physics and laws of nature, as well as between two-way time and one-way time. Time in Eq. (1.1) is two way. No implication that the past determines the future, or vice versa. "In going from the very simple science of mechanics to the very complex science of psychology, we must change from two-way to one-way time." We all get older and eventually die, a law of nature. But entropy is often referred to as "time's arrow" in that in all real processes (e.g. life) the entropy of the universe increases. Processes described adequately by Newtonian mechanics are reversible, whereas in thermodynamics we idealize (i.e. pretend) that reversible processes are possible. Lewis avers that "Gain of entropy always means loss of information and nothing more." This is preceded by a statement about entropy increasing in going from a known to an unknown system. Consider a system composed of three distinguishable molecules in a volume separated by a partition.

Nitrogen and oxygen are on one side, water vapor on the other. This is a known system, and we assume that there is only one way of specifying it. Thus, the multiplicity of the system is 1. Now punch a hole in the partition. All three molecules can be on one side or the other, or two on one side, one on the other. There are eight possibilities. We have gone from a known system to an unknown system. The *multiplicity* has increased, and hence, its entropy has increased. This is one of the "many faces of entropy" mentioned at the beginning of this chapter.

4.2 Entropy Changes of Liquids and Isotropic Solids

We obtained a simple analytical expression for the entropy of an ideal gas as a function of temperature and pressure (or volume) because of the ideal gas law relating pressure, temperature, and volume. We cannot follow the same path for liquids and solids because there is no simple, universal equation relating their state variables. Nevertheless, we can obtain simple approximate expressions for entropy changes of solids and liquids in the kinds of processes of interest to us.

The rate of change of entropy of a homogeneous substance undergoing a constant-volume process is

$$\frac{dS}{dt} = \frac{Q}{T} = \frac{1}{T}\frac{dU}{dt} = \frac{C_V}{T}\frac{dT}{dt}. \tag{4.76}$$

Similarly, for a constant-pressure process

$$\frac{dS}{dt} = \frac{1}{T}\frac{dH}{dt} = \frac{C_p}{T}\frac{dT}{dt}. \tag{4.77}$$

With a bit of algebra we can rewrite Eq. (4.73) as

$$\frac{C_V}{C_p} = 1 - \frac{T\alpha^2}{\rho c_p \kappa_T} = \frac{1}{\gamma}. \tag{4.78}$$

We showed in the previous section that $C_p = C_V$ at the temperature of maximum density of water ($\approx 4\,°C$), so let's pick a somewhat higher temperature, say, 20 °C. At this temperature the compressibility of water is about 5×10^{-10} Pa^{-1}, its coefficient of thermal expansion is 2.1×10^{-4} K^{-1}, its density is 1000 kg m^{-3}, and its specific heat capacity at constant pressure is 4200 J kg^{-1}K^{-1}. With these values in Eq. (4.78) we obtain 0.994 for the ratio of heat capacities. Thus, the two heat capacities of liquid water are the same within less than one percent. What about ice? The density of ice is slightly less than, and its specific heat capacity about half that of, water. The coefficient of thermal expansion of ice is about 1.7×10^{-4} K^{-1} and its compressibility is about 1.2×10^{-10} Pa^{-1}. With these values in Eq. (4.78) we obtain 0.966 for the ratio of heat capacities of ice (0 °C).. Thus, the two heat capacities of ice are the same within a few percent.

Because $C_p \approx C_v$ for a *few* liquids, most notably water, and probably *most* solids (at normal terrestrial temperatures), Eqs. (4.76) and (4.77) suggest that for many processes the rate of change of entropy is approximately

$$\frac{dS}{dt} = \frac{C}{T}\frac{dT}{dt},$$

(4.79)

and we need not worry about whether C is the heat capacity at constant pressure or volume. If C is approximately independent of temperature over the range of interest, we can integrate Eq. (4.79):

$$S - S_0 = C \ln\left(\frac{T}{T_0}\right),$$

(4.80)

where S is the entropy at temperature T and S_0 is the entropy at some reference temperature T_0. To understand why "few" is italicized, see the references at the end of Chapter 3.

The previous arguments suggesting that Eq. (4.79) can be used to determine entropy changes of liquids and solids were quick and dirty. For more convincing arguments, read on.

The change in entropy of a homogeneous system for any process in which temperature and pressure change is approximately

$$\Delta S \approx \frac{\partial S}{\partial T}\Delta T + \frac{\partial S}{\partial p}\Delta p.$$

(4.81)

This equation is essentially the first two terms in a Taylor series expansion of entropy. We rewrite Eq. (4.81) in a form more suitable for our purposes:

$$\Delta S = T\frac{\partial S}{\partial T}\frac{\Delta T}{T} + p\frac{\partial S}{\partial p}\frac{\Delta p}{p}.$$

(4.82)

The magnitude of the ratio

$$p\frac{\partial S}{\partial p}\Big/T\frac{\partial S}{\partial T}$$

(4.83)

tells us the relative contributions to the entropy change in the process for the same fractional changes in temperature $\Delta T/T$ and in pressure $\Delta p/p$.

From the first law of thermodynamics for a reversible process and the definition of entropy, we have

$$\frac{dS}{dt} = \frac{1}{T}\frac{dH}{dt} - \frac{V}{T}\frac{dp}{dt}.$$

(4.84)

By using the chain rule to expand dS/dt in Eq. (4.84) in terms of dT/dt and dp/dt and equating the coefficients of these two derivatives on both sides of Eq. (4.84), we obtain

$$\frac{\partial S}{\partial T} = \frac{1}{T}\frac{\partial H}{\partial T} = \frac{C_p}{T} \tag{4.85}$$

and

$$\frac{\partial S}{\partial p} = \frac{1}{T}\frac{\partial H}{\partial p} - \frac{V}{T}. \tag{4.86}$$

Applying the same arguments to Eqs. (4.85) and (4.86) as those leading from Eqs. (4.62) and (4.63) to Eq. (4.67) yields

$$\frac{\partial H}{\partial p} = V - T\frac{\partial V}{\partial T}. \tag{4.87}$$

This equation, the enthalpic counterpart to Eq. (4.67), combined with Eq. (4.86) gives

$$\frac{\partial S}{\partial p} = -\frac{\partial V}{\partial T}. \tag{4.88}$$

Now we have all the ingredients to determine the ratio Eq. (4.83), which with a bit of algebra becomes

$$p\frac{\partial S}{\partial p}/T\frac{\partial S}{\partial T} = -\frac{p\alpha}{\rho c_p}. \tag{4.89}$$

Does the term on the right side of this equation look familiar? We obtained it in Eq. (3.149) as the ratio of working to heating in a constant-pressure process. We also showed that the magnitude of this ratio for water at one atmosphere is around 10^{-5}. This implies that for any atmospheric process undergone by a liquid or solid (but *not* a gas) in which both temperature and pressure change, the entropy change is overwhelmingly dominated by the temperature change. This provides the justification for using Eq. (4.79) for determining entropy changes of liquids and solids. By analogy with defining an ideal gas by Eq. (4.69), we may define an *ideal solid* by zero compressibility. And why not? The magnitude of departures of real solids from ideality is comparable to departures of real atmospheric gases from ideality (see Section 5.4).

According to a law (usually, but incorrectly, called the Dulong–Petit law) stated by Petit and Dulong in 1819, "the atoms of all simple bodies have exactly the same capacity for heat." In modern terms this is the *molar* specific heat capacity, the value of which they measured by careful and clever cooling experiments to be $3R^* \approx 25$ J mol^{-1} K^{-1}, where R^* is the universal gas constant (Section 2.8). To convert this to J kg^{-1} K^{-1}, divide by the molar mass in kg per mole. The "simple bodies" were 13 elements, most of them metals. At or near room temperature, $3R^*$ is not a bad approximation for many solid elements. For ice the law yields a value about 65% of the measured value. Even the predicted value for salt (NaCl) is only half the measured value.

We now can extend the same trick we played in Section 4.1 for an ideal gas. We can forget that we derived Eqs. (4.85) and (4.88) by way of an ideal (i.e. nonexistent) process. They *define* the entropy change as a function of temperature and pressure. And from *measurements* of $C_p(T, p)$ and $V(p, T)$ we can integrate these equations *provided* there is no phase transformation. We show in Section 5.3 that there is an entropy discontinuity associated with a phase change (liquid to solid, vapor to liquid) at constant temperature.

4.3   Atmospheric Applications of the Second Law

The first law of thermodynamics, the energy of an isolated system is conserved, usually is emphasized more than the second law, the entropy of an isolated system cannot decrease. Yet, the first law by itself is a rudderless ship, adrift in a boundless sea. To show this, and to set the stage for an entropy maximization problem of relevance to the atmosphere, we consider what happens when two *finite* bodies, denoted 1 and 2, with initially different temperatures, are allowed to interact with each other but not with their surroundings (Fig. 4.5). We take this process to occur isobarically. When the composite system comes to equilibrium, what are the uniform temperatures T_{1f} and T_{2f} of its two components for given initial uniform temperatures T_{1i} and T_{2i}, where subscripts i and f denote initial and final? Until we invoke some kind of physical law, we cannot answer this question. All we can say is that the point (T_{1f}, T_{2f}) lies somewhere in the first quadrant in Fig. 4.6. The first law narrows this range of possibilities, but not as much as you might think.

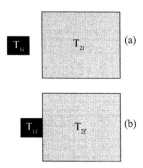

Fig. 4.5: (a) Two bodies with different initial temperatures T_{1i} and T_{2i} are (b) brought into contact. What are the final temperatures T_{1f} and T_{2f} of these bodies if the interaction between them is isobaric?

The actual process of isobaric interaction between the two bodies is irreversible. But each body proceeds from a definite equilibrium state to another such state. We may not know how the bodies traverse thermodynamic space, but we do know their points of departure and eventual destinations. Enthalpy is a state function that depends only on thermodynamic variables (temperature and pressure). The one

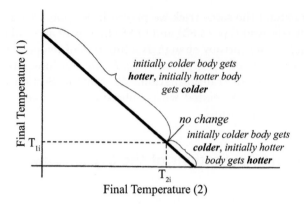

Fig. 4.6: The possible outcomes allowed by the first law of thermodynamics of the thought experiment depicted in Fig. 4.5.

body evolves by inscrutable processes from a state specified by (p, T_{1i}) to a state specified by (p, T_{1f}), and the other body evolves from a state specified by (p, T_{2i}) to a state specified by (p, T_{2f}). Set aside the actual process and assume that body 1 can undergo a reversible, isobaric process from the initial equilibrium state (p, T_{1i}) to the final equilibrium state (p, T_{1f}), and similarly for body 2. This could be achieved, for example, by placing body 1 (or 2) in contact with a series of many reservoirs having arbitrarily small temperature differences beginning with a reservoir at temperature $T_{1i}(T_{2i})$ and ending with a reservoir at temperature $T_{1f}(T_{2f})$. As far as body 1 is concerned, it could just as well have undergone this idealized process instead of the real irreversible process: the end result is the same. From the first law, it follows that for the idealized isobaric process, total enthalpy is conserved because the process undergone by the composite system is adiabatic:

$$\Delta H = 0. \tag{4.90}$$

We assume that the heat capacities (at constant pressure) for the two bodies, C_{p1} and C_{p2}, are independent of temperature. With this assumption, Eq. (4.90) yields

$$C_{p1}(T_{1f} - T_{1i}) + C_{p2}(T_{2f} - T_{2i}) = 0, \tag{4.91}$$

which can be solved for T_{2f}:

$$T_{2f} = \left(\frac{C_{p1}}{C_{p2}} T_{1i} + T_{2i} \right) - \frac{C_{p1}}{C_{p2}} T_{1f}. \tag{4.92}$$

Although we obtained this result for the idealized reversible process, it is also true for the real irreversible process because enthalpy is a state variable: it depends only on the state of the system (as determined by its temperature), not on how it evolved to that state.

The first law only shrinks the range of possible outcomes from any point in a rectangular domain to any point on the straight line Eq. (4.92). But we still are very much at sea. The first law allows for the possibility that the initially colder body will get hotter and the initially hotter body will get colder, or no change in temperature of either body, or the initially colder body will get colder, and the initially hotter body will get hotter. All these qualitatively different outcomes are equally likely if our only guide is the first law. Not very satisfying, is it? To find our way along a line of possibilities to a single point requires invoking the second law.

Again, the real process undergone by the two bodies in contact is irreversible. So, to determine the total entropy change we have to resort to reversible processes that take each body from its initial equilibrium state to its final equilibrium state. In a reversible, *isobaric* process, the heating rate is

$$Q = \frac{dH}{dt} = C_p \frac{dT}{dt}. \tag{4.93}$$

The corresponding entropy change is

$$\Delta S = \int_i^f \frac{C_p}{T} \frac{dT}{dt} dt, \tag{4.94}$$

which can be integrated under the assumption that C_p is independent of temperature over the range of interest:

$$\Delta S = C_p \ln\left(\frac{T_f}{T_i}\right). \tag{4.95}$$

The total entropy change in the idealized reversible process undergone by the composite system of two bodies is

$$\Delta S = C_{p1} \ln\left(\frac{T_{1f}}{T_{1i}}\right) + C_{p2} \ln\left(\frac{T_{2f}}{T_{2i}}\right). \tag{4.96}$$

But again, because entropy is a state variable, the entropy change Eq. (4.96) for the reversible process is the same as for the irreversible process connecting the same initial and final states.

To find the final state for which the entropy is an extremum, differentiate Eq. (4.96) with respect to T_{1f}, use Eq. (4.92) specifying T_{2f} as a function of T_{1f}, and set the result to zero:

$$\frac{d}{dT_{1f}} \Delta S = \frac{C_{p1}}{T_{1f}} - \frac{C_{p2}}{T_{2f}} \frac{C_{p1}}{C_{p2}} = 0, \tag{4.97}$$

which has the solution

$$T_{1f} = T_{2f}. \tag{4.98}$$

To determine if this extremum is a minimum or a maximum, take the second derivative

$$\frac{d^2}{dT_{1f}^2} \Delta S = -C_{p1}\left(\frac{1}{T_{1f}^2} + \frac{C_{p1}}{C_{p2}} \frac{1}{T_{2f}^2}\right) \leq 0, \tag{4.99}$$

from which it follows that the extremum is a maximum.

We conclude from this analysis that when the two bodies reach the same temperature, no further entropy increase is possible. And if we believe the second law, namely, that the entropy of an isolated system cannot decrease, we conclude that equality of temperature characterizes the final equilibrium state of the two bodies in contact. No further evolution (on average) of the composite system is possible because a change would entail a decrease in entropy, which is not allowed by the second law.

We have not explained anything. Our thermodynamic analysis tells us what happens, not why or how fast. Thermodynamics tells us nothing about the detailed mechanisms by which the initially hotter body cools and the initially colder body warms until they both reach the same temperature. Moreover, this equalization of temperature could occur in the blink of an eye or over the age of the universe. Thermodynamics by itself is powerless to predict the rate at which temperature equalization occurs.

At this point you may be saying to yourself, all that we have done with the first and second laws and a bit of mathematics is show what everyone knows (and what we previously assumed): bodies in contact eventually come to the same temperature. This is indeed true, and it is also not surprising given that thermodynamics is firmly rooted in observations of macroscopic systems. We should not be surprised when thermodynamics tells us what everyone knows because this is what went into the making of thermodynamics. But to be clear, the first law by itself is insufficient to show the temperature equalization derived here. We could start with the initial masses in the previous problem having the same temperature, and if by some magic the temperature in each mass were to revert to a different temperature, the first law would not be violated, but the second law would be.

To show that learning about entropy maximization is worth our effort, let us apply this principle to an atmospheric problem for which not everyone knows the answer. As a first step we generalize the previous analysis.

Maximum Entropy: Arbitrary Temperature Distribution in a Solid Slab

We can generalize the results of the previous subsection to determining the temperature profile that maximizes the entropy of an isolated solid slab much larger in lateral extent than its total thickness. We assume that the one-dimensional temperature profile depends only on horizontal distance along the normal to the slab and that the specific heat capacity per unit mass is independent of temperature. What temperature profile maximizes the total entropy? Rather than deal explicitly with integrals (which are limits of sums) we approximate the total entropy S by a finite sum over N layers each with the same mass δm, which from Eq. (4.80) is

$$S = c\delta m(\ln T_1 + \ln T_2 + \ldots + \ln T_N), \tag{4.100}$$

where T_i is the temperature in the ith layer. If the total enthalpy H is constant

$$H - c\delta m(T_1 + T_2 + \ldots + T_N) = 0. \tag{4.101}$$

Our aim is to maximize S subject to the constraint Eq. (4.101). We can solve Eq. (4.101) for the temperature T_k:

$$T_k = H/c\delta m - (T_1 + \ldots + T_{k-1} + T_{k+1} + \ldots + T_N),\qquad(4.102)$$

and hence, can write Eq. (4.100) as

$$S = c\delta m \ln T_1 + \ldots + \ln T_{k-1} + \ln T_{k+1} + \ldots + \ln T_N)$$
$$+ \ln[H - c\delta m(T_1 + \ldots T_{k-1} + T_{k+1} + \ldots + T_N)].\qquad(4.103)$$

Now take the derivative of S with respect to T_j and set it equal to zero:

$$\frac{\partial S}{\partial T_j} = \frac{1}{T_j} - \frac{1}{H - c\delta m(T_1 + \ldots + T_{k-1} + T_{k+1} + \ldots + T_N)} = 0.\qquad(4.104)$$

If each temperature is bounded, then for $N \gg 1$ and $\delta m/N\delta m \ll 1$, the sum in Eq. (4.104) is approximately an integral, and hence, removing a *single* temperature (e.g. T_j) cannot change the value of the limit of this sum. In other words, T_j in the first term contributes its full value to the derivative, whereas its contribution in the second term is swamped by a huge number of temperatures. Thus, each temperature that satisfies Eq. (4.104) in the limit is the continuous variable T. From Eq. (4.101) $cT = H/N\delta m = h$. The second derivative of S with respect to T is negative, so a uniform temperature distribution maximizes the total entropy. This is the simplest example in which an entropy increase is accompanied by energy spreading. The total internal energy of the slab is conserved but initially it is distributed arbitrarily. With time, this energy spreads until it is distributed uniformly and then does not revert to a nonuniform distribution, as predicted by the second law.

We can obtain the same result for the equilibrium temperature profile in an isolated slab by solving Eq. (7.13), which is both time and space dependent. For any arbitrary (bounded) initial temperature profile, the temperature becomes uniform, and we can determine how long this takes for any slab thickness and thermophysical properties. We can solve the same problem for an isolated sphere. Thermodynamics tells us only what the final state is, not how long it takes to get there.

We implicitly assume in our derivation of the equilibrium temperature profile that either the pressure of the slab is uniform, or its entropy is independent of pressure. But suppose that its one-dimensional temperature profile lies along a uniform gravitational field. What then? Solids in textbooks are "incompressible," but in the real world are compressible to a small but measurable degree, about 10 million times smaller than for air at one atmosphere. The entropy of an *ideal* incompressible solid at a fixed temperature cannot depend on pressure. Squeezing it cannot change its volume or internal energy and hence its entropy. To the extent that a *real* solid is incompressible its entropy is independent of pressure. The density of salt is 2170 kg m^{-3}, so $\rho g = 0.21$ atm m^{-1}. The pressure at the bottom of a 1000 m pillar of salt is 210 atm, whereas its compressibility is about 4×10^{-7} atm^{-1}. Thus, our proof by entropy maximization arguments holds to good approximation for isolated vertical solid columns. What about a column of isolated atmospheric air, the specify entropy of which also depends on pressure [Eq. (4.8)] and its pressure depends on height (Fig. 2.6)?

Our derivation of a uniform equilibrium temperature profile in an isolated solid slab is based on a thermodynamic argument, entropy maximization. Despite its name, there is nothing dynamic about thermodynamics. From the (dynamic) Fourier thermal conduction equation (7.11) for a solid body, one with no internal mass motion, which is less general than the second law, energy transfer driven by temperature gradients in an isolated solid body redistributes internal energy until all gradients vanish. But this equation is *not* valid for air.

The equilibrium temperature profile of an isolated column of air in Earth's gravitational field elicits arguments as heated as those in politics and religion. So, we must carefully examine this contentious issue from historical, physical, and experimental points of view.

In his 1871 *Theory of Heat*, James Clerk Maxwell wrote "We find that if a vertical column of gas were to be left to itself, till by the *conduction* [emphasis added] of heat it had attained a condition of thermal equilibrium, the temperature would be the same throughout." But he then goes on to state that

> this result is by no means applicable to the case of our atmosphere ... the effect of winds in carrying large masses of air from one height to another tends to produce a temperature distribution of a quite different kind ... In this condition of what Sir William Thomson [Lord Kelvin] has called the Convective equilibrium of heat, it is not the temperature that is constant, but the quantity ϕ, which determines the adiabatic curves.

That is, Kelvin favored the dry adiabatic lapse rate, Eq. (3.94), which is isothermal only in the limit $g \to 0$.

From the context it is evident that by convection Maxwell and Kelvin had in mind *perceptible* convection (i.e. winds). But following Eq. (7.13) we argue on physical grounds that there is always *imperceptible* convection, and in Section 7.3 we support this by observations. Zero convection in air exists only on paper. Idealizations inevitably lead to inconsistences, paradoxes, and errors. For example, if we insist that absolutely collisionless gases exist, two of them with different temperatures in an insulated container could never evolve to the same uniform temperature.

In what follows convection does not explicitly appear, which would seem to invalidate our conclusions. But this is no different from the Maxwell–Boltzmann equilibrium distribution of molecular speeds, Eq. (2.50), which does not depend on the collision cross section σ. Its *nonzero* value determines only *how long* it would take for a nonequilibrium distribution to evolve to an equilibrium distribution. And it is the same with imperceptibly small convection.

To insist, as some do, even vehemently, that the equilibrium profile of an isolated gas in a gravitational field is isothermal is to believe that gas molecules can magically have the same degree of immobility as molecules in a solid—this is the *solid-atmosphere fallacy*.

Entropy Maximization in the Atmosphere

The entropy of a gas, unlike a solid, depends on temperature *and* pressure. For atmospheric applications of the second law, it is useful to consider entropy as a function of the intensive variables T and p. The specific entropy (entropy per unit mass) can be written from Eq. (4.8) as

$$s - s_0 = c_p \ln \left(\frac{T}{T_0} \right) - R \ln \left(\frac{p}{p_0} \right), \tag{4.105}$$

where c_p is the specific heat capacity and R is the specific gas constant for air. Since potential temperature (Eq. 3.86) is

$$\Theta = T \left(\frac{p_0}{p} \right)^{R/c_p}, \tag{4.106}$$

it follows that

$$s - s_0 = c_p \ln \left(\frac{\Theta}{T_0} \right). \tag{4.107}$$

Further, the second law may be written as

$$\frac{dS}{dt} = C_p \frac{d \ln \Theta}{dt} = \frac{Q}{T}. \tag{4.108}$$

Thus, potential temperature is meteorologists' entropy. Contours of entropy (isentropes or constant Θ contours) are plotted on charts with other meteorological variables, such as pressure and winds, for isentropic analyses used to study atmospheric processes. Some atmospheric models use potential temperature as a vertical coordinate (called an isentropic vertical coordinate system).

In 1908, L. A. Bauer succinctly demonstrated the relationship between potential temperature and entropy and suggested that entropic temperature might be a more appropriate term. But he decided against pushing for this because by then potential temperature had come into common use.

In our previous treatment of entropy maximization in a solid slab, we considered isobaric processes. But this is not an option when dealing with the atmosphere because it is an ideal gas in which pressure varies vertically. To consider vertical entropy variations in air we can use potential temperature as a state variable because it depends on temperature *and* pressure.

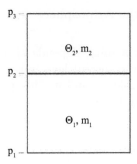

Fig. 4.7: Two air masses extending vertically and initially separated by an imaginary barrier. Potential temperatures Θ_1 and Θ_2 within these two masses are uniform.

The total mass (per unit area) of the atmosphere (on a locally flat Earth) between *any* two altitudes z_a and z_b is

$$\frac{m_{ab}}{A} = \int_{z_a}^{z_b} \rho\, dz. \tag{4.109}$$

Under the assumption of hydrostatic equilibrium, Eq. (4.109) becomes

$$\frac{m_{ab}}{A} = -\int_{z_a}^{z_b} \frac{1}{g}\frac{dp}{dz} dz = -\frac{1}{g}(p_b - p_a). \tag{4.110}$$

Thus, if we consider a layer of the atmosphere between two constant-pressure surfaces (planes), the total mass (per unit area) of this layer is constant. To illustrate entropy changes in atmospheric processes, we begin with a simple example of two masses with different potential temperatures that are aligned vertically as shown in Fig. 4.7. The two layers are initially separated by an insulating, impermeable barrier that is removed to allow the two masses to interact thermally and mix. If the vertical boundaries of our idealized system are constant pressure surfaces, then from Eq. (4.110) they shift in such a way that the mass in each layer remains constant and so does the total mass, $m_1 + m_2$.

The first law for this system is given by Eq. (3.88). Thus, for this isolated system the average potential temperature Θ_s will be conserved such that

$$C_p \frac{d\Theta_s}{dt} = 0 \tag{4.111}$$

and

$$\Theta_s = (m_1\Theta_1 + m_2\Theta_2)/(m_1 + m_2). \tag{4.112}$$

For this adiabatic process Θ_1 and Θ_2 will change when the barrier separating the two masses is removed, but Θ_s will remain constant. The second law provides the constraint needed to determine the final Θ distributions. Using potential temperature changes to calculate the entropy (from Eq. 4.101) for this adiabatic process we write

$$S = c_p m_1 \ln\Theta_1 + c_p m_2 \ln\Theta_2 + S_0. \tag{4.113}$$

We can determine the final state for which entropy will be a maximum by differentiating Eq. (4.113) with respect to Θ_1 and setting the result to zero to give

$$\frac{\partial S}{\partial \Theta_1} = c_p m_1 \frac{1}{\Theta_1} + c_p m_2 \frac{1}{\Theta_2}\frac{\partial \Theta_2}{\partial \Theta_1} = 0. \tag{4.114}$$

From Eq. (4.112), where Θ_s is constant,

$$\frac{\partial \Theta_2}{\partial \Theta_1} = -\frac{m_1}{m_2}. \tag{4.115}$$

Using this result in Eq. (4.114) gives

$$\frac{m_1}{\Theta_1} - \frac{m_2}{\Theta_2}\left(\frac{m_1}{m_2}\right) = m_1\left(\frac{1}{\Theta_1} - \frac{1}{\Theta_2}\right) = 0. \tag{4.116}$$

Thus, in the final equilibrium maximum entropy state

$$\Theta_1 = \Theta_2 = \Theta_s. \tag{4.117}$$

and the potential temperature distribution will be uniform for this vertical mixing case. But again, thermodynamics by itself does not tell us how the system will evolve in time to the equilibrium state of maximum entropy, although we do know that if the initial Θ_1 is greater than Θ_2, the system will be unstable and will quickly overturn to give a constant potential temperature profile.

To bring the entropy maximization problem to the real atmosphere, consider the potential temperature profile for a layer in the atmosphere. Figure 4.8 shows an idealized layer with a potential temperature profile that is generally statically stable but has smaller-scale variability at all heights. The layer is defined by a constant pressure surface at the top of p_2 and at the bottom of p_1. The mass in this layer

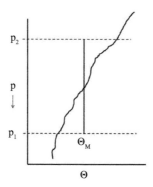

Fig. 4.8: An arbitrary potential temperature profile that has a mean potential temperature of Θ_M in the layer extending from pressure level p_1 to p_2.

(per unit area) from Eq. (4.110) is

$$\delta M = \frac{(p_1 - p_2)}{g} = \frac{\delta p}{g},$$ (4.118)

where $\delta p = p_1 - p_2$. The potential temperature averaged over the depth of this layer can be defined as

$$\Theta_M = \frac{1}{\delta p} \int_{p_2}^{p_1} \Theta dp.$$ (4.119)

The potential temperature profile in the layer can then be written as

$$\Theta(p) = \Theta_M + \Theta'(p)$$ (4.120)

where $\Theta'(p)$ is the deviation of Θ at any pressure from the mean and the integral of Θ' will be zero. Suppose that the layer is isolated from its surroundings. neither heated nor cooled by radiation or by interaction with the adjacent air (or ground). For this adiabatic process $(Q = 0)$ and the first law given by Eq. (3.88) gives

$$\frac{d\Theta_M}{dt} = 0.$$ (4.121)

To determine the equilibrium temperature profile in this layer, we must apply the second law. The total entropy (per unit area) of the layer is the integral of the specific entropy (entropy per unit mass):

$$S = \frac{c_p}{g} \int_{p_2}^{p_1} \ln \Theta \, dp,$$ (4.122)

where additive constants are omitted, and we are being sloppy by taking the logarithm of a quantity with dimensions. The mean specific entropy of the layer s_M from Eq. (4.122) using Eq. (4.120) can be written as

$$s_M = \frac{c_p}{\delta p} \int_{p_2}^{p_1} (\ln(\Theta_M + \Theta'(p))) \, dp.$$ (4.123)

The integrand in this equation can be rewritten as

$$\ln(\Theta_M + \Theta'(p)) = \ln \Theta_M \left(1 + \frac{\Theta'(p)}{\Theta_M}\right) = \ln \Theta_M + \ln \left(1 + \frac{\Theta'(p)}{\Theta_M}\right),$$ (4.124)

so that Eq. (4.123) can be written as

$$s_M = \frac{c_p}{\delta p} \int_{p_2}^{p_1} \left[\ln \Theta_M + \ln \left(1 + \frac{\Theta'(p)}{\Theta_M}\right)\right] dp$$ (4.125)

or

$$s_M = c_p \ln \Theta_M + \frac{c_p}{\delta p} \int_{p_2}^{p_1} \ln \left(1 + \frac{\Theta'(p)}{\Theta_M}\right) dp.$$ (4.126)

Because Θ'/Θ_M is small ($\Theta' \sim 3K; \Theta_M \sim 300K$), the ln function can be expressed as a Taylor series expansion

$$\ln(1+x) = x - \frac{1}{2}x^2 + \frac{1}{3}x^3 - \frac{1}{4}x^4 + \dots. \quad \text{for } -1 \le x \le 1. \tag{4.127}$$

With this expansion s_M, Eq. (4.126) is approximately

$$s_M \approx c_p \ln \Theta_M + \frac{c_p}{\delta p} \int_{p_2}^{p_1} \left[\frac{\Theta'(p)}{\Theta_M} - \frac{1}{2} \frac{\Theta'(p)^2}{\Theta^2_M} \right] dp, \tag{4.128}$$

where the higher-order terms in the series expansion have been dropped. But since by definition

$$\int_{p_2}^{p_1} \frac{\Theta'(p)}{\Theta_M} dp = 0, \tag{4.129}$$

specific entropy averaged over the layer can be written as

$$s_M \approx c_p \ln \Theta_M - \frac{c_p}{\delta p} \int_{p_2}^{p_1} \left[\frac{1}{2} \frac{\Theta'(p)^2}{\Theta^2_M} \right] dp. \tag{4.130}$$

But

$$\sigma_\Theta^2 = \frac{1}{\delta p} \int_{p_2}^{p_1} \Theta'^2 dp \tag{4.131}$$

is the potential temperature variance in the layer and it follows that

$$s_M \approx c_p \ln \Theta_M - \frac{1}{2} \frac{c_p}{\Theta^2_M} \sigma_\Theta^2. \tag{4.132}$$

Because the variance is positive, the entropy of the layer will be a maximum when the variance is zero and potential temperature in the layer is constant with pressure (height). It follows from Eq. (4.132) that

$$\frac{ds_M}{dt} \approx c_p \frac{d \ln \Theta_M}{dt} - \frac{1}{2} \frac{c_p}{\Theta^2_M} \frac{d\sigma_\Theta^2}{dt}. \tag{4.133}$$

Because the second law requires $ds_M/dt \ge 0$ and $d\ln \Theta_M/dt = 0$, it follows that $d\sigma_\Theta^2/dt < 0$. Thus, the potential temperature variance can only decrease in a dry isolated atmospheric column. A constant potential temperature maximizes the entropy of an isolated layer of the atmosphere. Could we have guessed this without appealing to entropy maximization? Stability considerations alone would not lead to this result. We probably would not guess that a neutral layer corresponds to maximum entropy, yet a constant potential temperature is merely the dividing line between stability and instability. Strictly by analogy with the uniform equilibrium temperature profile of an isolated slab we would predict that for a gas in a gravitational field the equilibrium profile would be uniform potential temperature.

Entropy maximization requires the equilibrium temperature of an isolated atmospheric layer subjected to mixing with no condensation or evaporation of water to decrease with height at the dry adiabatic rate. This result might appear to be at odds with our everyday experiences. Why isn't the equilibrium profile isothermal? If a solid is isolated from its environment and initially has a nonuniform temperature, conduction eliminates all temperature gradients. But in the atmosphere, energy transfer in an isolated layer is dominated by convection, rather than conduction. Lord Kelvin recognized this in an 1862 paper on the convective equilibrium of temperature in the atmosphere:

> The particles composing any fluid mass are subject to various changing influences, in particular of pressure, whenever they are moved from one situation to another. In this way they experience changes of temperature altogether independent of the effects produced by the radiation and conduction of heat. When all parts of the fluid are freely interchanged and not sensibly influenced by radiation and conduction, the temperature of the fluid is said to be in a state of convective equilibrium.

Energy transfer in the atmosphere and other fluids differs fundamentally from energy transfer in solids, in which mass motion is absent. Because pressure decreases with height in the atmosphere (a consequence of gravity) and air is fluid and compressible, ascending parcels expand (cool) and descending parcels compress (warm). Thus, the example of conduction in solids can impede our understanding of atmospheric convection. As noted previously, convection in the real atmosphere may be imperceptible, but it is never zero. Convection is not necessarily tornadic winds or even gentle breezes. Even Brownian motions are a form of micro-convection. The origin of the notion that the equilibrium temperature profile of an isolated column in a gas is the same as that in a solid reflects a failure to distinguish between a gas and a solid. We call this the *solid atmosphere fallacy*.

To show how dry convection produces a dry adiabatic lapse rate in the atmosphere, consider an isolated layer that is stable, and hence, has a lapse rate less than dry adiabatic. Suppose that a parcel moves from one level to another within this layer and then mixes with its surroundings. The temperature of parcels is not conserved as they move up or down. If a parcel moves to a higher level and then mixes with the air at that level, the layer is destabilized (i.e. the lapse rate increases) because a rising parcel cools at the dry adiabatic rate, and hence, will be cooler than its surroundings. Similarly, if a parcel moves to a lower level and then mixes with the air at that level, the layer is also destabilized because a descending parcel warms at the dry adiabatic rate, and hence will be warmer than its surroundings. Mixing therefore increases the entropy of the layer even though it is stable. Only in the absence of gravity would the dry adiabatic lapse rate be zero and the equilibrium temperature profile isothermal.

A few points must be added to this description of mixing in the atmosphere. The potential for mixing to occur in a layer depends on its stability and wind shear (variation of wind velocity with height). Mixing is much slower in a stable than in an unstable layer. In the free atmosphere we routinely observe stable layers but rarely observe unstable layers. Why? An unstable layer (if unsaturated) would quickly mix to the dry adiabatic rate. For a stable layer with little wind shear, however, mixing is suppressed, which allows other processes (e.g. radiation, advection) to play a larger role in the vertical temperature profile. In our idealized treatment of entropy maximization of an isolated layer we can ignore motions within the layer and the rate at which the equilibrium state is reached. We have all the time in the world. But we cannot ignore processes other than mixing if we attempt to explain observed temperature profiles. Globally averaged tropospheric and stratospheric temperature profiles, for example, can be explained as a consequence of radiative–convective equilibrium with the additional complication of condensation and evaporation of water. Furthermore, horizontal and vertical advection and energy transfer from the surface can substantially affect local vertical temperature profiles. But regardless of how other processes alter air temperature, local mixing can only increase entropy and, for dry processes, force the temperature profile closer to the dry adiabatic lapse rate.

Although our discussion has focused on the maximum entropy state of an isolated layer, well-mixed layers are most often observed near Earth's surface, where there may be substantial energy transport to and from the layer. Over land, for example, heating of the surface by solar radiation during the day can result in the formation of a shallow, very unstable layer near the surface. Convection generated in this layer can penetrate and stir the air above to form a well-mixed (constant Θ) layer. This convection can also transport energy into the layer from the surface and cause the mixed layer to warm and deepen as the temperature of the surface increases. This example illustrates the danger of invoking conduction or processes that depend on gradients to explain the vertical transfer of energy in the atmosphere because energy transports are large even though Θ (or equivalently, dry static energy) is constant with height in the mixed layer. Parcels close to the surface can be heated and thus be warmer than air near the bottom of the mixed layer. As these positively buoyant parcels rise, they cool at the dry adiabatic rate. But because temperature in the mixed layer also decreases at this rate, the rising parcels remain warmer than the mixed layer. Thus, as they mix with surrounding air, they increase the temperature of the mixed layer even though there is no gradient of Θ. This process is different from conduction in a solid. Although the example given here is associated with warming of the surface by solar radiation, mixed layers also can form and be maintained as cold air moves over a warmer surface. Figure 4.9 illustrates an example of a mixed layer maintained in this way, a potential temperature profile obtained from an instrumented aircraft flown off the coast of California where strong winds advected colder air over relatively warmer water. In oceanic regions like this the boundary layer is often fairly moist, resulting in the formation of clouds at the top of the mixed layer. Strong radiative cooling near the tops of

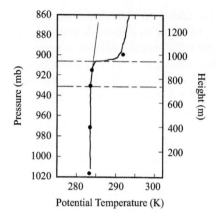

Fig. 4.9: Potential temperature profile calculated from temperature and pressure measurements obtained from an instrumented aircraft flown about 500 km off the coast of California. The solid circles are from ten-minute averages flown at constant-pressure levels in the same air mass but at different times. The sounding (solid line), from data taken during a slow descent of the aircraft over about five minutes, was obtained before the constant-level measurements were made. A shallow stratus cloud (indicated by dashed lines) was present in the top 200 meters of the mixed layer. Consequently, potential temperature in this part of the layer increases slightly because the lapse rate is less than the dry adiabatic lapse rate due to condensation of water vapor. The sea surface temperature is only about $14\,°C$ ($\Theta = 285.5$ K), but colder air advected into the study area resulted in a positive energy flux at the surface, which helped maintain a well-mixed boundary layer. Radiative cooling at the top of the thin cloud also may have enhanced mixing.

these clouds can further promote convection, which helps maintain a mixed layer even if surface heating becomes small.

As noted previously, the second law does not provide any information on the time rate of change to a maximum entropy state associated with adiabatic processes in the atmosphere. If a layer is absolutely unstable $(d\Theta/dz < 0)$, it will quickly mix adiabatically to a constant potential temperature state. But a very stable layer $(d\Theta/dz > 0)$ can maintain its structure for a week. Thus, from a prognostic perspective, the second law does not provide any time constraints. The second law, however, does have practical implications in maintaining mixed layers once they are formed. A mixed layer cannot be unmixed by adiabatic processes.

This point can be illustrated by considering deep mixed layers that form over the Saharan Desert due to strong surface heating and then are advected westward across the Atlantic by easterly winds in a deep layer. These layers are also exceptionally dry and often laden with dust. Deep mixed layers generated over the Sahara are called Saharan Air Layers (SALs). An example of a SAL and its persistence over time is shown in Fig. 4.10, where a deep mixed layer seen in atmospheric soundings made from Senegal has features that are observed over Barbados six days later. In the Barbados soundings of the elevated mixed layer observed from about 1.8 to 2.8 km can be tied to the deep mixed layer observed previously over Africa. The potential temperature in the layer has decreased due to infrared radiative cooling, but the

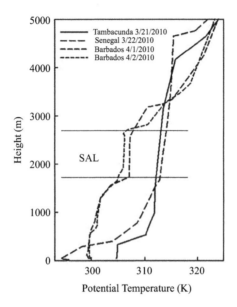

Fig. 4.10: Potential temperature profiles from upper-air soundings from 1) Tambacounda, Senegal on 21 March 2010 at 0900 UTC; 2) Dakar, Senegal on 22 March 2010 at 1200 UTC; 3) Grantley Adams International Airport (GAIA), Barbados on 1 April at 1200 UTC; and 4) GAIA, Barbados on 2 April at 12 UTC. The nearly constant potential temperature layers observed over Senegal were generated by dry convective mixing over the Saharan desert where surface temperatures were greater than 40 °C. Deep mixed layers generated over the Sahara are called Saharan Air Layers (SALs). The Tambacounda soundings shows modification of the SAL near the surface as it is advected over cooler woodland areas in this part of Senegal, and the Dakar sounding shows even cooler air near the surface, since Dakar is a coastal location where the sea surface is cooler than inland areas. This same air mass was then observed over Barbados (about 4500 km west) ten days later as it moved westward at a rate of about 5 ms^{-1}. The top of the SAL has descended about 1700 meters during the ten days, which corresponds to a vertical velocity of about 2 cms^{-1}. During this movement of the air mass across the Atlantic, the lowest 1.7 km was substantially modified by energy (and moisture) transports from the ocean surface and by radiation and dry and moist convection. The SAL between 1.7 and 2.7 km remains well mixed but has cooled by radiative processes during the ten days.

mixed layer structure remains. The potential temperature structure below this layer has been modified by energy transports from the surface and moist processes. SAL layers have important impacts on hurricane formation. In general, mixed layers generated over heated land surfaces are very prevalent and have important implications for weather forecasting since these layers can persist for days.

Thermodynamic Efficiency: The Carnot Cycle

The mass of the atmosphere per unit area of Earth's surface is approximately p_0/g, where p_0 is the surface pressure and g is the acceleration due to gravity. If we take the average surface pressure to be about 10^5 Pa, the mass of the atmosphere (per unit area) is about 10^4 kg m^{-2}. The mean radius of Earth is about 6400 km, which corresponds to a surface area of about 5×10^{14} m^2, and hence, the total mass of the atmosphere is about 5×10^{18} kg.

The specific heat capacity of air (at constant pressure) is about 1000 J kg^{-1}K^{-1}. Thus, every one-degree change in average temperature of the entire atmosphere corresponds to an enthalpy change of about 5×10^{21} J. This enthalpy change occurring over one year (3×10^7s) corresponds to about 2×10^{14} W, or 2×10^8 MW. A yearly change in the mean temperature of the atmosphere of only 0.1 °C is equivalent to 20,000 1000-MW power plants. And this number pales in comparison with what we obtain when we consider temperature changes of the oceans, with 1000 times the mass and four times the specific heat capacity per unit mass of the atmosphere.

Why do we bother to burn coal and oil in power plants when the atmosphere and oceans are vast reservoirs of energy? By lowering the temperature of the atmosphere a mere 0.1 °C per year, we could generate a staggering amount of power and at the same time mitigate the effects of global warming. Why aren't engineers beavering away at schemes for extracting energy from the atmosphere and the oceans? Is this a consequence of a giant conspiracy among OPEC, oil-refining companies, the coal industry, and automobile manufacturers to deprive the world of breakthroughs that would release enormous quantities of cheap but hitherto untapped energy? Why is all the energy in the atmosphere and oceans, free for the asking, going to waste? Scandalous, is it not?

It is indeed true that according to the first law, the atmosphere and oceans are vast reservoirs of energy, and this law places no restrictions on our ability to tap these reservoirs. The second law, unfortunately, throws cold water on grandiose schemes permitted by the first law. To show this, we must take a closer look at just what is meant by the efficiency of power generation.

Thermodynamic internal energy is disorganized energy, not in a form suitable for many activities in which humans engage. To get something useful to us, this disorganized energy must be transformed into organized energy. An automobile that bucked back and forth and sideways violently would possess kinetic energy but would not take its occupants anywhere. To be useful, the center of mass of the automobile must move mostly in one direction.

We consider the simplest conceptual device that transforms disorganized thermodynamic internal energy into organized energy capable of turning electric power generators, driving tractors, trucks, automobiles, airplanes, and an endless variety of machines, from giant steel mills to lawn mowers.

As our system we take a fixed mass of a working fluid, which could be a gas or a liquid. This system undergoes a *cycle*, by which is meant that its state changes in a series of processes such that the initial and final states are the same. The particular cycle of interest is made up of two isothermal and two adiabatic processes, all assumed to be reversible. To aid our thinking, it is helpful to represent this cycle on a pressure–volume (*pv*) diagram (Fig. 4.11). The working fluid is an ideal gas so that the diagram reflects reality, rather than artistic license. But the proof that follows is independent of the working fluid.

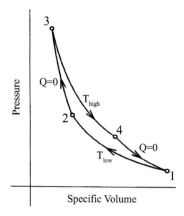

Fig. 4.11: Carnot cycle undergone by an ideal gas. The cycle is illustrated by a closed curve in *pv* space formed by the intersection of two isotherms (T_{high} and T_{low}) with two adiabats ($Q = 0$). This Fig. is an idealized indicator diagram (see Section 3.1) for a *zero-power* thermodynamic engine: finite work done in infinite time.

The system begins its cycle at point 1 and undergoes an isothermal compression at constant temperature T_{low}. We can imagine the system to be compressed while in contact with a reservoir at this temperature. At point 2, the system is further compressed, but this time adiabatically to point 3, where it then expands isothermally at temperature $T_{high} > T_{low}$. Again, this expansion may be imagined to occur while the system is in contact with a reservoir at this temperature. At point 4, the system is isolated from the reservoir and further expands adiabatically back to the starting point. And on and on forever. The implicit assumption that a working fluid can be compressed or expanded isothermally at the temperature of a reservoir and yet $Q \neq 0$ is an idealization.

Because the process is cyclic, the total working is, from the first law, the negative of the total heating:

$$W_{tot} = -Q_{tot}. \tag{4.134}$$

The total heating in the isothermal process between states 1 and 2 must be negative in order to maintain constant temperature in a compression. Thus, for this process

$$Q_{low} = -\int_1^2 Q \, dt \qquad (4.135)$$

where Q_{low} is positive. For the high-temperature isothermal process we have

$$Q_{high} = -\int_3^4 Q \, dt \qquad (4.136)$$

where Q_{high} is positive. Two of the processes are adiabatic, and hence, do not contribute to the total heating. The total heating over the cycle is

$$Q_{tot} = Q_{high} - Q_{low} \qquad (4.137)$$

which, from Eq. (1.79), yields

$$W_{tot} = -(Q_{high} - Q_{low}). \qquad (4.138)$$

Negative total working means that work is done by the system, which is certainly necessary if the cycle represents the operation of an engine.

What is the efficiency of this cyclic process? This depends on what is meant by efficiency. And it also depends on the desired output of the engine. We have to agree that this desired output is W_{tot}. To maintain a reservoir at an elevated temperature requires doing something that results in the expenditure of resources, for example, burning coal or gasoline or oil. Once a fuel has been burned, it cannot be burned again. The most economical use of fuels results when W_{tot} is as large as possible relative to Q_{high}. The low-temperature reservoir need not cost anything: it could be the atmosphere or a large body of water. We may define the efficiency of this thermodynamic engine as the ratio of the desired result, the magnitude of W_{tot}, divided by what it costs in energy units:

$$\eta = \frac{|W_{tot}|}{Q_{high}} = \frac{Q_{high} - Q_{low}}{Q_{high}}. \qquad (4.139)$$

The efficiency in the form of Eq. (4.139) is not expressed in terms of readily measurable thermodynamic variables. So let us go a step further.

Because entropy is a state variable, the change in entropy of the system is zero in a cycle. This total zero entropy change is made up of separate entropy changes:

$$\Delta S = \Delta S_{12} + \Delta S_{23} + \Delta S_{34} + \Delta S_{41} = 0. \qquad (4.140)$$

Neither of the two reversible adiabatic processes contributes to the total entropy change. The entropy change in the low-temperature isothermal process is

$$\Delta S_{12} = \int_1^2 \frac{Q}{T_{low}} dt = -\frac{Q_{low}}{T_{low}}. \qquad (4.141)$$

The entropy change in the high-temperature isothermal process is

$$\Delta S_{34} = \int_3^4 \frac{Q}{T_{high}} dt = \frac{Q_{high}}{T_{high}}. \tag{4.142}$$

From Eqs. (4.140), (4.141), and (4.142) it follows that

$$\frac{Q_{low}}{T_{low}} = \frac{Q_{high}}{T_{high}}. \tag{4.143}$$

With Eq. (4.143), the efficiency Eq. (4.139) becomes

$$\eta = 1 - \frac{T_{low}}{T_{high}}. \tag{4.144}$$

This efficiency is often called the *thermodynamic* or *Carnot efficiency*, and the corresponding cycle is called the *Carnot cycle*. The Carnot cycle on a temperature-entropy diagram (TS) is a rectangle, two isotherms and two isentropes (adiabats), from which Eq. (4.144) follows more readily. Although the work done in a cycle is proportional to the area of this rectangle, we have yet to see an instrument that measures it. Work is easier to measure from the area of a pressure-volume diagram (see Section 3.1).

The simplest explanation of why engines cannot be 100% efficient except in the unattainable limit $T_{low} \to 0$ is as follows. By definition, an engine operates in a cycle ($\Delta U = 0$). Consequently, Q is both positive and negative over different parts of the cycle, the exception being the limiting case in which $Q_{low} \to 0$. In this limit the low-temperature reservoir is in its lowest energy state and therefore energy cannot be transferred to it.

All the separate processes in the cycle were taken to be reversible. Real processes are irreversible. We showed previously that in at least one irreversible process (free expansion) work was required to restore the system to its original state. This result, together with our intuitive feelings about entropy gained from having considered its behavior, led us to postulate that the Carnot efficiency Eq. (4.144) for an idealized cycle is greater than that for any real cycle operating between the same two temperatures. And in fact, no one has ever constructed an engine more efficient than predicted by Eq. (4.144), which sets an upper limit to the efficiency of a thermodynamic engine that transforms disorganized energy into organized energy. The practical utility of the Carnot efficiency is that it puts an end to fruitless searches for a working fluid or cycle that will yield high efficiencies. The Carnot efficiency is independent of the working fluid and the cyclic process it undergoes.

If only the good die young, Sadi Léonard Nicolas Carnot (b. Paris, 1796; d. 1832) must have been very good. Dispatched to an early grave by cholera, Carnot lived not more than two years longer than Mozart, another candle that burned briefly but brightly. Unlike Mozart, Carnot's fame was almost entirely posthumous. During his life he published but one short monograph, *Reflections on the Motive Power of Fire and on Machines Fitted to Develop this Power*, which received favorable initial reviews (1824) and then was mostly forgotten for ten years until Clapeyron (whom we shall meet in Chapter 5) resuscitated it, putting Carnot's mostly qualitative arguments in a more analytical form. Yet although Clapeyron was himself a steam-engine engineer, and surely understood and appreciated the implications of Carnot's work for engineering practice, neither Clapeyron nor any other French engineers were markedly influenced by it. Only when this work was resuscitated for a second time in the middle of the nineteenth century by William Thomson (later Lord Kelvin) and Rudolf Clausius were the originality and profundity of Carnot's ideas fully appreciated, by then too late for him to enjoy the fruits of fame, even though he would have been comfortably middle-aged. Carnot's monograph, which he took pains to make accessible to as wide an audience as possible, still repays reading today, even though its theoretical underpinning is the discredited caloric theory according to which heat is a weightless fluid that can be neither created nor destroyed.

The pv diagram Fig. 4.11 provides a graphical means of visualizing the work done by a thermodynamic engine in a cyclic process. From the definition of the rate of working in a reversible process we have

$$w_{tot} = \int_1^2 p\frac{dv}{dt}dt + \int_2^3 p\frac{dv}{dt}dt + \int_3^4 p\frac{dv}{dt}dt + \int_4^1 p\frac{dv}{dt}dt, \tag{4.145}$$

where w_{tot} is the (positive) work done per cycle per unit mass of working substance. Equation (4.145) is often written compactly as

$$w_{tot} = \oint p\,dv, \tag{4.146}$$

where the circle intersecting the integral sign denotes integration around a cycle (in a clockwise sense for w_{tot} to be positive). According to Eq. (4.145) or (4.146) the work done in any cyclic process is proportional to the area enclosed by the closed curve representing the process on a pv diagram, called an indicator diagram (Section 3.1).

Now we have all that is necessary to answer the question posed at the beginning of this section: Why can't we extract energy from the atmosphere and oceans? The naive answer is that they would have to serve as both high- and low-temperature reservoirs. According to Eq. (4.144), the Carnot efficiency of an engine operating between these two reservoirs would be zero. Reality, although bleak, is not quite this bleak. There are finite temperature *differences* in both the atmosphere and the oceans. For example, the difference between the surface air temperature and that at 5 km might be 30–40 K. For an average temperature of 300 K, the corresponding Carnot efficiency is about 10%, which is the maximum attainable. A real engine exploiting this temperature difference would be less efficient. Similarly for the oceans. From the surface to a depth of about 1 km, the temperature may decrease by about 20 K, which corresponds to a Carnot efficiency less than 10%. It also has been pointed out by G. F. R. Ellis that the clear night sky is a low-temperature reservoir. The equivalent blackbody temperature of the clear sky is often taken to be about 250 K. If the mean surface temperature is 280 K, a 30 K difference could be exploited, but the associated Carnot efficiency would be low.

Devices could be constructed to exploit natural temperature differences. But how? This could be done only with turbines and pipes made of steel or aluminum, which would have to be designed, built, and maintained. Given the low Carnot efficiency of such devices, the investment in resources is not likely to yield a sufficient return. Schemes for generating power by exploiting ocean temperature differences have been promoted for over half a century. These schemes even have inspired their own acronym: OTEC (Ocean Thermal Energy Conversion). Presumably, adherents of OTEC hope to break the back of OPEC, but this has yet to occur, and we suspect that at least part of the reason is the fundamental limit imposed by the Carnot efficiency.

The atmosphere acts as a thermodynamic engine. Although Earth is in apparent radiative equilibrium, at low latitudes there is a net gain of radiant energy of about 100 Wm^{-2} and at high latitudes there is a net loss of 100 Wm^{-2}. This differential heating drives atmospheric motions. Surface temperatures near the equator are about 300 K and temperatures where the energy loss occurs at higher latitudes are about 260 K, giving a Carnot efficiency of about 13%. Because the net energy into the system at higher temperatures is about 100 Wm^{-2}, the rate of kinetic energy generation at this efficiency would be 13 Wm^{-2}. If kinetic energy were generated at this rate, the wind speeds in the atmosphere would be substantially higher than observed (see Problem 17). In the atmosphere the global kinetic energy dissipation rate, which balances the generation rate, is about 2–5 Wm^{-2}. Thus, the efficiency of the atmospheric thermodynamic engine is closer to 5%. The Carnot efficiency is not reached because on average the expansions and compressions in the global engine are not fully adiabatic. Hurricanes (tropical cyclones) can also be considered as thermodynamic engines where energy comes into these systems from inflow (expansion) at the ocean surface at temperatures of about 300 K. Energy loss from hurricanes occurs in the outflow (compression) at the tropopause at temperatures close to 200 K. Thus, the Carnot efficiency of these systems is about 33%. In very strong hurricanes, the ascent (expansion) in the eyewall and the descent (compression) in the outflow area are adiabatic so that efficiencies in hurricanes can approach the Carnot value.

Refrigerators and Coefficient of Performance

Equation (4.138) describes an ideal engine by means of which work is done *by* a system in a cycle. Run the cycle backwards and work is done *on* the system, which becomes a refrigerator. Its *coefficient of performance* is defined as the ratio of what is wanted relative to what it costs. Q_{low} is the energy transferred from the low-temperature reservoir to the system in one part of the cycle, whereas Q_{high} is the energy transferred from the system to the high-temperature reservoir (i.e. the surroundings at a higher temperature than the space to be refrigerated) in another part of the cycle. The cost of Q_{low} is the work done on the system, and hence, the coefficient of performance is

$$\xi = \frac{Q_{low}}{W_{tot}} = \frac{Q_{low}}{Q_{high} - Q_{low}} = \frac{1 - \eta}{\eta}. \qquad (4.147)$$

Rewrite this as

$$W_{tot} = \frac{Q_{low}}{\xi}. \tag{4.148}$$

The aim is to minimize W_{tot} for a given Q_{low}, and hence, ξ to be as large as possible. It can be greater than 1 if $\eta < 0.5$ and *infinite* if $\eta = 0$. At first glance this seems absurd. The lower the efficiency of the system run as an engine, the greater the coefficient of performance of the system run backwards as a refrigerator. A refrigerator allows for the possibility of doing what would not happen spontaneously. If we were to put the high-temperature reservoir in thermal contact with the low-temperature reservoir, the energy transport would be from high to low. By definition reservoirs are infinite (an idealization) so consider finite reservoirs that are brought into contact. Without intervention the temperature of the high-temperature reservoir decreases and the temperature of the low-temperature reservoir increases. A refrigerator can reverse this at the expense of doing work that contributes to monthly electricity bills.

Q_{high} is the energy transferred (in part of the cycle) from the system to the high-temperature surroundings (reservoir). Q_{low} is the energy transferred (in another part of the cycle) from the low-temperature reservoir to the system. The net effect is energy transfer from a low-temperature reservoir to a high-temperature reservoir. This does not violate the second law because the system is not isolated.

What is the significance of an infinite coefficient of performance? All it means is that $Q_{high} = Q_{low}$. In other words, it didn't cost anything to do nothing—hardly earth-shattering or miraculous.

The cold compartment of an ordinary refrigerator is an insulated box. Other than the resident anaerobic elf who turns the light on or off when the door is opened or closed, no one lives in a refrigerator, so it doesn't have to exchange air with its surroundings (unlike a livable house). The cold reservoir is the compartment, the hot reservoir is the kitchen. During part of the cycle, the system (the refrigerant) must be at a lower temperature than the space to be kept cold but during the other part of the cycle must be at a higher temperature than the surroundings. This does not violate the second law. The energy transfer in each part is in the approved direction. But the overall effect is *as if* a cold object were heating a hot object. This apparent violation of the second law is because of the work done. This law is often misunderstood. It doesn't say that the entropy of a system can only increase, it says that the entropy of the *universe* can only increase. Driving a refrigerator is an unseen distant power plant heedlessly increasing the entropy of the universe.

This analysis does not tell us how to design and make refrigerators, only that they are possible. Unlike perpetual motion machines, refrigerators are not forbidden by any physical laws.

The coefficient of performance of a refrigerator is difficult to understand because when it is infinite the refrigerator appears to not perform at all. For a better understanding consider the coefficient of performance ξ_c of a Carnot refrigerator:

$$\xi_c = \frac{T_{low}}{T_{high} - T_{low}}. \tag{4.149}$$

The lower the temperature T_{low} of the refrigerator compartment *required* to keep food from spoiling and the higher the temperature T_{high} of the surroundings (kitchen), the lower the Carnot coefficient of performance, which makes sense: the more you want, the more you have to pay for it. $T_{high} = T_{low}$ corresponds to a food storage bin, not a refrigerator. For any refrigerator worthy of the name, $T_{low} < T_{high}$. Confusion about the coefficient of performance is a consequence of ignoring what refrigerators are intended to do. If $T_{low} = 276$ K and $T_{high} = 295$ K, $\xi_c = 14.5$. To determine how close actual coefficients of performance are to this ideal value requires comparing the ratio of *measured* Q_{low} to *measured* W_{tot}. Measurements we have seen show that this ratio is appreciably lower than the theoretical value.

Thermodynamic Efficiencies of Real Engines

We may define a thermodynamic engine as a device that does work in a cycle by harnessing two constant temperature reservoirs with temperatures T_{high} and T_{low}, where $T_{low} < T_{high}$. An engine so defined is still an idealization, but less so than the Carnot engine, which sometimes is used in this extended sense. But this is misleading because this is a *specific* engine defined by the cycle (Fig. 4.11) in which the reservoir temperature and the working fluid temperature are *equal* during diabatic compression and expansion.

The Carnot engine is a marvelous device that does the maximum possible cycle work for a given ratio T_{low}/T_{high}. But there is a catch: the cycle time is infinite, and hence, the average *power* over a cycle is zero. The Carnot efficiency is only a first estimate of the limits of the possible. What are the limits of finite-power engines?

It is plausible that the efficiency of a thermodynamic engine depends *monotonically* on T_{low}/T_{high}. We probably would be astonished if the efficiency had local maxima and minima, which are not the same as the greatest value (1) and the least value (0). Because the efficiency must be less than the Carnot efficiency for a given temperature ratio, an educated guess for the simplest expression that satisfies a minimum set of requirements is $1 - (T_{low}/T_{high})^n$, where $n < 1$. And indeed, Curzon and Ahlborn obtain $n = 1/2$ for the efficiency at *maximum* power by analyzing a cycle similar to the Carnot cycle but occurring over a finite time. They assume that all processes are reversible, and a cycle composed of two isotherms—but for temperatures different from the two reservoir temperatures—and two adiabats. Even if you don't understand this analysis or agree with the assumptions underlying it, $n = 1/2$ is consistent with physical reasoning. And it lies halfway between the exponent for the least efficient ($n = 0$) and the most efficient ($n = 1$) cycles, and so is the best guess given the assumed form of the efficiency. That is, $p = 1/2$ minimizes the average square of $n - p$ for equally probable n between 0 and 1. More important, calculated efficiencies for three *real* power plants are remarkably close to observed efficiencies, which are appreciably less than the Carnot efficiencies.

The ideal Rankine cycle rather than the ideal Carnot cycle more adequately approximates real cycles in which high-temperature (600 °C and higher) and high-pressure (250 atm and higher) steam drives turbines to generate electrical power. The Rankine cycle is discussed in engineering thermodynamics textbooks.

Lapse Rate in Water

We can generalize Eq. (3.93) to terrestrial bodies of water. From the first law for an adiabatic process,

$$\frac{dH}{dt} - V\frac{dp}{dt} = 0. \tag{4.150}$$

Divide by the mass M and replace t by z, the distance into the fluid, the positive z-axis upward for air, downward for water. Using the hydrostatic equation in the form $dp/dz = \rho g$, we obtain

$$\frac{dh}{dz} = g. \tag{4.151}$$

In general,

$$\frac{dh}{dz} = g = \frac{\partial h}{\partial T}\frac{dT}{dz} + \frac{\partial h}{\partial p}\frac{dp}{dz} = c_p\frac{dT}{dz} + \rho g\frac{\partial h}{\partial p}, \tag{4.152}$$

and hence,

$$\frac{dT}{dz} = \frac{g}{c_p}\left\{1 - \rho\frac{\partial h}{\partial p}\right\}. \tag{4.153}$$

For an ideal gas $\partial h/\partial p = 0$, and except for sign, Eq. (4.153) is the dry adiabatic lapse rate in Earth's atmosphere. From the definition of enthalpy and the isothermal compressibility

$$1 - \rho\frac{\partial h}{\partial p} = \kappa_T\left(\frac{\partial U}{\partial V} + p\right) \tag{4.154}$$

From Eqs. (3.51) and (4.73)

$$\kappa_T\left(\frac{\partial U}{\partial V} + p\right) = T\alpha, \tag{4.155}$$

and hence, the adiabatic lapse rate is

$$\frac{dT}{dz} = \frac{g}{c_p}T\alpha. \tag{4.156}$$

As a check, for a gaseous atmosphere in a uniform gravitation field, this reduces to the dry adiabatic lapse rate (except for sign) because the isobaric coefficient of thermal expansion of an ideal gas is $1/T$.

We apply this to a cold lake well below the surface, where we assume a temperature of 278 K at which the thermal expansion coefficient of water is 16×10^{-6} K^{-1}. This corresponds to a lapse rate (strictly a rate of increase) 1.6×10^{-4} Km^{-1}. To no surprise, this is much smaller than the dry adiabatic lapse rate in air. Maybe this is only an accident, although we chose a plausible temperature without preconceptions, this rate of increase of temperature is remarkably close to what oceanographers seem to take for granted. For example, an oceanographic paper begins with the assertion that "at sufficient ocean depths (several kilometers) there is an increase of temperature with depth corresponding to the adiabatic rate $\Gamma \approx 10^{-6} \,^\circ\mathrm{C}\ \mathrm{cm}^{-1}$." The lapse rate in ocean water is complicated by salinity.

Annotated References and Suggestions for Further Reading

A rebuttal by Neil Postman to our complaint at the beginning of this chapter about how the term entropy has been mangled by popular use is in Neil Postman (1995) *The end of education.* New York: Knopf, p. 77: "The physicists describe all this [entropy] in mathematical codes and do not always appreciate the ways in which the rest of us employ their ideas of entropy and negentropy. Still, the universe is as much our business as theirs ..."

Two books devoted exclusively to entropy are Henry A. Bent (1965) *The Second Law.* Oxford University Press, and J. D. Fast (1968) *Entropy.* 2nd edn. Philadelphia: Gordon and Breach.

For criticisms of the now-widespread, and hence, virtually ineradicable notion that entropy is synonymous with disorder, see Karl K. Darrow (1944) "The concept of entropy," *American Journal of Physics,* 12, pp. 183–96; M. L. McGlashan (1966) "The use and misuse of the laws of thermodynamics," *Journal of Chemical Education,* 43, pp. 226–32; P. G. Wright (1970) "Entropy and disorder," *Contemporary Physics,* 11, pp. 581–8; Frank L. Lambert (1999) "Shuffled cards, messy desks, and disorderly dorm rooms—examples of entropy increase? Nonsense!" *Journal of Chemical Education,* 76, pp. 1385–7; Frank L. Lambert (2002) "Disorder—A cracked crutch for supporting entropy discussions," *Journal of Chemical Education,* 79, pp. 187–92; Dan Styer (2019) "Entropy as disorder: the history of a misconception," *The Physics Teacher,* 57, pp 454–8.

Before accepting that there is a unanimous opinion about entropy, see Robert H. Swendsen (2011) "How physicists disagree on the meaning of entropy," *American Journal of Physics,* 79, pp. 342–8. He "encountered a remarkable diversity of opinion for a subject that is well over a century old." What he says about entropy applies to much of science: "when people discuss the foundations of statistical mechanics, the justification of thermodynamics, or the meaning of entropy, they tend to assume that the basic principles they hold are shared by others."

For a very readable treatment of the molecular interpretation of entropy, requiring little mathematics, see Eric Johnson (2018) *Anxiety and the equation: understanding Boltzmann's entropy.* Cambridge, MA: MIT Press.

For Planck's account of his hypothesis about the statistical form of entropy see Max Planck (1950) *Scientific autobiography and other papers.* London: Williams and Norgate, pp. 41–2.

The relation between potential temperature and entropy was first pointed out by Louis Bauer (1908) "On the relationship between 'Potential Temperature' and 'Entropy'," *Physical Review*, 26, pp. 177–83.

Our discussion of entropy maximization in the atmosphere was inspired by F. K. Ball (1956) "Energy changes involved in disturbing a dry atmosphere," *Quarterly Journal of the Royal Meteorological Society*, 82, pp. 15–29.

A translation by R. H. Thurston of Carnot's monograph under the title *Reflections on the motive power of heat* was published by The American Society of Mechanical Engineers in 1943.

For a sketch of the life of Sadi Carnot, see James F. Challey's entry in the *Dictionary of scientific biography*. An insightful article on Carnot's work is by F. C. Frank (1966) "Reflections on Sadi Carnot," *Physics Education*, 1, pp. 11–18. The curious lack of impact Carnot's monograph had on the work of steam engine engineers, even Clapeyron, who did as much as anyone to bring Carnot's ideas to a wider audience, is discussed by Milton Kerker (1960) "Sadi Carnot and the steam engine engineers," *Isis*, 51, pp. 257–70.

For the use of the night sky as a low-temperature reservoir, see G. F. R. Ellis (1979) "Utilization of low-grade thermal energy by using the clear night sky as a heat sink," *American Journal of Physics*, 47, pp. 1010–11.

Ocean thermal energy conversion (OTEC) is discussed by Roger H. Charlier and John R. Justus (1993) *Ocean energies: environmental, economic and technological aspects of alternative energy sources*. New York: Elsevier.

For a derivation of the efficiency of a finite-power thermodynamic engine, see F. L. Curzon and B. Ahlborn (1975) "Efficiency of a Carnot engine at maximum power output," *American Journal of Physics*, 43, pp 22–4. For a simpler derivation, see D. C. Agrawal (2009) "A simplified version of the Curzon–Ahlborn engine," *European Journal of Physics*, 30, pp. 173–9. His expression for the efficiency, a different monotonic function of $x = T_{low}/T_{high}$, is 2/3 at $x = 0$ and zero at $x = 1$. And yet it yields efficiencies close to those calculated by Curzon and Ahlborn for real power plants. As a practical matter both efficiencies are the same over a realistic range of x. Expansions of both efficiencies in powers of $\delta = 1 - x$ agree to terms of order δ^2.

For the rate of increase of temperature deep within the ocean, see Mark Wimbush (1970) "Temperature gradient above the ocean floor," *Nature*, 227, pp. 1042–3.

For the coefficient of thermal expansion of water, see George S. Kell (1975) "Density, thermal expansivity, and compressibility of liquid water from 0 °C to 150 °C. Correlations and tables for atmospheric pressure and saturation reviewed and expressed on 1968 temperature scale," *Journal of Chemical and Engineering Data*, 20, pp. 97–105.

For more about the Petit–Dulong law, see Roberto Piazza (2019) "The strange case of Dr. Petit and Mr. Dulong," *Quaderni di Storia della Fisica*, 21, pp. 79–110. [You can find this by searching the Internet].

The effects of the hydrological cycle on the efficiency of Earth's atmospheric thermal engine are considered by Oliver Pauluis (2010) "Water vapor and mechanical work: a comparison of Carnot and steam cycles," *Journal of the Atmospheric Sciences*, 68, pp. 81–102.

A full treatment of hurricanes as Carnot engines is given by Kerry A. Emanuel (1988) "The maximum intensity of hurricanes," *Journal of the Atmospheric Sciences*, 45, pp. 1143–55. A tutorial on this topic is provided by Kerry A. Emanuel (2006) "Hurricanes: tempests in a greenhouse," *Physics Today*, 59, pp. 74–5.

About 20 years ago Harvey Leff and Frank Lambert independently began promoting the interpretation of entropy as an energy spread function, which meshes with the examples in this book. In a 2012 series of five highly recommended short expository articles about entropy ("Removing the mystery of entropy," *The Physics Teacher*, Vol. 50, pp. 28–31, 87–90, 170–2, 215–17, 274–6) Leff says much of what we say about entropy, but sometimes in different ways. Understanding is often enhanced by the same message expressed differently. A longer, more technical paper is from 2007: "Entropy: its language and interpretation," *Foundations of Physics*, 37, pp. 1744–66. See also his 2021 *Energy and entropy: a dynamic duo*. Boca Raton: CRC Press.

For a student laboratory measurement of the coefficient of performance of a real refrigerator, see Albert A. Bartlett (1975) "Introductory experiment to determine the thermodynamic efficiency of a household refrigerator," *American Journal of Physics*, 44, pp. 55–8.

Problems

1. The boilers for the earliest steam engines were not much different from ordinary tea kettles. Estimate the maximum efficiency of such steam engines.

2. Suppose that you want to increase the efficiency of an engine operating between two temperatures, T_{low} and T_{high}. You have two choices: You can decrease T_{low} by a given amount ΔT or you can increase T_{high} by the same amount. Which would you choose in order to obtain the greatest increase in efficiency? *Hint*: This question is easy to answer by applying simple calculus.

3. In Chapter 1 we asserted without explanation that the so-called energy crisis is really an entropy crisis. You now should be able to explain what we meant by this.

4. The atmosphere has been likened to a giant thermodynamic engine in which disorganized energy is transformed into the organized kinetic energy of the winds. The general circulation of the atmosphere is driven by temperature differences between the polar and equatorial regions. Estimate the maximum efficiency of the atmospheric engine. Do this for both summer and winter in

the Northern Hemisphere. On the basis of your calculations, what would you predict for the relative strength of winds in summer and winter? Keep in mind that in the equatorial region there are no seasonal changes comparable to those at mid-latitudes.

5. The specific entropy of liquid water is greater than that of solid water (ice) at the same temperature. This is true at $0\,°C$, and would also be true at any temperature. That is, the entropy of ice at, say, $-20\,°C$ is less than that of liquid water at $-20\,°C$. Why, therefore, doesn't ice in surroundings at $-20\,°C$ spontaneously melt? What is special about $0\,°C$ from a thermodynamic point of view? What is wanted here is an argument based on the *proper* use of the second law of thermodynamics. *Hints*: The enthalpy of water is greater than that of ice at the same temperature. You may assume that this enthalpy difference is approximately independent of temperature. Assume that this hypothetical melting of cold ice takes place at constant pressure in surroundings at constant temperature. You may also assume that the entropy difference between ice and water is approximately independent of temperature.

6. Many years ago one of the authors used to drive regularly through Yuma, Arizona. Outside town he would pass a strange orange grove. Scattered among the orange trees were large towers onto which were mounted propellers driven by engines. The scene was that of a flight of silent airplanes mysteriously hovering over the orange grove.

 The purpose of these propellers (huge fans) is to prevent frost damage on clear, cool nights. During such nights, air near the surface is cooled as the ground cools because of net longwave radiation. Since air higher in the atmosphere may be relatively unaffected by surface cooling, air near the surface can become very stable. How are fans used to keep the air near the surface warmer than it would be otherwise? At first glance, this seems crazy: aren't fans used for cooling?

 To show that these fans can yield the desired results, you must support your arguments with calculations. Consider a situation where at sunset air temperature near the surface is $10\,°C$ and the lowest several hundred meters of the atmosphere is well mixed. At sunrise, many hours later, air near the surface has cooled to $-5\,°C$ while the air temperature at 50 m and above is unchanged (you may assume that temperature varies linearly with height between the surface and 50 m). Now turn on the fans. Can this result in a surface temperature above freezing? *Hints*: This is a potential temperature problem. Do not be disturbed if your simple analysis leads to a temperature discontinuity. This corresponds to a very thin inversion layer.

7. In his delightful book *The Simple Science of Flight*, Henk Tennekes asserts on pp. 117–18 that

> The colder the air the better. The efficiency of jet engines improves as the difference between the intake temperature and the combustion chamber increases . . . This fact has far-reaching consequences for jetliners. The coldest air is found in

the lower stratosphere, about 10 kilometers ... The temperature there is about 55 °C below zero.

Estimate the change in efficiency of jet engines solely as a consequence of flying at 10 km instead of, say, at 3 km. You will have to guess or look up the temperatures in the combustion chamber of a jet engine. Keep in mind that if the intake air is colder at the higher altitude, so will be the air in the combustion chamber.

8. We are so accustomed to thinking of energy as being conserved and entropy as increasing that we may lose sight of the fact that the converse is possible. In what kind of a general process is the entropy of a system conserved, but its internal energy is not? This is a simple and straightforward question, not a trick.

9. You may have encountered the assertion that according to the second law of thermodynamics you cannot unscramble an egg. Yet, with a bit of thought, you should be able to devise a very simple method for unscrambling an egg. After you have done so, try to reconcile the supposedly impossible unscrambling of an egg with the second law of thermodynamics.

10. The so-called heat pump is essentially a refrigerator. In winter a heat pump functions as a heater, the low-temperature reservoir being the outside environment, the high-temperature reservoir being the inside of a house. In summer, a heat pump functions as an air conditioner, the low-temperature reservoir being the inside of a house, the high-temperature reservoir being its outside environment. Obtain an expression for the coefficients of performance of a heat pump acting as a heater and as an air conditioner. Estimate values for these coefficients of performance. See the previous problem.

11. You sometimes encounter assertions that a refrigerator "transfers heat from a cold body to a hot body" (in seeming defiance of the second law of thermodynamics). Is this assertion literally true? *Hint*: The easiest way to answer this question is to carefully consider each step of a Carnot cycle on a pv diagram. For simplicity, take the working fluid to be an ideal gas. Ask yourself if, at any point of the cycle, energy is transferred from a cold to a hot body by virtue of direct interaction between the two.

12. The author of a newspaper article on heat pumps (see Problem 20 in Chapter 1) asserts that "new models are so efficient they produce up to four units of heat for every one unit of electricity they use." This statement seems false but is in fact true. Convince yourself that it is true (see Problem 11), but misleading. Someone reading this article and taking it at face value might conclude that burning fuel oil (or natural gas) to heat a house in winter is madness. Why burn oil when you can use a heat pump and get four times as much heating? This statement is true, but incomplete. Please discuss.

13. We derive the Carnot efficiency by entropy arguments without assuming a particular working fluid. We can derive it more easily, although less generally, by assuming that the working fluid is an ideal gas. All we need is the first

law, the assumption that C_v is independent of temperature, which is *not* a requirement for a gas to be ideal, and the ideal gas law. With these assumptions derive the Carnot efficiency. It helps to use Fig. 4.11 as a guide.

14. Divide Eq. (4.73) by mass to express the specific heat capacities on a per unit mass basis and calculate the ratio c_p/c_v for a few solids at room temperature to convince yourself that it is (or is not) close to 1. Assume that reported heat capacities for solids are c_p because this is likely what was measured. Pick a few metals and a few insulators. You can find all the properties you need online.

15. We state in Section 1.1 that "Our entropy, presumably lower than it would be if we were reduced to a uniform gas, is paid for by an increase in the entropy of our surroundings because of our existence." The entropy of a human can neither be measured (who wants to volunteer?), nor calculated. What simple argument supports (but does not prove) that live humans are in a low entropy state from which they will ultimately ascend?

16. Consider a solid object in free space, far from any stars (including the Sun). Assume that its specific heat capacity is independent of temperature and that its rate of sublimation is negligible (constant mass). Because the object radiates, its temperature must drop, and hence, its entropy must *decrease*. This appears to be a violation of the second law. What is the only physically plausible way in which the second law can be saved? This problem is intended to stimulate you to make a discovery solely by careful thinking and believing that the second law is valid.

17. Consider a column extending through the depth of the atmosphere where kinetic energy is generated at the rate of 10 Wm^{-2}. If air in the column is initially at rest, what would be the average wind speed after five days if there were no kinetic energy dissipation?

18. In Problem 3.26 we state that to determine the speed of sound in a gas, the appropriate compressibility is the adiabatic (isentropic) compressibility κ_S, rather than the isothermal compressibility κ_T. This is also true for liquids. Show that $\gamma = \kappa_T/\kappa_S$. *Hints*: Eq. (4.85) is an expression for C_V as a function of the rate of change of S with respect to T in a constant pressure process. From this you should be able to guess at the rate of change in a constant volume process. This is a problem in which subscripts on partial derivatives are helpful. You need to use Eq. (3.60) but with different variables. This is the kind of boring mathematical problem we try to avoid because it doesn't teach you much about the physical world. It is only tedious manipulation of partial derivatives. The result may be useful because from the speed of sound in any liquid, its density, and its isothermal compressibility, one can determine the ratio of specific heat capacities.

5

Water and Its Transformations

Once upon a time, a well-meaning but misguided soul concocted a simplistic explanation of cloud formation: air rises, cools, and because cold air can't "hold" as much water vapor as hot air can, the vapor condenses, and a cloud forms. This explanation, designed to save those to whom it is directed from the burden of thought, has spread like wildfire, consuming the power to think wherever its flames have reached. Now it is difficult to find anyone who does not believe that air is like a sponge with a variable capacity that increases with temperature. Misleading terms such as *saturation vapor pressure* strengthen this belief.

Once an explanation, no matter how faulty, has appeared in print three times it becomes an immutable truth, and a minor industry keeps it alive. For example, a colleague sent us a paragraph from a book in which a physicist—who ought to have known better—carefully explains why hot air can hold more water vapor: as the air is heated, it expands, the separation between molecules increases, thereby allowing more room for water molecules. And a former student showed us a book in which air was represented as a set of boxes. The higher the air temperature, the bigger the boxes.

Examine this nonsense in light of what can be found on the first page of Wilhlem Ostwald's *Solutions* (1891): "two gases can always form homogeneous mixtures with one another in all proportions, provided they do not combine together chemically." Dalton's law of partial pressures (1802) also implies that air is not like a sponge with a limited capacity (Section 2.8). The sponge analogy is faulty and has been known to be so for more than two centuries.

Air is not some kind of sponge with a limited carrying capacity for water vapor because at normal temperatures and pressures the average separation between air molecules is about ten molecular diameters. Only about one part in a thousand of the volume of air is occupied by something palpable (Section 2.5). Air, hot or cold, is so thinly populated with molecules that it can easily accommodate many more, especially water vapor, which is only $\approx 1\%$ of air.

Consider what you must believe or be perplexed about if you uncritically accept that air is a sponge with a carrying capacity that increases with temperature:

Atmospheric Thermodynamics. Second Edition. Craig F. Bohren and Bruce A. Albrecht, Oxford University Press.
© Craig Bohren and Bruce Albrecht (2023). DOI: 10.1093/oso/9780198872702.003.0005

1. At a relative humidity of 100% all pores in the air sponge are filled with water vapor. Therefore, gasoline, benzene, alcohol, and thousands of other volatile compounds cannot evaporate into air with 100% relative humidity because there is no room for them. NO VACANCY.

2. At a given temperature, air above a solution in equilibrium holds less water vapor than air of the same temperature above pure water. Air somehow senses when it is above a solution and shrinks its pore sizes accordingly.

3. The air around a small droplet in equilibrium holds more water vapor than that around a large droplet (in Section 5.10, we clarify what is meant by "small" and "large"). Air somehow senses the size of droplets and can shrink or expand its pores accordingly.

4. Although air is a sponge that can hold only so much water vapor, by some mysterious alchemy it sometimes can increase the size and number of its pores. This is the phenomenon of *supersaturation*, essential to the formation of clouds (Section 5.11). Adherence to the sponge theory of air makes it difficult to understand (or even accept) supersaturation.

Confused? If so, blame the sponge theorists who have passed on nonsense dressed up as knowledge. The notion that water vapor "dissolves" in or is absorbed by air also is misleading and helps perpetuate the sponge analogy. Water vapor in air is *not* like salt dissolved in liquid water—the agent that *causes* salt to dissolve. Its solubility in different liquids varies by a factor of a million. But all the different kinds of molecules in air flit about in the same space with plenty of room to spare. In what follows, we attempt to demolish nonsense and tell the truth about water vapor in air.

5.1 Evaporation and Condensation of Water Vapor

We begin our discussion of water vapor in air with an analogy that does not break down under scrutiny. Consider a large cylindrical vessel of constant cross-sectional area A (Fig. 5.1). Wine from a huge cask flows from a spigot at a steady rate into the vessel and at the same time leaks from a hole in its bottom. The vessel is so tall that no matter how much we open the spigot, the vessel never overflows. Initially, there is no wine in the vessel, so none can leak from it. But as wine accumulates, the rate of leakage L increases with the height h of the column of wine.

Because the cross-sectional area of the vessel is constant, the rate of change of the volume of wine in it at any instant is A times dh/dt. The volume of the wine column increases at the volumetric inflow rate I and decreases at the leakage rate L:

$$A\frac{dh}{dt} = I-L(h).\tag{5.1}$$

The condition for (dynamic) equilibrium is that h does not change with time. Wine flows into and out of the vessel at the same rates. Denote by h_e the height of the

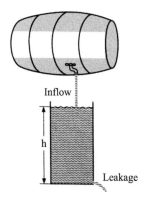

Fig. 5.1: Wine flows at a constant rate from the large cask into a leaky smaller vessel. When h is such that inflow balances leakage, the height of the wine column is constant.

wine column when equilibrium is reached. From Eq. (5.1) we obtain an implicit relation between h_e and I:

$$L(h_e) = I. \tag{5.2}$$

As I increases, so does h_e, because L is a monotonically increasing function of h. If the height of the wine column were steady and we were to open the spigot wider, the column would climb to a higher level until inflow and outflow again balance. With this analogy in mind, we turn to water vapor in air.

Consider a beaker partly filled with liquid water and covered tightly (Fig. 5.2). The space above the water contains no air and, initially, no water vapor. The molecules in the water are moving randomly in all directions, but not all with the same kinetic energy. The forces on a molecule within liquid water are repulsive at very short distances, attractive at greater distances, the range of both forces of order molecular dimensions. On average, the resultant of both forces is zero, but there are always fluctuations, as there must be if molecules jostle around, exhibiting small-scale accelerations driven by fleeting net forces. At the surface, however, attraction

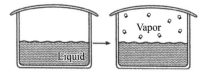

Fig. 5.2: Initially no vapor is above the liquid, but because molecules escape from the liquid, the concentration of vapor molecules increases until the rate of return (condensation) of vapor molecules to the liquid balances their rate of escape (evaporation).

wins out over repulsion (if it did not, the liquid would not be cohesive) but not completely, only on average. Liquids, like gases (Section 2.4), are composed of molecules with energies *distributed* around an average. Cohesion is not absolute. Every now and then a water molecule near the surface acquires enough kinetic energy in collisions with its neighbors to escape their attraction. Unlike molecules within water, those at its surface are not completely surrounded by neighboring molecules impeding their flight from the liquid. Water molecules that shake off their bonds enter the space above the liquid. This process is *evaporation*; the reverse process, the return of water vapor to the liquid, is *condensation*. To meteorologists evaporation usually means *net* evaporation: the *difference* between evaporation and condensation. This is understandable given that usually net evaporation is measured. For example, the level of water in a pan can be monitored over time, and a decrease in this level is a consequence of net evaporation. But failure to distinguish between net evaporation and evaporation is, we believe, a mistake partly responsible for all the nonsense about water vapor in air. Evaporation and condensation are physically distinct processes, essentially independent of each other. Evaporation depends on the state of the liquid; condensation depends on the state of the vapor above it.

The mass of water vapor M_v in the space above the liquid at any instant increases because of evaporation at a *fixed* rate E (if the water temperature is constant) and decreases because of condensation at a *variable* rate C that depends on M_v:

$$\frac{dM_v}{dt} = E - C(M_v). \tag{5.3}$$

All we need to know about C is that it increases monotonically with increasing M_v. C increases with time because evaporated water molecules enter the space above the water, thereby increasing M_v. Eventually, evaporation and condensation balance:

$$C(M_{ve}) = E, \tag{5.4}$$

where the subscript e indicates equilibrium. Equation (5.4) is an implicit relation between the equilibrium water vapor content in the space above the liquid as a function of evaporation rate. M_{ve} increases monotonically with E because C increases monotonically with M_v:

$$\frac{dM_{ve}}{dE} > 0. \tag{5.5}$$

M_{ve} is analogous to h_e, the equilibrium height of wine in the leaky vessel; the evaporation rate E is analogous to I, the rate of inflow of wine; and the condensation rate C is analogous to L, the rate of leakage of wine.

Evaporation is a spontaneous *sorting* process. Purely by chance some molecules acquire sufficient kinetic energy to escape from the liquid phase, at least temporarily. The higher the temperature T of the *liquid*, the greater the fraction of molecules with sufficiently high kinetic energies to escape, and hence, the higher the evaporation rate (i.e. $dE/dT > 0$). From Eq. (5.4)

$$\frac{dC}{dM_{ve}}\frac{dM_{ve}}{dT} = \frac{dE}{dT},\tag{5.6}$$

from which it follows that

$$\frac{dM_{ve}}{dT} > 0.\tag{5.7}$$

Misinterpretations of Eq. (5.7) are behind all the drivel about hot air having the capacity to hold more water vapor. There is *no* air in this thought experiment. What happens if air is added to the space above the liquid? *Nothing*, to very good approximation (Sections 5.8 and 5.10). The equilibrium concentration of water vapor is almost identical to what it would be if water molecules had evaporated into a vacuum. The presence of air, which is not a sponge, is largely immaterial. Equation (5.7) implies only that the concentration of water vapor *in equilibrium* with liquid water increases with its temperature because evaporation (E) increases exponentially with temperature (Section 5.3).

An implicit assumption underlying Eq. (5.7) is that the temperature of the liquid water remains constant during evaporation. But even if it is insulated, the more energetic (on average) molecules leave the liquid and the less energetic (on average) molecules are left behind, which is manifested by a temperature decrease of the liquid and a lower evaporation rate. Evaporation is inherently a self-limiting process.

Like a ball thrown upward, molecules that escape liquid water gain potential energy and lose some of their kinetic energy. Within the water they must have a minimum kinetic energy in order to escape. We also implicitly assume that those relatively few molecules that do escape from liquid into air quickly come into equilibrium with it at a nearly constant temperature. That is, the air is an infinite, constant temperature bath. Reality is more complicated.

If evaporation is a cooling process ($dT/dt < 0$), condensation is a warming process ($dT/dt > 0$). As water vapor molecules approach within a few molecular diameters of water, they are attracted by it and increase their kinetic energy as they enter the liquid. The difference between evaporation and condensation (net evaporation) determines if there is net warming or cooling of the liquid water. A thermodynamic analysis of (idealized) evaporative cooling and condensational warming (Section 5.3) supports the molecular interpretation.

5.2 Measures of Water Vapor in Air

The water vapor content of air may be specified in different ways. We could give the *absolute humidity*, expressed as a number density n_v of water molecules or as a mass density ρ_v. But the numbers so obtained are expressed in units that lack the soothing comfort of familiarity. Meteorologists are inordinately fond of temperatures and pressures, which is unsurprising given that they are readily measured, whereas number density and mass density are not. As a consequence, the concentration of water vapor in air is often expressed as a *vapor pressure* and given the symbol e. From vapor pressure (and temperature) absolute humidity follows:

$$n_v = \frac{e}{kT}, \quad \rho_v = \frac{e}{R_v T}, \tag{5.8}$$

where R_v is the gas constant of water vapor. Vapor pressure and absolute humidity are equivalent humidity variables (for a given T), although the units of e are more familiar.

If the concentration of water vapor in air is equal to the equilibrium value (at the temperature of the air), the corresponding vapor pressure is given a special symbol e_s and a special name: *saturation vapor pressure*. No significance should be attached to "saturation." This is a poorly chosen term that has led countless people astray. If we could burn all the meteorology textbooks and begin anew, we would call e_s the *equilibrium vapor pressure*, which signals its true meaning. But the damage has been done, and we have to live with it. In some fields, equilibrium (or saturation) vapor pressure is denoted simply as vapor pressure, it being understood that the vapor is in equilibrium with its condensed phase. But in meteorology there is a good reason for distinguishing between vapor pressure and equilibrium vapor pressure. Water vapor in atmospheric air is usually *not* in equilibrium with liquid or solid water. The term "saturated liquid" sometimes is encountered, which may be jarring because liquids can saturate porous bodies (e.g. sponges). Saturated water is liquid water *in equilibrium* with its (saturated) vapor.

Even more convenient than expressing water vapor concentration as a pressure is to dispense with dimensions altogether. There are at least three dimensionless measures of water vapor: *specific humidity q*, *mixing ratio w* (or humidity mixing ratio), and *relative humidity r*. The first two are almost the same:

$$q = \frac{\rho_v}{\rho}, \quad w = \frac{\rho_v}{\rho_d}, \tag{5.9}$$

where ρ_v is the density of the vapor, ρ is the density of air, and $\rho_d = \rho - \rho_v$ is the density of dry air. These two quantities are related:

$$q = \frac{w}{1+w}, \quad w = \frac{q}{1-q}. \tag{5.10}$$

Because both w and q are $\ll 1$ in the atmosphere, $w \approx q$. If all densities in Eq. (5.9) were expressed in the same units, q and w would be of order 0.01. But it is customary to multiply them by 1000, which is equivalent to expressing them in g/kg.

The third dimensionless measure of water vapor concentration, *relative humidity*, is defined in two slightly different ways. The traditional definition (for a given temperature) is the ratio of vapor pressure to saturation vapor pressure (of vapor in equilibrium with a *flat surface of pure water*):

$$r = \frac{e}{e_s}. \tag{5.11}$$

Relative humidity is often expressed as a percent ($r \times 100\%$). All was well until the World Meteorological Organization (WMO) recommended that relative humidity be defined as the ratio of the mixing ratio to the saturation mixing ratio:

$$r = \frac{w}{w_s}. \tag{5.12}$$

From the ideal gas law for water vapor and dry air,

$$e = \rho_v R_v T, \quad p_d = \rho_d R_d T, \tag{5.13}$$

and hence,

$$w = \frac{\epsilon e}{p_d} = \frac{\epsilon e}{p - e}, \tag{5.14}$$

where

$$\epsilon = \frac{R_d}{R_v} = 0.622, \tag{5.15}$$

and p is the total pressure. From Eqs. (5.11), (5.12), and (5.14) we obtain

$$\frac{w}{w_s} = \frac{e}{e_s} \left(\frac{p - e_s}{p - e} \right). \tag{5.16}$$

If the relative humidity is 100% according to the traditional definition ($e = e_s$), it is 100% according to the WMO definition. And 0% is the same for both definitions. But at intermediate relative humidities, the two definitions disagree although by only a small amount because $e \ll p$ in the atmosphere. We prefer the traditional definition, not because we are too hidebound to change, but because it is more in tune with the physics of evaporation and condensation. How humidity is measured is treated in Section 6.4.

We showed in Section 1.8 that the flux density of any quantity is a product of number density, velocity, and the quantity that is flowing. Thus, the flux density of water vapor molecules incident from all directions onto liquid water is proportional to n_v and to $\langle v \rangle$, the mean molecular speed. All that we lack is the constant of proportionality. The factor multiplying $n_v \langle v \rangle$ must be at least $1/2$ because only half the water molecules in any small volume of air above the water surface are directed toward it. An additional obliquity factor $1/2$ comes from integrating over all directions of incidence. Suppose that all molecules have the same speed v. The flux of these molecules onto a surface A is determined by the average of the velocity component $v \cos \theta$ in the direction of the normal to A, where θ is the angle between the velocity vector and this normal. The average of $\cos \theta$ is $1/2$, the ratio of the area of a hemisphere to its area projected onto a plane. Average over all speeds v to obtain $n_v \langle v \rangle / 4$ for the vapor flux density (if direction and speed are uncorrelated).

Consider a water surface with vapor above it (Fig. 5.3). Water molecules evaporate from the surface at the rate E. The total upward flux $F_\uparrow$ of water vapor at the surface is the sum of E and a term that is a consequence of water vapor impinging on the surface. A plausible assumption is that this term is proportional to the downward flux $F_\downarrow$. With this assumption the total upward flux of water vapor at the surface is

$$F_\uparrow = E + fF_\downarrow, \tag{5.17}$$

where f is a proportionality factor (which may depend on such quantities as the temperature of the surface). E is the evaporation rate into a vacuum, and we may interpret f as a kind of reflection coefficient.

Fig. 5.3: $F_\uparrow$ is the outward flux of vapor molecules *from* and $F_\downarrow$ the inward flux *to* its liquid.

For water vapor to be in equilibrium with liquid water, the upward and downward fluxes of vapor at the surface must be equal:

$$F_\downarrow = F_\uparrow. \tag{5.18}$$

The downward flux at the surface is

$$F_\downarrow = \frac{1}{4} n_{ve} \langle v \rangle A, \tag{5.19}$$

where n_{ve} is the equilibrium vapor density and A is the surface area. By combining Eqs. (5.17)–(5.19), we obtain

$$\frac{E}{A} = \tfrac{1}{4} n_{ve} \langle v \rangle (1 - f) = \tfrac{1}{4} n_{ve} \langle v \rangle \alpha_M, \tag{5.20}$$

where $\alpha_M = 1 - f$ is the *mass accommodation coefficient* (sometimes called the sticking probability). Thus, the evaporation rate per unit area at temperature T can be determined from the equilibrium vapor density, the mean speed of vapor molecules at this temperature, and α_M.

Now consider water vapor at temperature T with number density n_v *not* necessarily equal to the equilibrium value. We suddenly insert into this vapor a water surface at temperature T. The difference between the instantaneous rate of condensation C onto this surface and the rate of evaporation E from it is the difference between the downward and upward vapor fluxes: $C - E = F_{\downarrow} - F_{\uparrow}$. With this result and Eq. (5.17)

$$C = F_{\downarrow} - fF_{\downarrow}, \tag{5.21}$$

where

$$F_{\downarrow} = \frac{1}{4}n_v\langle v\rangle A. \tag{5.22}$$

From Eqs. (5.20)–(5.22) it follows that the ratio of condensation to evaporation is

$$\frac{C}{E} = \frac{n_v\langle v\rangle\alpha_M}{n_{ve}\langle v\rangle\alpha_M} = \frac{n_v}{n_{ve}}. \tag{5.23}$$

By using the ideal gas law, Eq. (5.23) can be put in the form

$$\frac{C}{E} = \frac{e}{e_s}. \tag{5.24}$$

This is the traditional definition of relative humidity, Eq. (5.11), which now has a simple physical interpretation as the ratio of the rate of condensation of water vapor (assuming something onto which it can condense) to the rate of evaporation of water vapor (assuming liquid or solid water from which water can evaporate).

A subtle but important difference between Figs. 5.2 and 5.3 is that the former depicts a closed system in equilibrium, whereas the latter depicts an open system *not* in equilibrium. The moment the lid is taken off a pan of water, we should not apply equilibrium thermodynamics to it. But shameless miscreants that we are, we do, and it seems to work well enough for our purposes.

C/E is for an *idealized* process in which *net* evaporation $E - C$ is $\langle v\rangle(n_{ve} - n_v)\alpha_M/4$, but only for an instant. This expression, a form of what is often called the *Hertz–Knudsen relation*, embodies the essential requirement for net evaporation, a *gradient* in the vapor number density above the water, but cannot tell us its constantly changing rate. Real gradients are more complicated and determined by, among other factors, wind speed, air temperature, solar and terrestrial radiation. Individual water molecules near the surface continually flit back and forth between the liquid and the vapor. Accommodation, like adsorption (Section 2.2), is often temporary. Evaporating water is a dynamic system subject to feedbacks. Net evaporation causes water to cool, which *reduces* evaporation. Net condensation causes water to warm, which *increases* evaporation.

Evaporation is *prima facie* evidence that molecules exist. A thin layer (≈ 1 mm) of water in a metal lid for a jar disappeared in about ten hours on a warm humid day in a house without air conditioning. Yet nothing perceptible left the water. This and unseen wind and smells were noted by Lucretius more than 2000 year ago: "Nature therefore works by unseen bodies."

We have measured net evaporation rates in a Pennsylvania forest between 0.06 mm/h and 0.20 mm/h over a day; the higher value is the yearly average "normal" evaporation from open water surfaces in hotter, sunnier Arizona.

Dew, Frost, Defrosters, Dehumidifiers, and Swamp Coolers

With a better understanding of condensation and evaporation we can address some questions about ordinary observations. For example, why does dew form? Because, you may be told, air temperature has dropped below the *dew point* (temperature following dew point is redundant). This is not only misleading, but also it is *not* an explanation. It is a *definition* of the dew point masquerading as an explanation. The dew point is *defined* as the temperature at which dew forms. But temperature of what? Dew forms on surfaces (or else it would not be visible as dew), and hence, the temperature of surfaces must drop below the temperature at which condensation of atmospheric water vapor equals evaporation from the surface. When this occurs, net condensation *may* become visible as dew.

Water drops at the tips of blades of grass or the edges of leaves seen early in the morning are often mistakenly called dew, whereas it is *guttation*. Atmospheric wave vapor condensed *onto* plants is dew. Guttation is drops of exuded (watery) sap that originate from *within* plants. Once you observe dew and guttation together you are unlikely to confuse them. Both may appear on the same blade of grass and can freeze, which is *not* frost.

How to determine the dew point T_d (in principle, at least) of air with partial pressure e of water vapor follows from Eq. (5.24):

$$e_s(T_d) = e. \tag{5.25}$$

We may consider Eq. (5.25) to define the dew point. This equation implies that the dew point of air is the temperature to which it must be cooled, keeping its vapor pressure constant (or its pressure and mixing ratio constant), such that this vapor pressure becomes equal to the equilibrium vapor pressure. To determine dew points we need to know how equilibrium vapor pressure depends on temperature. We defer this to the following section. For the moment, we do not need the precise dependence of equilibrium vapor pressure on temperature. We only need to know that it increases with T.

Having defined the dew point by Eq. (5.25), we may ask if this is the temperature at which dew forms. To be truthful, the answer to this question would have to be "No." The exact temperature at which dew forms on a surface is not well defined. In a critical and detailed discussion of dew-point hygrometry, R. G. Wylie and colleagues assert that "the concept of a sharply defined temperature, at which condensation begins abruptly, represents only an approximation to the truth." They also note that "in dew-point hygrometry, the usual simple concept of condensation is inadequate when the accuracy sought is higher than about 0.2 °C." Although we may define the dew point by Eq. (5.25), we cannot expect net condensation to occur exactly at this temperature. It is merely our best guess.

All our assertions about liquid water could have been made about ice. Even though solid, it evaporates, like everything, including us and all the objects around us. Fortunately, their rates of evaporation are exceedingly small (but not zero) at normal terrestrial temperatures. Evaporation of ice is called *sublimation*, and condensation of water vapor directly to ice is called *deposition*. But these two terms should not mislead you into thinking that sublimation and deposition are processes fundamentally different from evaporation and condensation.

Frost is ice formed directly from the vapor phase. The *frost point* is the counterpart of the dew point at temperatures below $0\,^{\circ}C$, but the distinction is not always made.

When discussing the formation of dew and frost, we must give careful thought to where e and e_s in Eq. (5.25) are determined. We are often careless in our use of air temperatures and dew points. When we say *the* air temperature, where in the air do we have in mind? Air temperature varies near surfaces, and hence, so does dew point. Consider, for example, a water surface. The vapor concentration in air just above this surface, within a thin stagnant boundary layer, is the equilibrium value. The dew point of this air is therefore the surface temperature. Whether there is net evaporation or net condensation depends on the dew point of the air *above* the thin boundary layer. By "above" is meant above the steep gradient of water vapor concentration near the surface, within a few millimeters, which is what drives diffusion through this layer (Section 7.3). Water is in equilibrium with its vapor only if there is no gradient in vapor pressure above the water. This is possible if the water is in a *closed* container, which it never is in the atmosphere and not often in our everyday lives.

Let e_a be the vapor pressure in air where the temperature is T_a. According to Eq. (5.25) the corresponding dew point is

$$e_s(T_d) = e_a \,. \tag{5.26}$$

For there to be a net flux of water vapor toward a surface, the vapor flux from the air toward the surface must be greater than the vapor flux from the surface toward the air. Fluxes are proportional to the product of number density, mean velocity, and area. Mean velocity varies as the square root of absolute temperature, whereas number density is e/kT, and hence, fluxes are proportional to $e/\sqrt{T}$. The condition for a net flux toward the surface is therefore

$$\frac{e_s(T_s)}{\sqrt{T_s}} < \frac{e_a}{\sqrt{T_a}} = \frac{e_s(T_d)}{\sqrt{T_d}} \sqrt{\frac{T_d}{T_a}} \,. \tag{5.27}$$

If $T_d \leq T_a$, Eq. (5.27) is satisfied if

$$\frac{e_s(T_s)}{\sqrt{T_s}} < \frac{e_s(T_d)}{\sqrt{T_d}} \,. \tag{5.28}$$

Because $e_s/\sqrt{T}$ increases monotonically with temperature (see the following section), Eq. (5.28) is equivalent to

$$T_s < T_d \,. \tag{5.29}$$

Thus, dew can form on a surface if its temperature drops below the dew point of the surrounding air. This is only a sufficient condition. Is it also necessary? Strictly, only if the square root of the ratio of (absolute) dew point to (absolute) air temperature is one. Under the conditions in which dew forms, this quantity is approximately one, so we can say that for dew to form on a surface *its* temperature must drop below the (approximate) dew point.

By the surface temperature of water is meant that in a *very* thin layer. This temperature can be measured with an infrared thermometer. A layer of water only about 1 μm thick transmits less than 1% of radiation at wavelengths of around 10 μm. Thus, the thermometer responds almost entirely to radiation corresponding to the temperature of this layer with about 3×10^{19} molecules per cm^2.

Students are sometimes taught that air temperature *cannot* drop below the dew point. Yet, no physical law supports this belief. When air *not* purged of particles is cooled to its dew point, the *approximate* temperature at which net condensation (onto particles) begins, two processes occur: Water vapor is removed from the air, which lowers its dew point, and net condensation gives rise to a temperature increase. Both processes act to prevent air temperature from dropping below the dew point. But there are at least two examples in which air temperature can drop below the dew point: in moist air filtered to remove particles onto which vapor can condense, and in air rising so rapidly that condensational warming doesn't keep up with adiabatic cooling. Everything takes time, including condensation.

Forecasters are taught that, on clear days with light winds, the late afternoon dew point is a good estimate of the minimum overnight temperature. Notice the caveats. Rather than say that air temperature *cannot* drop below the dew point, a more defensible assertion is that under many conditions air temperature *does not* drop below the dew point. The best way to convince yourself is to look at many early morning dew points and air temperatures.

If air temperature does drop *below* the dew point, called *undercooling*, the density of the vapor is *greater* than it would be if it were in equilibrium with the condensed phase, called *supersaturation* (Section 5.4). Not only is this possible in natural air, but it is also necessary for cloud formation (Section 5.11).

Having established that air temperature does not usually drop below the dew point, and also that a surface must drop below this temperature for dew to form, we face the unpalatable (to some) conclusion that for dew to form on a surface its temperature must drop below air temperature. This is not, as might appear, impossible. Surfaces can, and frequently do, cool *radiatively* to temperatures well below that of the surrounding air. All surfaces at all times, day and night, emit infrared radiation and absorb it from their surroundings (see Section 1.8). During the day, surfaces are heated by solar and infrared radiation from their surroundings, including the sky. At night, solar radiation ceases to warm surfaces, and *net* infrared radiative cooling of surfaces can be sufficient, especially on clear nights, to drive surface temperatures well below air temperature, sometimes even below the dew point (see Section 7.1 for more about dew and frost formation).

Automobiles are equipped with defrosters, a slight misnomer in that they are capable of removing water droplets as well as frost from windshields. A defroster directs a stream of warm air over a windshield, which is warmed to the point where evaporation exceeds condensation. The evaporation rate from a windshield increases with its temperature, whereas the condensation rate onto it is determined by the vapor density in the windshield's environment.

Those of us who live where winters can be cold have learned from experience that it is advisable to humidify our houses during winter to prevent cracking of wooden furniture and drying of our skin. Another way to achieve the same end would be to turn off our furnaces, an uncomfortable solution to problems caused by heating interior air. Outside air leaks into houses, is heated, and then leaks out. The total pressure p inside and outside is about the same, as evidenced by the failure of windows to be blown out. If the inside temperature is greater than that outside, the inside air density is correspondingly less because all habitable houses leak, which is required by building codes. From the ideal gas law

$$p = \rho_i R T_i = \rho_o R T_o, \tag{5.30}$$

where subscripts i and o denote inside and outside air, respectively. The ideal gas law applied to inside and outside vapor pressures yields

$$e_i = \rho_{vi} R_v T_i, \quad e_o = \rho_{vo} R_v T_o, \tag{5.31}$$

from which, together with Eq. (5.30),

$$\frac{e_i}{e_o} = \frac{\rho_{vi}}{\rho_{vo}} \frac{T_i}{T_o} = \frac{\rho_{vi}}{\rho_{vo}} \frac{\rho_o}{\rho_i}. \tag{5.32}$$

If negligible water vapor is added to the air inside a house because of breathing, sweating, cooking, and washing by its inhabitants, or removed by condensation, the mole fraction of water vapor in inside air is equal to that in outside air. With these assumptions the ratio of the vapor density to the total density is the same inside and outside:

$$\frac{\rho_{vi}}{\rho_i} = \frac{\rho_{vo}}{\rho_o}. \tag{5.33}$$

Equations (5.32) and (5.33) yield

$$e_i = e_o. \tag{5.34}$$

Although inside and outside vapor pressures are the same, the equilibrium (saturation) vapor pressure inside a heated house is considerably greater than that in frigid outside air. Thus, net evaporation from water-bearing objects (e.g. wood and skin) inside is also greater, and hence, the need to humidify [see Eq. (5.24)].

The reverse of a defroster is a dehumidifier, which is not much more than a cold surface onto which water vapor condenses because the evaporation rate from a sufficiently cold surface is less than the condensation rate of vapor from the surrounding air. A dehumidifier in the parlance of vacuum science is a *cold trap*—a cold surface by means of which vapors are removed from air by condensation.

In the hotter, drier regions of the Western United States, an inexpensive method of air conditioning is common but unknown to inhabitants of the humid East. These simple air conditioners, called *swamp coolers*, are louvered boxes containing panels of an absorbent material kept continuously wet. Outside air circulated through the wet material is cooled by (net) evaporation and directed inside. During the early summer months in southern Arizona, for example, air temperatures can be quite high yet accompanied by low relative humidities, and hence, swamp coolers can be inexpensive air conditioners. But during the thunderstorm season in late summer relative humidities rise to where swamp coolers provide little relief. In the Eastern United States, where hot summers usually are humid, swamp coolers likely would be worse than useless. But in a different form based on the same idea, they may be used for roof cooling. A water-saturated porous medium (e.g. sand) on a roof is cooled by net evaporation but the resulting humidified air remains outside. As the world attempts to mitigate the consequences of climate change with energy-efficient cooling, roof cooling by evaporation may become more attractive in some areas.

5.3 The Clausius–Clapeyron Equation

A key thermodynamic relation in Section 5.2 is the dependence of saturation vapor pressure on temperature, without which we cannot determine dew points and other meteorological variables. To proceed, we must determine this dependence.

Consider a fixed mass of water vapor (unmixed with dry air) that is compressed *isothermally*. Imagine the water to be enclosed in a cylinder fitted with a piston, the entire apparatus immersed in a huge constant-temperature ($> 0\,^{\circ}\mathrm{C}$) bath (reservoir). Initially, only water vapor is in the cylinder. As the piston is depressed very slowly (quasi-statically), the vapor pressure e increases, whereas the saturation vapor pressure e_s remains constant because temperature is constant. Eventually a point is reached at which $e = e_s$. With further compression liquid water appears in the cylinder because condensation slightly exceeds evaporation (see Eq. 5.24). Instead of a change in pressure, vapor is converted to liquid. The constant pressure during this process is the saturation vapor pressure (water vapor is in equilibrium with liquid water) at the temperature T of the bath [but see comments following Eq. (5.51)]. We showed in Section 4.1 that for any system (liquid, vapor, or a mixture of the two)

$$\frac{\partial U}{\partial V} = T\frac{\partial p}{\partial T} - p. \tag{5.35}$$

For an ideal gas, this partial derivative vanishes, but a system composed of liquid water in equilibrium with its vapor is *not* an ideal gas. During the phase change from vapor to liquid, only the relative amounts of vapor and liquid change, and hence, the total internal energy and volume of the system at any point in the compression process are given by

$$U = \tau U_w + (1 - \tau)U_v, \quad V = \tau V_w + (1 - \tau)V_v, \quad 0 \le \tau \le 1, \tag{5.36}$$

where τ is the mass of liquid relative to the (constant) mass of liquid and vapor. U_v is the internal energy of the system at the beginning of the phase change (entirely vapor), and U_w is the internal energy of the system at the end of the phase change (entirely liquid); the corresponding volumes are denoted by V_v and V_w. From Eq. (5.36) we obtain

$$\frac{\partial U}{\partial V} = \frac{\partial U}{\partial \tau} \bigg/ \frac{\partial V}{\partial \tau} = \frac{U_v - U_w}{V_v - V_w}. \tag{5.37}$$

We define the *enthalpy of vaporization* L_v as the difference between the enthalpy of a fixed mass of water vapor and that of the same mass of liquid:

$$L_v = H_v - H_w = U_v - U_w + p(V_v - V_w). \tag{5.38}$$

If we combine Eqs. (5.35), (5.37), and (5.38), we obtain

$$\frac{U_v - U_w}{V_v - V_w} = \frac{H_v - H_w}{V_v - V_w} - p = \frac{\partial U}{\partial V} = T\frac{\partial p}{\partial T} - p, \tag{5.39}$$

from which follows

$$T\frac{\partial p}{\partial T} = \frac{L_v}{V_v - V_w}. \tag{5.40}$$

The pressure p in Eq. (5.40) is the saturation vapor pressure e_s. And because this pressure (for pure water) depends only on temperature, the partial derivative in Eq. (5.40) is an ordinary derivative, and Eq. (5.40) becomes

$$\frac{de_s}{dT} = \frac{1}{T}\frac{L_v}{V_v - V_w}. \tag{5.41}$$

To deal with specific quantities (intensive variables), we divide the numerator and denominator on the right side of Eq. (5.41) by the total mass M of water to obtain

$$\frac{de_s}{dT} = \frac{1}{T}\frac{\ell_v}{v_v - v_w}, \tag{5.42}$$

where

$$\ell_v = \frac{L_v}{M} = h_v - h_w \tag{5.43}$$

is the enthalpy of vaporization per unit mass and

$$v = \frac{V}{M} \tag{5.44}$$

is the *specific volume* (inverse density). Because $v_v > v_w$ and e_s increases with increasing temperature, $\ell_v > 0$, which we could have surmised on physical grounds.

   Equation (5.42) is called the *Clausius–Clapeyron equation* or sometimes simply Clapeyron's equation. To determine if and how credit for this equation should be shared, we dug into the original writings of both men. Benoit-Pierre-Émile

Clapeyron (b. Paris, France, 1799; d. Paris, 1864), a French steam engine engineer, applied principles of Sadi Carnot (see Chapter 4) to obtain the equation

$$\frac{de_s}{dT} = \frac{1}{C}\frac{\ell_v}{v_v - v_w},\qquad(5.45)$$

where C is an unknown function of temperature. Clapeyron's 1834 paper containing this equation also translated Carnot's verbal descriptions into mathematical language and showed the Carnot cycle (Fig. 4.11), now a staple of thermodynamics textbooks, graphically on a pV or indicator diagram (see Section 3.1). Despite its theoretical and practical importance, Clapeyron's paper was ignored for almost 40 years, even by Clapeyron himself, until it was resuscitated by Rudolf Clausius (b. Köslin, Pomerania, 1822; d. Bonn, Germany, 1888) who, as noted in Chapter 4, coined the term *entropy*. Clausius deduced Eq. (5.42) in his book *The Mechanical Theory of Heat*, being scrupulous in citing Clapeyron's earlier paper. Clausius knew and appreciated not only what Clapeyron had done, but also what remained to be done: unveiling that pesky unknown function C and comparing theory with experiment. Unlike the inaccurately named Boyle–Mariotte and Charles–Gay–Lussac laws, Eq. (5.42) does truly deserve to bear the names of two scientists separated by age and nationality but forever linked by a fundamental equation to which both contributed.

For our purposes, $v_v \gg v_w$ and Eq. (5.42) is

$$\frac{de_s}{dT} \approx \frac{\ell_v}{Tv_v}.\qquad(5.46)$$

Although liquid water is far from being an ideal gas, water vapor is to good approximation, from which it follows that

$$e_s v_v = R_v T.\qquad(5.47)$$

If Eqs. (5.46) and (5.47) are combined, we obtain

$$\frac{1}{e_s}\frac{de_s}{dT} = \frac{\ell_v}{R_v T^2}.\qquad(5.48)$$

Although we show in a following subsection that the enthalpy of vaporization depends on temperature, ℓ_v often can be taken to be independent of temperature without serious error, especially over narrow temperature ranges. With this assumption, Eq. (5.48) can be integrated:

$$\frac{e_s}{e_{so}} = \exp\left(\frac{\ell_v}{R_v T_o}\right)\exp\left(-\frac{\ell_v}{R_v T}\right).\qquad(5.49)$$

We can rewrite this in a physically more transparent form by using $R_v = k/m_v$, where m_v is the mass of a water molecule:

$$e_s = C_0 \exp\left(-\frac{m_v \ell_v}{kT}\right),\qquad(5.50)$$

and C_0 is a constant. Does this equation look familiar? It has the same form as Eq. (2.48) for the decrease of density with height. The numerator in the exponent can be interpreted as the energy (strictly, enthalpy) required to break a water molecule free from its neighbors; the denominator is the average molecular kinetic energy available. The greater the value of kT, the greater the evaporation rate, and hence the greater the saturation vapor pressure. And the rate increases exponentially with kT because so does the fraction of molecules with sufficient energy to escape the liquid. We could have guessed Eq. (5.50) strictly by physical reasoning.

We prefer the term *enthalpy of vaporization* (sublimation, fusion) instead of the more familiar *latent heat of vaporization* (sublimation, fusion). More than half a century (1968) ago Defoe Giddings wrote that "the older term latent heat is rarely used in calorimetry," and instead recommended enthalpies. He made his living from calorimetry so needed to know exactly what he was measuring. Latent heat is an archaic term, a relic from the days in which it was thought that there were two kinds of heat: sensible, the kind you can feel, and latent, the kind you cannot. We now know better: there are not two kinds of heat—there is *no* kind of heat. Heating is a *process*, not a substance, a process that occurs because of a temperature difference between a system and its surroundings. Yet the term *latent heat* lingers on, often causing confusion, for example, the ascent of parcels of moist air is often taken to be an adiabatic process ($Q = 0$). But if the water vapor in such parcels condenses, there is an associated enthalpy change, which is often described as the "release of latent heat." This enthalpy change because of a phase transformation is accompanied by a temperature increase, *as if* the parcel exchanged energy with its surroundings. How do we reconcile this supposedly adiabatic process (no heating) with the fact that so-called latent heating occurs? The advantage of the term *enthalpy* is its unfamiliarity, and hence, it carries no misleading baggage. Unlike energy and entropy, the meaning of enthalpy has not been debased by frequent misuse in nonscientific contexts. Evidence of this is that some spell-checkers do not recognize enthalpy as a legitimate word. We encourage you to adopt the term *enthalpy of vaporization* instead of *latent heat of vaporization*, which is still used but has been losing ground since publication of the first edition of this book. To paraphrase Planck, progress in scientific terminology and notation advances "one funeral at a time." More and more chemistry textbooks have replaced all kinds of "heats" with enthalpies. Chemists use the unit $kJmol^{-1}$ for molar enthalpies, which are more revealing from a molecular point of view. They tell us the energy changes because of rearrangements of atoms and molecules (e.g. phase transitions and chemical reactions). To convert Jkg^{-1} to $kJmol^{-1}$, multiply by the molecular weight and divide by 10^6.

Joseph Black coined the term "latent heat," by which he meant a virtual temperature increase ΔT not observed in a phase change from liquid to vapor at apparently constant temperature; in modern terms, $c\Delta T = \ell_v$. This is the origin of denoting ΔT as latent, not observable as a temperature increase. The fundamental reason why $h_v > h_w$ (at the same temperature) is that the *internal potential energy* of the vapor is greater than that of the liquid because of the greater separation between vapor molecules, similar to the potential energy increase of a ball tossed upward in

Earth's gravitational field. We can observe this increase directly but not increases in potential energy at the molecular level. They are hidden, another term for which is "latent," as in latent fingerprints hidden until dusted with powder. We could call *all* potential energy latent energy. Indeed, in an 1853 paper entitled "On the General Law of the Transformation of Energy," William Rankine asserts that "All conceivable forms of energy [not heat] may be distinguished into two kinds; actual or sensible, and potential or latent." With no loss, latent can be tossed into the dustbin of archaic terms along with actual and sensible (i.e. kinetic). By recognizing that all potential energies at the molecular level are "latent," we unify, with a resultant economy of thought, what would otherwise be disparate phenomena. We need all the help we can get. As Ernst Mach wrote in a popular lecture on the economical nature of physical inquiry, "When the human mind, with its limited powers, attempts to mirror in itself the rich life of the world, of which it is itself only a small part, and which it can never hope to exhaust, it has every reason for proceeding economically."

The enthalpy of vaporization L_v can be associated with heating by integrating the first law of thermodynamics for a constant-pressure process in which a given mass of liquid water is transformed entirely into vapor at constant temperature:

$$\int_w^v Q\, dt = \int_w^v \frac{dH}{dt}\, dt = H_v - H_w = L_v. \tag{5.51}$$

The limits of integration indicate that initially the system is entirely liquid water and finally is entirely water vapor. This equation gives us some insight into the origins of the notion of some kind of strange and wonderful heat different from that usually encountered. According to Eq. (5.51), total heating is nonzero, which is inconsistent with the derivation of Eq. (5.38) based on an *assumed* isothermal, isobaric process. Every point on the line representing it is a possible state of a system composed of a liquid and its vapor. Each is an equilibrium state for a different mass fraction of vapor and cannot spontaneously make a transition to another state at the same temperature and pressure. Change requires a nudge from the surroundings. For Q to be nonzero there *must* be a small but finite temperature difference between a system and its surroundings. And for a liquid–vapor system to change from all vapor to all liquid, or vice versa, it cannot be *exactly* at the temperature of its surroundings. We snuck in the kind of idealization often swept under the rug in thermodynamics. Our derivation is logically indefensible.

Suppose that water vapor *and* dry air is in the hypothetical cylinder with which we began our derivation, and the initial *total* pressure is atmospheric pressure; p in the definition of enthalpy of vaporization is *not* this pressure. Moreover, the internal energy of the water is a function of T and the total pressure. Some air is dissolved in the water, and the internal energies of pure water and water containing dissolved air are not exactly equal. We doubt that a real transformation of water vapor entirely into liquid is an isobaric, isothermal process. We begin with a cylinder of only moist air containing water vapor at a partial pressure less than the saturation vapor pressure at T. If the partial pressure of the vapor increases, so does that of the nitrogen and oxygen. Compress further and water vapor condenses, which removes water vapor from the space above the water. The nitrogen and oxygen do not condense but are dissolved more in the water (Section 5.9). With further compression the *total* pressure increases even though the partial pressure of the water vapor changes only slightly (Section 5.7).

If we blissfully ignore all these niceties, we have an expression for Q for an *ideal* process in which water vapor, unmixed with dry air, is completely transformed into liquid water isobarically and isothermally. The extent to which this expression applies to real processes must be determined by experiment. Or we just have to use it and hope that Providence will save us from serious errors.

Other Enthalpy Differences

Although in deriving Eq. (5.42) we took the condensed phase of water to be liquid, it could have been solid (ice). If we take the constant temperature bath in our derivation to be $< 0\,^\circ\mathrm{C}$ we obtain the *enthalpy of sublimation*

$$\ell_s = h_v - h_i, \tag{5.52}$$

where the specific enthalpy of ice h_i replaces that of water. Ice in equilibrium with liquid water leads to the *enthalpy of fusion*

$$\ell_f = h_w - h_i. \tag{5.53}$$

Equations (5.51), (5.52), and (5.52) exemplify *first-order phase transitions*, defined by nonzero enthalpy differences between two phases at the same temperature and pressure. These three enthalpies (strictly, enthalpy differences) are not independent:

$$\ell_f + \ell_v = \ell_s. \tag{5.54}$$

Although we obtained Eq. (5.54) by adding Eqs. (5.38) and (5.52), this equation also follows from physical arguments. The enthalpy cost (total heating) in the transformation of ice to vapor must be the same if this occurs directly or by way of an intermediate liquid phase.

The enthalpy of fusion of ice is about eight times *smaller* than the enthalpy of vaporization (or sublimation) because intermolecular separations in the ice and liquid are roughly equal. As a consequence the potential energy difference between these two phases is much smaller than that between the vapor and condensed phases. A nonzero enthalpy of fusion arises primarily from a difference in *structure*. Liquids do have structure in that there are correlations in the positions of their molecules over short distances, whereas in crystalline solids this distance is indefinitely large.

To understand ℓ_f imagine a container of water at a fixed pressure and immersed in a constant-temperature $(0\,^\circ\mathrm{C})$ bath. By analogy with Eq. (5.51) the complete transformation of water to ice (per unit mass) is a result of integrated cooling. Alert readers may spot a contradiction in Eq. (5.51), which implies melting of ice at constant and uniform temperature. But $Q \neq 0$ requires a temperature *difference*. This is what happens when idealized (i.e. impossible) processes are invoked. There must be at least a small temperature difference between the ice and water. Ice cubes at *exactly* $0\,°\mathrm{C}$ in water at *exactly* $0\,°\mathrm{C}$ cannot melt.

Ice melts and water evaporates in the free atmosphere in a finite time. Moreover, real water contains dissolved air, and its pressure is ambient pressure, not solely or even mostly its vapor pressure. Net evaporation of real water is not isothermal and its enthalpy of vaporization changes with temperature [see Eq. (5.63)]. Thus, we cannot expect our idealized expressions to be exactly applicable to reality. They only have to be good enough for our purposes.

Except at very high pressures, the enthalpy of fusion is to good approximation the difference in internal energies of the two phases. Increasing the temperature of ice is a comparatively small perturbation of its thermodynamic state. Melting completely disrupts it. To melt ice takes 160 times more energy transfer from its surroundings than to raise its temperature by $1\,°\mathrm{C}$.

Mantras and Misconceptions about Phase Transitions

Mantras about the "release of latent heat" lack explanatory power. To go beyond them, consider a leaf on a fruit tree on an evening when subzero morning temperatures are forecast. Air temperature is slightly above freezing as is the temperature of the leaf. Water is sprayed on the leaf. The water on the leaf is the system, the leaf and the air its surroundings. As the night grows colder, air temperature drops below $0\,°\mathrm{C}$ and the water freezes. Thus, the enthalpy of the water must *decrease*: $H_i - H_w = -L_f = Q_{tot}$, which is *negative* for the water, and hence, *positive* for its surroundings. It is difficult to apportion this exactly between leaf and air, but the lion's share may go to the leaf because the water freezes on it. The net result is that the leaf, contrary to "common sense," is heated by the freezing of water on it, thereby nudging its temperature up a bit. This scheme is used by the owners of orchards, who may not understand why it works, only that it does (within limits). If it didn't, they wouldn't do it. They are not stupid. The thermal conductivity of ice is about four times that of water at around $0\,°\mathrm{C}$ and so the ice does *not* provide increased insulation, just the reverse (see Section 7.1 about insulation).

Because of this, the widespread term "frost damage" applied to plants is misleading. This conveys that deposition of frost on plants is what kills them. But according to the previous paragraph, deposition of frost on plants gives them a bit of protection from "frost damage." They may be free of frost but still die if temperatures drop sufficiently low. The distinction here is between temperatures below $0\,°\mathrm{C}$ ("frost") and ice deposited on surfaces from the vapor phase (frost). How does one explain that ice formed on plants by deposition kills them and yet ice formed on plants by freezing of liquid water protects them? Ice is ice. Why is one form lethal and the other beneficial?

Explanations in the previous paragraphs are based on macroscopic thermodynamics. At the molecular level, when liquid water freezes, there is a decrease of potential energy, which to balance the energy budget requires a comparable increase of kinetic energy in the surroundings, manifested by a temperature increase. This echoes a point made at the end of Chapter 1: energies are kinetic or potential. But because the density of ice is *less* than that of liquid water, both at the same temperature, the average separation between the molecules in ice *increases*. And yet ice is more cohesive. Thus, there is more to the decrease of intermolecular potential energy than simply a change in separation. Because of its *structure*, ice is more cohesive *despite* the increase in average separation. Water is not the only substance that expands upon freezing, but there aren't many.

Entropy and Enthalpy Differences in Phase Changes

From Eq. (4.63) we have

$$T\frac{\partial S}{\partial V} = \frac{\partial U}{\partial V} + p. \tag{5.55}$$

We apply the same arguments to Eq. (5.55) that were made when we derived Eq. (5.37). For liquid water and water vapor in equilibrium at temperature T and pressure p, Eq. (5.55) yields

$$T\frac{S_v - S_w}{V_v - V_w} = \frac{U_v - U_w}{V_v - V_w} + p, \tag{5.56}$$

from which it follows that

$$T(S_v - S_w) = U_v - U_w + p(V_v - V_w) = H_v - H_w = L_v. \tag{5.57}$$

If we divide both sides of Eq. (5.57) by the total mass of the system, we obtain

$$T(s_v - s_w) = \ell_v. \tag{5.58}$$

Because the enthalpy of vaporization is positive, the specific entropy of water vapor is greater than that of liquid water, both at the same temperature. A transition from the liquid phase to the vapor phase is accompanied by an increase in entropy of the system. But the entropy of the *universe* does not change. During the phase change the liquid–vapor system is in contact with a reservoir with constant temperature T. The entropy change of the reservoir is

$$\Delta S_{res} = \int_w^v -\frac{Q}{T}\,dt = \frac{1}{T}\int_w^v -Q\,dt = \frac{H_w - H_v}{T} \tag{5.59}$$

where the minus sign indicates that heating of the liquid–vapor system corresponds to cooling of the reservoir. The entropy change Eq. (5.59) of the reservoir is the negative of the entropy change of the liquid–vapor system (Eq. 5.57). Thus, the entropy of the universe does not change and the phase change is reversible—in this *idealized* process. All real phase transitions are irreversible.

Why is there heating (or cooling) of the liquid–vapor system and cooling (or heating) of the reservoir if the phase transition is isothermal and heating (cooling) is a consequence of temperature differences between a system and its surroundings? The answer to this riddle is that the phase change cannot be strictly isothermal except in the limit of an idealized infinitely slow process. Consider what happens when the vapor is compressed just a hair. For a brief moment, condensation exceeds evaporation, and hence, net condensation occurs. This causes the temperature of the system to increase slightly. Now the system is slightly warmer than its surroundings (the reservoir), and hence, the system is cooled, and the reservoir is heated (but its temperature does not increase because it is infinite) until the temperature of the system is again that of the reservoir. Now compress the vapor again a hair, wait for temperature to equilibrate, compress again, and so on *ad infinitum* until all the vapor has been transformed into liquid. Equations (5.51) and (5.59) apply only to an idealized process in which the system continuously passes through a succession of equilibrium states. Equation (5.58) can be extended to other phase changes by using Eqs. (5.52) and (5.53):

$$T(s_v - s_i) = \ell_s, \tag{5.60}$$

and

$$T(s_w - s_i) = \ell_f. \tag{5.61}$$

Phase transitions are characterized by *discontinuities* of entropy or, equivalently, enthalpy. There always are such discontinuities in liquid–solid transitions, unlike in liquid–vapor transitions, which vanish above the critical point (Section 5.4). Liquids always are closer to being vapors than solids.

Although probably not greatly relevant to atmospheric science, transitions exist between chemically identical phases with different structures, for example, the many different crystalline forms of ice.

Temperature Dependence of Enthalpy of Vaporization

Our derivation of the enthalpy of vaporization, Eq. (5.38), applies strictly to laboratory measurements made under controlled conditions: liquid water at the saturation vapor pressure for each temperature T and free of dissolved air. Natural liquid water, otherwise pure, contains some air. Moreover, the liquid water pressure is the sum of the (dry) air pressure and the saturation vapor pressure. To determine the temperature dependence of ℓ_v we assume that the presence of air is of no consequence. Justification for this is that the specific volume of liquid water is thousands of times smaller than that of water vapor, which to good approximation is an ideal gas with properties such as specific enthalpy that are not changed if mixed with nitrogen and oxygen at pressures of one atmosphere or less.

If we differentiate Eq. (5.38) with respect to T and use Eq. (5.43) we obtain

$$\frac{d\ell_v}{dT} = \frac{\partial h_v}{\partial T} - \left[\frac{\partial h_w}{\partial T} + \frac{\partial h_w}{\partial p} \frac{dp}{dT} \right]. \tag{5.62}$$

From Eq. (3.63) the specific enthalpy of the vapor h_v does not depend on pressure, but the specific enthalpy of the liquid h_w depends on temperature *and* pressure. Here, p is the saturation vapor pressure, and dp/dT is given by Eq. (5.42). From Eq. (4.87) and the definitions of the specific heat capacities, we obtain

$$\frac{d\ell_v}{dT} = -(c_{pw} - c_{pv}) - \frac{v_w(1 - T\alpha_w)\ell_v}{(v_v - v_w)T}, \tag{5.63}$$

where α_w is the coefficient of thermal expansion of liquid water; $T\alpha_w \ll 1$. Because $v_v \gg v_w$ even at $100\,°C$, the second term is negligible compared with $\Delta c_{wv} = c_{pw} - c_{pv}$. Thus, Eq. (5.63) becomes

$$\ell_v \approx \ell_{v0} - \int_{T_0}^{T} \Delta c_{wv} dT, \tag{5.64}$$

where the reference temperature T_0 is 273 K. Because $c_{pw} \geq c_{pv}$, ℓ_v decreases with increasing temperature, as it must because ℓ_v vanishes at the critical temperature (647 K) where the distinction between liquid and vapor vanishes (Section 5.4) and $c_{pw} = c_{pv}$. But if we limit temperatures to those between $-40\,°C$ and $60\,°C$ and take the integrand in Eq. (5.64) to be the difference in the specific heat capacities at $0\,°C$, 2350 Jkg^{-1} K^{-1}, then $\ell_v = 2.5 \times 10^6 - (2350)(T - T_0)$ agrees with values in the *Smithsonian Meteorological Tables* (p. 343) to within less than 0.3%. Moreover, the variation of ℓ_v over this range is about 10%. $c_{pw} \to c_{pv}$ with increasing temperature because the liquid and vapor become more like each other. If the second term in Eq. (5.63) is omitted, the result is sometimes called the *Kirchhoff equation*.

An equation for the enthalpy of fusion of water is similar to Eq. (5.63), but the second term is *not* negligible compared with the first. We obtain a differential equation $d\ell_f/dT - \ell_f/T = \Delta c_{wi}$, the solution to which is $\ell_f = \ell_{fo}(T/T_0) + \Delta c_{wi}T \ln(T/T_0)$ if we ignore the temperature dependence of $\Delta c_{wi} = c_{pw} - c_{pi}$. This correctly predicts that ℓ_f decreases with decreasing temperature, as it must, because liquid water becomes more like ice. The difference between calculated and tabulated values increases from 3% at $-10\,°C$ to 11% at $-40\,°C$. The variation of ℓ_f from $0\,°C$ to $-40\,°C$ is about 30%.

For many meteorological applications the temperature dependence of the enthalpies of vaporization and fusion is likely to be negligible. Scientists and engineers interested in liquids, including water, at high temperatures and pressures aren't so lucky.

Temperature Dependence of Saturation Vapor Pressure: A More Accurate Equation

As shown, to good approximation the enthalpy of vaporization of water is a linear function of temperature over the temperature range of greatest meteorological interest. With such a temperature dependence, Eq. (5.48) is not much more difficult to integrate than if ℓ_v were independent of temperature. If we combine Eqs. (5.48) and (5.64), we obtain

$$\frac{1}{e_s}\frac{de_s}{dT} = \frac{\ell_{vo} + (c_{pw} - c_{pv})T_0}{R_v T^2} - \frac{c_{pw} - c_{pv}}{R_v T}, \tag{5.65}$$

the solution to which is

$$\ln\frac{e_s}{e_{so}} = \frac{\ell_{vo} + (c_{pw} - c_{pv})T_o}{R_v}\left(\frac{1}{T_o} - \frac{1}{T}\right) - \frac{c_{pw} - c_{pv}}{R_v}\ln\frac{T}{T_o}. \tag{5.66}$$

For convenience we chose the reference temperatures (constants of integration) for the enthalpy of vaporization and for the saturation vapor pressure to be 273 K. If all the previous values for the physical parameters in Eq. (5.66) are combined, we obtain

$$\ln\frac{e_s}{e_{so}} = 6808\left(\frac{1}{T_o} - \frac{1}{T}\right) - 5.09\ln\frac{T}{T_o}. \tag{5.67}$$

Equation (5.49) with the enthalpy of vaporization at $0\,°C$ yields

$$\ln\frac{e_s}{e_{so}} = 19.83 - \frac{5417}{T}. \tag{5.68}$$

An equation in which the logarithm of the relative vapor pressure is written as $A - B/(C + T)$, where A, B, and C are determined by fitting measurements to temperatures over a limited range, is sometimes called the Magnus equation (1844), sometimes the Antoine equation (1888). In Eqs. (5.67) and (5.68) $e_{so} = 6.11$ hPa, the saturation vapor pressure at $0\,°C$. Figure 5.4 shows Eqs. (5.67) and (5.68) over the temperature range $-30\,°C$ to $30\,°C$. Equation (5.67) yields values nearly identical with tabulated values. Equation (5.68) is a good approximation, especially at the lower temperatures. Because the scale in Fig. 5.4 is such that errors in the two equations are not evident, the percentage difference between them and tabulated vapor pressures is shown on an expanded scale in Fig. 5.5. The gap between the two equations widens with increasing temperature. At $100\,°C$, the nominal boiling point, the saturation vapor pressure predicted by Eq. (5.67) is in error by less than 3%, whereas that predicted by Eq. (5.68) is in error by about 20%.

A change of notation transforms Eq. (5.66) into an expression for the temperature dependence of the saturation vapor pressure of ice:

$$\ln\frac{e_{si}}{e_{so}} = \frac{\ell_{so} + (c_{pi} - c_{pv})T_o}{R_v}\left(\frac{1}{T_o} - \frac{1}{T}\right) - \frac{c_{pi} - c_{pv}}{R_v}\ln\frac{T}{T_o} \tag{5.69}$$

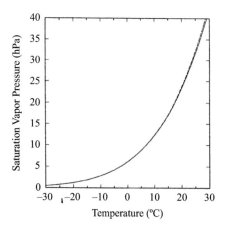

Fig. 5.4: Saturation vapor pressure of pure water versus temperature calculated using Eq. (5.67) (solid) and Eq. (5.68) (dashed). For an expanded view of how well these two equations agree with measurements, see Fig. 5.5.

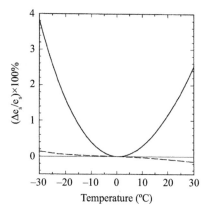

Fig. 5.5: Percent difference between calculated and measured (tabulated) saturation vapor pressures of pure water. The dashed curve shows the percent difference for Eq. (5.67) and the solid curve shows the percent difference for Eq. (5.68). The standard for comparison is saturation vapor pressure tabulated in the *Smithsonian Meteorological Tables.*

where ℓ_{so} is the enthalpy of sublimation at some reference temperature T_0 and c_{pi} is the specific heat capacity of ice. At $0\,°\text{C}$, the enthalpy of sublimation of ice is $2.834 \times 10^6\,\text{Jkg}^{-1}$ and its specific heat capacity is $2106\,\text{Jkg}^{-1}\text{K}^{-1}$. These values in Eq. (5.69) yield

$$\ln\frac{e_{si}}{e_{so}} = 6293\left(\frac{1}{T_o} - \frac{1}{T}\right) - 0.555\,\ln\frac{T}{T_o}. \tag{5.70}$$

The enthalpy of sublimation depends less on temperature than does the enthalpy of vaporization because of the smaller difference between the specific heat of ice and of water vapor compared with that between liquid water and vapor. A constant enthalpy of sublimation (at $0\,°C$) in Eq. (5.49) yields

$$\ln\frac{e_{si}}{e_{so}} = 22.49 - \frac{6142}{T}. \tag{5.71}$$

In Eqs. (5.70) and (5.71) $e_{so} = 6.11$ hPa.

Saturation vapor pressures over liquid water from $-40\,°C$ to $40\,°C$ in increments of $0.1\,°C$ are tabulated following the section on selected physical constants.

Difference between the Saturation Vapor Pressure above Ice and above Subcooled Water at the Same Temperature

We do not shrink from applying Eq. (5.67) for liquid water to temperatures below the nominal freezing point. To meteorologists, subcooling (or undercooling) is commonplace rather than exotic or "paradoxical." Figure 5.6 shows saturation vapor pressures above subcooled water (from Eq. 5.67) and above ice from (Eq. 5.70) for temperatures down to $-30\,°C$. These two vapor pressure curves intersect at $0\,°C$, whereas at lower temperatures they diverge, the vapor pressure for subcooled water always being greater than that for ice at the same temperature. This is consistent with the interpretation of saturation vapor pressure as a measure of evaporation rate. We expect water in the less tightly bound liquid state to evaporate more rapidly than ice at the same temperature. On this figure the two vapor pressure curves are almost indistinguishable, so we show their difference separately (Fig. 5.7). Note the maximum at about $-12.5\,°C$. Although the vapor pressure difference between liquid and solid water at the same temperature may seem small, it has profound consequences. As we show in Section 5.10, cloud droplets form on solid particles (condensation nuclei). Moreover, pure liquid water, especially in the form of tiny cloud droplets, does not invariably freeze when its temperature drops below $0\,°C$ without the help of foreign particles (ice nuclei) to initiate freezing. Nature is profligate with condensation nuclei but stingy with ice nuclei. Almost any particle in the atmosphere can serve as a nucleus for condensation of water vapor, but only certain substances can nucleate the freezing of liquid water. Because of this disparity, clouds of subcooled water droplets are the rule, rather than the exception. A cloud at a temperature below $0\,°C$ is not necessarily entirely ice. It may be mostly subcooled water droplets. But if a droplet happens to form on an ice nucleus or encounters one in its trajectory and becomes an ice crystal it would evaporate more slowly than a water droplet at the same temperature. An ice crystal in an environment of mostly subcooled water droplets grows at the expense of its neighbors shrinking. As the ice grain grows, it falls faster than its smaller neighbors and collides with them. They freeze upon contact (riming). This makes the ice grain even larger. As it accretes subcooled water droplets on its descent, it falls ever faster with an ever-larger cross-sectional area, both of which serve to increase its ability to grow even larger. Thus, high in a cloudy sky, unseen by human eyes, a drama unfolds, with the rich (ice grains) getting richer and the poor (subcooled water droplets) getting poorer. This is one process (often called the *Bergeron process*), driven by the small vapor pressure difference between liquid and solid water, for yielding precipitation. Raindrops are not just cloud droplets that have gorged on water vapor, but rather are agglomerations of cloud particles. If they melt before reaching the ground, the result is rain. If they melt and then subcool before reaching the ground, the result is freezing rain—subcooled rain drops that freeze on contact, sealing car doors shut and glazing streets and sidewalks with slippery ice.

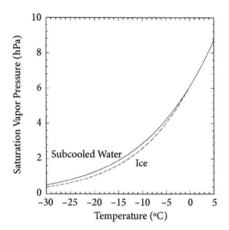

Fig. 5.6: Saturation vapor pressure of ice and subcooled water (pure water existing as liquid at a temperature below the nominal freezing point).

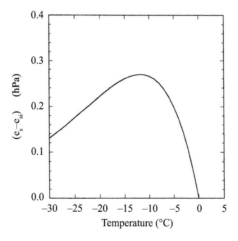

Fig. 5.7: Difference between the saturation vapor pressure of subcooled water e_s and that of ice e_{si}. The maximum is at $\approx -12\,°C$.

The Bergeron process is important for the formation of precipitation in *cold clouds*—those that exist above the 0 °C level in the atmosphere. In *warm clouds*, that is, those completely below the 0 °C level, water droplets can grow by the collision and coalescence process to form rain. But this requires that the droplet size distribution be broad so that larger droplets can fall relative to smaller ones and accumulate mass. The best conditions for warm cloud precipitation exist in warm marine environments where low condensation nuclei concentrations give rise to broad size distributions. Conversely, warm clouds with narrow size distributions do not precipitate. But even in the tropics, the Bergeron process is alive and well in deep cumulonimbus clouds and hurricanes where raging snowstorms exist in their upper levels.

Although water can exist as a liquid at temperatures below 0 °C, there is a limit below which pure liquid water cannot be subcooled. This limit, about -40 °C, is sometimes called the *Schaefer point* to honor Vincent Schaefer, who first established this temperature experimentally. Below the Schaefer point, pure water freezes by *homogeneous nucleation* of the ice phase.

Everything that you need to demonstrate subcooling can be found in a kitchen. On a cold (≈ -8 °C), calm morning well before sunrise, we placed small water drops (1–3 mm) from the end of a matchstick onto a bare "tin" can and one with a thin black plastic coating on its bottom, more than a dozen drops on each. The unfiltered well water initially was at 3 °C. The cans were left outside overnight in a plastic bag. Their surfaces were not cleaned. This was a severe test because the initial water temperature was close to 0 °C and the drops were placed on metallic objects a thousand times more massive and at subfreezing temperatures. We did everything possible to make the drops freeze quickly. And yet after more than an hour, five large drops on the bare can remained liquid. They glistened in the light of a flashlight, but to make sure they were liquid we probed some with the matchstick. The frozen droplets were cloudy. The smaller drops were on the plastic surface and most of them were unfrozen, especially the smallest ones. Drops that froze, some quickly, either must have contained a freezing nucleus or were deposited on one. And there had to be particles in the water because it is impossible to get rid of all of them (we know from the experience of trying). But all particles are not equally effective as nuclei. Moreover, their effectiveness increases with decreasing temperature. We tried this experiment in the freezing compartment of a refrigerator (≈ -19 °C) and all the drops froze. Do an experiment only once and you will know for the rest of your lives that water does not necessarily freeze at < 0 °C regardless of what is in thousands of textbooks; because of this, 0 °C is more accurately called the melting point of ice, rather than the freezing point of water. Ponds and lakes don't subcool because the probability of a sample of water having a freezing nucleus increases with the sample volume—and all it takes is one.

Undercooling is sometimes used as a synonym for the more widely used "supercooling," even though they are antonyms, which is why we prefer subcooling. "Sub" means under, "super" means over. Lake Superior is geographically above the other Great Lakes, not better than them.

Dew Point Depression and Human Comfort in Hot, Humid Weather

Genesis pronounces that "in the sweat of thy face shalt thou eat bread." Today, even sedentary office workers experience sweat on their faces, if only in gymnasiums. Sweat is *sensible perspiration*. The ceaseless transport of water vapor from *apparently* dry skin is about half of total *insensible perspiration*, the other half from respiration. The amount is not negligible, at least about 0.4 kg per day from the skin of an adult. This mass times an enthalpy of vaporization 2400 kJkg^{-1} (at $\approx 27\,°C$) is ≈ 960 kJ, ≈ 230 kcal, a substantial fraction of basal metabolism, the kilocalories required to do no more than vegetate.

Appearances to the contrary, water accreting on a can of cold beer in a humid environment is condensed ambient water vapor, *not* sweat. If it is, change brands. Sensible perspiration is produced by sweat glands in the skin, as much as ten times the amount of insensible perspiration. But absent excretion by them, there is still a *net* transport of water vapor from skin to ambient air, which depends on its humidity.

Because relative humidities are reported in newspapers and by radio and television meteorologists, summer discomfort has become associated with high air temperatures and high relative humidities. But meteorologists often take dew point (strictly, dew point depression, the difference between temperature and dew point) as a better measure of discomfort.

From the definitions of relative humidity and dew point

$$r = \frac{e}{e_s(T_a)} = \frac{e_s(T_d)}{e_s(T_a)}, \tag{5.72}$$

where the temperature of the air is T_a and its dew point is T_d. Evaporative cooling requires a *net* vapor flux from skin to surrounding air. The greater this flux, determined by the difference between the saturation vapor pressure at skin temperature, which is not uniform over the body, and the vapor pressure in the surrounding air, the greater the cooling. For simplicity, we take skin and air temperature to be the same. This likely underestimates the net flux because skin temperature is often higher than air temperature (which can be verified with an infrared thermometer). We assume that the evaporative flux is proportional to

$$e_s(T_a) - e = e_s(T_a) - e_s(T_d) = e_s(T_a)(1 - r). \tag{5.73}$$

All else being equal, an increase in relative humidity from 80% to 90% halves the evaporative flux. An even more dramatic example is the change from a relative humidity of 90% to 98%, a fivefold decrease in the evaporative flux. Relative humidity *per se* is not a good index of human comfort in hot, humid weather. We need a better index. From Eq. (5.49)

$$r = \frac{\exp(-\ell_v/R_v T_d)}{\exp(-\ell_v/R_v T_a)} = \exp\left[-\frac{\ell_v(T_a - T_d)}{R_v T_a T_d}\right]. \tag{5.74}$$

Because absolute air temperatures and dew points do not vary greatly, for our purposes we can take the coefficient of $(T_a - T_d)$ in Eq. (5.74) to be approximately constant:

$$r \approx \exp[-C(T_a - T_d)], \tag{5.75}$$

and hence,

$$1-r \approx 1 - \exp[-C(T_a - T_d)]. \tag{5.76}$$

Relative humidities affect human comfort in hot weather mostly when they are fairly high, which requires that $C(T_a - T_d)$ be small compared with unity. We therefore may approximate the exponential function in Eq. (5.76) as $1 - C(T_a - T_d)$, which yields

$$1-r \approx C(T_a - T_d). \tag{5.77}$$

Thus, at high relative humidities, the difference between air temperature and dew point is approximately proportional to the evaporative cooling flux. This is why when you ask a knowledgeable meteorologist for a forecast of how comfortable a hot summer's day in the humid Eastern United States will be, you are likely to be given air temperature and dew point, rather than air temperature and relative humidity. Mark Lawrence derived a value $C = 0.05$, where the temperature difference is in Celsius, which yields fairly accurate results for $r > 0.5$.

If we do not assume that skin and air temperature are equal, Eq. (5.73) predicts that net vapor transport for a fixed skin temperature decreases linearly with increasing vapor pressure of ambient air. But measurements made on live subjects indicate that the relation is *nonlinear* and beyond a certain vapor pressure, water loss is *greater* than predicted by a linear relation. The same investigators obtained a linear relation for a water surface, evidence that their results for skin were not because of a faulty technique. Breakdown of the linear theory is hardly a surprise given that skin is a much more complicated three-dimensional structure than an air–water interface. Skin and other organs are too complex to conform to idealized physical theories applied to simple conceptual models.

Except in nudist colonies, humans are usually partly clothed. Even absent evaporative cooling, clothing reduces cooling. The extent to which clothing impedes net vapor transport from clothed humans is a subject unto itself.

We live where summers are hot and humid and wish for the summer temperatures and dew points in Tonopah, Nevada and Ridgecrest, California, where peak temperatures are around 90 °F, but dew points (strictly, frost points) can be in the 20s. We have experienced firsthand that despite these high air temperatures we can be surprisingly comfortable. In contrast, when air temperature is 90 °F in the Eastern United States accompanied by dew points around 70 °F we are miserable. But if the dew point drops to 50 °F the next day, life becomes good again.

The lowest temperature attainable by a wet object (e.g. a damp dishrag) because of evaporative cooling is *not* the dew point but the somewhat higher wet-bulb temperature (see Section 6.4).

Lapse Rate of the Boiling Point

We defer a more complete discussion of boiling until Section 5.10 (see also comments about boiling in Section 5.4). For the moment, we use the boiling point of water to show yet another result obtained with the Clausius–Clapeyron equation. The boiling point T_b of water is *defined* as the temperature at which its saturation vapor pressure is equal to local atmospheric pressure:

$$e_s(T_b) = p. \tag{5.78}$$

Differentiate this equation to obtain the lapse rate of the boiling point:

$$\frac{de_s(T_b)}{dT_b}\frac{dT_b}{dz} = \frac{dp}{dz}, \tag{5.79}$$

from which it follows that

$$\frac{1}{e_s(T_b)}\frac{de_s}{dT_b}\frac{dT_b}{dz} = \frac{1}{p}\frac{dp}{dz}. \tag{5.80}$$

We may take pressure to be an exponentially decreasing function of altitude z with scale height H:

$$p = p_0 e^{-z/H}. \tag{5.81}$$

Combine Eqs. (5.48), (5.80), and (5.81) to obtain

$$\frac{dT_b}{dz} = -\frac{R_v T_b^2}{\ell_v H}. \tag{5.82}$$

With the assumption of a constant enthalpy of vaporization, Eq. (5.82) can be integrated

$$\frac{T_b}{T_{bo}} = \left(1 + \frac{z}{H_b}\right)^{-1}, \tag{5.83}$$

where T_{bo} is the boiling point at the surface and the *scale height for boiling* is

$$H_b = \frac{H\ell_v}{T_{bo}R_v}. \tag{5.84}$$

For $T_{bo} = 373$ K, $\ell_v = 2.26 \times 10^6$ Jkg^{-1} at this temperature (from Eq. 5.64), and $H = 8$ km, the scale height for boiling is about 107 km. From sea level to the summit of Mt. Everest we may approximate Eq. (5.84) by

$$T_b \approx T_{bo}\left(1 - \frac{z}{H_b}\right), \qquad z \ll H_b \tag{5.85}$$

and the corresponding lapse rate of the boiling point of water becomes

$$\frac{dT_b}{dz} \approx -\frac{T_{bo}}{H_b} \approx -3.5\,°\text{Ckm}^{-1}. \tag{5.86}$$

With a bit of care, boiling points can be measured to within about $0.2\,°C$ even using an ordinary thermometer. This uncertainty in temperature corresponds to an uncertainty in elevation of about 60 m. Although this is not sufficiently precise for landing an airplane, estimating your elevation on a mountain backpacking trip could provide you with a bit of fun over a campfire while boiling water for a cup of tea. We have inferred from boiling point measurements and Eq. (5.86) an elevation increase of 160 m (from 420 m) to within 25%.

Evaporative Cooling and Condensational Warming

Consider an idealized two-phase, single-component system composed of water (no air) in a rigid, closed, insulated container at an initially uniform temperature T_i. Water vapor is above the liquid. If the vapor pressure is less than the saturation vapor pressure at this temperature, liquid water will evaporate until evaporation and condensation rates are equal. The evolution of this system is neither a constant pressure nor a constant temperature process, but it *is* an adiabatic, isochoric process contrived to obtain a definite result.

Because it is easy to forget that internal energy is defined only to within an additive constant, we should write the specific internal energy of liquid water and water vapor as $u_\xi(T,p)+C$, where C is a constant; ξ indicates the phase ($\xi = v$ for vapor, w for liquid). At the same temperature and pressure, u is different for the two phases because of potential energy differences at the molecular level, without which there would be no distinction between them. The total internal energy of the system is constant, and hence,

$$m_{wi}u_w(T_i,p_i) + m_{vi}u_v(T_i,p_i) = m_{wf}u_w(T_f,p_f) + m_{vf}u_v(T_f,p_f), \qquad (5.87)$$

where i and f denote initial and final states. Underlying this equation is the *assumption* that the internal energies are *additive*. That is, the *interaction energy* between the two phases is negligible. But it not always is for sufficiently small systems (see Section 5.10). The terms involving C cancel. Because total mass $m = m_v + m_w$ is conserved

$$m_{wi} - m_{wf} = m_{vf} - m_{vi} = \Delta m_v. \qquad (5.88)$$

The final uniform pressure is determined solely by the final temperature. If the entire system were at a uniform temperature and the pressure were the saturation vapor pressure at that temperature, there would be no net evaporation, and hence, no temperature change.

We assume that the specific internal energy of the vapor and liquid do not depend on pressure, which is less valid for the vapor. If we expand the specific internal energy of both liquid and vapor in powers of the temperature difference $\Delta T = T_f - T_i$,

$$u(T_f) = u(T_i) + \frac{\partial u}{\partial T}\Delta T + \dots, \qquad (5.89)$$

and truncate this series after the second term, and use Eq. (5.88), Eq. (5.87) becomes

$$(m_{wi} - m_{wf})[u_w(T_i) - u_v(T_i)] = (m_{vf}c_v + m_{wf}c_w)\Delta T. \tag{5.90}$$

Because the density of liquid water is so much greater than that of water vapor, $u_v(T_i) - u_w(T_i) \approx \ell_v(T_i) - R_v T_i$, which is ℓ_v to within $\approx 5\%$. If we ignore this small difference, the final result is

$$\Delta T \approx \frac{-\Delta m_v \ell_v}{m_{vf}c_v + m_{wf}c_w} = \frac{-\Delta m_v}{m} \frac{\ell_v}{\bar{c}}. \tag{5.91}$$

$\bar{c}$ is a mass-weighted specific heat capacity (per unit mass):

$$\bar{c} = xc_v + (1 - x)c_w, \tag{5.92}$$

where $x = m_{vf}/m$ and $0 \leq x \leq 1$.

For net *evaporation*, $\Delta m_v > 0$ and $\Delta T < 0$, which is consistent with our experience of *evaporative cooling*. The temperature decrease depends on the ratio of the enthalpy of vaporization to the mass-weighted specific heat capacity, the dimensions of which are temperature. Solely by dimensional arguments, the temperature decrease must depend on this ratio. The unknown factor is the ratio of the mass change of the water vapor to the total mass, which is small if $\Delta T \bar{c}/\ell_v$ is. The final temperature of liquid and vapor is the same, and lower than the initial temperature. Total energy is conserved, so the kinetic energy decrease must be balanced by the potential energy increase of the molecules that escape the liquid. This is manifested macroscopically by a decrease of temperature (cooling).

For net *condensation*, $\Delta m_v < 0$ and $\Delta T > 0$. Although a temperature increase is not readily experienced, it can be observed experimentally. Attributing this to the "release of latent heat" is a shopworn, thought-stopping cliché. A better term is *condensational warming*, which is more closely linked to a molecular interpretation (see the end of Section 5.1). Warming implies a temperature increase, cooling a temperature decrease, whereas Q in the first law can be positive or negative. If $\Delta T < 0$, should we say that this is the "capture of latent cold," the logical counterpart of the "release of latent heat?"

Equation (5.91) contains the unknown ratio $\Delta m_{vf}/m$. But $\bar{c}$ is around $2 \times 10^3 \, \text{J kg}^{-1}\text{K}^{-1}$, ℓ_v is around $2 \times 10^6 \, \text{J kg}^{-1}$, and hence, $\ell_v/\bar{c} \approx 1000$ K. Because of this large multiplicative temperature, not much evaporation or condensation can result in cooling or warming of a few degrees in this idealized problem. What happens in reality? To answer this, we saturated a piece of absorbent paper with water and directed air over it with a fan. The initial temperature of the water, dry paper, and its substrate were about air temperature. Within seconds, the temperature of the wet paper dropped by $\approx 5\,°\text{C}$. And it did not appear any less wet—evidence that not much water evaporated. This experiment is complicated by energy transfer to the wet paper from the higher temperature air as well as a variable and indefinite total mass. Nevertheless, we observed an appreciable temperature drop of a small amount of water (not a bucketful) because of a much smaller amount of evaporation, as predicted. This is an example in which a surface temperature can drop *below* air temperature because of *net* evaporation.

Using two thermocouples that agreed to within $0.1\,^\circ\text{C}$, we simultaneously monitored the temperature of water about 5 mm deep in a small, shallow, ceramic bowl and the temperature of adjacent still air in an unheated room. The water temperature was about $1\,^\circ\text{C}$ lower. For both of these real processes, the systems were *not* isolated.

This analysis, like all others in this book, is based on idealizations and approximations. We assumed a constant mass system in a rigid, insulated container with zero thermal conductivity so that it does not affect the temperature change. Atmospheric processes in unconstrained air are messier and not under our control. We are masters of the universe only on paper. Although Eq. (5.91) can help us understand a class of real processes, and even give approximate quantitative results, it is not rigorous and cannot be expected to describe reality exactly.

5.4   van der Waals Equation of State

Evaporation of liquid water is an example of a *phase transition*. Water exists in three phases—vapor, liquid, and solid (ice also exists in several different crystalline phases)—and can pass from one phase to another. We begin our discussion of phase transitions and phase diagrams by way of the van der Waals equation of state.

The ideal gas law, $pV = NkT$, and its many variants is an example of a *thermal equation of state*, the qualifier "thermal" often omitted, an equation relating measurable thermodynamic variables. Although air is to good approximation an ideal gas, one of its components is water vapor, which at normal terrestrial temperatures can condense into liquid (or solid) water. Neither liquid nor solid water even approximates an ideal gas. Moreover, there is nothing in the ideal gas equation of state to indicate the abrupt transition of water from one phase (liquid, say) to another phase (solid, say). You can swim (uncomfortably) in water at a temperature slightly above freezing; a tiny drop in temperature—or a miracle—allows you to walk on water. Phase transitions are of great meteorological importance. Rain is born from clouds, which in turn forms when water changes phase from vapor to liquid (or solid). To understand these phase changes, we need a better equation of state than the ideal gas equation.

In an irreverent piece entitled "Memorial Service," H. L. Mencken asks:

> Where is the graveyard of dead gods? ... What has become of Sutekh, once the high god of the whole Nile Valley? ... What has become of Resheph, Anath, Ashtoreth, Nebo, Melek, Ahijah, Isis, Ptah, Baal, Astarte, Hadad, Dagon, Yau, Amon-Re, Osiris, Molech? ... they all have gone down the chute ... They were gods of the highest dignity—gods of civilized peoples—worshipped and believed in by millions. All were omnipotent, omniscient and immortal. And all are dead.

In a similar vein, we may ask "where is the graveyard of dead equations of state?" What has become of those of Boyle and Gay–Lussac, Rankine, Joule and Thomson, Dupré, Hirn, Recknagel, van der Waals, Clausius, Lorentz, Walter, Amagat,

Sarrau, Thiesen, Natanson, Sutherland, Lagrange, Violi, Antoine, Tait, Swart, Schiller, Jäger, Battelli, Brillouin, Weinstein, Rose Innes, Boltzmann, Boltzmann and Mache, Reiganum, Dieterici, Starkweather, Berthelot, Tumlirz, Mie, Goebel, Smoluchowsky, Batschinski, Planck, Leduc, Onnes, Peczalsi, Dalton, Kam, Wohl, Shaha and Basu, Chapman and Appleby, Schrieber, Porter, Boynton and Bramley, and Fouché? Once the darling of their creators, they all, like the gods listed by Mencken, are dead—with two exceptions. We already have considered one, that of Boyle and Gay–Lussac (under the name ideal gas law). The only other surviving equation of state still worshipped by many followers is that of van der Waals.

The late, great folklorist of the American Southwest, J. Frank Dobie, opined that a doctoral thesis is the transfer of bones from one graveyard to another. Alas, this is probably true of many theses, but a notable exception is that of Johannes Diderik van der Waals (b. Leiden, The Netherlands, 1837; d. Leiden, 1923). On June 14, 1873, van der Waals defended his thesis "On the Continuity of the Gaseous and the Liquid State." This event had such far-reaching consequences that a century later an international conference celebrating it was held at the University of Amsterdam, where van der Waals had been professor of physics until 1907.

The equation of state in van der Waals's thesis is remarkable in that it is readily derivable, physically transparent, and accounts at least qualitatively for much of the observed behavior of water (and other liquids).

To begin, we write the ideal gas equation of state in the form

$$pv_m = R^*T, \tag{5.93}$$

where v_m is the molar specific volume N_aV/N. This equation cannot be correct because it predicts that at constant pressure the molar specific volume is zero at $T = 0$. This defect is easily remedied by subtracting from v_m a quantity b to account for the finite volume of gas molecules:

$$p(v_m - b) = R^*T. \tag{5.94}$$

The quantity $v_m - b$ may be interpreted as the volume in which molecules are free to roam. Van der Waals took b to be four times the volume of the molecules, but also recognized that b might have to be decreased with increasing density. The precise relation of b to molecular volume is largely irrelevant because a molecule does not have a precisely defined volume. Each experimental method for determining molecular volumes (diameters) yields different results. Indeed, one method makes use of the van der Waals equation (see Problem 23). Another interpretation of b is that it is a consequence of the strong repulsive forces between molecules at very short distances from each other. Given that molecules do not have well-defined volumes, this interpretation is more in accord with reality. When we say that molecules have finite volumes, what we mean is that they do not appreciably penetrate each other because of strong repulsion at very short distances.

Equation (5.94) yields a nonzero volume at absolute zero, a requirement of any equation of state faithful to reality. We rewrite this equation in the form

$$p = \frac{R^*T}{v_m - b} \qquad (5.95)$$

so that we can turn our attention to how pressure is modified when intermolecular interactions are accounted for.

The ideal gas law is obtained under the assumption that intermolecular forces are negligible, which is strictly true only for average intermolecular separation greater than several molecular diameters. As this separation decreases, interactions make their presence felt to the extent that they no longer are negligible. With increasing number density, average intermolecular separation decreases, and the molecules exert attractive forces on each other. These forces give solids and liquids their cohesiveness. For very small intermolecular separations, the force between molecules is repulsive, as evidenced by the (relative) incompressibility of liquids and solids. We restrict ourselves to average intermolecular separations such that the force between molecules is attractive (as pointed out previously, b may be interpreted as accounting for repulsion). It is plausible that because of attractive forces, the (measured) pressure would be less, absent these forces. So again, we modify the ideal gas law by subtracting from the pressure in Eq. (5.95) the total force (per unit area) that molecules on one side of an imaginary bounding plane in a gas exert on molecules on the other side. We take the attractive force between molecules to decrease with distance such that beyond a certain fixed distance of separation, the radius of a sphere of influence, the attractive force exerted by one molecule on another, is negligible. Consider a single molecule on the left side of the boundary. The total attractive force experienced by this molecule is the volume of the sphere of influence times the number density of molecules (assumed uniform throughout the gas) times the average force between molecules. Only those molecules on the left side of the boundary in a layer of thickness equal to the radius of the sphere of influence experience attractive forces exerted by molecules on the right side. The total force on all the molecules in this layer is the product of its thickness, area, and number density of molecules. Thus, the total force per unit area that molecules on one side of the boundary exert on molecules on the other is proportional to the square of the number density of molecules *provided* that the average force between molecules does not depend on number density. Because the number density is proportional to $1/v_m$, the subtractive correction to Eq. (5.95) is proportional to $1/v_m^2$, and hence, we may write

$$p = \frac{R^*T}{v_m - b} - \frac{a}{v_m^2}, \qquad (5.96)$$

where a is a positive constant. From Eq. (2.13) and (2.78) it follows that $v_m^2 \propto d^6$, where d is the average separation between molecules.

This equation is considered so important in physics that The Netherlands issued a postage stamp bearing it and its inventor.

To give us some confidence in Eq. (5.96), we use it in Eq. (5.35):

$$\frac{\partial U}{\partial V} = T\frac{\partial p}{\partial T} - p = \frac{a}{v_m^2}. \tag{5.97}$$

According to this result, the internal potential energy of a van der Waals gas decreases (becomes more negative) with decreasing average separation, consistent with the assumption that gas molecules, on average, exert only attractive forces on each other. Repulsive forces were unacceptable to van der Waals. Even in their absence molecules would not necessarily clump because they are moving, as evidenced by our solar system.

Consider now the van der Waals isotherms. For constant T, p approaches infinity as v_m approaches b (for an ideal gas, p becomes infinite only in the limit of zero molar specific volume). For $v_m \gg b$, and because the first term in Eq. (5.96) is inversely proportional to v_m whereas the second term is inversely proportional to v_m^2, the second term is negligible compared with the first for sufficiently large v_m. For such large values of v_m, the van der Waals equation of state reduces to the ideal gas equation. Between these two limits, however, the van der Waals equation exhibits behavior quite unlike that of the ideal gas equation. For example, both the first and second derivatives of p with respect to v_m may vanish

$$\frac{\partial p}{\partial v_m} = 0, \quad \frac{\partial^2 p}{\partial v_m^2} = 0. \tag{5.98}$$

Because Eqs. (5.96) and (5.98) are three equations in three unknowns, we can solve for the pressure, temperature, and molar specific volume (p_c, T_c, v_c) for which these equations are satisfied:

$$p_c = \frac{a}{27b^2}, \quad T_c = \frac{8a}{27bR^*}, \quad v_c = 3b = \frac{3R^*T_c}{8p_c}. \tag{5.99}$$

The three quantities a, b, and R^* can be solved for p_c, v_c, and T_c, in order to write Eq. (5.96) as

$$p_r = \frac{8T_r}{3v_r - 1} - \frac{3}{v_r^2}, \tag{5.100}$$

where p_r, T_r, and v_r are called the *reduced* pressure, temperature, and molar specific volume:

$$p_r = \frac{p}{p_c}, \quad T_r = \frac{T}{T_c}, \quad v_r = \frac{v_m}{v_c}. \tag{5.101}$$

If pressure, temperature, and molar specific volume are expressed as reduced values, the van der Waals equation has the same form for all substances, called the *law of corresponding states*.

We now examine the conditions under which a van der Waals isotherm has minima and maxima. For $\partial p_r / \partial v_r$ to vanish requires that

$$\frac{(3v_r - 1)^2}{4v_r^3} = T_r . \tag{5.102}$$

The (nonnegative) function

$$f(v_r) = \frac{(3v_r - 1)^2}{4v_r^3} \tag{5.103}$$

vanishes at $v_r = 1/3$ (the lower limit of its domain) and in the limit as v_r becomes infinite. The derivative of f vanishes for $v_r = 1$, and hence, the maximum value of $f(v_r)$ is $f(1) = 1$. This in turn implies that Eq. (5.102) cannot be satisfied for $T_r > 1$, or, equivalently, for $T > T_c$. The temperature above which van der Waals isotherms have no minima or maxima is called the *critical temperature* (why we chose the subscript c); the corresponding pressure and molar specific volume are called the *critical pressure* and the *critical molar specific volume*.

Now turn to van der Waals isobars. From Eq. (5.100) we obtain

$$T_r = \left(\frac{3v_r - 1}{8} \right) \left(p_r + \frac{3}{v_r^2} \right) . \tag{5.104}$$

For T_r to have an extremum (minimum or maximum) requires that

$$\frac{3}{v_r^2} - \frac{2}{v_r^3} = p_r. \tag{5.105}$$

The values of the function on the left side of Eq. (5.105) never exceed one, from which we conclude that T_r has no extrema if $p_r > 1$ (i.e. $p > p_r$).

To probe the significance of critical points, we must carefully examine van der Waals isotherms (Fig. 5.8). For T less than T_c, an isotherm has one maximum and one minimum between which the slope is positive, unlike for ideal gases, the isotherms of which always have negative slope. What is the significance of a positive slope?

Imagine that a gas is placed in a cylinder fitted with a piston, the entire apparatus immersed in a constant-temperature bath below the critical temperature. The weight of the piston is adjusted so that the gas pressure just balances it. We choose the weight such that p lies in a region where the slope of the isotherm is positive. All is well, until a gnat lands on the piston. Because the total weight is now slightly greater than that of the gas pressure (times the piston area), the piston sinks slightly, which causes the gas volume to decrease slightly. In a region where the slope of an isotherm is positive, a decrease in volume implies a decrease in gas pressure. Thus, the imbalance increases, and the piston descends even more. The same kind of argument holds if the gnat had initially been on the piston with the total weight of piston and gnat balancing the gas pressure (times piston area). If the gnat flies away, the piston moves upward slightly, the gas volume increases, and if the slope of the isotherm is positive the gas pressure increases, which makes the

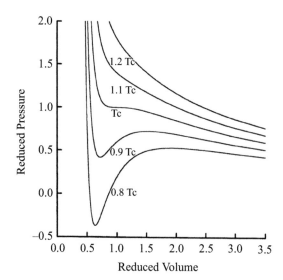

Fig. 5.8: Van der Waals isotherms near the critical temperature T_c. Reduced pressure is pressure relative to the critical pressure; reduced volume is specific volume relative to the critical specific volume. Above the critical temperature isotherms have no maxima or minima.

imbalance greater. We therefore conclude that where the slope of an isotherm is positive, the system cannot exist in stable equilibrium. Small displacements grow, rather than shrink.

The existence of a region of instability between a maximum and a minimum of a van der Waals isotherm is a signal that something happens in this region. This event is a *phase transition*, the transformation of vapor to liquid. Although the van der Waals equation does not yield the correct shape of an isotherm through a phase transition, this equation goes far beyond the ideal gas law, from which the existence of a phase transition is not even suspected. Moreover, the van der Waals equation predicts a critical temperature, above which no phase change occurs. About 12 years before van der Waals defended his thesis, Thomas Andrews (b. Belfast, Ireland, 1813; d. Belfast, 1885) had discovered the critical temperature (although not called such by him until 1869) of carbon dioxide, one of the few common gases with a critical temperature ($\approx 31\,°C$) near room temperature.

Although the van der Waals isotherm signals a phase transition in the region of the loop, this isotherm is not strictly correct for the transition, which occurs at constant pressure. Two years after van der Waals defended his thesis, the great Scottish physicist James Clerk Maxwell showed how to obtain from the van der Waals isotherm the (approximately) correct isotherm over the phase transition. Figure 5.9 shows a van der Waals isotherm below the critical temperature. At point E, the system is entirely vapor. The volume of the system is decreased at constant pressure p' until the system is entirely liquid. Along the straight line from

E to A, the system is an inhomogeneous mixture of liquid and vapor. According to the van der Waals isotherm, however, the phase transition proceeds along the loop $EDCBA$, for which the system is always homogeneous. Consider a reversible, isothermal process beginning and ending at the same state E. This cyclic process follows the van der Waals isotherm along $EDCBA$ and then returns along the constant pressure path ACE. Because this process is cyclic, the internal energy change is zero, which from the first law implies that the sum of total heating and working is zero:

$$\Delta U = \int (Q + W)\, dt = 0\,. \tag{5.106}$$

The entropy change in this cyclic process is also zero, and because the process is reversible and isothermal, we have

$$\Delta S = \int \frac{Q}{T}\, dt = \frac{1}{T} \int Q\, dt = 0\,. \tag{5.107}$$

From Eqs. (5.106) and (5.107) it follows that the total working is zero:

$$\int_E^C p\frac{dV}{dt}\, dt + \int_C^A p\frac{dV}{dt}\, dt + \int_A^C p'\frac{dV}{dt}\, dt + \int_C^E p'\frac{dV}{dt}\, dt = 0\,. \tag{5.108}$$

A simple geometrical interpretation of this equation is obtained by rewriting it

$$\int_A^C (p' - p)\frac{dV}{dt}\, dt = \int_C^E (p - p')\frac{dV}{dt}\, dt\,. \tag{5.109}$$

According to the Maxwell construction, the constant-pressure part of the isotherm over the region of a phase transition is obtained from the van der Waals isotherm by constructing a line of constant pressure p' intersecting the loop such that the areas of the two hatched regions in Fig. 5.9 are equal. The pressure p' so obtained is the saturation vapor pressure at the temperature T. In a following starred subsection, we show that the Maxwell construction applied to the van der Waals equation gives a functionally correct dependence of saturation vapor pressure on temperature (well below the critical temperature).

To understand both why the phase transition between E and A in Fig. 5.9 is likely to occur at constant pressure, and why this does not always occur, consider a vapor enclosed in a cylinder fitted with a movable piston, the entire apparatus immersed in a constant-temperature bath. Slowly move the piston downward, compressing the vapor. Eventually, a point (E) is reached where the pressure is equal to the saturation (equilibrium) vapor pressure. Now move the piston down further a hair. The vapor pressure rises above the equilibrium value, and hence, some of the vapor condenses in order to reduce the vapor pressure to the equilibrium value (provided there is a surface onto which vapor can condense). Because some of the vapor condenses to liquid, the volume of the system decreases. Now push the piston downward another hair. Again, for a brief moment the vapor pressure exceeds the equilibrium value until vapor condenses, again restoring the vapor pressure to the

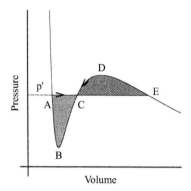

Fig. 5.9: A van der Waals isotherm showing the Maxwell construction. The pressure p' is such that the area enclosed by ABC is equal to that enclosed by CDE. This pressure is the saturation vapor pressure for the isotherm.

equilibrium value. Because some vapor condenses, the volume of the system further decreases. And so on and on, into the long hours of night, until all the vapor has been transformed into liquid. The fundamental reason why the isotherm is flat between 100% vapor at E and 100% liquid at A is that any increase in pressure results in an excess of condensation over evaporation, causing vapor to condense, thereby decreasing the pressure—a negative feedback.

When all the vapor has condensed, further compression yields a small change in volume for a large change in pressure. Because of the relative incompressibility of liquids, the isotherm on the liquid side of the phase change should be much steeper than on the gas side, as indeed it is according to the van der Waals equation.

Now go back to point E on Fig. 5.9. Suppose that, as the vapor is compressed, no condensation were to occur. This could happen within the vapor far from surfaces (including microscopic ones) onto which vapor condenses. The pressure would increase above the equilibrium value, and the vapor would be *supersaturated*, which means that its pressure is greater than the stable equilibrium value. This state of supersaturation is *metastable*: The vapor can exist in a supersaturated state, more or less indefinitely, until someone introduces a free surface into the vapor, which results in condensation, and hence, a lowering of the pressure to the stable equilibrium value. According to the van der Waals equation of state, supersaturation cannot rise above the point D, where the system becomes unstable.

Now consider the phase transition in the opposite sense: from liquid to vapor. Begin at point A, where the system is entirely liquid. As the volume of the system increases beyond A, vapor begins to appear (provided a free surface exists). If the piston is raised a hair, the vapor expands, and its pressure drops below the equilibrium value. This causes condensation to be less than evaporation (which is determined by temperature), and hence, net evaporation increases the vapor pressure back to the equilibrium value. And so on and on, until all the liquid has been transformed into vapor, just as in the reverse process of compression of vapor to liquid.

Suppose that as the water were expanded no vapor were to appear. This could occur within the liquid far from any surfaces, even microscopic ones. The pressure would drop below the equilibrium value and the liquid would be *superheated*, which means that its pressure is less than the equilibrium value. This state of superheating is metastable: the liquid can exist in a superheated state, more or less indefinitely, until someone introduces a free surface into the liquid, which results in evaporation, and hence, an increase of pressure to the stable equilibrium value. According to the van der Waals equation of state, superheating cannot descend below the point B, where the system becomes unstable.

Metastable states between A and B are said to be states of superheating because the boiling point of a substance is defined as the temperature at which the equilibrium vapor pressure is equal to the surrounding pressure. Liquid pressures in the region between A and B in Fig. 5.9 are lower than the equilibrium vapor pressure. Stated another way, the temperature along the metastable isotherm AB is higher than for all equilibrium isotherms (which are horizontal) intersecting AB.

Given the brief mention, at best, of superheating and supersaturation in physics textbooks, one might reasonably conclude that these are exotic phenomena observed only in well-equipped laboratories by scientists wearing white coats. Yet, you need go no farther than your kitchen to be in the presence of both phenomena occurring simultaneously: Just heat a kettle filled with water. Soon the water will boil, and a mixing cloud (Section 6.9) will emerge from the spout of the kettle. As we shall see in Section 5.11, a slight degree of supersaturation must occur for mixing clouds to form. Almost from birth, students are told that at sea level pure water boils at a temperature of exactly 100 °C. This is repeated so many times that it becomes a sacred truth, which to even question is to risk being burned at the stake as a heretic. It is indeed amusing how people so passionately defend a temperature—few of whom are ever likely to have measured it. Easy you say. Just shove a thermometer into boiling water. Would you be willing to bet a large sum of money that if, at an elevation where the pressure was ≈ 1013 hPa, you were to immerse the bulb of a perfect thermometer into pure boiling water, the thermometer would read exactly 100 °C? Don't argue about the outcome of a hypothetical experiment; perform a real one. If you do, you are likely to obtain a boiling temperature *above* 100 °C. A few degrees of superheat in water boiling in a kettle are almost inevitable. Observe for yourselves.

Measuring the boiling point of a liquid is not trivial. Plunging the bulb of a thermometer into a boiling liquid results in erroneously high values. In light of our discussion of superheating, we can understand why. Within a liquid there are no free surfaces other than perhaps those associated with microscopic suspended particles or cracks and pits on the walls of the containing vessel. If these particles are in short supply, if the vessel walls are smooth, considerable superheating within the liquid is possible. But the liquid–vapor interface is a free surface, which is our clue how to measure accurately the boiling temperature: suspend a thermometer just above the surface of the boiling liquid. The temperature so obtained is sometimes called the *condensation temperature.*

It has been known for over 200 years that the temperature of boiling pure water at sea level is *not* 100 °C. In perhaps another 200 years this readily verified experimental fact may have made its way into textbooks.

Now imagine air containing vapor to be cooled at constant pressure to a temperature below the dew point. The air is very clean and far from any surfaces. Thus, there is nothing onto which vapor can condense, and supersaturation is possible.

Merely because we do not directly observe superheating and supersaturation does not mean that they do not occur. There is a world of difference between the frequency of occurrence of an event and the frequency of its observation. Some textbooks state that pure water at sea level *cannot* be heated above 100 °C and that the relative humidity of air *cannot* rise above 100%. But Nature defiantly thumbs her nose at these textbooks and flagrantly violates all their proscriptions.

For some temperatures, van der Waals isotherms dip into regions of negative pressure (Fig. 5.8), which at first sight may be disconcerting. Yet, prediction of negative pressures is a success of the van der Waals equation, not one of its failures. Although negative pressures (in water, say) are not obviously part of our everyday lives, liquid water can indeed exist in *tension*, rather than compression. Because the region of negative pressures occurs where the van der Waals equation predicts a superheated state, we would suspect that tensile water is unstable, as indeed it is. It takes care to produce such water, but it can be done. In fact, it has been done for almost 150 years. When Lyman Briggs retired as director of the U.S. National Bureau of Standards, he devoted his sunset years to measuring ever lower negative pressures in water, reaching the lowest yet recorded, almost -300 atmospheres (and more than -400 atmospheres in mercury). As with superheating and supersaturation, negative pressures are everyday occurrences. For example, tension in plant fluids is quite common.

The phase transition from liquid to vapor (or vice versa) occurs only at temperatures below the critical temperature. Above this temperature, isotherms are continuous curves. During compression there is no sharp dividing line between gas and liquid. Although we may arbitrarily define a particular specific volume as the point at which a gas has been liquefied, this does not define a sharp boundary on one side of which lies liquid and on the other side of which lies gas. Although we approached the concept of critical temperature (and pressure) by way of the van der Waals equation of state, the existence of critical points in no way depends on the strict validity of this equation. For an experimentally determined phase diagram we can define the critical temperature T_c as the lowest temperature above which $\partial p/\partial v_m$ is continuous and the critical pressure p_c as the lowest pressure above which $\partial T/\partial v_m$ is continuous. The critical specific volume v_c is the measured value at (p_c, T_c). A vapor once was defined as a gas that can be liquefied easily. But what is meant by "easily" depends on how well one's laboratory is equipped and how much effort one is willing to expend. Now that all gases have been liquefied, it no longer makes sense to define a vapor to be an easily liquefied gas. Instead, a vapor may be defined as a gas below its critical temperature. Above this temperature, no gas can be liquefied because there is no sharp distinction between a gas and a liquid.

The following table shows critical temperatures and pressures for the major constituents of dry air and for water vapor (from *Smithsonian Physical Tables*).

Of these gases, only water vapor has a critical temperature well *above* temperatures of even the hottest regions of the atmosphere (within the lowest few hundred kilometers). And the critical temperatures of the other gases are well *below* the lowest temperatures in the atmosphere. For these reasons, we refer to water vapor, but not nitrogen or oxygen vapor, and must greet (with a horse laugh) assertions about air "condensing." The air of our everyday lives does not condense (for which we should be grateful), although one of its constituents may.

Gas	$t_c(°C)$	$p_c(\text{atm})$
Argon	−122	48.1
Nitrogen	−147	33.6
Oxygen	−119	49.7
Water	374	218

We note in Section 2.1 that the only molecular property in the ideal gas law is mass, whereas the van der Waals equation adds the molecular properties a and b. They are not obviously correlated with molecular structure and mass. For example, the molecular weight, critical temperature, and critical pressure of carbon dioxide are 44, 31.2 °C and 7.4 MPa, whereas for ozone they are 48, −12.3 °C, and 0.55 MPa.

We point out in Sections 3.7 and 7.1 that the thermophysical and transport properties of water are not greatly different from those of other atmospheric gases. But there is one glaring exception, absent which water vapor would be just another *permanent* atmospheric gas. There would be no clouds, rain, or oceans on Earth— nor humans to bemoan this. If you want a lucky number to tattoo on an ankle, consider 374. It is why we are on Earth, not motes in interstellar clouds.

Because of the great success of the van der Waals equation in yielding critical points, van der Waals isotherms are almost invariably shown for temperatures near the critical temperature, as in Fig. 5.8. But such a figure can be highly misleading for temperatures well below the critical temperature. Figure 5.10 shows such an isotherm. Note that reduced volume is plotted on a logarithmic scale and that the reduced volume of the vapor is more than 10,000 times greater than that of the liquid. The mere existence of a critical temperature and its physical significance gives us insight into the temperature dependence of some physical variables. For example, in Section 5.3 we derived the temperature dependence of the enthalpy of vaporization of water, which, for the temperature range considered, decreases linearly with increasing temperature, implying that at some temperature the enthalpy of vaporization of water should vanish. But we don't need equations to realize that at the critical temperature the enthalpy of vaporization must vanish. If there is no phase change above the critical temperature, there can be no enthalpy difference between liquid and vapor, and the enthalpy of vaporization is zero. Extrapolating Eq. (5.64) to zero yields a critical temperature for water of about 1340 K, more than twice its actual value. This error is no surprise, given that Eq. (5.64) was obtained under the assumption of constant specific heat capacities for liquid water and water vapor, whereas on physical grounds these two heat capacities must converge as the temperature approaches the critical temperature. Surface tension (Section 5.10) also must decrease with increasing temperature, vanishing at the critical temperature.

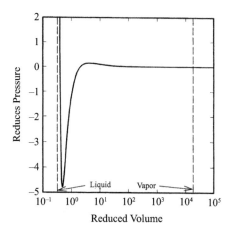

Fig. 5.10: At temperatures well below the critical temperature, Fig. 5.8 is misleading. For such temperatures, reduced volume should be plotted on a logarithmic scale. Shown here is a van der Waals isotherm for $T = 0.453T_c$, which corresponds to water at 20 °C.

Must a Liquid Boil in Order to Evaporate?

One can find in many books the assertion that water is raised to its boiling point and "then" begins to evaporate, or that the boiling point is the temperature at which the liquid-to-vapor phase transition occurs. And the latent heat of vaporization is sometimes called the latent heat of boiling. In light of everyday experiences, these assertions make no sense. Water spilled on an impervious kitchen floor eventually disappears. Water evaporates even though its temperature never rises above that of the kitchen. Even the hottest kitchen is well below 100 °C. Does water really have to be heated to its boiling point in order to evaporate?

To answer this, we must distinguish between what happens in textbooks and what happens in reality. Consider once again a cylinder fitted with a piston and partly filled with pure liquid water. Initially the piston rests on its surface. Now imagine that the water is slowly heated (at constant pressure equal to the weight of the piston divided by its cross-sectional area). At a definite temperature the piston rises slightly, and a thin gap appears between the water surface and the bottom of the piston. This gap contains water vapor—and nothing but water vapor—that has evaporated from the liquid. At this point the liquid is in equilibrium with its vapor. Moreover, the pressure of the vapor is the same as that of the liquid. Further heating is not accompanied by a temperature increase until all the liquid water has been transformed into vapor.

Because the boiling point of any substance is *defined* as the temperature at which its saturation vapor pressure equals its pressure, for the idealized experiment just considered it is indeed true that water begins to evaporate at its boiling point. But this does not mean that water in the open air must boil in order to evaporate. Evaporation and boiling are separate processes, and the distinction between them must be maintained to avoid needless confusion. Evaporation can and does take place without boiling, but not the converse. To fix in your mind the difference between evaporation and boiling, recall the witches' incantation from *Macbeth*: "Double, double toil and trouble; Fire burn and cauldron bubble." This is a vivid description of boiling, not evaporation. Can you imagine witches cackling over evaporating water? Boiling means bubbles, and hence, to understand boiling you have to understand the birth, life, and death of vapor bubbles (see Section 5.10).

Misconceptions about the abruptness of the phase transition from liquid water to vapor only at one definite temperature may originate from confusion with the transition from ice to liquid water. In our thought experiment with water, the transition occurs suddenly at a temperature that *depends strongly* on the applied pressure. But if liquid water is replaced with ice at, say −25 °C, and heated, liquid abruptly appears at about 0°C *almost independent* of this pressure.

Can a Solid Boil Before It Melts?

We have made heretical assertions about boiling, so one more probably won't condemn us to even lower depths of Hell. The boiling point of a substance is often treated as one of its fundamental properties. It is not. The (normal) boiling point of a substance is *defined* as the temperature at which its equilibrium vapor pressure is equal to one atmosphere. The "atmosphere" here is that on a particular planet, the one on which we happen to live. If we lived on a different planet, or if our atmosphere were different from what it is, boiling points would change, possibly dramatically, whereas melting points, specific heat capacities, and enthalpies of vaporization would hardly change at all. The saturation vapor pressure of water— liquid or ice—is about 6.11 hPa at 0 °C. If we lived on a planet with a surface atmospheric pressure of, say, 2 hPa, ice at −1 °C would be at a temperature above its boiling point—and yet be just as solid.

At this point, you might be saying to yourselves, "We don't live on a planet where one atmosphere is 2 hPa; so your argument, although amusing, is irrelevant. On our planet, the center of the universe, all solids melt before they boil." This is indeed true of most solids, but there is at least one exception: phosphorous. This element exists in a bewildering variety of forms called allotropes. Chemically they are identical, but structurally they are different. Red phosphorous (the color of which varies) exists in six forms, with melting points between 585 °C and 600 °C. But the temperatures at which the equilibrium vapor pressure of various forms of red phosphorous is one atmosphere range from 397 °C to 431 °C Thus, red phosphorous can truly be said to be a solid that boils before it melts.

Departures from Ideality According to the van der Waals Equation

From Eq. (5.99) it follows that

$$\frac{p_c}{T_c} = \frac{3}{8} R \rho_c, \tag{5.110}$$

where ρ_c is the density at the critical point. The ideal gas law predicts that

$$\frac{p}{T} = R \rho \tag{5.111}$$

at all temperatures and pressures. Thus, the error in the ideal gas law at the critical point is large. How much in error is the ideal gas law far from the critical point? To answer this, we begin with the van der Waals equation in the form

$$p = \frac{R^*}{v_m} \left(\frac{T}{1 - v_c/3v_m} - \frac{9}{8} T_c \frac{v_c}{v_m} \right), \tag{5.112}$$

which is obtained from Eqs. (5.96) and (5.99). Far from the critical point where $v_m \gg v_c$ we may approximate Eq. (5.112) by

$$p \approx \frac{R^*T}{v_m} \left[1 + \frac{v_c}{v_m} \left(\frac{1}{3} - \frac{9}{8} \frac{T_c}{T} \right) \right]. \tag{5.113}$$

If we transform from molar specific volume to density by way of

$$\frac{1}{v_m} = \frac{\rho}{M}, \tag{5.114}$$

where M is the molecular weight, Eq. (5.113) becomes

$$\frac{p}{RT} \approx \rho \left[1 + \frac{\rho}{\rho_c} \left(\frac{1}{3} - \frac{9}{8} \frac{T_c}{T} \right) \right]. \tag{5.115}$$

The quantity on the left side of Eq. (5.115) is the density according to the ideal gas law. This equation still isn't in a sufficiently transparent form. With Eq. (5.110) and more algebra we finally obtain

$$\frac{\rho_{vdW}}{\rho_{ideal}} \approx 1 + \frac{3}{8} \frac{p}{p_c} \frac{T_c}{T} \left(\frac{9}{8} \frac{T_c}{T} - \frac{1}{3} \right). \tag{5.116}$$

This equation answers the question posed at the beginning of this section. The quantity

$$\frac{9}{8} \frac{T_c}{T} - \frac{1}{3} \tag{5.117}$$

is positive for all temperatures less than T_c. Thus, Eq. (5.116) predicts that at a given temperature and pressure (well below the critical point), the density of a real gas is larger than that of the corresponding ideal gas. This is expected on physical grounds. To counteract intermolecular attractions, a given pressure requires a greater number density of molecules.

The following table shows the magnitude of the density correction from the van der Waals equation for nitrogen, oxygen, and argon, at a temperature of $0\,^\circ$C and a pressure of one atmosphere, and for water vapor at this temperature but at its saturation vapor pressure (6.11 hPa):

	A	N_2	O_2	H_2O
ρ_{vdW}/ρ_{ideal}	1.0012	1.00096	1.0013	1.000058

According to the van der Waals equation, we need not fret over departures from ideality of the common atmospheric gases at normal temperatures and pressures. At the critical point, however, $\rho_{vdW}/\rho_{ideal} = 1.3$, and although above this point the distinction between gas and liquid disappears, whatever inhabits this fluid region is definitely *not* an ideal gas.

* The Maxwell Construction and Saturation Vapor Pressure

According to the Maxwell construction, the saturation vapor pressure p' at temperature T is obtained by solving the three equations:

$$p' = \frac{8T}{3v_A - 1} - \frac{3}{v_A^2}, \tag{5.118}$$

$$p' = \frac{8T}{3v_E - 1} - \frac{3}{v_E^2}, \tag{5.119}$$

and

$$p'(v_E - v_A) = \frac{8T}{3} \ln\left(\frac{3v_E - 1}{3v_A - 1}\right) + 3\left(\frac{1}{v_E} - \frac{1}{v_A}\right). \tag{5.120}$$

Pressure, temperature, and molar volume are reduced values, although to avoid clutter we omit the subscript r. Equations (5.118) and (5.119) are the requirements that p' intersect the van der Waals isotherm at A and E, whereas Eq. (5.120) is the equal-area requirement (see Fig. 5.9). These equations relate four variables, and hence, in principle can be solved to obtain p' as a function of T. Perhaps this can be done in principle, but we never have seen it done, and we failed to do it in complete generality. So we lowered our ambitions and settled for a solution at temperatures well below the critical temperature. As evidenced by Fig. 5.10, at such temperatures

$v_E \gg 1$ and $v_E \gg v_A$. With these assumptions we may approximate Eqs. (5.119) and (5.120) by

$$p' \approx \frac{8T}{3v_E} \qquad (5.121)$$

and

$$p'v_E \approx \frac{8T}{3} \ln\left(\frac{3v_E}{3v_A - 1}\right) - \frac{3}{v_A}. \qquad (5.122)$$

Rewrite Eq. (5.118) as

$$p' + \frac{3}{v_A^2} = \frac{8T}{3v_A - 1}. \qquad (5.123)$$

To first approximation $v_A = 1/3$. We can't use this approximation on the right side of Eq. (5.123), but we can use it on the left side to obtain

$$3v_A - 1 \approx \frac{8T}{27}, \qquad (5.124)$$

where we also assumed that $p' \ll 27$ (which is true at pressures well below the critical pressure). If we approximate the second term on the right side of Eq. (5.118) by 9 and combine all these approximate equations, we obtain

$$p' \approx 27 \exp\left(-1 - \frac{27}{8T}\right). \qquad (5.125)$$

Recall that pressure and temperature in Eq. (5.125) are relative to critical values; so, our final step is to rewrite this equation in terms of absolute pressure and temperature:

$$\frac{p'}{p_c} \approx 9.93 \exp\left(-\frac{27T_c}{8T}\right). \qquad (5.126)$$

This equation cannot be correct for all temperatures because at the critical temperature p' must equal p_c, and this equation predicts that it is about $1/3\, p_c$. Nevertheless, Eq. (5.126) has the same form as the integrated Clausius–Clapeyron equation, Eq. (5.49). For water, the slope of the $\ln p'$ versus $1/T$ curve is about -2200 from Eq. (5.126), whereas the correct value (assuming a constant enthalpy of vaporization) is about -5400. The Maxwell construction applied to the van der Waals equation yields an exponential dependence of saturation vapor pressure on $1/T$ (well below the critical temperature), but the exponent is incorrect by about a factor of two.

An Overview of the Many Successes of the van der Waals Equation

To leave you in a state of proper awe of the van der Waals equation, we summarize some of its many successes:

1. Phase change.
2. Slope of isotherm on the liquid side of the phase change greater than on the vapor side.
3. Critical temperature above which there is no distinction between liquid and gas.
4. Metastable states of superheating and supersaturation.
5. Limits to superheating and supersaturation.
6. Negative pressures.
7. Molecular diameters.
8. Law of corresponding states.
9. Departures from ideality (density of a real gas greater than that of an ideal gas).
10. Functional dependence of saturation vapor pressure on temperature (from Maxwell construction).

An impressive list, is it not? If you do Problems 25 and 26, you should become even more impressed by the van der Waals equation.

5.5 Phase Diagrams: Liquid–Vapor, Liquid–Vapor–Solid, Triple Point

The solid, liquid, and vapor equilibrium states of a substance can be illustrated with both pv and pT diagrams. In Section 5.4 we showed with pv diagrams the success of the van der Waals equation in describing phase changes at constant temperature. And the relationship between saturation vapor pressure and temperature for water was illustrated in Section 5.3 with pT diagrams. pv and pT diagrams can be combined to form a pvT equilibrium surface for water in all three phases. That is, the function $p = p(v, T)$ is the equation of a surface in a three-dimensional pvT space. Many textbooks show such surfaces for water that include its critical point and the triple point (a point at which all three phases can exist in equilibrium). Look carefully at these diagrams. They often have no units, nor is there any indication that variables should be on a logarithmic scale in order to depict anything close to reality. Although rarely stated, these diagrams are schematic. They may have been generated (or copied from earlier texts) to empha-size key features of phase transitions but are distorted substantially. We attempted to construct for water a real thermodynamic surface that includes the critical point and the triple point but abandoned this project once we understood the true nature of the equilibrium surface. To provide an accurate portrayal of this surface for water, we therefore show combinations of pv and pT diagrams, labeled with units and drawn to correct scale, and use these to illustrate key features of phase transitions.

The range of liquid–vapor equilibrium pv states and selected isotherms are shown in Fig. 5.11. Only the critical point, the 100 °C and the 0 °C isotherms are shown together with the envelope of equilibrium points for the vapor-to-liquid transition at all temperatures ranging from the critical temperature to the triple point. Interpretation of these isotherms and how they relate to condensation is the same as that for van der Waals isotherms. As in Section 5.3 we consider the variation in specific volume of a fixed mass of water vapor as it is compressed isothermally. At temperatures and pressures well below the critical point the change in specific volume associated with isothermal compression before condensation occurs is described to good approximation by the ideal gas law. At higher temperatures and pressures, however, the ideal gas law is inadequate. But regardless of the exact form of the equation of state, isotherms in the vapor phase are similar for a wide range of temperatures. Along any isotherm below the critical temperature, sufficient compression results in the formation of liquid drops (if suitable condensation nuclei are present) as the pressure reaches the saturation vapor pressure. Once liquid begins to form, further isothermal compression converts more vapor to liquid with no change in pressure, and both phases coexist. As compression continues, all the vapor is eventually converted to liquid. Beyond this point, further compression results in

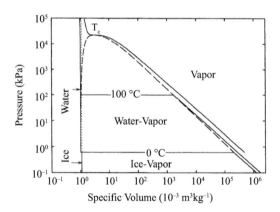

Fig. 5.11: Equilibrium pressure and specific volume states of water in all three phases. The dashed envelope is the set of all states for which the transition from vapor to liquid, or vice versa, occurs. On a logarithmic scale, isotherms for a wide range of temperatures, from the triple point to the critical point, are closely packed together. At pressures greater than the triple-point pressure, the ice-phase line is indicated by a dotted line.

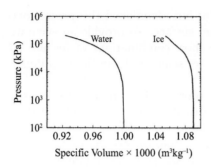

Fig. 5.12: Specific volume versus pressure for water and ice at the melting temperature (the temperature at which ice and water are in equilibrium).

little change in specific volume compared with the volume changes associated with compression of the vapor.

Figure 5.12 shows the variation of the specific volume of both ice and water at the melting point, which ranges from $0\,°C$ to $-20\,°C$ for pressures between 1 and about 2000 atmospheres. At pressures less than about 100 atmospheres, ice and water are nearly incompressible, and the difference between their specific volumes is almost constant. At one atmosphere $v_j - v_w = 0.091 \text{ m}^3\text{kg}^{-1}$, and at 2000 atmospheres (melting temperature $-20\,°C$), this difference increases to $0.135 \text{ m}^3\text{kg}^{-1}$. Measurements show no indication of a critical point (like that for the vapor and liquid phases) where the ice and liquid phases become indistinguishable.

Although the pv diagram shown in Fig. 5.11 can be used to describe phase changes associated with isothermal compression, this is not a process that occurs naturally in the atmosphere. In general, pT diagrams provide a more useful depiction of atmospheric processes. A detailed discussion of the variation of the saturation vapor pressure of water with temperature appears in Section 5.3. Because many atmospheric processes involving water occur near its triple point, we examine the equilibrium surface near it.

In Section 5.3 we derived the Clausius–Clapeyron equation for the relation between pressure and temperature of liquid water in equilibrium with its vapor. The same arguments that led to this equation apply to ice in equilibrium with vapor and with liquid water. Thus, Eq. (5.40) can be extended to ice–vapor $(i - v)$ and ice–water $(i - w)$ equilibria merely by a change of notation:

$$\left(\frac{dp}{dT}\right)_{v-w} = \frac{\ell_v}{T(v_v - v_w)}, \tag{5.127}$$

$$\left(\frac{dp}{dT}\right)_{i-v} = \frac{\ell_s}{T(v_v - v_i)}, \tag{5.128}$$

$$\left(\frac{dp}{dT}\right)_{i-w} = \frac{\ell_f}{T(v_w - v_i)}. \tag{5.129}$$

Equation (5.127) is Eq. (5.40) written in a slightly different form. Here, p is the pressure and T the temperature of the two phases in equilibrium, and v_v, v_w, v_i are the specific volumes of vapor, water, and ice.

Near the triple point the specific volume of water vapor is much greater than that of water or ice, and hence, it follows from Eqs. (5.127) and (5.128) that the slopes of the pT curves are positive, although the ice–vapor equilibrium curve is slightly steeper than the liquid–vapor equilibrium curve because $\ell_s/\ell_v = 1.13$. At the triple point, the slope of the liquid–vapor curve is about $0.4\,\mathrm{hPaK}^{-1}$ and the slope of the vapor–ice curve is about $0.5\,\mathrm{hPaK}^{-1}$. Although the difference between these slopes is small (see Fig. 5.13), as noted in Section 5.3, this difference has important consequences for cloud and precipitation physics. For ice–water equilibrium, the slope of the pT curve (Eq. 5.129) is negative because the specific volume of ice is greater than that of water. Furthermore, because these specific volumes ($1.091 \times 10^{-3}\,\mathrm{m}^3\,\mathrm{kg}^{-1}$ for ice and $1.000 \times 10^{-3}\,\mathrm{m}^3\,\mathrm{kg}^{-1}$ for water) are not greatly different, the slope of this curve ($-1.34 \times 10^5\,\mathrm{hPaK}^{-1}$) is large compared with those of the water–vapor and ice–vapor equilibrium curves, even though ℓ_f is smaller than ℓ_v and ℓ_s (Fig. 5.14).

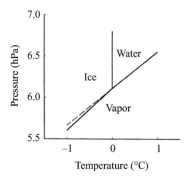

Fig. 5.13: Equilibrium pressure versus temperature for water near its triple point. The dashed line is an extrapolation to temperatures below $0\,^{\circ}\mathrm{C}$ (subcooled water). On this scale the dependence of melting temperature on pressure (vertical line) is not evident. See Fig. 5.14 for a display of this dependence.

A negative slope of a solid–liquid equilibrium curve means that the substance expands upon freezing. Water is unusual in this respect but not unique. The slope is positive for substances (e.g. carbon dioxide) that contract upon freezing. One consequence of the negative slope of the liquid–solid equilibrium curve for water is that its melting point decreases with increasing pressure, about 0.0076 K per atmosphere near the triple point. The simplest explanation for this is that the density of ice is *less* than that of water, and squeezing ice increases its density, facilitating its transition to liquid. The triple point (273.16 K) is the temperature at which pure ice, water vapor, and pure liquid water are in equilibrium (at a pressure of about 6.11 hPa); the ice point (273.15 K) is the temperature of air-saturated water (exposed to air at a pressure of one atmosphere) in equilibrium

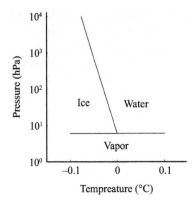

Fig. 5.14: Similar to Fig. 5.13, but with pressure plotted on a logarithmic scale, which makes the dependence of melting temperature on pressure evident. On this scale the ice–vapor and water–vapor curves are almost horizontal lines, but with slightly different slopes.

with ice. Air dissolved in water (see Section 5.9) lowers its melting point by about 0.0023 K. This lowering combined with that resulting from a pressure increase of about one atmosphere is why the ice point is 0.01 K lower than the triple point.

Lowering of the melting point with increased pressure is sometimes called *pressure melting*. This can be demonstrated by attaching weights (5 kg or more) to the ends of a thin (metal) guitar string draped over a block of ice. With time the string cuts its way through the ice. Although pressure melting is often invoked to explain the thin layer of water that acts as a lubricant for skiing or ice skating, the pressures that can be obtained are insufficient to account for the observed melting unless the ice is very close to 0 °C. For example, a 75-kg skier equipped with skis each having a geometric surface area (length times width) of $0.25\,\mathrm{m}^2$ would increase the average pressure on the bottom of skis by about 0.015 atm. This would result in a lowering of the melting point of the underlying snow by only 0.0001 °C. The actual contact area between the ski and the snow, however, may be only 1/1000 of the geometric area, and hence, the pressure at the contact points would increase correspondingly. But even then, the melting point would be lowered by only about 0.1 °C. A hollow-ground skate blade might have a geometric area only 10^{-3} as large as that of a ski. But the geometric area of the skate is likely to be closer to the actual contact area, so that at best one might expect a lowering of the melting point by 1 °C. Friction, not pressure, causes the melting needed to lubricate skis and skates. Pressure melting may, however, provide a melting layer that helps lubricate the interface between glaciers and the underlying land. But the weight of a glacier alone cannot explain the lowering of the melting point. Lateral forces associated with the movement of glaciers across irregular surfaces can produce local pressure increases sufficient to appreciably lower the melting temperature.

We focused attention on the triple point of water because normal terrestrial temperatures are well above the critical temperatures of the other principal constituents of the atmosphere (nitrogen, oxygen, and argon). What about the trace gas carbon dioxide? It has a critical temperature of 304.2 K (and a critical pressure of 7.33×10^5 hPa), which is well above temperatures encountered in most of the troposphere. Why, therefore, don't we have to consider the liquid and solid phases of carbon dioxide in Earth's atmosphere? Fortunately, the triple point of carbon dioxide is at 216.5 K and 5172 hPa, and hence, because of its low concentration carbon dioxide exists only in the vapor phase. Weather forecasters would be further challenged if constituents of the atmosphere other than water existed naturally in more than one phase.

Although in this and other sections we refer to ice without qualification, ordinary ice is strictly called ice I, which is one particular crystalline form (hexagonal) of ice. Several other stable high-pressure forms exist. By high pressure is meant thousands of atmospheres, which is why we do not encounter these *polymorphs* of ice in our everyday lives. These polymorphs differ from ordinary ice in their crystal structure, and hence, also in their thermodynamic properties. P. W. Bridgman, whom we quote in the References in Chapter 1 and the introduction to Chapter 3, acquired fame from pioneering studies of ice in its high-pressure forms. Missing from this list of polymorphs is (metastable) ice IV, which has been observed only by Bridgman as D_2O ice (D indicating deuterium, an isotope of hydrogen twice as massive as ordinary hydrogen but chemically identical). To this list we may add ice Ic, a metastable cubic form of ice I, and vitreous ice, a glassy ice about which little is known. As far as anyone knows there is only one kind of water vapor whereas solid water exists in several forms. For our purposes, there is only one kind of liquid water, although there is recent (2017) evidence for metastable high-density and low-density phases of water.

A triple point is a point in pvT space where three phases coexist in equilibrium. When we refer to the triple point, we usually mean the point where ice I, liquid water, and vapor coexist, but there are other possible triple points for water because all three phases could be ice. To date, *eighteen* different phases of ice have been identified.

We become so fixated on water that it is easy to forget that every pure substance (except helium) has a triple point. Water just happens to have a triple point near room temperature. The triple point of mercury is $-39\,°C$. Unlike the boiling point, the triple point is an intrinsic property of a pure substance independent of its environment.

The existence of ices up to ice VIII inspired the novel *Cat's Cradle* by Kurt Vonnegut, Jr. The central character of this novel is a hypothetical form of ice, "ice–nine," a room-temperature form of ice. It supposedly is not observed because of the lack of a naturally occurring freezing nucleus to provide the template necessary for liquid water to congeal into ice-nine. This novel is remarkable in that its discussion of nucleation and the various forms of ice is scientifically correct, except that ice–nine did not exist—although it does now, but not at ordinary temperatures and pressure, so the world's liquid water is safe. This correctness is not surprising, given that Kurt's brother, Bernie, was an atmospheric scientist with considerable knowledge of cloud physics. *Cat's Cradle* is a delightful story of the quest for the ice-nine freezing nucleus and the consequences of its discovery. This is one of two novels every meteorologist should read, the other being George Stewart's *Storm*, which gave rise to the practice of naming tropical storms.

Almost as exotic as room-temperature ice–nine, but less fanciful, is superheated or "hot" ice, the existence of which has been debated for over a century. By superheated ice is meant water existing as a solid above 0 °C. Thomas Carnelly created a storm in the 1880s by claims of ice existing at temperatures above 100 °C. Recent claims are more modest, a mere few degrees of superheat. But even if ice can be superheated just a bit, and only under special circumstances, the existence of superheated ice is intellectually satisfying. Liquid water, as we have seen, is readily superheated (raised to temperatures above 100 °C without boiling) and subcooled (lowered to temperatures below 0 °C without freezing). Although the normal range at which liquid water exists (at atmospheric pressure) is 0 °C to 100 °C, the actual range is about −40 °C to 280 °C. And water vapor can exist at relative humidities appreciably greater than 100% (supersaturation) without condensing. Thus, it is only fitting that to complete this set of metastable states of water we should have superheated ice. The barrier to superheating ice is the liquidlike layer on its surface (see Section 1.7 and Annotated References in Ch. 1). Because of the rarity or even nonexistence of superheated ice and the millions of tons of subcooled water in clouds, a more accurate term for freezing point of water is melting point of ice.

5.6 Free Energy

When in doubt, begin with something you know, the first law of thermodynamics for a reversible process:

$$\frac{dU}{dt} = T\frac{dS}{dt} - p\frac{dV}{dt} \, . \tag{5.130}$$

If we add to both sides of this equation $V\,dp/dt$ and $S\,dT/dt$, we obtain

$$\frac{d}{dt}(H - TS) = V\frac{dp}{dt} - S\frac{dT}{dt} \, . \tag{5.131}$$

In a constant-pressure *and* -temperature process the quantity $H - TS$ is conserved. Experience has shown us that conserved quantities are islands of stability in a sea of change, friendly faces in a hostile world. So, we had better give $H - TS$ a name, *Gibbs free energy* (also called *Gibbs function*), and a symbol:

$$G = H - TS. \tag{5.132}$$

With this definition the first law of thermodynamics for a reversible process is

$$\frac{dG}{dt} = V\frac{dp}{dt} - S\frac{dT}{dt}\,.$$

(5.133)

From the chain rule for differentiation, we obtain

$$\frac{\partial G}{\partial p} = V, \qquad \frac{\partial G}{\partial T} = -S.$$

(5.134)

Why we would be interested in a quantity that is conserved only in a constant temperature and pressure process? If these two thermodynamic variables don't change, what does? Have you forgotten already that pressure and temperature are constant in the phase change from liquid water to vapor? It would seem that the Gibbs free energy might be handy when we are interested in phase changes. To show that this is so, let us tackle by a different approach a problem we already solved.

Denote by g_v and g_w the respective specific free energies of water vapor and liquid water at the same temperature and pressure. During a transition from one phase to the other, the total free energy does not change:

$$\frac{d}{dt}(M_v g_v + M_w g_w) = 0,$$

(5.135)

where M_v is the mass of water vapor and M_w is the mass of liquid water. Because pressure and temperature are constant in this process, the specific free energies do not change, although the masses of liquid and vapor do:

$$g_v \frac{dM_v}{dt} + g_w \frac{dM_w}{dt} = 0\,.$$

(5.136)

Total mass is conserved

$$\frac{dM_v}{dt} = -\frac{dM_w}{dt}\,,$$

(5.137)

which leads to

$$\frac{dM_v}{dt}(g_v - g_w) = 0\,.$$

(5.138)

In general, dM_v/dt does not vanish, and hence, Eq. (5.138) implies that

$$g_v(p, T) = g_w(p, T)\,.$$

(5.139)

Equation (5.139) sometimes takes the form of equality of molar free energies in the two phases. What is true for molar free energies is true for specific free energies because molecular weight does not change from the vapor to the liquid phase.

The equal sign (=) has two quite different meanings, a source of possible confusion. For example, the same symbol appears in the equations

$$\sin^2 x + \cos^2 x = 1 \tag{5.140}$$

and

$$x^2 + 2x + 1 = 0. \tag{5.141}$$

In the first equation, = represents the declarative "is." For *all* x, the sum of the squares of its sine and its cosine *is* 1. But in the second equation, = represents the interrogative "is there?" Is there any number x such that when it is squared, added to twice itself and to 1, the result is zero? Equation (5.140) is always true (identity), whereas Eq. (5.141) may be true only for certain values of x, which are called the solutions to this equation (answers to a question). In Eq. (5.139) the equal sign is of the second type, a sign that asks the question, are there values of p and T such that the specific free energy of water vapor is equal to that of liquid water? If such values exist (we assume they do), the pressure p is the saturation vapor pressure e_s at the temperature T. Stated another way, from the infinitude of number pairs (p, T), only those pairs for which Eq. (5.139) is satisfied are physically realizable pressures and temperatures of a vapor in equilibrium with its liquid.

With a better understanding of the meaning of Eq. (5.139), we can use it to determine the dependence of saturation vapor pressure on temperature. If the temperature is changed to $T + \Delta T$ and the pressure is changed to $p + \Delta p$, the condition for equilibrium becomes

$$g_v(p + \Delta p, T + \Delta T) = g_w(p + \Delta p, T + \Delta T). \tag{5.142}$$

Expand Eq. (5.142) in a Taylor series and retain only the first two terms:

$$g_v(p, T) + \frac{\partial g_v}{\partial p}\Delta p + \frac{\partial g_v}{\partial T}\Delta T = g_w(p, T) + \frac{\partial g_w}{\partial p}\Delta p + \frac{\partial g_w}{\partial T}\Delta T. \tag{5.143}$$

This equation and Eq. (5.139) yield

$$\left(\frac{\partial g_v}{\partial T} - \frac{\partial g_w}{\partial T}\right)\Delta T = \left(\frac{\partial g_w}{\partial p} - \frac{\partial g_v}{\partial p}\right)\Delta p. \tag{5.144}$$

We now can write

$$\frac{dp}{dT} = \lim_{\Delta T \to 0} \frac{\Delta p}{\Delta T} = \left(\frac{\partial g_v}{\partial T} - \frac{\partial g_w}{\partial T}\right) \Big/ \left(\frac{\partial g_w}{\partial p} - \frac{\partial g_v}{\partial p}\right). \tag{5.145}$$

From Eq. (5.134) we have

$$\frac{dp}{dT} = \frac{-s_v + s_w}{v_w - v_v}. \tag{5.146}$$

If we multiply the numerator and denominator of the right side of Eq. (5.146) by T, we obtain

$$\frac{dp}{dT} = \frac{T(s_v - s_w)}{T(v_v - v_w)} = \frac{\ell_v}{T(v_v - v_w)}. \tag{5.147}$$

With a notation change from p to e_s, Eq. (5.147) is the Clausius–Clapeyron equation (5.42). Thus, by free energy arguments we obtained a familiar result, which indicates that free energy might be a useful thermodynamic variable when phase changes engage our attention.

Unfortunately, the term *free energy* means different things to different people and is denoted by different symbols. Be prepared to be confused, or at least uncertain, when you read papers and other books. A survey of books on thermodynamics indicates that no two authors quite agree.

The quantity $U - TS$ is also sometimes called free energy, or to be more specific, *Helmholtz free energy*, and assigned the symbol F. But don't be surprised to find A chosen for $U - TS$ and called the *work function*. In yet another incarnation, $U - TS$ is assigned the symbol ψ and called the *Helmholtz potential* or *Helmholtz free energy* or *free energy at constant volume* or *thermodynamic potential at constant volume*. Confused? Read on. What we called the Gibbs free energy $H - TS$ and denoted by G is also called *free energy without qualification* and denoted by F, the same symbol sometimes used for $U - TS$. But $H - TS$ also is denoted by Φ and called *the thermodynamic potential* or, to be more specific, *the thermodynamic potential at constant pressure*. Other names are *free energy at constant pressure* and *free enthalpy*. To add further to the confusion, $U + pV$, $U - TS$, and $H - TS$ are referred to indiscriminately as thermodynamic potentials. This welter of symbols and names should cause you to never take for granted that you know what someone else means by free energy or thermodynamic potential.

From Eq. (1.75) and Eq. (4.5), the work done in an isothermal process is the change in the Helmholtz free energy, $W_{tot} = \Delta(U - TS)$. If this free energy decreases, work is done *by* the system, presumably the maximum work because it is based on ideal processes.

5.7 Effect of Air Pressure on Saturation Vapor Pressure

When we derived Eq. (5.147), the system of interest consisted entirely of a single substance—water—in two phases, liquid and vapor. But this is a contrived system. Water in nature almost always coexists with water vapor and dry air, the latter component dominating the total pressure. How does this air change the saturation vapor pressure at a given temperature? We derived the Clausius–Clapeyron equation for a single-component system (pure water) by free energy arguments, so let us apply them to a two-component (dry air and water) system.

Consider liquid water in equilibrium with its vapor mixed with dry air. The total free energy of a mixture of ideal gases is the sum of the free energies of its components. Thus, the total free energy of our system is

$$M_a g_a(p, T) + M_v g_v(e_s, T) + M_w g_w(p_t, T),$$ (5.148)

where p, M_a, and g_a are the partial pressure, mass, and specific free energy, respectively, of dry air, and the other quantities are as defined in the previous section. The pressure of the liquid water is equal to the total pressure $p_t = p + e_s$ of the air with which the liquid coexists. In a constant (total) pressure and temperature process the total free energy does not change:

$$M_a \frac{dg_a}{dt} + g_a \frac{dM_a}{dt} + M_v \frac{dg_v}{dt} + g_v \frac{dM_v}{dt} + M_w \frac{dg_w}{dt} + g_w \frac{dM_w}{dt} = 0.$$ (5.149)

Neither the mass of air nor the total mass of water substance changes:

$$\frac{dM_a}{dt} = \frac{d}{dt}(M_v + M_w) = 0.$$ (5.150)

We also have

$$\frac{dg_w}{dt} = \frac{\partial g_w}{\partial p_t} \frac{dp_t}{dt} = 0$$ (5.151)

because total pressure is constant:

$$\frac{dp_t}{dt} = \frac{d}{dt}(p + e_s) = 0.$$ (5.152)

The rates of change of the specific free energies of the air and vapor are

$$\frac{dg_a}{dt} = \frac{\partial g_a}{\partial p} \frac{dp}{dt} \quad \text{and} \quad \frac{dg_v}{dt} = \frac{\partial g_v}{\partial e_s} \frac{de_s}{dt}.$$ (5.153)

It follows from Eq. (5.134) that

$$\frac{\partial g_a}{\partial p} = \frac{1}{\rho_a}, \quad \frac{\partial g_v}{\partial e_s} = \frac{1}{\rho_v},$$ (5.154)

where ρ_a and ρ_v are the densities of air and water vapor. We need one more result:

$$\frac{M_a}{\rho_a} = \frac{M_v}{\rho_v} = V,$$ (5.155)

where V is the common volume occupied by air and vapor, before we can combine all these equations to obtain

$$\frac{dM_v}{dt}[g_v(e_s, T) - g_w(p_t, T)] = 0.$$ (5.156)

Because dM_v/dt is arbitrary, we must have

$$g_w(p + e_s, T) = g_v(e_s, T),$$ (5.157)

which is the condition for equilibrium between liquid water and its vapor mixed with air. We could have guessed at this result or simply postulated it on the basis of our previous success at deriving the Clausius–Clapeyron equation.

When the air pressure is changed (isothermally) to $p+\Delta p$, the saturation vapor pressure changes to $e_s + \Delta e_s$, and the new condition for equilibrium is

$$g_w(p+\Delta p+e_s+\Delta e_s, T) = g_v(e_s+\Delta e_s, T). \tag{5.158}$$

If we expand Eq. (5.158) in a Taylor series and retain only the first two terms, we obtain

$$g_w(p_t, T) + \frac{\partial g_w}{\partial p_t}(\Delta p + \Delta e_s) = g_v(e_s, T) + \frac{\partial g_v}{\partial e_s}\Delta e_s . \tag{5.159}$$

Divide both sides of this equation by Δp and take the limit as Δp goes to zero:

$$\frac{de_s}{dp} = \frac{\rho_v}{\rho_w - \rho_v}. \tag{5.160}$$

In deriving this equation we also used Eq. (5.154). At temperatures of interest (well below the critical temperature), the density of liquid water ρ_w is much greater than that of water vapor ρ_v, and hence, Eq. (5.160) can be approximated:

$$\frac{de_s}{dp} = \frac{\rho_v}{\rho_w}. \tag{5.161}$$

This equation sometimes is called *Poynting's formula*.

To judge by his obituary, John Henry Poynting (b. Monton, England, 1852; d. Birmingham, 1914) was a man after our own hearts. Sir Joseph Larmor writes of Poynting that

> his rebellion against an excessive anthropomorphism which had begun to cling around the notion of natural laws, as if they were really legal enactments to be obeyed or disobeyed by inert matter almost as if it possessed will-power and could exercise choice, some substances being praised as good radiators while others are stigmatized as bad—most gases being admittedly unable to reach a standard of perfection held out to them by Boyle's law, though a few of excessive merit might surpass it,—Poynting's revolt against this kind of attitude to laws of nature, though doubtless more than half humorous, was in itself wholesome.

Poynting's name is best known to students of physics by way of the eponymous Poynting vector, which specifies the magnitude and direction of the transport of electromagnetic energy. You are reading these lines courtesy of the Poynting vector. This contribution to electromagnetic theory overshadows Poynting's other contributions to science, among which is counted Eq. (5.161).

Because the density of liquid water is vastly greater than that of water vapor in air, the presence of air only slightly changes the saturation vapor pressure. But note that this change is *positive*. The greater the pressure, the greater the saturation vapor pressure. This puts one more nail in the coffin of the sponge analogy. The denser the air, the *more* water vapor it can "hold." To be fair, we have neglected dissolved air in water, which lowers its saturation vapor pressure. But in Section 5.9 we show that the *net* consequence of air above water is to increase (slightly) its saturation vapor pressure at a given temperature.

We obtained Poynting's formula by mathematical arguments. A physical argument helps to cement it in our minds, especially since an increase in saturation vapor pressure with an increase in total pressure on a liquid may seem counterintuitive. After all, you would find it more difficult to escape if pressed to the mat by a heavyweight wrestler than by a featherweight. And yet saturation vapor pressure is a measure of the ease with which molecules can escape from a liquid. Why do they escape more readily (although by a tiny amount) when the liquid is subjected to increased pressure?

For molecules to escape from a liquid requires them to have sufficient kinetic energy to overcome the forces binding them together into a liquid. The greater these attractive binding forces, the less likely a molecule is to have sufficient energy to escape. The change in volume of liquid water subjected to increased pressure is exceedingly small because at short separations the force between molecules is strongly repulsive. Water resists being compressed, but when it is, the average distance between molecules must decrease slightly. In turn, this means that some water molecules experience greater repulsive forces, which aid them in escaping from the liquid. An analogy is provided by the escape velocity of molecules from a planetary atmosphere (see Section 2.4). The greater the density of the planet (for fixed diameter), the greater its attractive gravitational force on molecules in its atmosphere, and hence, the greater the velocity they must have to escape. Reducing this attractive force decreases the escape velocity. When molecules in a liquid are squeezed closer together, the average repulsive force between them increases, and an increase in repulsion is equivalent to a decrease in attraction.

Water vapor is to good approximation an ideal gas. Thus, the vapor density in Eq. (5.161) is $e_s/R_v T$, which yields the differential equation

$$\frac{1}{e_s}\frac{de_s}{dp} = \frac{1}{R_v T \rho_w},$$
(5.162)

the solution to which is

$$\ln\left(\frac{e_{s2}}{e_{s1}}\right) = \frac{p_2 - p_1}{R_v T \rho_w},$$
(5.163)

under the assumption (a good one) that the density of water is independent of pressure. From this equation it follows that at $20\,^\circ$C, a pressure increase $(p_2 - p_1)$ of one atmosphere corresponds to a saturation vapor pressure increase of less than 0.1%.

5.8   Lowering of Vapor Pressure by Dissolution

Dissolving anything in water lowers its saturation vapor pressure (at a given temperature). The physical reason for this is not difficult to find. As we have seen, saturation vapor pressure increases with increasing rate of evaporation. This rate depends on the temperature of the evaporating substance. But it also depends on its concentration (molecules per unit volume). When anything is dissolved in water, the concentration of water molecules decreases, and hence, we expect the evaporation rate to also decrease.

An *ideal solution* is one in which the solute and solvent are essentially identical except for their names. In general, when one substance is dissolved in another, the resulting system has properties that are not simply weighted sums of the properties of the two substances. Suppose, for example, we dissolve substance 1 in substance 2 (e.g. salt in water or alcohol in water). If this process of dissolution (or mixing if both substances are liquids) is carried out adiabatically ($Q = 0$) and isobarically, total enthalpy is conserved:

$$H_1(T_i) + H_2(T_i) = H_{12}(T_f), \tag{5.164}$$

where H_1 and H_2 are the enthalpies of substances 1 and 2 and H_{12} is the enthalpy of the solution formed from the two. The common temperature of the two substances before mixing is T_i and the temperature of the solution after dissolution is complete is T_f.

It is sometimes said that heat is "absorbed" or "generated" in dissolution, which is surely nonsense given that this process is carried out adiabatically, and hence, total heating is zero. How can we make some sense out of nonsense?

Suppose that a solution, immediately after it is formed, is allowed to interact (isobarically) with its surroundings at temperature T_i until the two are in thermal equilibrium. This is *not* an adiabatic process, and hence, total heating

$$\int_f^i Q \, dt = H_{12}(T_i) - H_{12}(T_f) \tag{5.165}$$

is not zero in general. Because of enthalpy conservation, Eq. (5.164), we can write

$$\int_f^i Q \, dt = H_{12}(T_i) - [H_1(T_i) + H_2(T_i)] = \Delta H_{\text{sol}}, \tag{5.166}$$

where the enthalpy difference ΔH_{sol} between the enthalpy of a solution and the sum of enthalpies of its constituents (all at the same temperature and pressure) is called the *enthalpy of solution* (mixing) or heat of solution. Its physical interpretation is that it is the total heating required to restore the solution to the common temperature of its separate components before dissolution. If the heat capacity C_{p12} of the solution is independent of temperature over the range (T_i, T_f), we have

$$T_f - T_i = \frac{-\Delta H_{\text{sol}}}{C_{p12}}. \tag{5.167}$$

When the enthalpy of solution is positive, the temperature of the solution is less than that of its components before dissolution, and hence, the solution must be heated in order to restore it to the initial temperature. When the enthalpy of solution is negative, the temperature of the solution is greater than that of its components before dissolution, and hence, the solution must be cooled in order to restore it to the initial temperature. Strictly speaking, these conclusions depend on the assumption of an (approximately) temperature-independent heat capacity. When this temperature dependence is not negligible, we still may define the enthalpy of solution as an enthalpy difference between two states of the solution. However, we cannot say with certainty that a temperature increase upon dissolution corresponds to a negative enthalpy of solution, and vice versa.

In addition to possible temperature changes upon dissolution, the volume of a solution may not be equal to the sum of the volumes of its separate components (Section 2.8). Mass is conserved, but volume need not be. A solution with zero enthalpy of solution and zero volume change is said to be ideal. In such a solution, the environment of a solute molecule surrounded by solvent molecules is essentially the same as in the pure solute, and similarly for the solvent molecules.

Consider now an ideal solution composed of two components, which we denote as A and B. The equilibrium vapor pressure p_A of component A, say, does not depend on the absolute number of molecules of both type provided that the system is sufficiently large (for what is meant by "sufficiently," see Section 5.10). Thus, p_A can depend on only the temperature T of the solution, and the two mole fractions x_A and x_B

$$x_A = \frac{N_A}{N_A + N_B}, \quad x_B = \frac{N_B}{N_A + N_B}, \tag{5.168}$$

where N_A is the number of molecules of type A and N_B is the number of molecules of type B. But the two mole fractions are not independent,

$$x_A + x_B = 1, \tag{5.169}$$

consequently, p_A depends on only two variables

$$p_A = p_A(T, x_B). \tag{5.170}$$

Similarly, p_B, the equilibrium vapor pressure of component B, is a function of only two variables

$$p_B = p_B(T, x_A). \tag{5.171}$$

For an ideal solution, the total pressure p must be constant:

$$p = p_A + p_B \tag{5.172}$$

because components A and B differ by name only (they are physically interchangeable). We may imagine, for example, that A is radioactive whereas B is not, but the two are otherwise identical. Because p is constant,

$$\frac{\partial p}{\partial x_B} = \frac{\partial p_A}{\partial x_B} + \frac{\partial p_B}{\partial x_B} = \frac{\partial p_A}{\partial x_B} + \frac{\partial p_B}{\partial x_A}\frac{dx_A}{dx_B} = 0, \tag{5.173}$$

from which and Eq. (5.169) it follows that

$$\frac{\partial p_A}{\partial x_B} = \frac{\partial p_B}{\partial x_A}. \tag{5.174}$$

The pressures must also satisfy the boundary conditions

$$p_A(T,0) = p_{Ao} = p, \qquad p_A(T,1) = 0, \qquad (5.175)$$

$$p_B(T,0) = p_{Bo} = p, \qquad p_B(T,1) = 0, \qquad (5.176)$$

where p_{Ao} is the equilibrium vapor pressure above a pure solution of component A and p_{Bo} is the equilibrium vapor pressure above a pure solution of component B (which for an ideal solution are both equal to p). The functions

$$p_A = p_{Ao}(1 - x_B), \qquad p_B = p_{Bo}(1 - x_A) \qquad (5.177)$$

satisfy Eqs. (5.172) and (5.174) together with the boundary conditions Eqs. (5.175) and (5.170). We may rewrite the expression for p_A as

$$p_A = p_{Ao} x_A = p_{Ao} \frac{N_A}{N_A + N_B}. \qquad (5.178)$$

This is *Raoult's law* or rather the modern textbook version of it. François Marie Raoult (b. Fournes, France, 1830; d. Grenoble, 1901), who had a long and distinguished career as professor of chemistry at the University of Grenoble, expressed his law, which he obtained experimentally, not theoretically, somewhat differently

$$p_{Ao} - p_A = p_{Ao} \frac{N_B}{N_A}. \qquad (5.179)$$

To underscore the difference between what Raoult said and what his successors have said he said, we rewrite Eq. (5.179) as

$$p_A = p_{Ao}\left(1 - \frac{N_B}{N_A}\right) \qquad (5.180)$$

and Eq. (5.178) as

$$p_A = p_{Ao}\left(1 + \frac{N_B}{N_A}\right)^{-1}. \qquad (5.181)$$

Raoult's law is strictly valid only for ideal solutions, yet often gives good results even for *dilute* nonideal solutions ($N_B \ll N_A$). For such solutions Eqs. (5.180) and (5.181) give *almost* the same results. Raoult's law describes a *colligative* property of solutions. Such properties depend (ideally) only on the fraction of solute molecules dissolved in a solvent. Figure 5.15 shows measured relative humidity above aqueous solutions of two salts, sodium chloride, NaCl, and ammonium sulfate, $(NH_4)_2SO_2$, compared with predictions from Raoult's law. To obtain better agreement between theory and experiment, the number of moles of the salt is multiplied by two. That is, x_B in Eq. (5.177) is obtained by replacing N_B in Eq. (5.168) by $2N_B$, where N_B is the number of moles of the salt added to water. When sodium chloride, ammonium sulfate, and other salts are dissolved, they dissociate into ion pairs. For example, each mole of NaCl dissolved in water yields one mole of positive sodium ions and one mole of negative chlorine ions. A large measure of Raoult's fame rests on this necessary fiddling with his law to obtain agreement with experiment. This law applied to salt solutions provided experimental evidence for dissociation, a controversial hypothesis proposed in 1887 by Arrhenius. At first sight, the notion that a solution of sodium chloride consists of sodium and chlorine ions in water seems ludicrous, if not insane. Sodium is highly reactive, and chlorine gas is poisonous. If you were to ingest either of these elements, you would not likely survive the experience. Yet, coursing through our bodies is blood in which sodium and chlorine ions benignly swim. And we could not function without them.

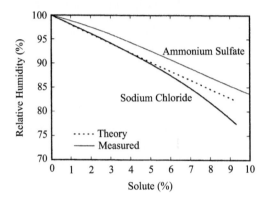

Fig. 5.15: Relative humidity above aqueous solutions at 20 °C in equilibrium with water vapor. The dashed line shows theory according to Raoult's law. The two solid lines are measurements, taken from *International Critical Tables*, for sodium chloride and ammonium sulfate.

At low concentrations, both solutions obey Raoult's law. As the concentration increases, departures from Raoult's law increase.

To determine the extent to which a salt solution is ideal, we added table salt to water. Although we measured no perceptible volume change upon dissolution, we did measure the temperature of a concentrated salt solution to be about 1 °C *lower* than the (room) temperature of the pure water and salt before dissolution (see Section 3.9 for details). Despite this measurable enthalpy of solution, low-concentration sodium chloride solutions still obey Raoult's law. At low concentrations, only *how much* is dissolved determines the vapor pressure lowering, but at higher concentrations, *what* is dissolved increasingly affects the lowering. As evidenced from Fig. 5.15, departures from Raoult's law are greatest at high concentration.

The temperature drop when sodium chloride is dissolved in water (adiabatically and isobarically) makes the term *heat of solution* seem a bit laughable. Why not cold of solution? And where does this heat or cold come from? Certainly not the surroundings, from which the system is insulated. We have seen the so-called heat of solution said to be "heat absorbed," but from where is a mystery. Perhaps the heat is absorbed from the ether, another outmoded concept. The example of a temperature drop upon dissolution (of an exceedingly common substance) is a further argument for replacing the misleading term heat of solution with enthalpy of solution, which has the advantage of being free of encrusted misconceptions. A heat of solution is simply a difference between the enthalpies of a solution and those of its separate constituents, both at the same temperature. Why not call a spade a spade?

You might think that although Raoult's law is relevant to chemistry, it has no relevance to meteorology. You'll learn otherwise in Section 5.11.

5.9   Air in Water: Henry's Law

Water exposed to air is unrelentingly bombarded at an intense rate by air molecules. This rate (per unit area) for any component of air is (see Section 5.2)

$$\frac{1}{4}n\langle v \rangle, \tag{5.182}$$

where n is the number density of the component with mass m and mean speed

$$\langle v \rangle = \sqrt{\frac{8kT}{\pi m}}. \tag{5.183}$$

Molecules that strike the water will be captured by it and retained for at least a brief time before possibly returning to the air. Because of this loitering of molecules at the water surface, their concentration there increases above that in the air. This is the phenomenon of *adsorption* (see Section 2.1), an accumulation of gas molecules at a surface resulting in a higher concentration than in the free gas. This is analogous to a pileup of cars in a crash on a congested highway. Adsorption is similar to condensation.

Suppose that initially pure water is exposed to air, molecules of which accumulate at the surface. Because of a concentration gradient of dissolved air molecules from the surface into the water, they migrate (diffuse) into the body of the water, called *absorption*, which results in a concentration decrease. Adsorption and absorption considered as a single process is called *sorption*. As the concentration of molecules increases, so does their rate of return to the air. Eventually, molecules dissolved in the water will come into equilibrium with those in the air. As with evaporation and condensation, this equilibrium is dynamic: Air molecules continually enter and leave the water at the same rate. Sorption is to be distinguished from the dissociation of solid salt into sodium and chlorine ions by interaction with water.

Let c be the concentration of molecules in the water at any time. We assume that it is continually stirred to maintain a more or less uniform concentration. The rate of change of concentration is given by

$$\frac{dc}{dt} = C - E, \tag{5.184}$$

where E is the escape rate and C is the capture rate, which are like the evaporation and condensation rates considered in Section 5.1. The only difference is that neither the escaping (evaporating) nor captured (condensed) molecules are water. By arguments similar to those underlying Eq. (5.17), it follows that the capture rate is given by

$$C = \frac{1}{4}n\langle v\rangle\alpha, \tag{5.185}$$

where α is the mass accommodation coefficient. By using the ideal gas law we can write Eq. (5.185) as

$$C = \frac{p\alpha}{\sqrt{2\pi mkT}}, \tag{5.186}$$

where p is the partial pressure of the component. For fixed temperature and partial pressure, C is constant. Attainment of an equilibrium concentration $(dc/dt = 0)$ therefore requires that E increase with increasing concentration. The simplest such function is

$$E = \beta c. \tag{5.187}$$

where β is independent of concentration (although it may depend on temperature). We combine Eqs. (5.184), (5.186), and (5.187) to obtain

$$\frac{dc}{dt} = \frac{p\alpha}{\sqrt{2\pi mkT}} - \beta c. \tag{5.188}$$

When equilibrium is reached, c ceases to change with time. Thus, the equilibrium concentration c_∞ is, from Eq. (5.188),

$$c_\infty = \frac{\alpha}{\beta}\frac{p}{\sqrt{2\pi mkT}}. \tag{5.189}$$

This can be written more compactly by combining the various factors in Eq. (5.189)

$$c_\infty = \frac{p}{K(T)}, \tag{5.190}$$

where

$$K = \frac{\beta\sqrt{2\pi mkT}}{\alpha}. \tag{5.191}$$

Equation (5.190) is one of the forms of *Henry's law*. As could have been predicted with near certainty, this is not the form of the law enunciated in 1803 by William Henry (b. Manchester, England, 1774; d. Manchester, 1836): "under equal circumstances of temperature, water takes up, in all cases, the same volume of condensed gas as of gas under ordinary pressure." Let us try to reconcile Henry's law as stated by Henry with the modern version.

Take the dimensions of c_∞ to be moles per unit volume. The total number of moles n_m of a gaseous component of air in equilibrium with this component dissolved in a volume V_w of water is $c_\infty V_w$. From the ideal gas law it follows that the volume V_g of this dissolved gas at the partial pressure p and temperature T of the gas is

$$V_g = \frac{n_m R^* T}{p}. \tag{5.192}$$

With this equation and Eq. (5.190) it follows that

$$\frac{V_g}{V_w} = \frac{R^* T}{K}. \tag{5.193}$$

An implicit assumption that crept into our analysis is that the volume of the water is essentially the same with and without the dissolved gas. This is usually a good approximation, but it is an approximation nonetheless and is not strictly true.

According to Eq. (5.193), the equilibrium concentration of a gas dissolved in water (or any liquid), expressed as the volume of this gas at the pressure and temperature of the gas exposed to the liquid relative to the liquid volume, is fixed for a given temperature and independent of pressure (the same for a "condensed gas" as for a gas at "ordinary pressure").

"'When I use a word,' Humpty Dumpty said, in rather a scornful tone, 'it means just what I choose it to mean—neither more nor less'." *Solubility* is a Humpty Dumpty word. The solubility—ability to dissolve—defined by Eq. (5.193) is sometimes called the *Ostwald absorption coefficient*, although the great physical chemist Wilhelm Ostwald himself used simply "solubility" for this quantity. The *Bunsen absorption coefficient* at temperature T is defined as the amount of gas, expressed as a volume at $0\,°C$ and a pressure of 1 atm, dissolved in unit volume of liquid in equilibrium with the gas at a partial pressure of 1 atm. These two absorption coefficients differ by the ratio of the absolute temperature of interest to 273.15 K. The quantity $c_\infty/p = 1/K$ has every right to be called a solubility in the sense that it is a property only of the gas–liquid system (at a given temperature), not of the partial pressure of gas to which the liquid is exposed (to the extent that Henry's law is valid). To add to the confusion, concentrations (and hence, solubilities) can be expressed as moles (of dissolved gas) per unit volume (of liquid), moles per unit mass, mass per unit volume, mass per unit mass, and so on. Moral: When anyone tries to sell you a "solubility," read the fine print on the contract.

We wrote Henry's law in the form Eq. (5.190) to emphasize the causal relation between the concentration of gas dissolved in a liquid and the partial pressure of gas to which the liquid is exposed. The origins of this relation are curiously absent from many treatments of Henry's law, as if it were just an inscrutable mathematical law with no simple physical interpretation. Why does the equilibrium concentration (at a given temperature) of gas in a liquid depend on the partial pressure of the gas? Because at a given temperature, pressure is proportional to the rate at which gas molecules bombard the liquid, which is *how* the gas enters it. Why is the equilibrium concentration different (at a given temperature) for the same gas in different liquids or for different gases in the same liquid? Because the escaping tendency of gas molecules in a sea of liquid molecules depends on the interaction between the two, which can be quite different for different gas–liquid combinations. A gas molecule that enters a liquid and finds itself in the grip of strong attractive forces is not likely to escape. The stronger these attractive forces, the greater the equilibrium concentration of gas in the liquid. To convince yourself of this, consider the limiting (unrealistic) example of infinite attractive forces. All gas molecules that enter the liquid never return to the gas, and hence, the concentration of gas in the liquid increases without limit as long as there is a fresh supply of gas molecules striking the liquid.

An important example of gas concentrations in liquids is the dissolved oxygen that fish "breathe." They are the major beneficiaries of Henry's law. Oxygen concentrations in water are often expressed in $mgL^{-1} \approx 0.56\,ppm$; 12 mgL^{-1} is often taken as the maximum. Fish are literally dead in the water at 1–2 mgL^{-1}. The concentration (m^{-3}) of oxygen in sea-level air is about 25 times the maximum concentration of oxygen in water.

Pressure per se is not the physically relevant quantity in Henry's law. Don't be misled by the form of this law into thinking that gas molecules are pressed into a liquid like nails driven into a board with a hammer. Pressure is, at a fixed temperature, merely a convenient measure of the random collision rate of gas molecules with the liquid surface. These collisions, and nothing else, cause gas molecules to enter a liquid. Intermolecular forces retard them from exiting.

With a bit of algebra we can write Henry's law Eq. (5.190) as

$$p = \frac{K\rho_w}{M_w}\frac{n_m}{n_w}, \tag{5.194}$$

where ρ_w is the density of water, M_w its molecular weight, and n_w the number of moles of water mixed with n_m moles of gas (where $n_w \gg n_m$). Note the similarity between Raoult's law and Henry's law Eq. (5.194): a linear relation between a pressure and a mole fraction. This is hardly a surprise. Raoult's law and Henry's law are two sides of the same coin. Both laws apply to (dilute) solutions. It is customary to denote as solvent the component of greatest abundance in a solution, and as solute the component of least abundance. Raoult's law gives the pressure of solvent in equilibrium with solution when the mole fraction of solute is small, whereas Henry's law gives the pressure of solute in equilibrium with solution when the mole fraction of solute is small.

The proportionality constants are quite different in Raoult's and Henry's laws. Again, this is hardly surprising. Under the conditions for which Raoult's law is valid, solvent molecules escape from an environment dominated by solvent molecules. But under the conditions for which Henry's law is valid, solute molecules escape from an environment dominated by solvent molecules. Thus, Henry's law is at its best when Raoult's law is at its worst, and vice versa.

Raoult's law and Henry's law are not fundamental laws of nature, merely good approximations for many gases and many liquids—but not all. Henry's law states that at a given temperature, the equilibrium concentration of a gas in a liquid is proportional to the partial pressure of that gas above it. Natural water contains dissolved oxygen, nitrogen, argon, carbon dioxide, and so on, each gaseous component of air separately satisfying Henry's law as if it alone were present, which was shown by Dalton only a few years (1807) after Henry published his findings. We note in passing that because carbon dioxide is in the atmosphere in concentrations of about 419 ppm (2021 global average) and is highly soluble in water, precipitation is naturally acidic, a weak carbonic acid solution. Acidity of a solution is measured by pH, the negative of the logarithm (base 10) of the hydrogen ion concentration in moles per liter. Pure water has a pH of 7. Water in equilibrium with atmospheric carbon dioxide has a pH of about 5.6, the precise value depending on temperature because solubility is temperature dependent. Rainwater is naturally acidic. Acid rain as a consequence of the malodorous activities of humans has a pH of around 4.5. That of the oceans is around 8, and hence, they are basic (alkaline). Ocean "acidification" is a trend toward neutral.

Change in Saturation Vapor Pressure with Total Pressure

We now have all the tools for answering a question left hanging in Section 5.7: what is the *net* effect of air on the saturation vapor pressure of water? According to Poynting's formula (Eq. 5.161), saturation vapor pressure increases with increasing total pressure in the ratio of the densities of the vapor and liquid phases:

$$\frac{de_s}{dp} = \frac{\rho_v}{\rho_w}. \tag{5.195}$$

At $0\,°C$, the saturation vapor pressure of water is about 6.11 hPa. The density of liquid water ρ_w is about 1000 kg m^{-3} According to the ideal gas law, the density of water vapor at 6.11 hPa and $0\,°C$ is about 0.00485 kg m^{-3}. Thus, increasing the *total* pressure on liquid water from 6.11 hPa to 1013 hPa increases the saturation vapor pressure by approximately

$$\Delta e_s \approx \Delta p \frac{0.00485}{1000} = 4.88 \times 10^{-3} \text{ hPa}. \tag{5.196}$$

But air cannot overlie water without entering it, and anything dissolved in water *lowers* its saturation vapor pressure (at a fixed temperature). Air fits into the category of "anything," and hence air dissolved in water must lower its saturation vapor pressure. About 29 cm^3 of air is dissolved in one liter (1000 cm^3) of water at $0\,°C$ in equilibrium with air. The total number of air molecules in this volume at a pressure of one atmosphere can be obtained from the ideal gas law

$$N = \frac{pV}{kT} = 7.8 \times 10^{20}. \tag{5.197}$$

The number density of molecules in water is

$$\frac{\rho_w N_a}{M_w} = 3.35 \times 10^{28} m^{-3}, \tag{5.198}$$

where N_a is Avogadro's constant and M_w is the molecular weight of water, and hence, one liter (10^{-3}m^3) of water contains 3.35×10^{25} water molecules. The mole fraction of dissolved air is therefore 2.32×10^{-5}. According to Raoult's law, the saturation vapor pressure is lowered by this fraction, about 1.4×10^{-4} hPa. This decrease in vapor pressure because of dissolved air is approximately 35 times smaller than the increase because the total pressure of the water is greater than it would be if it coexisted solely with its vapor. Thus, the net effect of air is to *increase* the saturation vapor pressure of water. This increase is small, but it provides another brick to hurl at the sponge analogy, which predicts a decreasing vapor pressure with increasing air pressure. After all, as the air molecules become more tightly packed, shouldn't there be less room for water molecules? To determine how air changes the saturation vapor pressure of water, we appealed to both Henry's and Raoult's laws.

In 1802, Dalton asserted that the vapor pressure of a liquid in a gas is the same as in a vacuum (not absolutely true but a very good approximation). Yet almost 200 years later the notion is widespread and virtually ineradicable that air is a kind of sponge with a limited holding capacity for water and that the pores in this gaseous sponge expand and contract with temperature changes. Dalton is turning over in his grave.

5.10   Size Dependence of Vapor Pressure: Water Droplets, Solution Droplets, and Bubbles

The boundary between a liquid and a vapor is not as sharp as it appears. Near a boundary there is a more or less continuous change, although abrupt, from one phase to the other. An intermediate phase, neither liquid nor vapor, inhabits a narrow transition region, often ignorable—but not always. Let δ be its thickness for a system with volume V and surface area A. Assume that the number density n of molecules is the same everywhere throughout the volume (not strictly true). With these assumptions, the ratio of the number of molecules in the transition region to the total number of molecules is

$$\frac{nA\delta}{nV} = \frac{A\delta}{V}.$$ (5.199)

The surface-to-volume ratio A/V is the inverse of a characteristic linear dimension of the system. For example, A/V for a sphere of diameter d is $6/d$, and Eq. (5.199) is

$$\frac{6\delta}{d}.$$ (5.200)

The transition region must be at least a molecular diameter and is likely to be several. If we take $10d_m$ as a rough guess for the thickness of the transition region, where d_m is the molecular diameter, the ratio of molecules in the transition region to the total number is

$$\frac{60d_m}{d}.$$ (5.201)

Assume that if transition molecules make up about one-tenth the total we can no longer ignore the transition region if

$$d \leq 600 \, d_m \, . \tag{5.202}$$

Diameters of molecules such as water and oxygen are about $3 \times 10^{-4} \mu$m, which gives the criterion

$$d \leq 0.2 \ \mu\text{m}. \tag{5.203}$$

On the basis of admittedly crude reasoning we conclude that finite size may be of consequence to systems (particles) smaller than about 0.2 μm. As we shall see, this is supported by more detailed analysis. Although the common objects of our everyday lives are much larger, the atmosphere contains hordes of particles this size and smaller, and hence, the consequences of finite size are of special relevance to meteorology.

Droplet and Bubble Vapor Pressure: Physical Interpretation

In Section 5.1 we give a physical argument *why* the equilibrium vapor pressure over a *flat* surface should increase with temperature. For a molecule to evaporate (escape) from a liquid (or solid) it must have sufficient energy to overcome the attraction of its nearest neighbors, and the fraction of molecules with this energy increases with increasing temperature. In Section 5.3 we derived Eq. (5.49), which supports this argument and shows *how much* the equilibrium vapor pressure increases and how it depends on enthalpy of vaporization and gas constant. We follow a similar procedure for the equilibrium vapor pressure of droplets and bubbles, and first give a physical argument for why it is different from that for flat surfaces.

By equilibrium vapor pressure of a droplet of a given diameter we mean its environmental or *external* vapor pressure such that the droplet neither grows nor shrinks. By equilibrium vapor pressure of a bubble of given diameter we mean its *internal* vapor pressure such that the bubble neither grows nor shrinks assuming that the only gas in the bubble is vapor from the surrounding liquid.

A water molecule, say, on a flat surface has an indefinitely large number of neighbors attracting it, and hence, impeding its escape, although only a few nearest neighbors are effective because intermolecular attractions drop off rapidly with distance on the molecular scale. Consider a molecule on the surface of a small droplet. The smaller it is, the fewer its nearest neighbors relative to what it would be if it were on a flat surface, the more likely it is to escape, and hence, the greater the equilibrium vapor pressure. To determine how small the droplet has to be for this vapor pressure to be appreciably greater than for a flat surface requires detailed analysis, but Eq. (5.203) gives us a rough estimate.

For a bubble, we reverse the argument. A molecule at the surface of a tiny bubble has *more* nearest neighbors than it would if it were on a flat surface, a smaller evaporation rate, and hence, a smaller equilibrium vapor pressure.

Spherical water droplets and bubbles have the same radius of curvature at every point. Although we cannot get inside a bubble or outside a droplet to observe differences in vapor pressure, we can extend our physical explanation to any surface with a varying radius of curvature, and hence, a rate of evaporation that varies accordingly. The radius of curvature of a flat surface is infinite, that of an (ideal) sharp edge is zero. Although snowflakes may be born as delicately beautiful dendrites, if isolated at temperatures well below freezing for days or weeks they evolve (metamorphose) into lumps with all the aesthetic appeal of mashed potatoes. This is not because of melting, but rather due to differential rates of evaporation (sublimation to be picky), increasing with decreasing radius of curvature. As a consequence, flatter parts grow at the expense of sharper parts shrinking. Left to evolve on its own, a snowflake would end its life as a more or less spherical blob, a shape with no gradients of radii of curvature. This is compelling evidence that vapor pressure (evaporation rate) depends on radius of curvature if sufficiently small.

Mechanical Equilibrium of Balloons, Corneas, Droplets, Bubbles: The Young–Laplace Equation and the Road to Surface Tension

To proceed we must grapple with the concept of surface tension, a subject generally treated badly in textbooks. To make matters worse, eminent scientists have argued that surface tension is fictitious. That is, it is impossible for a static liquid to support a tensile stress parallel to its surface because this is a shear force and "everyone knows" that an essential property of static liquids is that they cannot support shear stresses. This is a legitimate argument but happens to be wrong. To ease into this we begin with an example of surface tension obviously *not* fictitious, that in the skin of an inflated balloon. As evidenced by Chapter 3, we are obsessed with balloons.

Figure 5.16 depicts an inflated spherical balloon. The pressure inside is p_i and that outside is p_o. We apply a force balance to half the balloon. If σ is the force per unit length along the circumference of the skin of the balloon, mechanical equilibrium requires that

$$p_i = p_o + \frac{2\sigma}{r}, \tag{5.204}$$

where r is the radius. This is sometimes called the Young–Laplace equation. As expected, the pressure inside is greater than that outside, and you may have noticed that a balloon becomes easier to inflate as it increases in size, as predicted. The quantity $2\sigma/r$ is sometimes called the *capillary pressure*. It can be measured by inserting a smart phone with a barometer into a balloon and inflating it (see Vandermarlière, 2016). But we have measured σ directly (see Problem 79) and it agrees within a factor of 2 with measurements of $p_i - p_o$.

Capillary pressure may be closer to home than you realize, just with a different name: *intraocular pressure*. A normal part of an eye examination by an optometrist is a measurement of this pressure. A topical anesthetic is applied to your cornea, then a probe gently pressed against it. By measuring the force required for a given

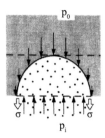

Fig. 5.16: A force balance applied to an inflated spherical balloon in mechanical equilibrium.

flattening, the intraocular pressure is inferred. Measured in millimeters of Hg, the normal range is 12–22 mm. Pressures outside this range may indicate glaucoma.

An example of what appears to be the consequence of a real surface tension of an air–water interface like the tension in the skin of a drum, is provided by water striders, insects that make dimples on water because of the downward force of their weight. This is balanced by an upward force, a consequence of a nonzero divergence of the normal to the depressed surface. For example, the quantity $2/r$ in the capillary pressure is the divergence of the normal to a sphere; the divergence of the normal to a flat surface is zero. But depress an elastic surface and you will feel a resisting force because the normal to the surface is not uniform. The pressure difference across the surface is proportional to this divergence, similar to a force being the gradient of a potential energy.

We are accustomed to thinking of pressure in a fluid as positive and isotropic, as it implicitly is elsewhere in this book. By isotropic is meant that the pressure force on any surface in the fluid is normal to the surface. This is indeed true *within* a *static* fluid, but not in a moving fluid or at the surface of a static fluid. A velocity gradient or a surface destroys isotropy. Well within a static fluid (several molecular diameters) every direction is indistinguishable from every other direction. Within an interfacial region, however, this no longer holds, the fundamental reason for the pressure anisotropy in this region.

Now if we are so bold as to apply Eq. (2.204) to a liquid droplet (not necessarily water) in air at a pressure $p_a = p_o$,

$$p_i = p_a + \frac{2\sigma}{r}. \tag{5.205}$$

p_a is not necessarily solely the equilibrium vapor pressure of the droplet. For a bubble in the same liquid

$$p_g = p_o + \frac{2\sigma}{r}, \tag{5.206}$$

where $p_g = p_i$ and the subscript g denotes a gas *in* the bubble, not necessarily entirely vapor from the surrounding liquid. The differences between the liquid and gas pressures are equal in magnitude but opposite in sign. But for both droplet and bubble the internal pressure must be greater than the external pressure. At 20 °C, the surface tension of a gas–water interface is 0.073 Nm^{-1}, which we assume is independent of the gas. A small spherical droplet holds its shape because of surface tension regardless of the surrounding gas. For a radius of 1 μm, the capillary pressure of water is about 1500 hPa.

Without knowing it, you are likely to have used Eq. (5.205). For an infinite radius of curvature (flat surface), this equation gives the condition for pressure equality across an interface, which seems so obvious we take it for granted. The barometer (see Section 2.1) is based on this equation. In deriving the expression for the height of liquid in a barometer, we assume pressure equilibrium between this liquid and the surrounding air, which is Eq. (5.205) for "large" r, by which is meant $r \gg 2\sigma/p_a$. For $p_a = 10^5$ Pa and σ for water, a radius of curvature appreciably greater than 1 μm can be said to be "large."

Those who argue that surface tension is fictitious do not deny the existence of σ, but aver that it is "really" specific surface free energy. σ does indeed have the dimensions of energy per unit area, and we know that it takes energy to make a surface. You can glare at a piece of chalk until the end of time, and it will not break. A bit of work is required to break it and create two surfaces. Work done on a system increases its energy, and we can consider it to reside in the surface, which is no more mysterious than assigning an internal energy to a volume. Decompose the working rate W into two components: $-p\,dV/dt$ and everything else, denoted by W'. By arguments similar to those leading to Eq. (5.131) we obtain

$$W' = \frac{d}{dt}(H - TS) + S\frac{dT}{dt} - V\frac{dp}{dt}.\tag{5.207}$$

Thus, for a process at constant pressure and temperature

$$W' = \frac{d}{dt}(H - TS) = \frac{dG}{dt}.\tag{5.208}$$

The *coefficient of surface tension* σ (usually shortened to surface tension) is often defined by way of the rate at which work is done in changing the area of a surface at constant temperature and pressure:

$$W' = \sigma\frac{dA}{dt}.\tag{5.209}$$

From Eqs. (5.208) and (5.209), it therefore follows that the surface free energy is

$$G = \sigma A.\tag{5.210}$$

Equilibrium Vapor Pressure of Droplets and Bubbles: A Physical Interpretation

The Young–Laplace equation tells us nothing about the dependence of the vapor pressure of a droplet in equilibrium with its vapor or the vapor pressure within a bubble. This equation is only a condition for mechanical equilibrium. For example, the pressure inside a bubble must satisfy this equation or it will either expand or contract. This is an important point we return to.

Before deriving an expression for the dependence of vapor pressure on droplet and bubble size, we can test and sharpen our intuition and understanding by obtaining an approximate expression by analogy and physical reasoning. The exponent in Eq. (5.50) is an energy per molecule divided by kT, the same form as in the barometric formula. We therefore can make an educated guess that the dependence of vapor pressure on size is also exponential with an analogous exponent. Consider a droplet of radius r. If surface tension can be interpreted as an energy per unit area, $4\pi r^2\sigma$ is the surface energy of a droplet. Divide this by the total number of molecules in the droplet, $(\rho_w N_a/M)(4\pi r^3/3)$, to obtain the energy per molecule, and hence, the quotient

$$\frac{3\sigma}{\rho_w r k T(N_a/M)} = \frac{3\sigma}{\rho_w r R_v T},\tag{5.211}$$

where $m = M/N_a$ is the mass of a molecule and R_v is the gas constant for the vapor. Because we expect the vapor pressure to increase with decreasing size, we take the exponent to be positive, and hence, our best guess for the vapor pressure is

$$e_s = C\exp(3\sigma/r\rho_w R_v T).\tag{5.212}$$

This is the more rigorous Kelvin equation if we replace 3 by 2. And if we had chosen $3kT/2$ for the mean kinetic energy of a molecule [Eq. (2.10)] we would have obtained the Kelvin equation. Not bad for a one-line derivation! For a bubble, we expect that changing the curvature from $1/r$ to $-1/r$ in this equation will do the trick, which does indeed give a vapor pressure that decreases with decreasing size.

An implicit assumption is that σ depends only on temperature, not size, but at some sizes this cannot be strictly true. We expect *all* thermodynamic properties of sufficiently small particles to depend on size. An example is the appreciable melting point depression of metallic particles smaller than tens of nanometers. Below some size, the concept of a melting point ceases to be valid. A single molecule does not have a melting point or surface tension. What about a particle composed of 10, 100, 1000 molecules? At what point does the particle become "large"? The size dependence of σ has been a subject of theoretical investigation for many years but we need not concern ourselves about this.

The Kelvin Equation and the Difficult Birth of Cloud Droplets

A simple and clever derivation of the Kelvin equation by Galvin is based on the thought experiment depicted in Fig. 5.17. A thin capillary tube of radius r is inserted into a liquid. The inside of the tube is imagined to be a nonwetting surface, by which is meant that a droplet on it would be a perfect sphere. This is described by saying that the contact angle is 180°, an idealized upper limit resulting in what appears

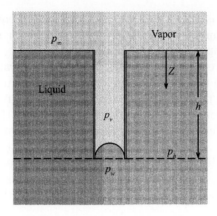

Fig. 5.17: This depicts a thought experiment in which a non-wetting capillary tube is inserted into a liquid.

to be an unrealistic drop in the liquid level in the tube. A liquid hemisphere is at its bottom. The pressure at a depth h in the liquid is

$$p_h = p_\infty + \rho_w g h, \tag{5.213}$$

from which it follows that

$$h = \frac{p_h - p_\infty}{\rho_w g}. \tag{5.214}$$

The vapor pressure in the tube increases with depth

$$p_v(z) = p_\infty \exp(gz/R_v T). \tag{5.215}$$

Assume that the pressure in the hemisphere p_w is uniform and equal to p_h. From the Young–Laplace equation we have

$$p_w = p_v(h) + p_c, \tag{5.216}$$

where p_c is the capillary pressure. Combining these equations yields

$$p_v(h) = p_\infty \exp\left[\frac{p_v(h) - p_\infty + p_c}{\rho_w R_v T}\right]. \tag{5.217}$$

g serves a purpose and then vanishes from the stage. The vapor pressure $p_v(h)$ is what we are after. At first glance it appears that this would require an iterative solution. This is not necessary if $p_v(h) - p_\infty \ll p_c$. We already showed that for $r = 1$ µm, the capillary pressure for water is more than 1 atm. As a consistency check, approximate Eq. (5.217) by

$$p_v \approx p_\infty \exp\left(\frac{2\sigma}{r \rho_w R_v T}\right). \tag{5.218}$$

For water for $T = 293\,\mathrm{K}$, $\rho_w R_v T = 1.35 \times 10^5 \mathrm{kPa}$, $r = 1\,\mu\mathrm{m}$, and $p_\infty = 20\mathrm{hPa}$, $p_v(h) \approx p_\infty = e_\infty$, whereas $p_c = 1500\,\mathrm{hPa}$. Thus, the inequality is satisfied with room to spare and

$$e/e_\infty = \exp(2\sigma/r\rho_w RT) \tag{5.219}$$

is a good approximation that gets better with decreasing size and probably is good for all sizes of cloud droplets. Equation (5.219) is the Kelvin equation, remarkably close to Eq. (5.212).

e_∞ is the saturation vapor pressure Eq. (5.50) for a surface with *zero* curvature $(1/r)$. But ℓ_v depends on temperature *and* size for sufficiently small sizes (large curvatures). This is often overlooked, but the error in doing so is small except for droplets in the earliest stages of growth, composed of, say, 20–50 molecules. At some size σ, ℓ_v, ρ_w, and T cease to have a meaning because they characterize only aggregations of many molecules. The only uncertainty is, exactly how many?

We can approximate h by

$$h \approx \frac{p_c}{\rho_w g}. \tag{5.220}$$

For $g = 9.8\,\mathrm{m\ s^{-2}}$, $h = 14\,\mathrm{m}$ for $r = 1\,\mu\mathrm{m}$, and $h = 1400\,\mathrm{m}$ for $r = 0.01\,\mu\mathrm{m}$. As we said, this is a "thought experiment."

There are more complicated routes to the Kelvin equation, but it is more fun to obtain it by simple and clever arguments. And taking a detour around mathematical derivations leads to better physical understanding.

We now have all the ingredients to calculate how equilibrium vapor pressure depends on size. The surface tension of water (in air) at $20\,^\circ\mathrm{C}$ is approximately $0.073\,\mathrm{Nm^{-1}}$, the density of liquid water is $10^3\,\mathrm{kgm^{-3}}$, and the gas constant for water vapor is $461\,\mathrm{Jkg^{-1}K^{-1}}$. Thus, at $20\,^\circ\mathrm{C}$, the equilibrium vapor pressure from Eq. (5.219) is

$$\frac{e_{sr}}{e_{s\infty}} = \exp\left(\frac{0.00108}{r}\right), \tag{5.221}$$

where r is in micrometers. From this equation we generate the following table:

r (μm)	$e_{sr}/e_{s\infty}$
1	1.001
0.1	1.01
0.01	1.11
0.001	2.93

This confirms the correctness of our previous guess that when droplet diameters are less than about 0.2 μm, we cannot ignore the consequences of a finite surface-to-volume ratio. The ratio $e_{sr}/e_{s\infty}$ corresponding to a molecular radius 1.5×10^{-4} μm is about 1300. We shouldn't take this number seriously because Eq. (5.221) is based on macroscopic thermodynamics, which is of dubious applicability to single molecules. Nevertheless, our calculations show that very small droplets cannot exist except in environments with exceedingly high relative humidities. Recall that relative humidity is defined relative to a *flat* surface of pure water. Thus, this table indicates that a water droplet of radius 0.001 μm could not exist in an environment with relative humidity less than about 293%, appreciably higher even than that in muggy Washington, D.C. during the summer.

Equation (5.221) and the results obtained from it are at the root of why mature cloud droplets cannot begin their lives as single water molecules. If a cloud droplet of diameter 10 μm (a typical value) were merely one that had grown from a single molecule, its demands on the environment would have been insatiable in the early stages of its growth. Fortunately, nature provides an abundance of *condensation nuclei* in the atmosphere, which enable cloud droplets to avoid an awkward stage of growth in which they would be voracious consumers of water vapor.

In an atmosphere scrubbed clean of all particles (a task to daunt Hercules), cloud formation would not be impossible but would require relative humidities much higher than those to which we are accustomed. The consensus of opinion is that cloud droplets would form in air completely free of particles when the relative humidity reaches about 400%. Clean moist air would have to be cooled to well below its nominal dew point to lower the kinetic energy of water molecules sufficiently to increase the probability that they bind together in random collisions to form clusters that live long enough to serve as embryos for further growth. This is the process of *homogeneous nucleation*, in which water droplets form on clusters of their own kind, in contrast to *heterogeneous nucleation*, in which droplets form on foreign particles. Homogeneous nucleation is not forbidden in the atmosphere, but it need not, and rarely (if ever) does, occur because in all but the cleanest air there are so many suspended particles, many of natural origin, that cloud droplets form long before the relative humidity reaches the high value necessary for homogeneous nucleation.

Vapor Pressure of Solution Droplets

We showed in the previous subsection that very small droplets of pure water cannot exist in environments of 100% relative humidity or less because they would rapidly waste away by net evaporation. But if a droplet were to begin its existence as a *solution* droplet, the vapor pressure reduction by dissolution might compensate for the vapor pressure increase with decreasing size. Suppose that a droplet begins its existence on a *soluble* nucleus. Water vapor bombards this nucleus, and a thin solution layer is formed, which evaporates less rapidly than pure water at the same temperature and with the same radius of curvature. Initially, the droplet is a saturated solution around a solid core. As the droplet grows, it becomes entirely solution with ever decreasing concentration. Thus, the vapor pressure reduction is greater the smaller the droplet, in opposition to the vapor pressure increase with decreasing droplet size.

Consider a small salt (NaCl) grain of diameter d_s. The solubility of salt in cold water is about 36 g per 100 cm^3 of water. That is, when 36 g of salt is added to 100 cm^3 of pure water, all the salt dissolves to form a saturated solution in the sense that additional salt would not dissolve. The mass m_s of the salt grain is $\rho_s V_s$, where ρ_s is the density of salt (about 2.17 g cm^{-3}). The volume of water V_w required to dissolve the grain to form a saturated solution is obtained from $m_s/V_w = c$, where c is the concentration of the solution (mass per unit volume of water). When a given volume of any substance is dissolved in a given volume of a liquid, the volume of the resulting solution is not necessarily the sum of the volumes of its two components. We dissolved about 50 g of ordinary table salt in a few hundred cm^3 of water. As far as we could tell, the volume of the solution so obtained differed from the sum of the volumes of salt and water by at most a few percent. So, we shall assume no volume change upon dissolution of salt in water. With this assumption, the volume V_o of a saturated solution droplet is

$$V_o = \frac{\pi d_o^3}{6} = V_w + V_s = \frac{m_s}{\rho_s} + \frac{m_s}{c} = \frac{\pi d_s^3}{6}\rho_s\left(\frac{1}{\rho_s} + \frac{1}{c}\right), \tag{5.222}$$

where d_s is the diameter of the salt grain. The diameter d_o of a (saturated) solution droplet is therefore

$$d_o = d_s\left(1 + \frac{\rho_s}{c}\right)^{1/3} = 1.91\, d_s\,. \tag{5.223}$$

For diameters d between d_s and $1.91 d_s$, the droplet is a salt core coated with a layer of saturated salt solution. As the droplet grows because of net condensation of water vapor, the salt concentration decreases. We assume that the salt solution obeys Raoult's law even though we know that this is not strictly true for high concentrations (see Fig. 5.15). To determine the dependence of equilibrium vapor pressure on droplet size because of the salt dissolved in it, we need only the mole fraction x_w of water in the droplet:

$$x_w = \frac{m_w/M_w}{m_w/M_w + 2m_s/M_s}, \tag{5.224}$$

where M_w and M_s are the molecular weights of water and salt, respectively, and m_w is the mass of water (which depends on droplet size, whereas the mass of salt is fixed). The factor 2 is required because salt dissociates into sodium and chlorine ions in aqueous solution. If we denote by V the total volume of the droplet, Eq. (5.224) can be written

$$x_w = \frac{V - V_s}{V + V_s\left(\frac{2\rho_s M_w}{\rho_w M_s} - 1\right)}. \tag{5.225}$$

We can rewrite Eq. (5.225) in terms of droplet diameter d:

$$x_w = \frac{1 - (d_s/d)^3}{1 + 0.334(d_s/d)^3}. \tag{5.226}$$

For $d \gg d_s$ Eq. (5.226) becomes

$$x_w \approx 1 - \left(\frac{d_s}{d}\right)^3.$$

(5.227)

Although the evaporation rate increases with decreasing size because of the greater ease with which a water molecule can escape from a pure water droplet, the evaporation rate decreases because of the increasing concentration of dissolved salt. We can combine Eqs. (5.221) and (5.226) to obtain an expression for the equilibrium vapor pressure (relative to that for an infinite pure water droplet) over a solution droplet of diameter $d \geq 1.91 d_s$ formed on a salt grain of diameter d_s:

$$\frac{e_s}{e_{s\infty}} = \exp\left(\frac{0.00216}{d}\right)\left(\frac{1 - (d_s/d)^3}{1 + 0.334(d_s/d)^3}\right), \qquad d \geq 1.91 d_s$$

(5.228)

where d is in micrometers. For diameters between d_s and $1.91\,d_s$, the vapor pressure equation is

$$\frac{e_s}{e_{s\infty}} = 0.817\exp\left(\frac{0.00216}{d}\right), \qquad d_s \leq d \leq 1.91 d_s.$$

(5.229)

Note that Eq. (5.228) is the product of two functions of diameter—one monotone increasing, the other monotone decreasing—which allows for the possibility of a maximum for $e_s/e_{s\infty}$. We may interpret e_s as the vapor pressure of the environment of a droplet that neither shrinks nor grows. A plot of e_s versus d is therefore a plot of all the equilibrium states of a solution droplet formed on a salt particle of diameter d_s.

The quantity $e_s/e_{s\infty}$ is sometimes called the *saturation ratio*, relative humidity without the factor 100%. Figure 5.18 shows the equilibrium saturation ratio as a function of size for a droplet formed on a salt (NaCl) nucleus of diameter 0.015 μm. This curve is called a *Köhler curve*, but it is a bit different from the Köhler curves you are likely to find elsewhere. We wanted you to see once in your life a Köhler curve for droplet diameters ranging all the way from that of a salt nucleus to many times this diameter. Figure 5.19 shows a set of more conventional Köhler curves, for salt nuclei of various diameters, in which the initial stage of droplet growth is suppressed. If it were not for the soluble nucleus incorporated in each droplet, these curves would climb to stupendous heights with decreasing droplet diameter. Instead, soluble nuclei greatly lower the equilibrium vapor pressure for the smallest droplets. All these curves rise above the 100% relative humidity level. Typical cloud droplets have diameters around 10 μm. Thus, soluble nuclei or solution droplets with diameters a few micrometers or less cannot grow to cloud droplet size unless they are exposed to a supersaturated environment. A nucleus in an environment with relative humidity 100% or less can grow only so much before it reaches an impenetrable barrier to further growth. To make this point clearer we appeal to Fig. 5.20, which shows a single Köhler curve (for a nucleus of diameter 0.015 μm) to avoid clutter.

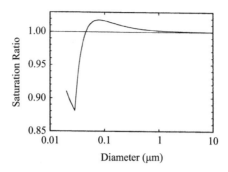

Fig. 5.18: Environmental saturation ratio (at 20 °C) such that a solution droplet of given diameter neither shrinks nor grows (no net evaporation). Saturation ratio is the vapor pressure relative to the saturation vapor pressure (over a *flat* surface of *pure* water). The salt (NaCl) nucleus on which droplets form has a diameter of 0.015 μm. The sharp change in slope signals total dissolution of the nucleus in water.

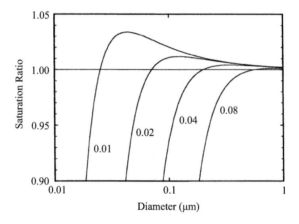

Fig. 5.19: A series of curves similar to the single curve in Fig. 5.18. Each curve is labeled by the diameter (in μm) of the salt (NaCl) nucleus on which the droplet forms. The lower parts of these curves, for the smallest diameters, are suppressed.

Figure 5.19 shows why a distinction is sometimes made between *condensation nuclei* and *cloud condensation nuclei*. A condensation nucleus is an *insoluble* particle on which droplets *can* form. A cloud condensation nucleus is a *soluble* particle on which droplets form *preferentially*. They begin to grow at lower saturation ratios and so have an edge over insoluble particles. Absent soluble particles in the atmosphere, clouds would still form but at slightly higher saturation ratios.

Suppose that a solution droplet is in equilibrium with its environment (represented by point A on the curve in Fig. 5.20). Further suppose that this droplet is suddenly brought into an environment with a higher saturation ratio, indicated by

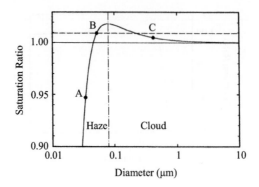

Fig. 5.20: Same curve as in Fig. 5.19 with the lower part of the curve suppressed. If a droplet of diameter shown at *A* is in equilibrium and suddenly brought into an environment with saturation ratio (relative humidity) shown by the dashed line, the droplet can grow only to the diameter shown at *B*. But if a droplet shown at *C* is brought into this same environment, the droplet can grow without limit (assuming an unlimited water vapor supply).

the dashed line. Disequilibrium is such that the droplet grows by net condensation, but as it does its rate of evaporation increases. When the droplet has grown to point *B*, it can grow no further. If it were to do so, its rate of evaporation would exceed its rate of condensation, and hence, the droplet would shrink. Consider another droplet in equilibrium with its environment, represented by point *C* on the curve. This droplet suddenly finds itself in an environment with a higher saturation ratio. Disequilibrium again causes droplet growth, but as this droplet grows its rate of evaporation decreases, and hence, the droplet continues to grow without limit (assuming the availability of water vapor, of course). Droplet *A* grows to a new stable state, whereas droplet *C* never achieves stability. This difference in stability provides the distinction between haze and cloud. A haze particle has a diameter lying to the left of the peak of the Köhler curve; a cloud particle has a diameter lying to the right. This curve is the key to understanding why visibility can be so poor under high humidities. When relative humidity rises above a critical value, say around 75% or so (still well below 100%), small soluble particles in the atmosphere begin to grow, becoming solution droplets. Although they do not grow as large as cloud droplets, they do become large enough to affect visibility. Scattering by particles small compared with the wavelength of illumination increases as the square of particle volume. Thus, a tiny soluble nucleus that grows by only a factor of two in diameter scatters 64 times as much light as it did when it was dry. More often than not, decreased visibility on humid days is attributed to scattering by water vapor. Yet, it is only when atmospheric water *ceases* to be vapor, condensing into solution droplets, that scattering markedly increases. Note carefully here the distinction between an observation and its interpretation. The observation that visibility often greatly deteriorates, all else being equal, on days of high humidity, is valid. The knee-jerk explanation of this observation as increased scattering by water vapor is cockeyed. Water vapor makes up only a few percent of air and scattering by a water vapor molecule is actually *less* than that by oxygen or nitrogen, the dominant species.

A line of constant saturation ratio may intersect a Köhler curve twice; for example, saturation ratios between about 0.88 and 0.91 and greater than 1.0 (Fig. 5.18). Each intersection point corresponds to an equilibrium diameter. What determines which is realized in nature? Although the intersection points correspond to two distinct equilibrium states, one is stable. the other unstable. A region of a Köhler curve with negative slope corresponds to states of unstable equilibrium. We can be reasonably sure that a droplet in a given environment will find its way to the stable equilibrium state just as a ball balanced on a protuberance will, if nudged a hair, roll into a neighboring depression.

Boiling Demystified and More Heresy

We now have all the ingredients necessary to understand the significance of the boiling point. Keep in mind that boiling requires *bubbles*. According to Eq. (5.206), a bubble in a liquid cannot exist without collapsing unless the pressure in the bubble is greater than that in the surrounding liquid. The pressure within a liquid is at least atmospheric pressure, increasing slightly with depth. Thus, for a vapor bubble in a liquid not to collapse requires the vapor pressure to be at least atmospheric pressure. The pressure in a vapor bubble in a pure liquid depends on temperature and size. By definition the nominal boiling point is the temperature at which the equilibrium vapor pressure (over a *flat surface*) is one atmosphere. This temperature is therefore the *minimum* temperature at which a vapor bubble can muster enough internal pressure to resist being crushed by its surroundings. Because this requires a vapor pressure higher than the surrounding pressure, becoming equal to this pressure only in the limit of infinite bubble radius, and because the Kelvin equation says that the vapor pressure inside a bubble is smaller than that over a flat surface, the temperature at which a vapor bubble will not collapse is slightly *higher* than the nominal boiling point.

If you put a metal pan filled with water on a stove to heat, before long you will hear pinging. This is the sound of collapsing vapor bubbles. They form on the hot bottom of the pan, break loose from their moorings, and rise into the cooler liquid above, which is not yet at the boiling point. Although the pressure of the surroundings of these bubbles decreases as they rise, this is not sufficient to compensate for the lower temperature, and hence, lower vapor pressure. As a consequence, the rising bubbles collapse, emitting a small shock wave that resounds against the sides of the pan. If you close your eyes and listen to water as it is heated, you can tell when the boiling point has been reached. Once the temperature of the water is uniformly at or slightly above the boiling point, vapor bubbles no longer collapse.

Tiny vapor bubbles in a liquid face the same awkward stage of growth as do tiny water droplets. According to Eq. (5.206), the pressure inside a vapor bubble necessary to prevent it from collapsing increases with decreasing size (for fixed surroundings). To make matters worse, saturation vapor pressure decreases with size. The only way a tiny vapor bubble can survive is by an increase in the liquid temperature to well above the nominal boiling point. As a rule, the extremely high temperatures necessary to sustain tiny embryo bubbles are not observed in boiling water. Thus, there must be a means by which vapor bubbles avoid the stage of their growth that would require high temperatures. Now Eq. (5.206) comes to the rescue. Note that p_g in this equation is the *total* gas pressure. A vapor bubble that begins its existence as a tiny bubble containing mostly air at a pressure of around one atmosphere can resist being crushed by its surroundings. A bubble in boiling water begins its life mostly as air but ends it mostly as vapor.

We did a simple experiment. We put snow into a food processor, added water, and turned the speed to HIGH. We added more water until all the snow had melted and the temperature was $1\,°C$. Churning water aerates it. After sitting in a pan for several hours the water temperature was $17\,°C$ and the side of the pan below the water level was covered with hundreds of tiny bubbles, about 0.3 mm in diameter, moored to microscopic pits and cracks. When we knocked a few bubbles loose, they rose and shrank to nothing. They were stabilized by attachment to tiny reservoirs of trapped air. These could provide the air bubbles required to nucleate boiling. Overnight the water temperature dropped to $12\,°C$ and most of the bubbles were gone. According to Eq. (5.206) the pressure in these mostly air-filled bubbles had to be just a bit higher than atmospheric pressure. From Henry's law and Fig. 5.18, the temperature increase from $1\,°C$ to $17\,°C$ resulted in a supersaturated (nonequilibrium) solution of air in water.

Invisible air bubbles are necessary for heterogeneous nucleation of boiling. They play the same role as condensation nuclei in the formation of clouds. A corollary of this is that filtered water (or any liquid) from which much of the air has been removed can be appreciably superheated in smooth-walled vessels. Small water droplets cannot exist (without help) because they evaporate too much whereas small bubbles in water cannot exist (without help) because they evaporate too little. Help comes in the form of nuclei to boost droplets and bubbles over a wall. Similarly, nuclei are necessary for water to freeze and carbon dioxide bubbles to form in freshly poured beer under conditions we have come to take for granted. And yet even mention of nucleation in physics textbooks is rare.

To show that there is *not* an abrupt huge increase in the evaporation rate of water at the boiling point preventing the water temperature from rising above $100\,°C$ as long as no bubbles form, e_s (which determines the evaporation rate) from Eq. (5.68) at $98\,°C$ is 933 hPa and at $99\,°C$ is 966 hPa, an increase of less than 4%.

Why can't the temperature of pure water exceed $100\,°C$ at an ambient pressure of one atmosphere? Actually, it can, which has been known for more almost 250 years. But there is no rigid temperature boundary that water is forbidden to cross by occult forces. Water in an open pan on a stove is heated from *below* and simultaneously cooled from *above* by net evaporation, net radiation, and convection accompanied by a small *decrease* in the mass of liquid water unless heated for sufficiently long. Heating usually dominates over cooling—*until* bubbles are formed in profusion. At that point cooling by evaporation *sharply* increases because bubbles transport water vapor out of the liquid as they burst at its surface. To overcome this increased evaporative cooling, turn up the gas on the stove. But this results in more vigorous boiling—a negative feedback opposing increased heating by the stove.

At *all* temperatures from $0\,°C$ to $100\,°C$, water evaporates, and the enthalpy of vaporization *decreases* by only $\approx 10\%$. The uniqueness of $100\,°C$ at a pressure of one atmosphere is that it is the temperature at which bubbles *can* exist in water without collapsing, *not* the temperature at which "latent heat" suddenly is aroused from its slumber. Because (net) evaporation is a cooling process, shouldn't it be called the "release of latent cold"? What *is* sudden is the formation of many water vapor bubbles. Q—its source the stove—raises the temperature of the water, resulting in water vapor being generated *within* it. Mantras about the "release" of latent heat are "explanation" by jargon.

Although it is difficult—but *not* impossible—to heat pure water at sea level to much above $100\,°C$, it may be difficult to heat water in an *open* container to above even *lower* temperatures. We note in Section 1.7 that it is difficult to treat open systems thermodynamically, but this doesn't stop us from trying. Suppose that a mass M of water decreases by ΔM because of net evaporation, where $|\Delta M| \ll M$. The more energetic molecules escape, leaving behind those less energetic. For this *local* energy decrease to be quickly and uniformly shared, the water must be well mixed by internal convection or stirring. $\Delta M \ell_v$ is approximately the decrease in internal energy of the water, which in a time interval Δt results in a temperature *decrease* ΔT_e, where $M c_w \Delta T_e \approx \Delta M \ell_v$. We may interpret $\ell_v dM/dt$ as a negative pseudo-heating term added to Q driven by temperature differences between the water and its surroundings (e.g. air and a gas stove burner). Q has a positive component from the burner and a negative component because of energy transfer to a room at a lower temperature (see Section 7.1). $M c_w dT/dt = Q + \ell_v dM/dt$ determines the approximate rate of increase of temperature of M, assumed to be almost constant over the time of interest. Because the second term is negative and evaporation increases with temperature, $dT/dt = 0$ is possible. To test this we did a kitchen experiment. We heated 0.67 kg of water in a coffee can on a "heat diffuser" on the burner of a gas stove at its *lowest* setting. The 5-cm diameter can was wrapped in foil-clad insulation to decrease the variable negative contribution to Q. The initial height of water was 8.5 cm, which decreased by ≈ 0.4 cm. For about 20 minutes the temperature 1 cm below the water surface increased linearly, as expected if Q is approximately constant and $\gg |\ell_v dM/dt|$ (a least-squares fit to linear yielded $R^2 = 0.91$). Then the curve began to flatten, was more or less flat in 70 minutes, and the temperature fluctuated fractions of a degree around $80\,°C$. Putting a loose-fitting lid on the can resulted in boiling in about 25 minutes. The difference between $M c_w \Delta T$ and $\Delta M \ell_v$ for $\Delta T \approx 60\,°C$ and $|\Delta M| \approx 0.04$ kg corresponds to an average Q of about 22 W (40 W–18 W) over 70 minutes. Despite its flaws, simple theory accords with measurements and supports the explanation that water temperature plateaus at about $100\,°C$ mostly because of an abrupt large increase in evaporation of boiling water. But if Q is too small the water may not reach this temperature until the decrease of mass is appreciable. The negative contribution to Q is difficult to quantify because it includes contributions from convection and net emission of infrared radiation (see Section 7.1), which increase with increasing temperature. We estimate that the *maximum* radiative cooling of the open water surface was 3.6 W, and we expect comparable cooling by convection. Insulation of the can decreased radiation and convection. Apparently, evaporation dominated cooling, consistent with the exponential increase of vapor pressure with temperature. The positive contribution to Q (burner) was approximately constant.

Having mentioned removing air from water opens up another heresy. We have encountered countless times the assertion that air can be *completely* removed from water by heating. Even Ostwald, in his beautiful book *Solutions*, refers more than once to "dissolved gas entirely removed by raising temperature." This doesn't make much sense *if* it is based on the temperature dependence of the solubility of air in water, especially if it is boiled in an open pan *exposed* to air. At the very least one would have to use a container with a narrow outlet to prevent ambient air from entering the water. The equilibrium concentration c_∞ of a gas dissolved in a liquid is given by Henry's law Eq. (5.190). The rate at which gas molecules enter the liquid is determined by p, the partial pressure of the gas in contact with the liquid; the rate at which they escape from the liquid is determined by K. Based on the form of the Clausius–Clapeyron and Kelvin equations, and our interpretation of them, the temperature dependence of K should have the form

$$K = A \, \exp\left(\frac{-E}{RT}\right), \tag{5.230}$$

where R is the gas constant of the dissolved gas, A is a constant, and E is some kind of enthalpy of solution, a measure of how tightly the dissolved gas molecules are bound to the liquid. Because the escape of a gas molecule from a liquid is similar to the escape of one of its own molecules, we even can hazard a guess that E should be comparable to the enthalpy of vaporization.

Our guess is verified in Fig. 5.21, which shows the measured solubility (Bunsen absorption coefficient) $R^*T_0/K(T)$ of air in water versus temperature together with a least-squares fit to the measurements. The quantity E so obtained is 3×10^5 Jkg^{-1}, about one-eighth the enthalpy of vaporization of water. Solubility does indeed decrease with increasing temperature, but not to zero. From room temperature to the boiling point, the solubility is halved, the factor by which the equilibrium concentration of air dissolved in water can drop, a far cry from reduced to zero.

Suppose that water at 20 °C in equilibrium with air is heated to the boiling point. The solubility of air in the water decreases, but the concentration of air dissolved in it does not instantaneously follow suit. At the higher temperature, the dissolved air is not in equilibrium with the surrounding air, but it takes time for the water to evolve to a new equilibrium state. This approach to equilibrium can be hastened by agitating the water, as in boiling, but regardless of how fast equilibration occurs, all or even most of the air in water cannot be removed solely by raising its temperature to 100 °C. The way to remove air from water, if you believe

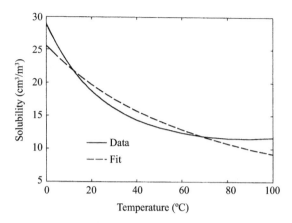

Fig. 5.21: Solubility of air in pure water, the amount of air expressed as a volume at $0\,^\circ$C and the pressure of the experiment divided by the volume of water in which the air is dissolved. The solid curve shows measurements taken from *International Critical Tables*. The dashed curve is an exponential fit to the measurements.

Henry's law, is to reduce the pressure of the air to which the water is exposed and allow it to come into equilibrium with its new low-pressure environment, called *vacuum degasification*. Boiling hastens this process (the reverse of aeration), but it is the turmoil of boiling, not the high temperature, that rids water of air. At sufficiently low pressures, water will boil at room temperature. If boiled at around $100\,^\circ$C, even in a whistling tea kettle, air can't get back into the water, but water is lost. You throw out the baby (water) with the bath water (air). To minimize loss of water by evaporation, vacuum degasification should be done at low temperatures.

Annotated References and Suggestions for Further Reading

Not only is the structure and properties of water in its several phases still a hot topic of research, it has been a fertile source of controversies, for example, the existence of "hot ice" more than a century ago (Section 5.5). A purportedly anomalous form of liquid water, eventually dubbed polywater, emerged from a Russian backwater in the early 1960s and attracted considerable worldwide scientific and popular attention, peaking by 1971, dead in the water by 1974. For an excellent history of the rise and fall of polywater see Felix Franks (1981) *Polywater*. Cambridge, MA: MIT Press.

A water-soluble controversy was sparked by a 1988 paper in *Nature* by Jacques Benveniste and collaborators claiming that a solution of biomolecules in water could be diluted to such an extent that although it could not have contained a single molecule it still retained a detectable memory of them, just as we can remember rubbing shoulders with people in their absence. Subsequently, it was claimed that signals from such empty water with a memory could be digitized and transmitted over telephone lines. If true, much of physics and chemistry would have to be drastically revised or abandoned. To date, this has not happened.

"Cold fusion" was announced in March 1989, in newspapers, on television news, and at a press conference in which electrochemists Stanley Pons and Martin Fleischmann claimed to have achieved room temperature fusion of deuterium nuclei using palladium electrodes immersed in a bath of heavy water. Up to that point, fusion of light nuclei was believed to be possible only at temperatures and pressures in the interiors of stars. Years of effort to achieve sustained fusion on Earth had failed. If cold fusion had been true, it would have been the greatest scientific discovery and benefit to humankind of all time—a source of clean energy for eons. But when something seems too good to be true, it usually isn't. Cold fusion was not an exception. For an account of the rise and fall of cold fusion, see Gary Taubes (1993) *Bad science: the short life and weird times of cold fusion.* New York: Random House.

For a clear, simple discussion of solutions of gases in gases and of gases in liquids, see the first two chapters of Wilhelm Ostwald (1891) *Solutions.* London: Longmans. Note in particular the italicized statement on page 5: "*the vapour-pressure of a liquid in a gas is the same as in a vacuum,*" in support of which Ostwald cites an 1803 paper by Dalton. Contrast what was known almost two centuries ago with the misconception that air is a variable-capacity sponge for soaking up water vapor.

The value of the mass accommodation coefficient of water is in dispute. Estimates range from a small fraction to 1. It appears, however, that the most likely value is 1. See P. E. Wagner (1982) "Aerosol growth by condensation," in W. H. Marlow (ed.) *Aerosol microphysics II: chemical physics of microparticles.* New York: Springer, pp. 129–78.

The difficulty of making precise dew point measurements is discussed by R. G. Wylie, D. K. Davies, and W. A. Caw (1965) "The basic process of the dew-point hygrometer," in Robert E. Ruskin (ed.) *Humidity and Moisture*, Vol. I. New York: Rheinhold, p. 125.

Our derivation of the Clausius–Clapeyron equation follows that given by Enrico Fermi (1956) *Thermodynamics.* Mineola: Dover, pp. 62–8.

See *Dictionary of Scientific Biography* for biographical sketches of Clausius (Edward E. Daub) and Clapeyron (Milton Kerker).

Clausius's 1879 book in which he gives the Clausius–Clapeyron relation is *The mechanical theory of heat*, W. R. Brown (trans.). London: Macmillan.

Translated papers by Clapeyron and Clausius as well as Carnot's monograph are given in E. Mendoza (ed.) (1960) *Reflections on the motive power of fire by Sadi Carnot and other papers on the second law of thermodynamics by É. Clapeyron and R. Clausius.* Mineola: Dover.

For a discussion of the role of rainfall in causing rapid melting of snow (Problems 4 and 5), see Craig F. Bohren (1995) "Rain, snow, and spring runoff revisited," *The Physics Teacher*, 33, pp. 79–81.

For simple but fascinating experiments with subcooling, see N. Ernest Dorsey (1948) "The freezing of supercooled water," *Transactions of the American Philosophical Society*, 38, 247–328.

For the history of the discovery of the process by which rain forms in clouds of subcooled droplets, see the series of short articles by Arnt Eliasson, Duncan C. Blanchard, and Tor Bergeron (1978) "The life and science of Tor Bergeron," *Bulletin of the American Meteorological Society*, 59, pp. 387–92.

For a discussion of the Schaefer point, see Duncan C. Blanchard (1996) "Serendipity, scientific discovery, and Project Cirrus," *Bulletin of the American Meteorological Society*, 77, pp. 1279–85. See also the 1979 paper by the same author: "Science, success, and serendipity," *Weatherwise*, 32, pp. 236–41.

In this chapter we laid the foundations for understanding attempts to make rain by seeding clouds. Not surprising, cloud seeding is shrouded in misconceptions, perhaps the foremost among them being that seeding creates rain clouds out of clear air. But rain requires clouds on a grand scale, and the best that puny humans can do is to coax existing clouds to release their water as precipitation. One way of doing so is by dropping dry ice (solid carbon dioxide) into clouds of subcooled water droplets, which was discovered by the late Vincent Schaefer. By the inexorable law of error propagation, the notion has spread that dry ice causes these droplets to freeze, which is not what Schaefer concluded. On the contrary, he provided evidence that the nucleation of ice crystals by dry ice occurs in the vapor phase. In a delightful article by Bernard Vonnegut (1981) "Misconception about cloud seeding with dry ice," *Journal of Weather Modification*, 13, pp. 9–11, the author quotes 16 authors who got the story wrong. Following this paper is a rebuttal by B. J. Mason, one of the authors listed.

The many equations of state at the beginning of Section 5.4 are listed by J. R. Partington and W. B. Shilling (1924) *The specific heats of gases*. London: Ernest Benn Limited.

For an echo of our assertion that molecules do not have well-defined volumes see John T. Edward (1970) "Molecular volumes and the Stokes–Einstein equation," *Journal of Chemical Education*, 47, pp. 261–70. He concludes that "for particles of this size [molecular] the notion of particle volume ... becomes hazy."

Among the many papers published in the *Proceedings of the van der Waals Centennial Conference on Statistical Mechanics*, C. Prins (ed.), Amsterdam: Elsevier, the following two do not require specialized knowledge: J. D. Boer (1974) "van der Waals in his time and the present revival—opening address," pp. 1–27; M. J. Klein (1974) "The historical origins of the van der Waals equation," pp. 28–47.

For a discussion of the equation of van der Waals including translations from his dissertation, see Stephen G. Brush (1961) "Development of the kinetic theory of gases. V. The equation of state," *American Journal of Physics*, 29, pp. 593–605.

Although we give the critical temperature of water as 374 °C, measured values reported by various investigators lie about a tenth of a degree on either side of this value. Similarly, measured critical pressures lie a few tenths of an atmosphere on either side of 218 atm. A summary of critical data for water (and much more) is given in G. S. Kell (1972) "Thermodynamic and transport properties of fluid water," in Felix Franks (ed.) *Water: a comprehensive treatise,* Vol. I: *The physics and physical chemistry of water.* New York: Plenum, pp. 363–412.

For a biographical sketch of Thomas Andrews, see E. L. Scott's entry in *Dictionary of Scientific Biography.*

James Clerk Maxwell, in his 1872 *Theory of Heat* (Boston: D. Appleton), discusses Andrews' measurements of critical points. Maxwell also notes that James Thomson (a brother of Lord Kelvin) suggested that the isotherms through a phase transition are not discontinuous (in slope), and he sketched isotherms remarkably similar to the isotherms later obtained theoretically by van der Waals.

Maxwell discusses van der Waals's dissertation in a paper entitled "van der Waals on the continuity of the gaseous and liquid states," reprinted as Paper 69 in (1952) *The Scientific Papers of James Clerk Maxwell.* Mineola: Dover. Paper 71, "On the dynamical evidence of the molecular constitution of bodies," contains a derivation of what is now called the Maxwell construction. For a treatise on the van der Waals fluid see David C. Johnston (2014) *Advances in the thermodynamics of the van der Waals fluid.* London: Morgan & Claypool. Among other things, he derives the caloric equation of state from a molecular point of view.

For a good history of the van der Waals equation, see A. Ya. Kipnis, B. E. Yavelov, and J. S. Rowlinson (1996) *Van der Waals and molecular science.* Oxford: Oxford University Press, especially ch. 3.

Superheating of liquids is discussed by Robert C. Reid (1976) "Superheated liquids," *American Scientist*, 64, pp. 146–56.

Boiling and melting points of various forms of phosphorous are given by John R. Van Wazer (1958) *Phosphorus and its compounds: Vol. I: Chemistry.* New York: Interscience, pp. 101–117.

For discussions of negative pressure, see Alan T. J. Hayward (1971) "Negative pressure in liquids: Can it be harnessed to serve man?" *American Scientist*, 59, pp. 434–43; P. F. Scholander (1972) "Tensile water," *American Scientist*, 60, pp. 584–90. The paper by Hayward is especially recommended for its clarity and simplicity.

For more about Lyman Briggs, see the entry by Stuart W. Leslie in *Dictionary of Scientific Biography.*

The only place we have seen the relation between vapor pressure and temperature according to the van der Waals equation is in his biographical sketch by J. A. Prins in *Dictionary of Scientific Biography*. But Prins gives only a figure (figure 4) with no details of how it was obtained and no references to where it may be found.

One of the few accurate presentations of the phase diagram for water can be found in Kenneth C. Young (1993) *Microphysical processes in clouds*. Oxford: Oxford University Press.

For references to the controversy created by Camelly about superheated ice, see N. Ernest Dorsey (1940) *Properties of ordinary water–substance*. New York: Reinhold, p. 643.

For more about the polymorphs of ice, see Peter V. Hobbs, (1974) *Ice physics*. Oxford: Oxford University Press, which surveys everything that was known about ice up to the publication of this massive compendium of physical and chemical properties. See Section 1.5 of this book and references cited therein for more on superheated ice.

A shorter treatise on water, including ice, is D. Eisenberg and W. Kauzmann (1969) *The structure and properties of water*. Oxford: Oxford University Press.

For arguments why "pressure melting cannot be responsible for the low friction of ice," see Samuel C. Colbeck (1995) "Pressure melting and ice skating," *American Journal of Physics*, 63, pp. 888–90. Colbeck's theoretical arguments subsequently were supported by what appears to be overwhelming experimental evidence. See Samuel C. Colbeck, L. Najarian, and H. B. Smith (1997) "Sliding temperatures of ice skates," *American Journal of Physics*, 65, pp. 488–92. See also the paper on premelting of ice by Dash *et al.* cited in the references at the end of Chapter 1.

For the role of pressure melting in glacier sliding, see W. S. B. Patterson (1981) *The physics of glaciers*. 2nd edn. New York: Pergamon, ch. 7.

The classic work on free energy is by Gilbert Newton Lewis and Merle Randall (1923) *Thermodynamics and the free energy of chemical substances*. London: McGraw-Hill. Read the preface of the first edition and marvel at its beauty. Lewis was a scientist of the first rank who also could write with clarity and style. Another lyrical book by a famous physical chemist is C. N. Hinshelwood (1951) *The structure of physical chemistry*. Oxford: Oxford University Press. Read the first chapter and compare it with more recent monographs. Try not to cry.

See Sanford A. Moss (1943) "American standard letter symbols for heat and thermodynamics," *American Journal of Physics*, 11, pp. 344–9 for a discussion of the confusing terms and symbols used for various free energies.

The obituary of Poynting from which we quoted can be found together with all of his contributions to science in John Henry Poynting (1920) *Collected scientific papers*. Cambridge: Cambridge University Press.

Raoult's paper is translated in the 1899 *The Modern Theory of Solutions*, Harry C. Jones (trans.) New York: Harper. For a biographical sketch of Raoult, see the entry by Louis I. Kuslan in *Dictionary of Scientific Biography*. For a highly readable account of Arrhenius and the controversy over ions in solution, see

Bernard Jaffe (1976) *Crucibles: the story of chemistry*, 4th edn. Mineola: Dover. Read the entire book. It's a treasure.

A molecular interpretation of vapor pressure reduction by dissolution is given by J. H. Hildebrand and R. L. Scott (1950) *The solubility of non-electrolytes*. 3rd edn. New York: Reinhold, pp. 17–18.

For a biographical sketch of William Henry, see the entry by E. L. Scott in *Dictionary of scientific biography*.

Our calculations in Section 5.9 of the small changes in the saturation vapor pressure of pure water at $0\,^{\circ}C$ because of dissolved air and a total pressure of 1 atm agree with those of James E. McDonald (1963) "Intermolecular attractions and saturation vapor pressure," *Journal of the Atmospheric Sciences*, 20, pp. 178–80.

An excellent elementary treatment of surface tension is given by Michael V. Berry (1971) "The molecular mechanism of surface tension," *Physics Education*, 6, pp. 79–84.

Modified Köhler curves (with kinks in them) are discussed by Jen-Ping Chen (1994) "Theory of deliquescence and modified Köhler curves," *Journal of the Atmospheric Sciences*, 51, pp. 3505–16.

For a simple demonstration of condensational warming, see Dale R. Durran and Dargan M. W. Frierson (2013) "Condensation, atmospheric motion, and cold beer," *Physics Today*, 66, pp. 74–5.

For the distinction between sweating and net condensation in the context of "sweating like a pig," see Craig F. Bohren (2016) "Sweating like a pig. Physics or irony?" *The Physics Teacher*, 54, pp. 142–4

In 1874 John Aitken published a paper entitled "On Boiling, Condensing, Freezing, and Melting" that begins: "That water does not always boil at the same temperature, when under the same pressure, has long been well known." He cites a report on thermometers published in 1777. It seems that what was once "well known" no longer is. This paper is in the *Collected Scientific Papers of John Aitken* or can be found online. His marvelous 1885 paper, "On dew" is worth reading. His papers are entirely descriptive, the fruit of keen observations. And he was a master of clear exposition.

Although we—and we are not alone—have jettisoned the term "latent heat" coined by Joseph Black, we greatly admire him and pay him the ultimate compliment of reading him and repeating some of his experiments (see problems). His *Lectures on the Elements of Chemistry, Vol. Part 1. General Effects of Heat* published after his death (1799) from his manuscripts by John Robison is the work of a first-class scientific mind, but we do not honor Black by treating it as sacred writ. Like us, he criticizes, sometimes acerbically, the terms, definitions, and explanations of his predecessors and contemporaries—by name. Contrary to what we have read, he did not clearly distinguish between "heat" and temperature. Nor did he use the term energy or understand its conservation. He embraced a particle "nature of heat ... the most probable of any that I know." Although he did not explicitly use the term caloric, he considered the "subtile elastic matter of heat is self-repelling matter ... while they are attracted by the other kinds of matter, and that with different degrees of force." He believed that "heat is the effect of a peculiar substance."

For the difficulties of accurately measuring boiling and condensation temperatures, see Wojciech Świętosławski (1945) *Ebulliometric measurements*. New York: Reinhold, ch. 1. He notes that "Numerous experiments have shown that the temperature measured inside a boiling liquid or solution does not correspond to its real boiling point. The temperature is always higher ... The main phenomenon that produces the increase is the superheating of the liquid."

For observations of the metamorphism of individual snowflakes, initially at $-5\,^{\circ}$C, over two months, see Si Chen and Ian Baker (2010) "Evolution of individual snowflakes during metamorphism," *Journal of Geophysical Research*, 115, D21114, figs. 2 and 3.

K. P. Galvin (2005) "A conceptually simple derivation of the Kelvin equation," *Chemical Engineering Science*, 60, pp. 4659–60

A basic treatment of precipitation processes in warm and cold clouds is in John M. Wallace and Peter V. Hobbs (2006) *Atmospheric science: an introductory survey*. 2nd edn. London: Academic, 2006). Chapter 6.

From analysis of data from the Tropical Rain Measuring Mission, K. Lau and H. T. Wu (2003) "Warm rain process over tropical oceans and climate implications," *Geophysical Research Letters*, 30(24) p. 2240, conclude that "warm rain accounts for 31% of the total rain amount and 72% of the total rain area in the tropics."

The first explanation of why marine warm clouds with broad droplet size distributions have a greater propensity to precipitate than warm continental clouds associated with narrow droplet size distributions was given by Patrick Squires (1958b) "The microstructure and colloidal stability of warm clouds. Part II: The causes of the variations in microstructure," *Tellus*, 10, pp. 262–71.

The effect that aerosols may have on precipitation from shallow warm clouds and their fractional coverage over the subtropical oceans was considered by Bruce. A. Albrecht (1989) "Aerosols, cloud microphysics, and fractional cloudiness," *Science*, 245, pp. 1227–30.

The physical explanation of evaporation and condensation in Section 5.1 is similar to that in Rudolf Clausius's justifiably famous paper (1857) "The Nature of the Motion which we call Heat." It is remarkably clear, mostly descriptive, with little mathematics, and physically insightful. Well worth reading. English translations are in Brush (1965), in his 2003 anthology, and can be found on the Internet.

For measurements of insensible perspiration from the skin of live human subjects, see A. B. Goodman and A. V. Wolf (1969) "Insensible water loss from human skin as a function of ambient vapor concentration," *Journal of Applied Physiology*, 26, pp. 203–07.

For values of typical daily amounts of insensible perspiration from skin and lungs, see Christopher J. Lote (2012) *Principles of renal physiology*. 5th edn. New York: Springer, p. 11.

For a simple evaporimeter made from scrap materials, see Craig F. Bohren (1990a) "You never miss the water until the well runs dry," *Weatherwise*, 43, pp. 342–47. This also shows the fourfold variation in evaporation rate over a day.

For more about evaporation see Frank E. Jones (1992) *Evaporation of water: with emphasis on applications and measurements*. Boca Raton: CRC Press.

For a simple experiment showing that grass cut off from its roots does not guttate even though surrounding rooted grass does, see Craig F. Bohren (1990b) "All that glistens is not dew," *Weatherwise*, 43, pp. 284–87.

For evaporation rates in Arizona, see Keith R. Cooley (1970) "Evaporation from open water surfaces in Arizona," *Agricultural Experiment Station and Cooperative Extension Service Folder 159*. Tucson: University of Arizona Press.

For a very detailed discussion of homogeneous nucleation, see James E. McDonald (1953) "Homogeneous nucleation of supercooled water droplets," *Journal of Meteorology*, 10, pp. 416–33. His aim was to give a theoretical explanation of why −40 °C is the lowest temperature at which subcooled droplets can exist. He uses the Kelvin equation to estimate the minimum size of an "embryo" to nucleate freezing. An embryo is an aggregation of water molecules with the potential to grow. He gives an example of a rarely mentioned pitfall of which all scientists should be aware. An incorrect theory may agree well with observations, but when corrected *not* as well. Errors can cancel each other.

For a lengthy review of nucleation of ice, see B. J. Murray, D. O'Sullivan, J. D. Atkinson, and M. E. Webb (2012) "Ice nucleation by particles immersed in supercooled cloud droplets," *Chemical Society Reviews*, 31, pp. 6519–54. They conclude that "ice nucleation below about −15 °C is dominated by soot and mineral dusts. Above this temperature the only materials known to nucleate ice are biological."

For a remarkably clear discussion of the physics of liquids see J. A. Pryde (1966) *The liquid state*. London: Hutchinson. This slim book is a gem, especially the first five chapters. We smiled when we read (p. 2) "Fire [one of the four Elements of antiquity] appears in modern dress as 'heat energy,' a concept arrived at historically via the notions of phlogiston and caloric."

Problems

1. Raindrops are water drops sufficiently large that they fall with appreciable speeds. A typical raindrop has a diameter of 1 mm. That of a typical cloud droplet is perhaps 100 times less. Although it might be thought that cloud droplets grow into raindrops by condensation, this can be shown to be false by simple thermodynamic arguments. calculate the maximum size to which a cloud droplet can grow solely by condensation. You will need the following:

The gas constant R_v for water vapor is $461.5 \mathrm{Jkg}^{-1}\,\mathrm{K}^{-1}$.

A useful approximation for the saturation vapor pressure (hPa) as a function of temperature is

$$\log e_s = 9.4041 - \frac{2354}{T},$$

where the logarithm is to the base 10.

You may make the following assumptions:

a. An air parcel at a temperature of $30\,^\circ$C is saturated at cloud base (500 m).

b. This parcel ascends adiabatically to the tropopause (16 km) without condensation (i.e. the parcel becomes supersaturated).

c. Then excess water vapor condenses onto the nuclei that were in the parcel at the beginning of its ascent.

d. Each nucleus acquires the same amount of water.

e. Initially, the concentration of nuclei in the parcel is 100 cm^{-3}, which corresponds to very clean air.

   Hint: Don't make this problem harder than it need be. You are only trying to show that cloud droplets cannot grow into raindrops by condensation of water vapor.

   Briefly explain why these assumptions, although they are reasonable (a cloud base temperature of $1000\,^\circ$C will give larger droplets, but this temperature is absurd), lead to a good estimate of the maximum cloud droplet size.

2. How much liquid water must evaporate before the process of (net) evaporation ceases, assuming no source of energy to heat the water? You can determine only the relative amount of water loss by evaporation. The enthalpy of vaporization of water is about 2.5×10^6 Jkg^{-1}. Assume that initially the water and air are at $20\,^\circ$C and that the air temperature above the surface remains constant. The relative humidity of the air above the water is 50% and also remains constant. By "above the surface" is meant well above the air–water interface. At this interface the air and water temperature are the same and the relative humidity is 100%. On the basis of your answer, what do you conclude about the process of (net) evaporation? *Hint*: Before jumping on your horse and riding off in all directions, first ask yourself why (net) evaporation ceases. If you can answer this, the rest is easy.

3. Suppose that net evaporation of water proceeded at the *maximum* possible rate; that is, every molecule that left did not return. What would this evaporation rate be? Express your result in a rate of change of water depth. On the basis of your calculation, what do you conclude about your own observations of the (net) evaporation of lakes, ponds, and water spilled on floors? *Hint*: All that is wanted here is a very rough estimate, so you need not waste time determining precise values. The mean speed of gas molecules at ordinary temperatures is of the order of hundreds of meters per second.

4. When rain falls on a snowpack, the snow is sometimes observed to melt more quickly than it would have if there had been no rain. The apparently obvious explanation for this is that the rain melts the snow. To check the validity of this explanation determine how much rain (in inches) is required to melt one inch (water equivalent) of snow. Assume that the temperature of the rain is $2\,^{\circ}$C and that of the snow is $0\,^{\circ}$C.

5. On the basis of your answer to the previous question, you should be inspired to look for a deeper explanation of why increased snowmelt seems to accompany rain. Please do so.

6. A student once asked us to explain the following. He had put some hot water into a plastic bottle in order to clean it. He screwed the cap on the bottle and then shook it, following which the bottle swelled slightly. When he unscrewed the cap, there was an audible outrush of gases from the bottle. Please explain. Estimate the magnitudes of any potential contributors to the pressure increase in the bottle. Assume that the air temperature was $20\,^{\circ}$C and that of the hot water was $60\,^{\circ}$C ($140\,^{\circ}$F).

7. A forecast of overnight frost in Florida causes the owners of orange groves to become nervous. Unfortunately, the term *frost* has (at least) two meanings: a night of frost could mean that the temperature dropped below the freezing point or that frost (ice deposited from the vapor phase) appeared on surfaces. Consequently, the term *frost damage* seems to be interpreted by some people as damage done to plants because frost is deposited on them. This question has two parts.
   a. Are subfreezing temperatures *necessary* for the deposition of frost? Are subfreezing temperatures *sufficient* for the deposition of frost?
   b. Explain why the appearance of frost on, say, orange trees might be a more welcome sight to the owner of an orchard than to frost-free trees on an otherwise identical cold night.

8. During the winter the wife of one of the authors cans soup. She pours freshly made hot soup into glass jars, not filling them completely, covers their mouths with metal lids and fastens them as tightly as possible with retaining rings. Then she puts the jars in a large pressure cooker and heats them for more than an hour. Following this, she sets all the jars aside to cool, during which time she listens carefully for a metallic pinging sound coming from each jar, which signifies that its lid has now bowed inward slightly, and the jar is sealed. If she doesn't hear a ping from a jar, she redoes it using a new lid. Explain the physical reason for the ping. Why should a jar be redone if it doesn't yield a ping?

9. Can (pure) liquid water be simultaneously subcooled and superheated? If so, under what conditions? This is not a trick question; it merely requires paying careful attention to the definitions of subcooled and superheated. Water is subcooled if it exists in the liquid phase at a temperature below the nominal freezing point. Water is superheated if it does not boil at a temperature above the nominal boiling point.

10. You probably have heard people say that something made them so mad that their blood boiled. This is not an entirely fanciful metaphor. If you were to

ascend high enough in the atmosphere, your blood would boil. Estimate the altitude at which this would occur.

11. A book on training dogs to track, search, and rescue contains the following assertion: "Humidity is the amount of water contained in the air. If the humidity is high, the air will carry a maximal amount of water. Because air's 'carrying capacity' is almost saturated, it is unable to absorb any great amount of other odorous materials, severely slowing the airborne spread of scent." Please discuss.

12. Standard sea-level atmospheric pressure is defined as 760 mm of Hg: the height of a column of mercury supported by standard pressure at its bottom end, with the pressure at the top of the column identically zero. When sea-level pressure is determined with a mercury-filled barometer, what is the error (percentage) in the measured pressure, at room temperature, resulting from the finite vapor pressure of Hg? The vapor pressure of Hg? The vapor pressure of mercury measured in millimeters of Hg is

$$\log_{10} p = 10.377 - 0.8256 \log_{10} T - \frac{3285}{T},$$

where T is absolute temperature.

13. Problem 6 in Chapter 2 was to determine the pressure in a pressure cooker at which its safety valve (a weight) releases gases. If you did this problem correctly, you should have obtained a value of about 3 atmospheres. Take the temperature in the pressure cooker just after it is sealed shut but not yet on the stove to be $20\,°C$ (293 K). The initial internal pressure is one atmosphere. Now heat the pressure cooker (which has a layer of water in it) to the point where the safety valve is just about to release gases. The pressure cooker is a closed system with a fixed volume. Thus, one might be tempted to calculate the maximum temperature in the pressure cooker by using the ideal gas law in the form

$$p_f/p_i = T_f/T_i,$$

where the subscripts i and f refer to initial and final. For our problem the ratio of pressures is about three; hence, the maximum temperature we calculate is $3(293\,\text{K}) = 879\,\text{K} = 606\,°C = 1123\,°F$. This seems too high. If you look at the oven dial on an ordinary kitchen stove, you are not likely to find a temperature much greater than about $550\,°F$. Moreover, $606\,°C$ is close to the melting point of aluminum. Do pressure cookers really operate at peak temperatures of over $1000\,°F$? Where is the error in the reasoning that led to this high temperature?

14. This problem is an extension of the previous one. If you understand why what seemed at first glance to be plausible reasoning gave an excessively high peak temperature in the pressure cooker, you should be able to estimate the correct peak temperature. Please do so.

15. People who live in the northeastern United States sometime notice dew on the windows of their homes during winter. Careful examination, however, reveals that this dew almost always is limited to *inside* surfaces. And yet the outer surface of a window, exposed to outside air, is certainly colder than its inner surface, exposed to warmer inside air. Moreover, we showed in Section 5.2 that the relative humidity inside a house in winter is likely to

be *lower* than that outside. All this is very puzzling—at first glance. Put on your thinking cap and unravel the puzzle.

16. In a class in which we discussed the previous problem, a student recounted his observation of the reverse phenomenon: dew on the outside of a window. When, where, and under what conditions might this have occurred?

17. Summers are hot and humid in the northeastern United States, because of which people may run dehumidifiers in the basements of their homes. Why basements? Why not the ground and first floors?

18. Students often have difficulty accepting that ice, a solid, sublimates. Solids seem so permanent. We suspect that at least part of this difficulty results from our everyday observations that solid objects do not perceptibly waste away by net evaporation. How long would it take the mass of a gold ring to decrease by 10% at room temperature? Assume that evaporation of the gold occurs at the maximum possible rate. For simplicity take the gold to be pure. Gold is about 19.3 times denser than water and the atomic weight of gold is 196.9. The dependence of the equilibrium vapor pressure p (in Pa) of gold on absolute temperature T is given by

$$p = 1.037 \times 10^{11} \exp\left(-\frac{40480}{T}\right).$$

This equation was obtained from data for metals in Saul Dushman (1962) *Scientific foundations of vacuum technique*, 2nd edn. New York: Wiley, table 10.2.

Now determine the temperature to which the gold must be raised so that the mass of the ring would be diminished by 10% during your lifetime. The melting point of gold is 1336 K.

Although we chose a gold ring, the same results are obtained for silver, aluminum, iron, and many other metals. On the basis of your results, can you state succinctly what it is about ice that makes it so different from a great many metals and other solids under the conditions in which we usually observe them?

19. Relative humidity usually decreases with height in the atmosphere even though temperature, and hence saturation vapor pressure, decreases. From this result you should be able to *estimate* an upper limit for the scale height H_v of water vapor in the atmosphere. This scale height may be defined by

$$\frac{1}{H_v} = -\frac{1}{e}\frac{de}{dz}.$$

Hint: Begin with the definition of relative humidity.

After you have obtained your estimate for the scale height of water vapor, try to explain why it is different from the scale height for total pressure.

20. The grandmother of one of the authors used to hang wet wash out to dry even when air temperatures were well below freezing. Was this just a waste of time? If so, why? If not, why not? Why didn't she just hang the wash to dry in the basement of her house?

21. An observant retired physics teacher in Ajax, Ontario, Chris Curran, sent us a sketch (see Fig. 5.22) of a pattern of ice and dew on a windowpane.

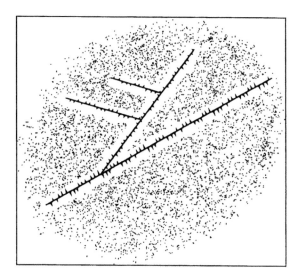

Fig. 5.22: A pattern of ice and dew on a windowpane.

Ice had formed in scratches in the glass, which also was covered more or less uniformly with dew. Flanking these ice-filled scratches were clear spaces, apparently of uniform width, devoid of both dew and ice. Please explain the reason for these clear spaces.

22. In Section 5.2 we stated, but did not prove, that $e_s/\sqrt{T}$ increases monotonically with absolute temperature T. This is plausible but not obvious because although e_s increases monotonically with T, so does $\sqrt{T}$. Please supply the missing proof (for water).

23. By using values for the critical temperature and pressure of water and the van der Waals equation, you can estimate the diameter of a water molecule. How does the diameter you obtain square with what you may have been told about the size of the water molecule? With critical temperatures and pressures of the other molecular components of air, you also can estimate the relative sizes of these molecules. Please do so. Again, check your results against what you know or can find.

24. We argued in this chapter that if you were to attempt to measure the boiling point of pure water at sea level (standard pressure) by immersing the bulb of a thermometer in boiling water, you would almost be certain to obtain a temperature a degree or more higher than $100\,°C$. Critics of this assertion might argue that the observed superheating is simply a consequence of the increased pressure with depth in the water. Show that this increased pressure cannot even begin to account for the magnitude of readily observed superheating in boiling water.

25. Problem 12 in Chapter 3 was to estimate the rise in sea level for a $1\,°C$ temperature rise because of thermal expansion of water. Show that if you use

the ideal gas law to determine the coefficient of thermal expansion of water, the result will be incorrect, and not by a small amount. Then show that the van der Waals equation yields a value much closer to the measured value. *Hint*: Results from the starred subsection in Section 5.4 will be useful.

26. Problem 16 in Chapter 3 was to estimate the change in sea level if water were absolutely incompressible. To do so, you needed the compressibility of water. Show that if you use the ideal gas law to determine this compressibility, the result is incorrect by a very large factor. Then use the van der Waals equation to estimate the compressibility of water and its temperature dependence. Compare your estimate with measured values. *Hint*: Results from the starred subsection in Section 5.4 will be useful. If you do this problem and the preceding one correctly, you'll gain even more respect for the van der Waals equation.

27. How does the boiling point of water change if you dissolve salt or sugar or anything in it? Before jumping on your horse and riding off frantically in search of an equation, try to answer this question solely by physical reasoning. That is, explain in words why the boiling point of water changes (or does not change), and in what direction (increase or decrease), when a substance is dissolved in it.

28. Do people add salt to boiling water for cooking pasta in order to change the boiling point of the water? This question, which is related to the previous one, must be answered quantitatively. If you are ambitious, you can test your predictions based on theory by carefully measuring the boiling temperature of a known mass of water to which salt is added in measured amounts. *Hint*: The molecular weight of NaCl is 58.4; the molecular weight of water is 18.

29. Toward the end of Section 4.1 we criticized the widespread notion that entropy is just a synonym for disorder. We supported our criticism with examples involving metastable states, specifically subcooled water and a supersaturated solution. These examples had occurred to chemists. Another example would more likely have occurred to a meteorologist. Please give such an example. That is, give an example of a change involving a metastable state, of great importance to meteorology, in which a necessary increase in entropy (of an isolated system) is not accompanied by what a reasonable person would immediately and unambiguously associate with increased disorder.

30. If you rummage in the cabinets in a chemistry laboratory, you are likely to find a curious bottle, marked Boiling Stones, containing small, porous, chemically inert stones. These stones are intended to be added to water when it is boiled. Does whoever puts rocks in water also have them in his head? Why on earth would anyone put stones in boiling water? After you have answered this question to your satisfaction, you can have some fun by asking chemistry students (and teachers) about boiling stones

31. You are faced with the task of determining elevation differences by measuring boiling point changes of a liquid. How would you select the liquid in order to obtain the greatest possible sensitivity of boiling point to elevation change?

32. If you bend over and open the door of a hot oven to take a peek at what is being cooked inside, and you are wearing glasses, your glasses may fog. This

seems to be at odds with the fact that in cold climates, where houses are heated during winter, humidifiers are often used. Please explain.

33. While you are hiking in mountains, the weather unexpectedly changes, becoming overcast, windy, and cold, followed by heavy rain. You have no rain gear (when you began your hike the sun was shining), so your clothes become sopping wet. After the rain ceases, they eventually dry. How much energy did your body need to provide in order to dry your wet clothes? If biochemical reactions (metabolism) did not act to keep your body temperature constant, how much would your body temperature change? Assume that your clothes take up about a pound of water (0.5 kg) and that your specific heat capacity is approximately that of liquid water. What does the answer to this question indicate that you should carry with you on mountain hikes?

34. To go further with the previous problem, determine how the temperature decrease (if uncompensated for by metabolism) is related to the size of the hiker. Assume that the amount of rainwater taken up by the hiker's clothing is proportional to his surface area. What lessons are to be learned from the answer to this problem?

35. To make Fig. 5.11, we used measured vapor pressures of aqueous solutions of varying concentration tabulated in *International Critical Tables*. These data were not given in the form we present. In particular, for ammonium sulfate, $(NH_4)_2SO_2$, vapor pressure was given as a function of the number of grams of the solute (ammonium sulfate) dissolved in 100 g of the solvent (water). The vapor pressure of pure water at $20\,°C$ is 17.539 mm Hg. Ten grams of ammonium sulfate added to 100 g of water lowers the vapor pressure to 17.1 mm Hg. From these data you should be able to estimate the molecular weight of ammonium sulfate. Compare your estimate with the accepted value.

36. If a pressurized aircraft flying at high altitude suffers rapid decompression, a cloud can form in the cabin. Explain. Consider an aircraft with cabin pressure equal to atmospheric pressure at 5000 ft and flying at 30,000 ft. If the temperature of the cabin air is $20\,°C$ and its relative humidity 20%, can a cloud form if there is rapid decompression? Below approximately what altitude would the aircraft have to be flying such that a cloud could not form upon cabin decompression?

37. We asserted in this chapter that if the surface pressure on Earth were quite different from its present value (say, 2 hPa instead of 1000 hPa), the boiling point of water would be quite different, but its enthalpy of vaporization would hardly change. Please verify this assertion. *Hint*: To do this problem you'll have to draw upon results in Chapters 3 and 4, specifically in Sections 3.2, 3.6, and 4.2.

38. Suppose that what we call modern science, including thermodynamics, had been developed in Tibet instead of Western Europe and Britain. At approximately what temperature would water boil in kitchens in the British Isles?

39. Cold water from a kitchen tap is poured into an old saucepan. No bubbles are seen in the water immediately after it is poured, but several hours later the bottom of the saucepan is carpeted with tiny bubbles. Please explain.

40. Estimate the boiling point at the bottom of the ocean. There are two answers to this question. One of them is obtained without much thought, the other

requires you to examine critically your assumptions and the first answer. *Hint*: Review the section on the van der Waals equation of state.

41. We argued that because we can see vapor bubbles in a liquid, their density must be much less than that of the surrounding liquid. What other simple observational evidence supports this assertion?

42. Every atom of oxygen is not like every other atom. Oxygen has two stable isotopes, ^{16}O (average abundance of 99.759%) and ^{18}O (average abundance of 0.204%). These two isotopes are essentially chemically identical (same number of electrons), but ^{18}O is heavier by virtue of two more neutrons in its nucleus. Water is a compound of oxygen, and hence, comes in (at least) two varieties, chemically identical but with different molecular weights (18 and 20). On p. 89 of *Climate History and the Future*, H. H. Lamb asserts that

> Of all the multifarious physical traces of former climate none offers the opportunity for more precise measurements or appears to yield more detailed information than the varying proportion of oxygen-18 in the ice at different depths in the polar sheet and in the carbonates deposited on the ocean bed. The variations are related to temperature.

The simplest problem to consider in order to understand how stable oxygen isotopes can be used to infer temperature differences is the relative variation in the ratio of oxygen-18 to oxygen-16 in saturated water vapor. Show how the difference between this ratio at temperature T and at some reference temperature T_o relative to the ratio at the reference temperature depends on T. How precisely must ^{18}O - to- ^{16}O ratios be measured in order to infer temperature changes of $1\,°C$?

43. Estimate the maximum possible percentage (by number) of water vapor in moist air in Earth's atmosphere. This number should be such that you could not rule out with absolute certainty that it could never occur in the atmosphere, but you would be highly suspicious of any higher values. After you have made your estimate, search climatological records for evidence that it has occurred. To help you in your search, first determine where on Earth the highest percentages of water vapor are likely to be found.

44. If the temperature of all the world's oceans were raised to $100\,°C$ or even a few degrees higher, would the ocean water (near the surface) boil? Why or why not? Ignore the salinity of sea water.

45. Backpackers have to carry all their food on long trips. Much of the weight of most food is contributed by water, and hence, smart (or lazy) backpackers carry dried food, in particular *freeze-dried* food. Why freeze-dried? To answer this question, you can go to a library and try to find something about freeze-drying, or you can put on your thinking cap and try to figure out the process of freeze-drying based on what you learned in this chapter.

46. In Problem 5 in Chapter 4 you were allowed to assume that the entropy and enthalpy differences between water and ice are independent of temperature. Now remove this assumption. That is, do Problem 5 again

but this time account for the temperature dependence of the entropy and enthalpy differences.

47. At the beginning of Section 2.1 we described a demonstration of atmospheric pressure in which a small amount of water is poured into a can, heated to boiling, then the can is capped, cooled, and crushed. Please explain.

48. This problem is an extension of Problem 39 in Chapter 2. Can the difference between the distances ridden by Rominger and Induráin in the same time be accounted for quantitatively by different air resistances? To answer this question you have to assume that both riders are capable of exerting the same force and have the same cross-sectional area and drag coefficient. That is, both riders are identical in every way but ride in different meteorological conditions.

49. Suppose that you were stuck on a remote tropical island, your only meteorological instrument a very precise thermometer (0.1 °C resolution from 80–100 °C). How would you estimate the pressure decrease associated with an approaching hurricane? What is the smallest pressure change that could be detected?

50. When you open a bottle of beer, you may notice a cloud form in its neck. The cloud was not there before the bottle was opened. Please explain. *Hints*: Beer is mostly water. The gas in the space above the beer is mostly carbon dioxide at a pressure greater than one atmosphere. This cloud is not a mixing cloud, but it is composed of water droplets.

    It is very likely that the cloud formed by homogeneous nucleation. Although there were undoubtedly particles in the carbon dioxide when the beer was bottled, the larger particles eventually settle into the beer, whereas the smaller ones diffuse to the sides of the bottle where they are captured. Under the assumption that the cloud in the neck forms by homogeneous nucleation, estimate the minimum pressure in the bottle before it was opened. When you obtain this pressure, ask yourself—or try to find out—if it makes sense. *Hint*: The ratio of specific heats of carbon dioxide is 1.3.

51. A colleague once suggested a scheme for cheap power generation in regions where winter temperatures drop well below freezing. The scheme is to amass ice that forms naturally (and at no cost) during winter, insulate this ice, and then use it during the warmer months to generate power. Please assess the feasibility of this scheme. One of the authors noted on his electricity bill that for a period of about one month he used about 1000 KWh of energy (1 KWh is a power of 1000 W for one hour). Estimate the *minimum* amount of ice needed to generate this much electrical energy. *Hints*: To answer this problem, you need results from Section 4.4 on the Carnot cycle. But don't just thoughtlessly apply the Carnot efficiency, the definition of which requires that we agree on what is meant by efficiency. For this problem, no fuel is burned, but instead ice (which was frozen at no cost) is melted. You still have to use results in Section 4.4, but to determine the maximum work that can be obtained when a given mass of ice melts. Also keep in mind that the melting ice is a low-temperature reservoir, the high-temperature reservoir being the relatively warmer spring, summer, and fall air.

52. Some students, in our experience, think that ice (especially in the form of snow) is readily warmed to temperatures appreciably greater than $0\,°C$. Their thinking is that common in the days before the notion of latent heat was put forward. The only way really to convince yourself that the temperature of ice does not rise above $0\,°C$ (with rare exceptions) is to observe this, which requires only a thermometer. Put some ice chips into cold water and stir until the mixture comes to equilibrium. Measure the temperature of the ice–water mixture at various times and verify that it does not rise above $0\,°C$ as long as ice is present and the mixture is stirred. Nor does the temperature drop below $0\,°C$. This experiment, which takes only a few minutes, may seem so trivial that we should be almost ashamed to suggest it. Perhaps, but how else can you truly know that ice does not rise above $0\,°C$? No doubt you have been told this countless times, yet our experience with hundreds of students is that some of them do not believe this, and who can blame them if they have not observed it for themselves?

53. Equation (2.17) defines the ratio of absolute temperatures at the boiling and ice points as the ratio of the corresponding pressures of a real gas in the limit as the pressure of the gas approaches zero. With the help of the van der Waals equation, you should be able to determine the shape of the curve p_b/p_i versus p_i (or, equivalently, gas density) and thus see how to extrapolate this curve to zero density. Please do so. *Hint*: Use the simplest correction to the ideal gas law obtained from the van der Waals equation.

54. If you did Problem 3 correctly, you learned that (net) evaporation would rapidly cease if water did not interact with its surroundings. The total amount of radiant energy incident on a surface outside Earth's atmosphere and pointed toward the sun is about $1365\text{W}/\text{m}^2$ This is a *maximum* flux of solar radiation. The actual flux at the surface is less because of attenuation by the atmosphere and because the direction of the sun is not, in general, perpendicular to the surface. For sake of argument, let's say that a representative value for the flux of solar radiation at the surface is one-third the maximum value. Further assume that all this solar radiation is absorbed by water and is its only source of heating (i.e. neglect emission and absorption of infrared radiation and also convective energy transfer from the atmosphere). Given these assumptions, estimate the rate of net evaporation of water exposed to sunshine on a clear day. Express this evaporation rate as a change of depth per unit time (e.g. mm/h or cm/h). After you have done so, compare your estimate with measurements of net evaporation.

55. This problem was sent to us by George Greaves, a number theorist with an interest in physics. A fisherman walks to the middle of a fairly large lake covered with ice one meter thick. He drills a hole in the ice. What happens? In particular, does water flow into the hole and, if so, how far? We could have given this question at the end of Chapter 2 but decided to give it at the end of this chapter because of its discussion of differences between ice and water.

56. In Section 2.1 we made the heretical statement that the number of molecules in given volume of a real gas and at fixed temperature and pressure is *not*

the same for all gases. Show that this follows from the van der Waals thermal equation of state.

57. Devise a simple method for measuring the enthalpy of fusion of water using a kitchen stove, thermometer, a stopwatch, and kitchen scales. You need the specific heat capacity of water per unit mass. Before tackling this measurement think about the basic physical principles and assumptions underlying it. Quickly and crudely we measured an enthalpy of fusion with only a 30% error (possibly by luck rather than skill). If you make this measurement, you will own the enthalpy of fusion. It will no longer be a mysterious number handed down from on high. Our method is based on the work of Joseph Black.

58. If the previous problem whets your appetite for measurements, try measuring the enthalpy of vaporization of water using the same equipment. We measured a value too low by about a factor of two. Not bad for a first hurried attempt. If you are ambitious, try to do better by taking more care.

59. In this chapter we note that we have often read the assertion that water must be heated or, worse, brought to a boil before it can evaporate. To convince yourself that this is nonsense, try a simple although not very exciting experiment. Put some water in a shallow concave dish. We added a drop of red food coloring to water in a small, white ceramic dish. The maximum depth of the water was no more than three millimeters. Set your dish inside where it won't be disturbed and examine the water level every day or so. The temperature of the water must be close to ambient so cannot be said to be heated. Thus, it should not evaporate. Does it? If so, do you believe your eyes or what you read in a book?

60. Estimate the maximum total fractional amount of water vapor in the *entire* clear atmosphere per unit area (on an assumed flat Earth). You will have to estimate a plausible maximum for the number fraction of water molecules at the surface and you will need a result from a problem in Chapter 2.

61. Estimate how much "dewfall" (a figurative term: dew does not fall) is possible overnight. You have to make an educated guess as to the mass density of water vapor near the surface and the thickness of the layer of air that you have to consider. Express your answer as a depth of a uniform layer of condensed water. Then search to find measurements to compare with your estimate. Keep in mind that because you have to make assumptions, you cannot expect your estimate to be very accurate. All you can hope for is the order of magnitude.

62. This is related to the previous problem. Explain why what you calculated and what measurements you find may seem too low on the basis of careful observations of real dew.

63. This is related to the previous two problems. You are unlikely to have ever observed a uniform layer of water of known thickness. Devise a simple experiment to create a layer comparable to what you estimated. *Hints*: A large shallow pan, preferably rectangular, and a syringe capable of measuring small volumes are useful. So that water will wet the pan you need to take a very, very tiny amount of liquid detergent and spread it uniformly over the pan. Make a layer of water by injecting water into the pan, little by little, and stop when the layer appears comparable with dew you have observed.

Keep in mind that you have made some rough approximations and also may have to depend on your memory of deposits of dew.

64. To show that the almost always unqualified assertion that hot air "holds" more water vapor than "cold" air is not true, dig through readily available temperature and humidity records. Only *one* example in which the water vapor density at a particular site with a certain air temperature is *less* than the water vapor density at a different site with a lower temperature refutes the assertion. This is a black swan observation. The assertion that *all* swans are white is refuted once and for all if a *single* black swan is found. If you can find one example in which colder air "holds" more water vapor than hotter air, try to find others.

65. If you found the previous problem interesting, try the following. What is the general condition under which "cold" air "holds" more water vapor then "hot" air. *Hint*: Determine temperatures and relative humidities such that the vapor density of the "cold" air is greater than that of the "hot" air. Interpret your equation and convince yourself that it makes sense. If you do this problem, it may help you to find observations.

66. The norm in many textbooks is p-V isotherms, often misleadingly distorted and without numerical values for p and V. p-T isochores are less common, T-V isobars rarer still. Try to sketch as best you can a $T - v_m$ isobar for water at a pressure well below p_c. Consider water vapor in a cylinder fitted with a movable piston to maintain constant pressure. The vapor can be cooled slowly. For sake of concreteness take the pressure to be 20 hPa and the initial temperature to be 300 K. A log scale for v_m and linear scale for T are recommended. Perhaps most difficult is to determine what happens when the temperature drops to the saturation temperature, although a hint is that the isobar should bear a family resemblance to the isotherm.

67. In a 2019 article in *The Economist* (November 30) on using liquid air for energy storage we find, without qualification, that "At a temperature of $-196\,°C$ all of air's component gases will be liquefied." What exactly does this mean? An important qualifier is omitted. What is it? *Hint*: The critical temperature of air is $-140\,°C$, and hence, air can be liquefied at *any* lower temperature. According to the article this method could compete with lithium-ion battery storage, which (if true) is well worth knowing about.

68. If we blow on cold hands with our "hot" breath to warm them, why do we blow on hot soup to cool it? *Hint*: There may be more than one reason. If so, rate them according to their importance.

69. This is related to the previous problem. Think of a simple demonstration you can do in seconds using only your own body to support your answer to this problem.

70. Does the way we blow on our hands affect the perceived warming? With a fast-response thermocouple you can do a simple experiment to support your answer. Try to explain what you measure.

71. By blowing on cold hands do we really gain much other than briefly? You can use an infrared thermometer to answer this question.

72. A few small ice cubes at $0\,°C$ are dropped into a glass of tap water. Obtain an expression for the temperature of the water after *all* the ice has melted. For simplicity assume that the ice–water mixture is in an insulated cylinder fitted with a movable piston resting on the water surface. Ignore the difference between the specific enthalpy of the water at its initial and final temperatures. Interpret your results. State any assumptions. Ask yourself if your expression makes physical sense. You know from experience that the water gets colder. Would you call this a "release of latent cold"? What assumptions underly your result? Interpret physically in words, not jargon, why the water cools.

73. Assume that the ice in the previous problem is initially at a temperature lower than $0\,°C$. Determine the temperature of ice cubes in the freezing compartments of typical refrigerators. How does this change your previous result and what do you conclude? After doing these problems, ask your fellow students to explain how ice added to water or a soft drink cools it. What is the disadvantage of using ice to cool a drink? How could you eliminate this?

74. For a variation on the simple experiment discussed in the text, try the following. Measure the temperature of a sheet of dry absorbent paper with an infrared thermometer. Measure air temperature. Then wet the paper with water at room temperature and measure the more or less steady temperature of the paper. Repeat this experiment using a fan to blow air over the paper. What do you conclude?

75. Why are electrical resistance heaters said to be 100% efficient? Discuss.

76. A common experience in the Northeastern United States in late winter is the sudden appearance of "sweat" on the floor of an unheated room or building following unexpected spring-like weather. Explain.

77. In Section 5.3 we show that the difference between the specific internal energy of water vapor and liquid at the same temperature is the enthalpy of vaporization to within about 5%. How close is the enthalpy of fusion of water at $0\,°C$ to the difference between the specific internal energies of liquid water and ice? Try to interpret your result, in particular, the sign of the difference.

78. The *flash point* of gasoline is about $-43\,°C$, by which is meant that its saturation vapor pressure (density) at this temperature is such that the vapor will ignite (in air) given a source of ignition (lit match, spark, cigarette). We have seen the following demonstration (on the advice of our attorneys we cannot suggest you do it). About 20 ml of gasoline is poured into a shallow dish. A lit long wooden match is held about 15 cm above it. Nothing happens. The match is lowered more and more down to about a centimeter until finally it is dropped into the gasoline, which then immediately bursts into flames. Explain. The flash point of diesel fuel is about $52\,°C$. If the same demonstration is done with diesel, nothing happens—at first. But if the burning match is left in the diesel for about 30 seconds it suddenly bursts into flames. Explain. How do these demonstrations help you understand vapor pressure gradients above water?

79. The pressure inside an inflated balloon can be measured using a manometer, bicycle pump, valve, and some plumbing. But this pressure (strictly, the

capillary pressure) can be measured more easily (but less accurately) using items found in many kitchens. Can you devise such a method? *Hint*: An ordinary kitchen scale is adequate.

80. If you did the experiment in the previous problem, you can determine the surface tension of the balloon.

81. With the average intraocular pressure (capillary pressure) in the "normal" eye and the average radius of curvature of the cornea (which you can find easily), calculate the surface tension of the human cornea and compare it with the surface tension of a balloon and of water. What do you conclude?

82. Suppose that we add a small amount of liquid water to a one-liter container of dry air. The temperature is kept constant. The water will evaporate. How much liquid is required such that all of it evaporates and the vapor pressure is the saturation vapor pressure corresponding to the constant temperature. Do this for temperatures of $0\,^\circ$C and $100\,^\circ$C. What do you conclude from your calculations?

83. This problem is related to the previous one. Early attempts to measure isobaric expansion of gases between the melting temperature of ice and the boiling point of water (see Section 2.1) gave highly variable results, which was attributed to small amounts of liquid water contamination in the gas container. An implicit assumption (indeed, a requirement) in these measurements is that the amount of gas is constant. What happens if this is not satisfied? Before doing any mathematical analysis, give a physical explanation of why water causes errors and their likely sign.

84. The most recent claim for the lowest barometric pressure at sea level is 860 hPa. What is the boiling point of water at this pressure? *Hint*: Because Eq. (5.68) is not sufficiently accurate you will have to use Eq. (5.67). But it does not have an algebraic solution, so you will have to solve it iteratively or by making approximations. What is the boiling point of water on Mars?

85. In the absence of nuclei, the temperature of water at a pressure of one atmosphere can be increased above $100\,^\circ$C without boiling. If nuclei are suddenly introduced into the water, what would happen? At what temperature might this result in dire consequences. *Hint:* You will have to do some calculations or find data online (not difficult), then exercise some judgment. What is important is to understand what happens and to what extent, not to determine an exact temperature.

86. In Section 4.1 (see Fig. 4.1) we assumed that that in a free expansion of an ideal gas in a rigid, insulated container, there is no temperature change, which is not exactly true. What is true from the first law is that there is no change in the internal energy. There can be a temperature change for a real gas. Based on simple physical arguments (at the molecular level) will the equilibrium temperature of a real gas increase or decrease? Is your answer supported by the van der Waals equation? For a given ratio of volumes, how would you expect the temperature change to depend on the final number

density of the gas? Physical reasoning is sufficient to answer this question. Mathematical acrobatics are unnecessary.

87. This is an advanced version of the previous problem. The internal energy of a monatomic van der Waals fluid given by Johnson (see References) is

$$U = \frac{3}{2}NkT - N^2\frac{\bar{a}}{V},$$

where $\bar{a}$ is independent of temperature. This is at least plausible and consistent with Eq. (5.97). In a free expansion for which $\Delta U = 0$, determine the temperature change. Then calculate it for xenon at an initial number density for an ideal gas at a pressure of one atmosphere and temperature $20\,°C$. Assume that the final volume is twice the initial volume. $\bar{a}$ for xenon is $1.2 \times 10^{-48}\,J\,m^3$. We chose xenon because it has the largest value of $\bar{a}$ for monatomic gases.

88. This is related to the previous problem. Assuming that you did the calculations correctly, do you think that this temperature change could be measured in a real laboratory. Consider how this could be done and what problems would be faced. How does the total heat capacity of the gas enter into this? What about the volume of the container and the area of its walls? Keep in mind that $Q = 0$ is possible only on paper. If the temperature of the gas changes, there must be energy exchange between it and its container

89. According to Eqs (6.95) and (6.104), the dry (meaning no change of phase) lapse rate depends on the relative amounts of water vapor and dry air. What is the possible realistic range of c_p? Is this range worth worrying about?

90. (Net) evaporation can cool liquids, but not all liquids equally. Saturate an initially dry absorbent paper towel with room-temperature water and measure the towel's temperature change with an infrared thermometer. Repeat this experiment in the same environment with isopropyl alcohol. You can buy 91% at drugstores. If there is a difference, why? Can you come up with a simple figure of merit, based on physical reasoning, to rate liquids according to their evaporative cooling power. You will need to find some liquid properties. *Hint*: Use molar properties. Intrinsic properties aside, what is different about water?

91. Dewar flasks (dewars) are very well-insulated (and expensive) containers for holding liquid nitrogen for months. The equilibrium vapor pressure of liquid nitrogen is about 100 kPa at 77 K, 1500 kPa at 110 K, and 3400 kPa at 126 K. Dewars are not perfectly insulating, because of which they must leak just a bit. Why? The knee-jerk answer is partly true. But the complete explanation requires a bit of thinking that will test your understanding of equilibrium vapor pressure, evaporation, and condensation. If you are familiar with pressure cookers, how they work safely is related to this problem.

92. Suppose that the enthalpy of fusion of ice were small, by which is meant $\ell_f \ll c\Delta T$, where ΔT is a few degrees, but its melting temperature remained around zero. How would this have affected the habitability of areas on Earth with heavy seasonal snowfall?

93. Joseph Black (c. 1760) added water at a temperature of 176 °F to an equal mass of ice (presumably at 32 °F). The final temperature of the mixture was 32 °F. At first glance, this seems impossible. What is the greatest initial temperature of water such that interaction with an equal mass of ice can result in water at 32 °F?

94. Suppose that air did not satisfy the ideal gas law sufficiently well that we had to use the van der Waals equation for real air. Convince yourself that there is not an analytical solution for $p(z)$ in an isothermal atmosphere even if the finite size of molecules is ignored. The hydrostatic equation still holds.

95. Compare the density of water at the critical point with the density of "normal" water so that you realize the extent to which water vapor becomes more like liquid water as it is compressed.

96. What is the capillary pressure of a typical cloud droplet? What is the capillary pressure of a typical raindrop? Compare these two pressures with atmospheric pressure. Take the surface tension of water to be 0.072 Nm^{-1}. What do you conclude?

97. We argue in Section 5.4 that surface tension must decrease with increasing temperature. Can you do a simple experiment to verify this? *HINT*: See Problem 8 in Chapter 2.

98. You may have heard the assertion that it is "too cold to snow," which sounds absurd until you think more about it. Detailed calculations are not necessary. Everything you need is in this chapter. *Hint*: Assume that the vapor pressure at the surface is the equilibrium vapor pressure at the surface temperature and that it decreases with a fixed scale height. Consider surface temperatures from 0 °C to −30 °C.

99. In absolutely clean air at 20 °C, its dew point this temperature, (a) how much would the temperature have to drop *at constant pressure* before the relative humidity was 400%? No detailed calculations are necessary. (b) If this air is expanded adiabatically, what expansion ratio is necessary for 400% relative humidity? This is *not* a constant pressure process. Before tackling this problem, ask yourself if the temperature drop will be smaller or larger. You may want to write a small computer program to solve part (b). We did not. You need results from Section 3 3.

100. We state in this chapter that "everything evaporates." To support this, the 99th edition of the *Handbook of Chemistry and Physics* (4–62) lists temperature-dependent vapor pressure data for many metals even at room temperature. How long would it take for aluminum foil just below its melting point, say 900 K, to evaporate completely assuming that every aluminum atom that evaporates does not return and the temperature does not change? To get you started, the vapor pressure of aluminum at 900 K is 8.3×10^{-8}Pa. You will need a few readily obtainable parameters.

101. Robert Boyle was frigophilic, as evidenced by *His New Experiments and Observations Touching Cold* (1664/5). He bemoaned that it is an "important subject ... hitherto ... almost totally neglected." This is a remarkable

collection of experiments and observations. For example, he filled a brass cylinder, five inches long, about an inch and three quarters in diameter, with water, inserted a wooden plug in the end, and placed a 74 lb object on the plug. This was set outside in subfreezing weather. The water froze and the object was lifted the "width of a barley corn" (0.333 in.). Estimate how much work was done and calculate the change in enthalpy of the water upon freezing. What do you conclude?

102. Santa Ana winds are associated with warm, dry air flowing off the shore in Southern California. These downslope winds originate from the mountainous high-desert areas to the east, and often stoke wildfires. To illustrate the mechanism responsible for these winds, consider a case in which the height of the source area is 1.5 km (~850 hPa), the temperature of the air is 15 °C, and its relative humidity is 50%. What will be the temperature and the relative humidity of this air if it descends to sea level? Discuss your results and provide a short description that could explain to the public why the Santa Ana winds are so warm and dry.

103. Problem 12 addresses the finite vapor pressure in a mercury barometer. What about a water barometer? Estimate the saturation vapor pressure correction to the height of the water column for a water barometer at a temperature of 20 °C. Give a rough estimate of the accuracy of the water temperature needed to give a pressure uncertainty of less than 0.5 hPa. To what precision would the height of the water column need to be measured to resolve a 0.5 hPa accuracy with a water barometer?

104. We show in Section 5.8 that the saturation vapor pressure of a solution is less than that of pure water, both at the same temperature. Thus, a given amount of solution at a given initial temperature in a given open container and in a given environment should evaporate more slowly than pure water. Is this necessarily true? This is a subtle problem that does not require difficult mathematics, but that *does* require careful thought, rather than jumping to conclusions. Ultimately, the answer requires simple but carefully done experiments. Yes or no might be equally correct. Try to understand this problem qualitatively and design an experiment to solve it. The answer may depend on the concentration of the solution. We made a first cut at such an experiment.

105. Estimate the upper limit of the amount of water vapor exhaled in 24 hours by an adult human. You will need the tidal volume of adults and breathing rate (or measure your own). Express your result in kilograms and liters of water.

106. Drinking a hot beverage on a hot, dry day may result in greater cooling. At first glance, this may seem counterintuitive. Explain. *Hint*: Estimate how much the average core temperature of an adult could increase because of drinking a very hot cup of coffee.

107. What is a reasonable (but not impossible) upper limit for the percentage of water molecules in air at sea level? Can you find weather data for anywhere in the world where this upper limit was approached within factor of two or three?

6

Moist Air and Clouds

The previous chapter is the longest and, in many ways, the central chapter of this book. The present chapter is a logical extension of Chapter 5, and even could have been combined with it. But the emphasis in the two is different. The theme of Chapter 5 is the principles of evaporation and condensation, with detailed analyses of how saturation vapor pressure depends on temperature, total pressure, dissolved air, and even droplet size. These principles are mostly applied to single cloud droplets (cloud microphysics). In this chapter we apply them to the macroscopic properties of clouds, emphasizing cloud formation and stability, and include a few topics (e.g. virtual temperature, wet-bulb temperature) that could have been incorporated in the previous chapter. In that chapter we viewed the atmosphere through a microscope; in this chapter we switch to a telescope.

6.1 Precipitable Water in the Atmosphere

If *all* the water vapor from the surface to the upper reaches of the troposphere were to suddenly condense and fall to earth (without evaporating), how deep would the resulting puddle be? We can estimate this depth, denoted by d_w, by invoking the ideal gas law and the Clausius–Clapeyron equation and making a few judicious approximations. Precipitable water is sometimes expressed in units of kgm^{-2} or gcm^{-2}, which divided by the density of water is d_w. The total depth of water is an integral from the surface upward

$$\rho_w d_w = \int_0^\infty \rho_v dz, \qquad (6.1)$$

where ρ_w is the density of liquid water and ρ_v is that of water vapor. Because water vapor in Earth's atmosphere is to very good approximation an ideal gas

$$d_w = \frac{1}{\rho_w R_v} \int_0^\infty \frac{e}{T} dz, \qquad (6.2)$$

where e is the vapor pressure. We obtain an upper limit for d_w by taking e at each altitude to be the saturation vapor pressure e_s at the corresponding temperature.

Atmospheric Thermodynamics. Second Edition. Craig F. Bohren and Bruce A. Albrecht, Oxford University Press.
© Craig Bohren and Bruce Albrecht (2023). DOI: 10.1093/oso/9780198872702.003.0006

The gradient of e_s is

$$\frac{de_s}{dz} = \frac{de_s}{dT}\frac{dT}{dz} = -\frac{de_s}{dT}\Gamma, \tag{6.3}$$

where Γ is the lapse rate. From the Clausius–Clapeyron equation (5.48)

$$\frac{e_s}{T} = \frac{R_v T}{\ell_v}\frac{de_s}{dT}. \tag{6.4}$$

For an assumed constant lapse rate, the previous three equations yield

$$d_w = -\frac{1}{\rho_w \ell_v \Gamma}\int_0^\infty T\frac{de_s}{dz}dz. \tag{6.5}$$

To obtain an upper limit for d_w, we take T in Eq. (6.5) to be constant and equal to the surface value T_0. With this assumption

$$d_w \approx \frac{T_0 e_{s0}}{\rho_w \ell_v \Gamma}. \tag{6.6}$$

The infinite upper limit of integration in Eqs. (6.1) and (6.5) is symbolic. By infinity is meant sufficiently high in the atmosphere that any water vapor at higher altitudes makes a negligible contribution to the total precipitable water. For a 25 °C surface temperature the saturation vapor pressure e_{s0} is about 32 hPa. For this temperature and vapor pressure and a lapse rate of 6.5 °C km^{-1}, d_w is about 6 cm, not a stunningly high value even though it is an overestimation because we assumed a saturated atmosphere and T in Eq. (6.5) equal to the surface value. The largest value for d_w inferred from satellite measurements that we could find is about 5 cm, to no surprise in the tropics; in mid-latitudes d_w is about 1 cm. Moreover, this amount of water is realizable only if *all* the atmospheric water vapor condenses, which is physically impossible.

Our estimate for d_w tells us something about storms. Rainstorms yielding more than a few centimeters of rain are not rare. This rainwater cannot simply be all the water vapor overhead condensed into cloud droplets that congregate into rain drops. Storms must be great engines for processing water vapor, gathering it from extensive surroundings, and concentrating it in a smaller area.

If you were faced with the task of calculating precipitable water from data, how would you go about doing so? You would have to remember that the definition of specific humidity q, Eq. (5.9), is the ratio of vapor density to air density. With this definition in Eq. (6.1)

$$d_w = \frac{1}{\rho_w}\int_0^\infty q\rho \, dz. \tag{6.7}$$

Then use the hydrostatic equation to obtain

$$d_w = -\frac{1}{g\rho_w}\int_0^\infty q\frac{dp}{dz}dz = \frac{1}{g\rho_w}\int_0^{p_o} q \, dp, \tag{6.8}$$

where p_0 is the surface pressure. Thus, integrating the specific humidity profile (as a function of pressure) yields the precipitable water.

6.2   Lapse Rate of the Dew Point: Level of Cloud Formation

Clouds can form in the atmosphere as air from a given level is lifted. Although we need not consider *how* it is lifted, we may define the *lifting-condensation level* (LCL) as the level to which a parcel of air would have to be lifted dry adiabatically to reach a relative humidity of 100%. The height of this level for a given parcel is a unique function of its temperature and humidity. To calculate this height, we imagine a parcel to be lifted from some initial height z_0 (or pressure p_0) to a height $z_{LCL}(p_{LCL})$ at which the parcel's temperature equals its dew point. As the parcel is lifted, its temperature T decreases (approximately) at the dry adiabatic rate, so that at any height z

$$T = T_0 - \Gamma_d(z - z_0),\tag{6.9}$$

where T_0 is the temperature of the parcel at z_0, and $\Gamma_d = g/c_p$ is the dry adiabatic lapse rate. The dew point T_d of the parcel at any height is

$$T_d = T_{d0} - \Gamma_{dew}(z - z_0)\tag{6.10}$$

provided that the lapse rate Γ_{dew} of the dew point of the parcel is constant. To obtain the lifting-condensation level we set T in Eq. (6.9) equal to T_d in Eq. (6.10):

$$z_{LCL} = z_0 + \frac{T_0 - T_{d0}}{\Gamma_d - \Gamma_{dew}}.\tag{6.11}$$

To determine the lapse rate of the dew point of a parcel, we begin by differentiating the equation defining the dew point, $e_s(T_d) = e$, with respect to z:

$$\frac{de_s}{dT_d}\frac{dT_d}{dz} = \frac{de}{dz}.\tag{6.12}$$

The vapor pressure e is proportional to the total pressure p:

$$e = \frac{w}{w + \epsilon}p,\tag{6.13}$$

where w is the mixing ratio and ϵ is the ratio of the gas constants of water vapor and dry air. We assume that until (net) condensation occurs, the mixing ratio of the parcel is constant, and hence,

$$\frac{de}{dz} = \frac{w}{w + \epsilon}\frac{dp}{dz}.\tag{6.14}$$

If we combine Eqs. (6.12)–(6.14)

$$\frac{1}{e_s}\frac{de_s}{dT_d}\frac{dT_d}{dz} = \frac{1}{p}\frac{dp}{dz}.\tag{6.15}$$

From Eq. (6.4) with $T = T_d$

$$\frac{1}{e_s}\frac{de_s}{dT_d} = \frac{\ell_v}{R_v T_d^2}.$$ (6.16)

For an adiabatic process

$$\frac{1}{p}\frac{dp}{dz} = \frac{c_p}{RT}\frac{dT}{dz} = -\frac{g}{RT}.$$ (6.17)

Equations (6.15)–(6.17) combined yield the lapse rate of the dew point:

$$\Gamma_{dew} = -\frac{dT_d}{dz} = \frac{g}{\epsilon \ell_v}\frac{T_d^2}{T},$$ (6.18)

where $g/\epsilon \ell_v \approx 6.3 \times 10^{-6}$ m^{-1}, nearly independent of temperature. Although T_d^2/T does depend on two (absolute) temperatures, its range in the troposphere is small. Typical values of the right side of Eq. (6.18) lie between about 1.7 and 1.9 °C km^{-1}. Take an intermediate value, 1.8 °C km^{-1}, as a rough approximation for the lapse rate of the dew point of a parcel. This lapse rate, together with a temperature lapse rate of 9.8 °C km^{-1} in Eq. (6.11), yields

$$z_{LCL} - z_0 \approx \frac{T_0 - T_{d0}}{8}.$$ (6.19)

Without having made any outrageous approximations we arrive at a useful rule of thumb: clouds can form by lifting air at height z_0 (assuming the existence of a lifting mechanism) a vertical distance (in km) one-eighth the dew point depression (difference between air temperature and dew point) at that height. Although this rule of thumb seems not to be widely known, it is, in a slightly less accurate form, over 150 years old. In his *Philosophy of Storms* (1841), the American meteorologist James P. Espy asserted that "The bases of all clouds forming by the cold of diminished pressure from upmoving columns of air, will be about as many hundred yards high as the dew point in degrees is below the temperature of the air at that time." Equation (6.19) in English units yields 75 yards in place of Espy's 100 yards.

A subtle point, often overlooked, is buried in this analysis: clouds can form on ascent only because the temperature of a parcel *happens* to decrease more rapidly than its dew point. This is not a universal law of nature, merely a consequence of the thermophysical properties of water and air. We return to this point in Section 6.9. Be prepared for a shock.

Often the lifting-condensation level of air near the surface is a good estimate of the height of the base of cumulus clouds with their roots in the planetary boundary layer. When this layer is well mixed, z_{LCL} of surface air will differ little from that of air at any height in the layer. Potential temperature and the mixing ratio of a parcel are conserved. If the boundary layer is well mixed, these two quantities are constant with height, and hence, parcels originating from any height in the layer will have the same lifting-condensation level.

Figure 6.1 shows observations of cloud base height, obtained from a ceilometer, compared with z_{LCL} calculated from surface observations of temperature and dew

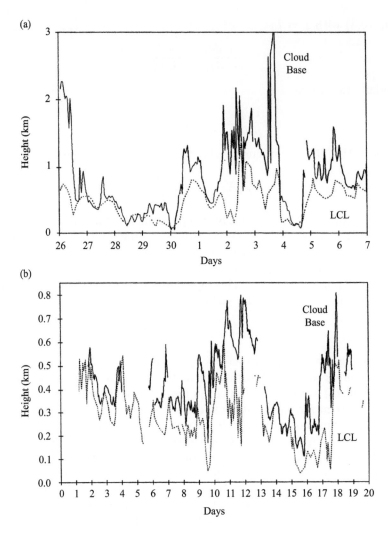

Fig. 6.1: Hourly-averaged cloud base height obtained with a laser ceilometer compared with the lifting-condensation level (LCL) calculated from hourly averaged
surface temperature and dew point. (a) Observations from a site near Penn State
University in late November and early December 1987. (b) Observations associated
with stratocumulus clouds during the first three weeks of July 1987, at San Nicolas
Island, about 100 km southwest of Los Angeles, California.

point. We chose data for two regions where low clouds are observed frequently.
When the boundary layer is well mixed and clouds are strongly coupled to surface
processes, agreement is good between z_{LCL} and observed cloud base. At other times
the agreement is not so good, indicating that the clouds formed by processes not
closely linked to the temperature and humidity of air near the surface.

The often good agreement between the observed height of cloud base and the calculated height at which a parcel becomes saturated upon adiabatic ascent should give us a bit of confidence that, despite the vagueness of parcel theory, it does yield results in accord with reality.

Although Eq. (6.19) is a useful approximation, for some applications a more accurate estimate may be needed. If we know the temperature T_{LCL} of a parcel at the lifting-condensation level, it follows from Eq. (6.9) that

$$z_{LCL} - z_0 = \frac{T_0 - T_{LCL}}{\Gamma_d} . \tag{6.20}$$

Both the temperature and dew point of a parcel with constant mixing ratio undergoing dry adiabatic ascent (or descent) depend only on height z, and hence there exists a functional relation between T and T_d. To obtain this relation we differentiate T_d with respect to z:

$$\frac{dT_d}{dz} = \frac{dT_d}{dT}\frac{dT}{dz} . \tag{6.21}$$

If this equation is combined with Eq. (6.15), we obtain

$$\frac{1}{e_s}\frac{de_s}{dT_d}\frac{dT_d}{dT}\frac{dT}{dz} = \frac{1}{p}\frac{dp}{dz} . \tag{6.22}$$

Now use the Clausius–Clapeyron equation, Eq. (6.16), and the equation relating pressure and temperature gradients in an adiabatic process, Eq. (6.17), to obtain a differential equation for T_d:

$$\frac{dT_d}{dT} = \frac{c_p T_d^2}{\epsilon l_v(T_d)T} . \tag{6.23}$$

The temperature dependence of the enthalpy of vaporization from Eq. (5.64) is

$$l_v(T_d) = l_r + (c_{pv} - c_{pw})T_d, \tag{6.24}$$

where

$$l_r = l_{vr} - (c_{pv} - c_{pw})T_r . \tag{6.25}$$

T_r is some reference temperature at which the enthalpy of vaporization is l_{vr}. Equation (6.24) combined with Eq. (6.23) yields the differential equation

$$\frac{d}{dT}\left(-\frac{l_r}{T_d} + (c_{pv} - c_{pw})\ln T_d - \frac{c_p}{\epsilon}\ln T\right) = 0, \tag{6.26}$$

the solution to which is

$$-\frac{l_r}{T_d} + (c_{pv} - c_{pw})\ln T_d - \frac{c_p}{\epsilon}\ln T = \text{const} . \tag{6.27}$$

At the lifting-condensation level $T = T_d = T_{LCL}$, and hence, from Eq. (6.27)

$$\frac{\ell_r}{T_{LCL}} - (c_{pv} - c_{pw}) \ln \frac{T_{LCL}}{T_{d0}} + \frac{c_p}{\epsilon} \ln \frac{T_{LCL}}{T_0} = \frac{\ell_r}{T_{d0}}. \tag{6.28}$$

With a bit more algebra, this can be written

$$\frac{1}{T_{LCL}} + A \ln \frac{T_{LCL}}{T_{d0}} = \frac{1}{T_{d0}} + B \ln \frac{T_0}{T_{d0}}, \tag{6.29}$$

where

$$A = -\left(\frac{c_{pv} - c_{pw}}{\ell_r} \frac{c_p}{\epsilon \ell_r}\right), \quad B = \frac{c_p}{\epsilon \ell_r}. \tag{6.30}$$

A is about 1.26×10^{-3} K^{-1} and B about 5.14×10^{-4} K^{-1} Equation (6.29) can be solved iteratively for a given temperature and dew point, but this is not necessary. Because the lapse rate of dew point is about $1.8\,°$C km^{-1}, T_{LCL} will not vary greatly from T_{d0}. Suppose, for example, that a parcel has to be lifted as much as 15 km to reach the lifting-condensation level. This corresponds to a temperature change of $27\,°$C, only about a 10% change in absolute temperature. Thus, for our purposes we may approximate the logarithmic term in Eq. (6.29) by

$$\ln \frac{T_{LCL}}{T_{d0}} \approx 1 - \frac{T_{d0}}{T_{LCL}}. \tag{6.31}$$

This approximation in Eq. (6.29) yields an explicit expression for T_{LCL} as a function of temperature and dew point at any height z_0:

$$T_{LCL} = \frac{1 - AT_{d0}}{\frac{1}{T_{d0}} + B \ln \frac{T_0}{T_{d0}} - A}. \tag{6.32}$$

This equation gives T_{LCL} values within 0.02 K of the iterative solution to Eq. (6.29) for temperatures between 250 and 300 K, even for a dew point depression as large as 20 K. A 0.02 K error in T_{LCL} results in a 2-m error in the estimate for $z_{LCL} - z_0$.

An even simpler estimate for T_{LCL} can be obtained from Eqs. (6.9) and (6.10). If we eliminate $z - z_0$ from these equations at $z = z_{LCL}$, where $T = T_d = T_{LCL}$,

$$\frac{T_0 - T_{LCL}}{T_{d0} - T_{LCL}} = \frac{\Gamma_d}{\Gamma_{dew}}, \tag{6.33}$$

which can be solved for T_{LCL}:

$$T_{LCL} = \frac{T_{d0}\Gamma_d - T_0\Gamma_{dew}}{\Gamma_d - \Gamma_{dew}}. \tag{6.34}$$

With $\Gamma_{dew} = 1.8\,°$C km^{-1}, this becomes

$$T_{LCL} = \frac{9.8T_{d0} - 1.8T_0}{8}. \tag{6.35}$$

This estimate of T_{LCL} differs by $0.3 - 1.0$ K from the solution to Eq. (6.29) for temperatures between 250 and 300 K and a dew point depression of 20 K; for a dew point depression of 10 K, the error drops into the range $0.1 - 0.5$ K.

6.3 Density of Moist Air: Virtual Temperature

We expressed the criterion for static stability (see Section 3.5) by way of temperature gradients, because at constant pressure, temperature is a surrogate for density. This approach always would be acceptable if the fraction of water vapor in air were constant. But because the atmosphere contains variable amounts of water vapor, making stability arguments indirectly by way of temperature gradients is not quite the same as invoking density gradients directly. The condition for static stability of the atmosphere in density terms is

$$\frac{d\rho_p}{dz} > \frac{d\rho_s}{dz},\tag{6.36}$$

where ρ_p is the density of a parcel undergoing adiabatic ascent or descent and ρ_s is the density of the parcel's surroundings. If the density of a parcel increases more rapidly than that of its surroundings, the parcel is negatively buoyant when displaced upward and positively buoyant when displaced downward. The condition for static instability is the converse of Eq. (6.36).

From the Poisson relation between temperature and volume in an adiabatic process (see Section 3.3), the relation between temperature and density follows:

$$\frac{T_p^{1/(1-\gamma)}}{\rho_p} = \text{const}.\tag{6.37}$$

The derivative of this equation with respect to height z is

$$\frac{1}{\rho_p}\frac{d\rho_p}{dz} = -\frac{1}{\gamma - 1}\frac{1}{T_p}\frac{dT_p}{dz}.\tag{6.38}$$

In this equation the temperature gradient of the parcel is the dry adiabatic lapse rate, and hence,

$$\frac{1}{\rho_p}\frac{d\rho_p}{dz} = -\frac{g}{\gamma R}\frac{1}{T_p}.\tag{6.39}$$

The ideal gas law for the surroundings can be written

$$p = \rho_s R T_s = \rho_s \frac{R}{R_d} T_s R_d,\tag{6.40}$$

where R_d is the gas constant of dry air. We define the *virtual temperature* T_v as

$$T_v = \frac{R}{R_d} T = \frac{M_d}{M} T,\tag{6.41}$$

where M_d is the mean molecular weight of dry air and M is that of moist air. Virtual temperature is the temperature that air of given pressure and density would have if the air were completely free of water vapor. Because the mean molecular weight of moist air is never greater than that of dry air, virtual temperature is never less than temperature.

Virtual temperature in Eq. (6.41) is not expressed in terms of familiar thermodynamic variables. From Eq. (2.102) the mean molecular weight is

$$M = f_d M_d + f_v M_v, \tag{6.42}$$

where M_d and M_v are the molecular weights of dry air and of water vapor. The corresponding number fractions f_d and f_v of these two components satisfy

$$f_d + f_v = 1. \tag{6.43}$$

From Eqs. (6.42) and (6.43)

$$\frac{M_d}{M} = \frac{1}{1 - f_v(1 - \epsilon)}, \tag{6.44}$$

where $\epsilon = M_v/M_d$. The specific humidity q can be written

$$q = \frac{f_v M_v}{f_v M_v + f_d M_d}, \tag{6.45}$$

which can be solved for f_v:

$$f_v = \frac{q}{\epsilon + q(1 - \epsilon)}. \tag{6.46}$$

By combining Eqs. (6.44) and (6.46), we obtain

$$\frac{M_d}{M} = 1 + q\frac{1 - \epsilon}{\epsilon} = 1 + 0.608q. \tag{6.47}$$

With an explicit expression for M_d/M, there is not a shadow of a doubt that virtual temperature is always greater than or equal to temperature:

$$T_v = T(1 + 0.608q). \tag{6.48}$$

From Eqs. (6.40) and (6.41), the ideal gas law can be written

$$p = \rho_s R_d T_v. \tag{6.49}$$

Take the derivative of both sides of Eq. (6.49) with respect to z:

$$\frac{1}{p}\frac{dp}{dz} = \frac{1}{T_v}\frac{dT_v}{dz} + \frac{1}{\rho_s}\frac{d\rho_s}{dz}. \tag{6.50}$$

By using the hydrostatic equation, Eq. (6.50) can be written

$$\frac{1}{\rho_s}\frac{d\rho_s}{dz} = -\frac{g}{T_s R} - \frac{1}{T_v}\frac{dT_v}{dz}. \tag{6.51}$$

Now subtract Eq. (6.51) from (6.39):

$$\frac{1}{\rho_p}\frac{d\rho_p}{dz} - \frac{1}{\rho_s}\frac{d\rho_s}{dz} = -\frac{g}{\gamma R T_p} + \frac{g}{R T_s} + \frac{1}{T_v}\frac{dT_v}{dz}. \tag{6.52}$$

The gradient of virtual temperature in Eq. (6.51) is that of the surroundings, not of the parcel.

Stability is determined by the sign of

$$\frac{d\rho_p}{dz} - \frac{d\rho_s}{dz}.\tag{6.53}$$

Both derivatives are evaluated where the density and temperature of parcel and surroundings are equal (i.e. at the height from which the parcel is imagined to ascend or descend). Thus, stability is determined only by the sign of

$$\frac{g}{R}\frac{\gamma-1}{\gamma} + \frac{T_s}{T_v}\frac{dT_v}{dz}\tag{6.54}$$

which is obtained from the right side of Eq. (6.52) for $T_p = T_s$. The first term is the dry adiabatic lapse rate, and hence, the criterion for stability can be written

$$\frac{dT_v}{dz} > -\frac{T_v}{T_s}\Gamma_d = -\frac{R}{R_d}\frac{g}{R} = -\frac{g}{R_d},\tag{6.55}$$

where g/R_d is the adiabatic lapse rate for dry air (which is not exactly the same as that for air containing water vapor). We won't invent a new symbol for this quantity. Instead, we'll just try to keep in mind that to determine stability we need to know if the gradient of virtual temperature rather than temperature is greater than, less than, or equal to the adiabatic lapse rate for dry air:

$$\frac{dT_v}{dz} > -\Gamma_d(\text{stable}), \quad \frac{dT_v}{dz} = -\Gamma_d \text{ (neutral) and } \frac{dT_v}{dz} < -\Gamma_d(\text{unstable}).\tag{6.56}$$

These static stability criteria are identical to those obtained previously with the temperature gradient replaced by the virtual temperature gradient (where Γ_d is for dry air).

Virtual temperature is just a convenient way to account for the variable water vapor content of air, not a temperature that can be measured with a thermometer. By differentiating Eq. (6.41) with respect to height

$$\frac{1}{T_v}\frac{dT_v}{dz} = \frac{1}{T}\frac{dT}{dz} + \frac{1}{R}\frac{dR}{dz}.\tag{6.57}$$

This shows that the virtual temperature gradient is compounded of a temperature and a moisture gradient.

If we take 0.01 as a representative value for the specific humidity of moist air, its virtual temperature is a bit more than half a percent greater than its temperature. But keep in mind that these are absolute temperatures. For normal terrestrial temperatures, a half percent or so increase in absolute temperature corresponds to a few degrees Celsius.

As might be expected, we can define a *virtual potential temperature* Θ_v as

$$\Theta_v = \left(\frac{p_0}{p}\right)^{R/c_p} T_v\tag{6.58}$$

and the stability criteria are identical to previous criteria in Section 4.3, with temperature replaced by virtual potential temperature.

6.4   Wet-Bulb Temperature

Wet-bulb temperature is more difficult to understand than dew point. The usual definition is something like the following: "The wet-bulb temperature is the temperature to which air may be cooled by evaporating water into it at constant pressure until saturation is reached." In this process, the mixing ratio is not constant, which is why dew point and wet-bulb temperature are different. This definition is somewhat misleading because it implies that air can be cooled merely by adding water vapor to it—that water molecules somehow are intrinsically colder than nitrogen and oxygen molecules. What really happens is that when liquid water undergoes net evaporation, it cools, as does air in contact with this water.

To understand better the distinction between dew point and wet-bulb temperature, let us consider how they might be measured in idealized experiments.

To determine the dew point of air, we enclose it in a cylinder fitted with a piston. This freely moving piston ensures a constant pressure. Cooling coils are wrapped around the cylinder so that we can cool its contents to any desired temperature. No water is in the cylinder other than the water vapor in the air. Now we cool this air to a temperature at which the saturation vapor pressure equals the partial pressure of the water vapor. This temperature is the dew point.

To measure the wet-bulb temperature of air, we can use the same cylinder, but we put some liquid water in it as well, get rid of the cooling coils, and wrap insulation around the cylinder. The initial temperature of the liquid water is that of the air. Because of net evaporation, the liquid water cools, and hence, so does the air in the cylinder. At the same time, water vapor is added to this air, which retards the rate of net evaporation. Eventually the system reaches an equilibrium temperature, and the equilibrium vapor pressure is the saturation vapor pressure at this temperature. This equilibrium temperature depends on how much liquid water was initially in the cylinder. The more water, the higher the temperature. So we successively reduce the amount of liquid water and measure the equilibrium temperature. The lowest temperature so obtained is the wet-bulb temperature. Thus, we can define the wet-bulb temperature as the *lowest* temperature to which moist air can be cooled (adiabatically and isobarically) solely by evaporative cooling. Don't let "wet" in wet-bulb temperature mislead you. It can drop below the freezing point.

The previous paragraph provides guidelines for a quantitative discussion of the wet-bulb temperature. In an adiabatic, isobaric process, enthalpy is conserved:

$$H_{di} + H_{vi} + H_{wi} = H_{df} + H_{vf} + H_{wf}, \tag{6.59}$$

where the subscripts i and f denote initial and final, d denotes dry, v denotes vapor, and w denotes liquid water. If we denote by $h(T)$, with appropriate subscripts, the specific enthalpies of the three components in the cylinder at temperature T, Eq. (6.59) becomes

$$M_d h_d(T_i) + M_{vi} h_v(T_i) + M_w h_w(T_i) = M_d h_d(T_f) + M_{vi} h_v(T_f) + M_w h_w(T_f), \tag{6.60}$$

where M with appropriate subscripts indicates initial and final masses of the three components (dry air, water vapor, and liquid water). From the definition of enthalpy of vaporization, we have

$$\ell_v(T_i) = h_v(T_i) - h_w(T_i), \quad \ell_v(T_f) = h_v(T_f) - h_w(T_f). \tag{6.61}$$

The total mass M_w of water is conserved in this closed system:

$$M_{vf} + M_{wf} = M_{vi} + M_{wi} = M_w. \tag{6.62}$$

The mass of dry air M_d is also conserved. Equations (6.60)–(6.62) can be rearranged

$$M_d[h_d(T_i) - h_d(T_f)] = M_w[h_w(T_f) - h_w(T_i)] + M_{vf}\ell_v(T_f) - M_{vi}\ell_v(T_i). \tag{6.63}$$

To proceed, we must make a few approximations. Because the specific heat capacities of air and liquid water are nearly constant with temperature over the range of meteorological interest, we can write

$$h_d(T_i) - h_d(T_f) = c_{pd}(T_i - T_f), \quad h_w(T_i) - h_w(T_f) = c_w(T_i - T_f). \tag{6.64}$$

And if T_i is not greatly different from T_f, the two enthalpies of vaporization are approximately equal:

$$\ell_v(T_i) \approx \ell_v(T_f) = \ell_v. \tag{6.65}$$

If Eqs. (6.64) and (6.65) are combined with Eq. (6.63), we obtain

$$(M_d c_{pd} + M_w c_w)(T_i - T_f) = (M_{vf} - M_{vi})\ell_v. \tag{6.66}$$

Divide both sides of Eq. (6.66) by M_d and use the definition of the mixing ratio w:

$$\left(c_{pd} + \frac{M_w}{M_d}c_w\right)(T_i - T_f) = \ell_v(w_f - w_i). \tag{6.67}$$

Before proceeding, we pause from our algebraic labors and ask ourselves where we are going and how to get there.

The properties of the air, its initial mixing ratio w_i and temperature T_i, are given. The final equilibrium temperature is obtained by solving Eq. (6.67) for T_f; the final mixing ratio w_f is the saturation mixing ratio $w_s(T_f)$ at this temperature. Note that T_f depends on the total mass of water M_w relative to that of dry air M_d. Because of evaporative cooling $T_f \leq T_i$; also, because water is added (isobarically) to the air, $w_f \geq w_i$. If the air is initially saturated, $w_i = w_s(T_i)$, and the solution to Eq. (6.67) is $T_f = T_i$, as expected on physical grounds. In general, this equation is solved by finding the intersection of the curves of two (dimensionless) functions of T_f:

$$f(T_f) = \frac{c_{pd} + (M_w/M_d)c_w}{\ell_v}(T_i - T_f) \tag{6.68}$$

and

$$g(T_f) = w_s(T_f) - w_i. \tag{6.69}$$

[g in Eq. (6.69) is *not* the specific free energy]. Both of these functions, shown in Fig. 6.2, are positive for $T_f \leq T_i$. $f(T_f)$ is zero for $T_f = T_i$ and increases for $T_f < T_i$; $g(T_f)$ is zero when $T_f = T_d$ and increases for higher temperatures. Thus, the point of intersection of the two functions (see Fig. 6.2) lies above the temperature axis and to the right of T_d (i.e., $T_f \geq T_d$). All else being equal, the temperature at the intersection point is greater the greater the value of M_w/M_d. The lowest value of T_f is obtained in the limit as this ratio of masses approaches zero, which requires that M_d become indefinitely large since M_w cannot be zero (if it were zero, neither the mixing ratio nor temperature would change). We call this minimum value of T_f the wet-bulb temperature T_w, which is determined by solving the equation

$$\frac{c_{pd}}{\ell_v}(T_i - T_w) = w_s(T_w) - w_i. \tag{6.70}$$

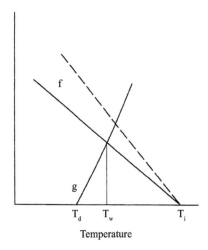

Fig. 6.2: The wet-bulb temperature is determined by the intersection of the functions f and g. T_i is the temperature of moist air, and T_d is its dew point. The temperature to which moist air can be lowered (adiabatically and isobarically) by evaporative cooling depends on the ratio of the total mass of water to the mass of dry air. The dashed curve is f for a finite value of this ratio; the solid curve is f in the limit of zero for this ratio. Thus, the wet-bulb temperature T_w is determined by the intersection of f (solid) with g.

According to Eq. (6.13)

$$w = \frac{\epsilon e}{p - e} \approx \frac{\epsilon e}{p}.$$ (6.71)

where $e \ll p$. With this approximation in Eq. (6.70) we obtain the *psychrometric equation*

$$e = e_s(T_d) = e_s(T_w) - \frac{p c_{pd}}{\epsilon \ell_v}(T_a - T_w),$$ (6.72)

where we replaced T_i in Eq. (6.70) with T_a to denote the temperature of the air of interest; the quantity

$$\frac{p c_{pd}}{\epsilon \ell_v}$$ (6.73)

is sometimes called the *psychrometric constant* (at sea level about 0.65 hPaK^{-1}). Thus, if we measure air temperature T_a (sometimes called the *dry-bulb temperature*) and the corresponding wet-bulb temperature, we can obtain the air's vapor pressure from Eq. (6.72) and one of the various approximations for the saturation vapor pressure at the wet-bulb temperature (see Section 5.3). It follows from Eq. (6.72) that $T_d \leq T_w \leq T_a$ and that $T_d = T_w$ if $T_a = T_w$. Before pocket calculators and personal computers were widely available, determining relative humidity and dew point from measurements required psychrometric charts or tables. Today they are unnecessary. Without much effort a short program can be written to calculate relative humidity and dew point as a function of pressure and dry- and wet-bulb temperatures. If you have a programmable calculator or personal computer, write such a program.

A simple approximation for the wet-bulb temperature follows from expanding the two vapor pressures in Eq. (6.72) about T_a and truncating to two terms:

$$e_s(T_d) \approx e_s(T_a) + \left(\frac{de_s}{dT}\right)_a (T_d - T_a), \quad e_s(T_w) \approx e_s(T_a) + \left(\frac{de_s}{dT}\right)_a (T_w - T_a).$$ (6.74)

Their difference is

$$e_s(T_d) - e_s(T_w) \approx \left(\frac{de_s}{dT}\right)_a (T_d - T_w) = \frac{e_s(T_a)\ell_v}{R_v T_a^2}(T_d - T_w) = -\frac{p c_{pd}}{\epsilon \ell_v}(T_a - T_w),$$ (6.75)

where we used the Clausius–Clapeyron equation and Eq. (6.72). T_w as a linear function of T_a and T_d follows from Eq. (6.75)

$$T_w \approx \left(\frac{a}{a+b}\right) T_d + \left(\frac{b}{a+b}\right) T_a,$$ (6.76)

where

$$a = \frac{e_s \ell_v}{R_v T_a^2}, \quad b = \frac{p c_{pd}}{\epsilon \ell_v}.$$ (6.77)

For a at $15\,°C$ and b at sea level

$$T_w \approx 0.6\,T_d + 0.4\,T_a. \tag{6.78}$$

At high relative humidity (60–90%) and for air temperatures from $0\,°C$ to $30\,°C$, the difference between the approximate wet-bulb temperature from Eq. (6.78) and the exact wet-bulb temperature is less than $1\,°C$ and the average difference over this range of temperature and humidity is $0.1\,°C$.

The wet-bulb temperature also has implications for atmospheric processes. For example, rain drops falling from the base of a cloud can evaporate, cool, and hence, cool (and moisten) the air beneath the cloud. In this process, the wet-bulb temperature in the layer remains constant. If the wet-bulb temperature is below $0\,°C$ a transition from rain to frozen precipitation is possible when the air cools to the wet-bulb temperature. Weather forecasters use the wet-bulb temperature in the lower layers of the atmosphere to determine if a precipitation changeover is possible.

The wet-bulb temperature is also the relevant temperature when considering heat stress on humans. Evaporation of perspiration, sensible and insensible (Section 5.3), is a major mechanism for cooling the human body in warm environments. Because evaporative cooling depends on a difference between wet skin temperature and the wet-bulb temperature of the environment, no cooling occurs if this temperature exceeds $35\,°C$, the approximate maximum skin temperature that can keep the body core temperature from exceeding $37\,°C$. Thus, $35\,°C$ can be considered the *killer wet-bulb temperature*. Wet-bulb temperatures greater than $25\,°C$ experienced during strenuous outdoor activities may also lead to fatalities. The temperature and relative humidity for wet-bulb temperatures of $35\,°C$ and $25\,°C$ are shown in Fig. 6.3 and illustrate how increasing temperature or relative humidity can give deadly wet-bulb temperatures. For an air temperature of $38\,°C$ and a relative humidity of 82% the wet-bulb temperature is $35\,°C$, but at $42\,°C$ and relative humidity of only 63% the wet-bulb temperature is the same. Wet-bulb temperatures detrimental to human activities outdoors are already encountered frequently in the tropics and subtropics and episodically during heat waves at higher latitudes.

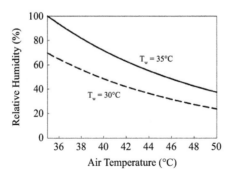

Fig. 6.3: Temperature and relative humidity for constant wet-bulb temperatures of $30\,°C$ and $35\,°C$.

As Earth warms, these areas and the frequency and intensity of heat waves will increase and with enough warming some areas may experience killer wet-bulb temperatures routinely and possibly become uninhabitable.

We have laid all the groundwork for determining yet another temperature, the (isobaric) *equivalent temperature* T_e, defined as the temperature to which air would rise if *all* its water vapor were to condense in an adiabatic, isobaric process. To determine this temperature, we could begin, as we did with the wet-bulb temperature, by appealing to conservation of enthalpy. But we don't need to do this because our result is already contained in Eq. (6.67). All we have to do is set $w_f = 0$ in this equation and change a few symbols:

$$T_e = T_a + \frac{\ell_v w}{c_{pd} + w c_w}, \tag{6.79}$$

where T_a is the temperature of the air and w its mixing ratio. Although equivalent temperature is easier to calculate than wet-bulb temperature, the price we pay for this is that the process defining equivalent temperature is fictitious, allowed by the first law of thermodynamics (enthalpy conservation in an isobaric, adiabatic process) but forbidden by the second law (entropy cannot decrease). Problem 14 shows this.

The difference between air temperature and equivalent temperature is surprisingly high. For a mixing ratio of 0.001(1 g/kg), which corresponds to very dry air, this temperature difference is about 2.5 °C ($c_{pd} \approx 1000$ J kg^{-1}K^{-1}; $\ell_v \approx 2.5 \times 10^6$ J kg^{-1}).

Is the Temperature of a Wet Bulb the Wet-Bulb Temperature?

Although we may define the wet-bulb temperature according to an equation, this quantity doesn't do us much good unless we can measure it. So, let us consider what happens when air is blown through cloth moistened with water. Energy transfer from the air to the wet cloth results in evaporation. Denote by T_a the temperature of the air and by T_w the steady-state temperature of the wet cloth once equilibrium is reached. The air is cooled and has water vapor added to it in an isobaric, adiabatic process.

Consider some time interval after the steady state is reached. During this interval a mass M_d of dry air passes through the cloth accompanied by a mass M_v of water vapor. Because this system composed of dry air, water vapor, and liquid water undergoes an adiabatic, isobaric process, total enthalpy is conserved:

$$M_w h_w(T_w) + M_d h_d(T_a) + M_v h_v(T_a) = M'_w h_w(T_w) + M_d h_d(T_w) + M'_v h_v(T_w), \tag{6.80}$$

where M'_w is the mass of liquid water in the cloth at the end of the interval and M'_v is the corresponding mass of water vapor mixed with the mass M_d of dry air that has passed through the cloth. Conservation of total water mass requires that

$$M_w + M_v = M'_w + M'_v. \tag{6.81}$$

The enthalpy of the water trapped in the pores of the cloth changes (in steady state), not because its temperature changes, but because its mass changes. If we use Eq. (6.61), assume, as previously, that the enthalpy of vaporization at T_a is approximately equal to that at T_w, and invoke

$$h_w(T_w) = h_w(T_a) + c_w(T_w - T_a) \tag{6.82}$$

as well as Eq. (6.81), conservation of enthalpy (Eq. 6.80) becomes

$$(c_{pd} + \frac{M_v}{M_d}c_w)(T_a - T_w) = \ell_v(w' - w). \tag{6.83}$$

We assumed that the outgoing air is cooled to the temperature of the wet cloth. With the additional assumption that the outgoing air is saturated (i.e. w' is the saturation mixing ratio at temperature T_w) and that $c_w M_v/M_d$ is negligible compared with c_{pd}, Eq. (6.83) is identical to Eq. (6.70), the defining equation for the wet-bulb temperature. We conclude that, subject to the various assumptions we made, the temperature of a wet bulb T_w is indeed the wet-bulb temperature. We now have a means to measure it: wrap the bulb of a thermometer with a piece of wet cloth, ventilate it, and read the lowest temperature (assuming that the cloth does not become dry).

Humidity Measurements

Atmospheric humidity is measured for meteorological and other applications, for example, to monitor and control indoor humidity, which affects human comfort and the physical properties of many organic materials. Visit an art museum and you will see humidity sensors in nearly every room. Although indoor measurements often need not be highly accurate, some meteorological applications may require higher accuracies.

One of the simplest ways to measure this temperature is with a sling psychrometer, a frame on which two liquid-in-glass thermometers are mounted. The bulb of one is covered with a cloth sleeve (wick) wetted with pure water before the thermometers are whirled about a handle attached to the frame. After a minute or so of whirling, readings are quickly taken of the wet- and dry-bulb temperatures. There are, however, several sources of error with this method, including radiative heating, inadequate ventilation of the wick, impure water, and difficulty reading the thermometers quickly. Keeping the wick wet for continuous operation of wet-bulb thermometers can be difficult, and measurements at temperatures below freezing are not practical. The accuracy of both the dry and wet-bub temperatures are about 0.5 °C, corresponding to a relative humidity accuracy of about 5%, although other sources of error can push this closer to 10%. The advantage of a sling psychrometer is that it requires no calibration, is portable, and can easily be used in remote locations.

Dew point measurements offer another possibility for determining atmospheric humidity. A common design for a dew point hygrometer incorporates a thermo-electric device to cool a small mirror. An optical sensor detects dew (or frost) on the mirror while its temperature is measured with a thermocouple or thermistor. Chilled-mirror hygrometers can be used to measure humidity over nearly the entire range of temperatures in the atmosphere and can measure the dew point to an accuracy of 0.1 °C. Consequently, chilled-mirror hygrometers are used as a reference standard in meteorology laboratories. They have a relatively slow response time compared with other electronic sensors.

Dew point and wet-bulb hygrometers exploit the physical properties of water and water–air systems. Another class of hygrometers is based on changes in the physical properties of materials with changes in humidity. One of the simplest instruments of this type is the hair hygrometer, which works on the principle that human hair (thin, blond hair is the best) elongates with increases in relative humidity. Mechanical mechanisms amplify changes in the length of the hair to provide a visual display. Hair hygrometers are slow to respond (about 20 min or more) and may not be stable when exposed to extreme temperature and humidity. The hair hygrometer was invented in the late eighteenth century and has been used since then for monitoring and displaying relative humidity. Hair hygrometers were also used in the early radiosondes, but they were not effective in this application due to their slow response time. Hygrometers using hair or artificial fibers are still used in hygrometers manufactured for analog displays that require no electrical input.

Various hygrometers make use of changes in electrical properties of materials with humidity. Typically, these hygrometers are based on either a resistance or capacitance that varies with humidity. Originally, thin films of carbon or a salt such as lithium chloride on a glass substrate were used as resistive humidity sensors. But these hygrometers suffered from hysteresis (a dependence on previous states, not just present state), which limited their performance on radiosondes. But advances in the development of resistive sensors have provided sensors that maintain their calibration and have a good response time. These resistive sensors consist of two electrodes separated by a thin hygroscopic polymer film that changes resistance with relative humidity changes. Capacitive sensors also make use of this thin film technology where the capacitance of these sensors changes with relative humidity. They are now in wide use in meteorological applications and generally have an accuracy of about 2% and a response time of about 1 s. The development of capacitive humidity sensors was an advancement in the sensors used in radiosondes. Humidity measurements from radiosondes are a particular challenge since low-cost but accurate sensors are needed and humidity can change substantially with height. Like other wet-bulb and dew point instruments, electrical elements used for humidity measurements may degrade in marine environments due to the deposition of salt on sensors. The development of small and inexpensive signal and digital processing circuitry needed to convert either capacitance or resistance to humidity has resulted in the mass production of inexpensive digital hygrometers. They typically have an accuracy of around 5%, which is sufficient for many home or commercial applications.

Another class of instruments is based on absorption by water vapor of radiation at different wavelengths. Although some of these instruments operate in the infrared and near-infrared, one instrument is based on absorption by water vapor at the Lyman-alpha line (0.12 μm). Lyman-alpha hygrometers are used where a fast response time (less than 0.1 s) is needed to measure turbulence-scale humidity fluctuations from research aircraft. Because these hygrometers do not measure absolute humidity accurately, the mean humidity is usually tied to chilled-mirror dew point measurements.

6.5 Lapse Rate for Isentropic Ascent of a Saturated Parcel

We derived the dry adiabatic lapse rate in Section 3.4 for an unsaturated process. But there may come a point when we can no longer assume that some of the water vapor in a rising parcel does not condense. When condensation occurs, the parcel becomes a mixture of dry air, water vapor (at saturation), and liquid water droplets. Liquid water is not an ideal gas so to determine lapse rate we appeal to entropy conservation. The derivation is tedious, and you are almost certain to make errors along the way. We derived the lapse rate for a saturated parcel several times, each time obtaining different results. Indeed, you are likely to find expressions similar but not identical to ours. Nevertheless, we think that what follows is correct.

The total entropy of a parcel is

$$S = M_v s_v + M_d s_d + M_w s_w, \tag{6.84}$$

where subscripts v, w, and d denote vapor, liquid water, and dry air, respectively. If we differentiate Eq. (6.84) with respect to z, use Eq. (5.58) for the entropy difference between liquid and vapor, and conservation of total water mass,

$$\frac{dM_v}{dz} = -\frac{dM_w}{dz}, \tag{6.85}$$

we obtain

$$\frac{dS}{dz} = \frac{dM_v}{dz}\frac{\ell_v}{T} + M_v \frac{ds_v}{dz} + M_w \frac{ds_w}{dz} + M_d \frac{ds_d}{dz}. \tag{6.86}$$

From the expression for the entropy of an ideal gas we have

$$\frac{ds_d}{dz} = \frac{c_{pd}}{T}\frac{dT}{dz} - \frac{R_d}{p_d}\frac{dp_d}{dz}, \tag{6.87}$$

$$\frac{ds_v}{dz} = \frac{c_{pv}}{T}\frac{dT}{dz} - \frac{R_v}{e}\frac{de}{dz} = \frac{c_{pv}}{T}\frac{dT}{dz} - \frac{R_v}{e_s}\frac{de_s}{dT}\frac{dT}{dz}, \tag{6.88}$$

where we took $e = e_s$, which enabled us to write

$$\frac{de_s}{dz} = \frac{de_s}{dT}\frac{dT}{dz}. \tag{6.89}$$

From Eq. (4.79) we have

$$\frac{ds_w}{dz} = \frac{c_w}{T}\frac{dT}{dz}.$$

(6.90)

If Eqs. (6.87), (6.88), and (6.88) are substituted into Eq. (6.86), we obtain

$$\frac{dS}{dz} = \frac{\ell_v}{T}\frac{dM_v}{dz} + (M_w c_w + M_v c_{pv} + M_d c_{pd})\frac{1}{T}\frac{dT}{dz} - \frac{M_v R_v}{e_s}\frac{de_s}{dT}\frac{dT}{dz} - \frac{M_d R_d}{p_d}\frac{dp_d}{dz}.$$

(6.91)

To proceed, we denote the (constant) total mass of water in the parcel as

$$M = M_w + M_v.$$

(6.92)

This equation, together with (see Section 5.3)

$$\frac{d\ell_v}{dT} = c_{pv} - c_w$$

(6.93)

and the Clausius–Clapeyron equation can be folded into Eq. (6.91) to obtain

$$\frac{dS}{dz} = (Mc_w + M_d c_{pd})\frac{1}{T}\frac{dT}{dz} + \frac{d}{dz}\left(\frac{\ell_v M_v}{T}\right) - \frac{M_d R_d}{p_d}\frac{dp_d}{dz}.$$

(6.94)

We can simplify this by defining

$$c_p = \frac{Mc_w + M_d c_{pd}}{M_d},$$

(6.95)

which leads to

$$\frac{T}{M_d}\frac{dS}{dz} = c_p\frac{dT}{dz} + T\frac{d}{dz}\left(\frac{\ell_v w_s}{T}\right) - \frac{R_d T}{p_d}\frac{dp_d}{dz},$$

(6.96)

where w_s is the saturation mixing ratio. The total pressure p is $p_d + e_s$, from which by differentiation with respect to z, the ideal gas law applied to dry air and water vapor, and $w_s = R_d e_s / R_v p_d$ it follows that

$$-\frac{R_d T}{p_d}\frac{dp_d}{dz} = -\frac{R_d T}{p_d}\frac{dp}{dz} + \frac{w_s \ell_v}{T}\frac{dT}{dz}.$$

(6.97)

This result substituted in Eq. (6.96) yields

$$\frac{T}{M_d}\frac{dS}{dz} = c_p\frac{dT}{dz} + \frac{d}{dz}(w_s \ell_v) - \frac{R_d T}{p_d}\frac{dp}{dz}.$$

(6.98)

With the assumption of hydrostatic equilibrium

$$\frac{dp}{dz} = -\frac{gp}{RT},$$

(6.99)

where R is the gas constant of the parcel's surroundings, the third term on the right side of Eq. (6.98) becomes

$$\frac{R_d T}{p_d}\frac{dp}{dz} = -g\frac{R_d p}{R p_d}.$$ (6.100)

From the definitions of the average molecular weight and gas constant for a mixture (see Section 2.8), we have for air with partial pressure p_d of dry air and e of water vapor

$$\frac{1}{R} = \frac{p_d}{p}\frac{1}{R_d} + \frac{e}{p}\frac{1}{R_w}.$$ (6.101)

This can be written as

$$\frac{R_d p}{R p_d} = 1 + \frac{e}{p_d}\frac{R_d}{R_w} \approx 1$$ (6.102)

if $e/p_d \ll 1$. Now we have to be careful. In Eq. (6.100), R is the gas constant for the surroundings of the parcel, p is the total pressure (of parcel or surroundings), but p_d is the partial pressure of dry air in the parcel. In Eq. (6.101), however, R, p, and p_d apply to the same chunk of air. So we have to make an additional assumption (a good one) that the difference between the partial pressure of water in the parcel and that in the surroundings is small compared with the total pressure. With these last approximations, Eq. (6.98) becomes

$$\frac{T}{M_d}\frac{dS}{dz} = c_p\frac{dT}{dz} + \frac{d}{dz}(w_s \ell_v) + g.$$ (6.103)

If the ascent (or descent) is isentropic (reversible, adiabatic), $dS/dz = 0$, and we obtain the *moist adiabatic lapse rate*,

$$\frac{dT}{dz} = -\frac{g}{c_p} - \frac{1}{c_p}\frac{d}{dz}(w_s \ell_v).$$ (6.104)

The first term on the right side of Eq. (6.104) is almost the dry adiabatic lapse rate Eq. (3.93). The small difference arises from the appearance of the specific heat capacity of liquid water rather than water vapor in c_p. For an unsaturated process Eq. (6.104) reduces to Eq. (3.93), which it must. For a rising parcel in which condensation occurs, the saturation mixing ratio w_s decreases with increasing z. Consequently, the lapse rate of such a parcel is *less* than the dry adiabatic lapse rate. Could you have guessed the form of the second term in the lapse rate? It has a simple physical interpretation. Water vapor does not condense into liquid water without other consequences. Although the details of this phase change are hidden in a thermodynamic analysis, its consequences are observable as an increase in temperature of the air–vapor–water system. This condensational heating is often referred to as "the release of latent heat," an archaic phrase coined in an era when it was thought that there were two kinds of heat, *sensible* and *latent* (see Section 5.3). The term *sensible heat* has more or less disappeared from physics but has been preserved in the geophysical sciences. Meteorology bears the same relation to physics as the French spoken in Quebec does to that spoken in Paris. French speakers settled in Canada several centuries ago. Language, like science, changes, but the French of Canada evolved separately from that of France. Consequently, after several centuries, Quebec French sounds archaic to Parisians, a language flavored with words and phrases of a France that no longer exists. And so it is with meteorology, which often preserves the physics at a time when the two sciences went their separate ways. This is why, for example, many meteorologists are taught that absolute zero is the temperature at which all motion stops, whereas most physicists know (or should know) that this is not true (see Section 2.1).

Why is the phrase "release of latent heat" objectionable? Because it is a mantra masquerading as an explanation. To accept an explanation of temperature rise because of net condensation as the result of the release of some mysterious ethereal fluid is like eating sawdust instead of a proper meal; the sawdust fills your stomach but has no nutritional value. And yet a proper explanation of condensational warming, based on sound physical principles, is readily available (see Section 5.1).

We may write Eq. (6.104) more succinctly by introducing the *moist static energy*

$$h = c_p T + gz + \ell_v w_s, \tag{6.105}$$

from which it follows that

$$\frac{dh}{dz} = 0. \tag{6.106}$$

In deriving Eq. (6.104) we used almost everything we have learned so far, the basic tools of atmospheric thermodynamics: the ideal gas law, Dalton's law of partial pressures, the Clausius–Clapeyron equation, the first and second laws of thermodynamics, entropies of vapor and liquid phases, enthalpy of vaporization and its dependence on temperature, mean molecular weights of mixtures, and the hydrostatic approximation.

We can go further with Eq. (6.104) by differentiating $w_s \ell_v$:

$$\frac{dT}{dz} = -\frac{g}{c_p} - \frac{w_s}{c_p} \frac{d\ell_v}{dz} - \frac{\ell_v}{c_p} \frac{dw_s}{dz}. \tag{6.107}$$

The enthalpy of vaporization changes with height z of the parcel because its temperature does:

$$\frac{d\ell_v}{dz} = \frac{d\ell_v}{dT} \frac{dT}{dz}. \tag{6.108}$$

With Eq. (6.108) in Eq. (6.107) and some rearranging, we obtain

$$\frac{dT}{dz} \left(1 + \frac{w_s}{c_p} \frac{d\ell_v}{dT} \right) = -\frac{g}{c_p} - \frac{\ell_v}{c_p} \frac{dw_s}{dz}. \tag{6.109}$$

Because of Eq. (6.93) for the temperature dependence of the enthalpy of vaporization, Eq. (6.109) becomes

$$\frac{dT}{dz}\left(1 - \frac{w_s(c_w - c_{pv})}{c_p}\right) = -\frac{g}{c_p} - \frac{\ell_v}{c_p}\frac{dw_s}{dz}. \tag{6.110}$$

The saturation mixing ratio w_s is of order 0.01 or less, and the relative difference of specific heat capacities is approximately 1.3, and hence, for our applications

$$0 < \frac{w_s(c_w - c_{pv})}{c_p} \ll 1, \tag{6.111}$$

which enables us to *approximate* Eq. (6.110) as

$$\frac{dT}{dz} = -\frac{g}{c_p} - \frac{\ell_v}{c_p}\frac{dw_s}{dz}. \tag{6.112}$$

We have not yet finished. We can go further. From Eq. (6.71) we have

$$w_s \approx \epsilon\frac{e_s}{p}, \tag{6.113}$$

which when differentiated yields

$$\frac{1}{w_s}\frac{dw_s}{dz} = -\frac{1}{p}\frac{dp}{dz} + \frac{1}{e_s}\frac{de_s}{dz} = -\frac{1}{p}\frac{dp}{dz} + \frac{1}{e_s}\frac{de_s}{dT}\frac{dT}{dz}. \tag{6.114}$$

By once again calling upon the hydrostatic approximation and the Clausius–Clapeyron equation, by now old and trusted friends, Eq. (6.114) becomes

$$\frac{1}{w_s}\frac{dw_s}{dz} = \frac{g}{RT} + \frac{\ell_v}{R_vT^2}\frac{dT}{dz}. \tag{6.115}$$

This equation in Eq. (6.112) yields

$$\frac{dT}{dz} = -\frac{g}{c_p}\frac{1 + \ell_v w_s/RT}{1 + \ell_v^2 w_s/c_p R_v T^2}. \tag{6.116}$$

For T around 280 K

$$\frac{\ell_v}{RT} \approx 30, \quad \frac{\ell_v}{R_vT} \approx 20, \quad \frac{\ell_v}{c_pT} \approx 9. \tag{6.117}$$

The saturation mixing ratio at this temperature is of order 0.01. In general, neither $\ell_v w_s/RT$ nor $\ell_v^2 w_s/c_p R_v T^2$ can be neglected compared with unity, and hence, we can approximate no further. We have to live with Eq. (6.116), cumbersome though it may be.

We now rest from our mathematical labors and contemplate the significance of our results. When a parcel of unsaturated air rises adiabatically, its temperature decreases at the dry adiabatic rate. When saturation is finally reached (at the lifting-condensation level, where temperature and dew point converge), water vapor begins to condense (assuming the presence of condensation nuclei). At this point the lapse rate changes discontinuously to that given by Eq. (6.116). For typical values of saturation mixing ratio and temperature, the lapse rate changes by roughly a factor of two, temperature abruptly decreasing at $\approx 5\,^\circ\mathrm{C\,km}^{-1}$ instead of $\approx 10\,^\circ\mathrm{C\,km}^{-1}$. As the parcel continues to rise, the saturation mixing ratio decreases, and the lapse rate gradually returns to the dry adiabatic value. An implicit assumption underlying our analysis is that any liquid water formed by condensation within the parcel is carried with it in its ascent. In the following subsection we remove this restriction.

Equivalent Potential Temperature and Wet-Bulb Potential Temperature

In the previous subsection we derived the moist adiabatic lapse rate, a rate at which temperature decreases with height at a given level in the atmosphere. In many applications, however, it is necessary to determine the temperature of a parcel as it ascends (or descends) moist adiabatically over some depth, which for cumulus convection can extend from near the surface to the tropopause. Although the expression for the moist adiabatic lapse rate could be integrated to obtain the temperature of a parcel as it ascends, Eq. (6.116) was derived using the hydrostatic approximation, which may not be valid for vigorous cumulus convection. Thus, we seek a relation, not limited by this approximation, between temperature and pressure during moist adiabatic ascent (or descent). In so doing we obtain a temperature parameter that is conserved in a saturated adiabatic process. By adiabatic process we mean an isentropic (reversible adiabatic) process.

Equation (6.91) can be written

$$\frac{1}{M_d}\frac{dS}{dt} = (w_t c_w + c_{pd})\frac{1}{T}\frac{dT}{dt} + \frac{d}{dt}\left(\frac{\ell_v w_s}{T}\right) - \frac{R_d}{p_d}\frac{dp_d}{dt}, \tag{6.118}$$

where the time derivative is a rate of change following a parcel and w_t is the total water (liquid plus vapor) mixing ratio, assumed to be constant. For isentropic motion $dS/dt = 0$, and Eq. (6.118) can be integrated to obtain T of the parcel as a function of p_d during ascent or descent for a given w_t. For a saturated parcel $p = p_d + e_s(T)$; thus, p_d and T uniquely determine p. Define

$$\Theta_d = T\left(\frac{p_0}{p_d}\right)^{R_d/c_p}, \tag{6.119}$$

where p_0 is some reference pressure (usually taken to be 1000 hPa) and $c_p = c_{pd} + w_t c_w$. Θ_d is approximately Θ, the difference between these two temperatures arising from p_d in Eq. (6.119) instead of p, R_d instead of R, and c_p slightly different from the specific heat capacity of moist air $(c_{pd} + wc_{pv})$. We now can write Eq. (6.118) as

$$\frac{1}{M_d}\frac{dS}{dt} = c_p\frac{d\ln\Theta_d}{dt} + \frac{d}{dt}\left(\frac{\ell_v w_s}{T}\right) = \frac{d}{dt}\left(c_p\ln\Theta_d + \frac{\ell_v w_s}{T}\right) \qquad (6.120)$$

because from Eq. (6.119)

$$c_p\frac{d\ln\Theta_d}{dt} = c_p\frac{d\ln T}{dt} - R_d\frac{d\ln p_d}{dt}. \qquad (6.121)$$

For an isentropic process, Eq. (6.120) can be integrated from an initial state at T and $p_d(p)$ to a final state denoted by the subscript f:

$$c_p\ln\frac{\Theta_{df}}{\Theta_d} = -\left(\frac{\ell_v(T_f)w_{sf}}{T_f} - \frac{\ell_v w_s}{T}\right), \qquad (6.122)$$

where Θ_d and w_s are functions of T and p. At a sufficiently high altitude (low pressure) where all vapor in the parcel has condensed $(w_{sf} = 0)$, Θ_{df} is defined as the *equivalent potential temperature* Θ_e:

$$c_p\ln\frac{\Theta_e}{\Theta_d} = \frac{\ell_v w_s}{T} \qquad (6.123)$$

or

$$\Theta_e = \Theta_d\exp\left(\frac{\ell_v w_s}{c_p T}\right). \qquad (6.124)$$

Equivalent potential temperature is approximately the potential temperature a saturated parcel would have if it were lifted to very low pressure so that all its water vapor condenses. This relationship is approximate because the exponent in the defining equation for Θ is not quite the same as that in the defining equation for Θ_e even for completely dry air $(w_s = 0)$.

For a saturated process Θ_e bears the same relation to entropy as potential temperature does for an unsaturated process. From Eq. (6.120) and the definition of equivalent potential temperature Eq. (6.123), we have

$$\frac{1}{M_d}\frac{dS}{dt} = \frac{c_p}{\Theta_e}\frac{d\Theta_e}{dt}. \qquad (6.125)$$

This has the same form as Eq. (4.108), and hence, the designation "equivalent" potential temperature. According to Eq. (6.125), Θ_e is conserved in an isentropic process, which provides the rationale for defining this quantity in the first place. This conservative property of Θ_e is particularly useful for treating processes in which condensation and evaporation occur.

Although we defined Θ_e for saturated air, this quantity also can be defined for unsaturated air with temperature T, pressure p, and mixing ratio w by considering it to be lifted to the lifting-condensation level (Section 6.2), where Θ_d is

$$\Theta_{LCL} = T_{LCL} \left(\frac{p_0}{p_{LCL} - e_s(T_{LCL})} \right)^{R_d/c_p} \tag{6.126}$$

and hence,

$$\Theta_e = \Theta_d(LCL) \exp \left(\frac{\ell_v(T_{LCL})w}{c_p T_{LCL}} \right) . \tag{6.127}$$

Because T_{LCL} and p_{LCL} of an unsaturated parcel are uniquely determined by its thermodynamic state (p, T, w) and are conserved during dry adiabatic ascent or descent, Θ_e for such a parcel is also conserved.

The *saturation equivalent potential temperature* of an unsaturated parcel is the equivalent potential temperature it would have if it were saturated, defined as

$$\Theta_{es} = \Theta_d \exp \left(\frac{\ell_v(T)w_s(T)}{c_p T} \right) . \tag{6.128}$$

Although Θ_{es} depends on the thermodynamic state of the air, this quantity is not conserved in unsaturated ascent or descent. For saturated air, of course, $\Theta_e = \Theta_{es}$. Although Eqs. (6.124) and (6.128) have the same form, they are interpreted differently. In Eq. (6.124), w_s is the mixing ratio of saturated air at temperature T, whereas in Eq. (6.128) it is the saturation mixing ratio corresponding to a temperature T of unsaturated air.

Calculating accurate values of Θ_e and Θ_{es} for unsaturated air from Eqs. (6.127) and (6.128) is somewhat tedious because Θ_d, not Θ, appears in these equations, and care must be taken to obtain correct values of c_p and ℓ_v. For applications where estimates within 0.5 K are adequate, Θ_e and Θ_{es} can be determined from the approximate relations

$$\Theta_e = \Theta \exp \left(\frac{2.675w}{T_{LCL}} \right), \tag{6.129}$$

and

$$\Theta_{es} = \Theta \exp \left(\frac{2.675w_s}{T} \right), \tag{6.130}$$

where w and w_s are in g/kg and Θ is used instead of Θ_d.

Further insight into the nature of Θ_e can be obtained by rewriting Eq. (6.127) as

$$\Theta_e \approx \Theta \exp \left(\frac{\ell_v w}{c_p T_{LCL}} \right) . \tag{6.131}$$

This is a good approximation because for typical conditions Θ_d is not more than about 5 K greater than Θ. Thus, Θ_e is always greater than Θ, and as w approaches zero, Θ_e approaches (approximately) Θ. Because the argument of the exponential function in Eq. (6.131) is usually less than 0.2, it can be further approximated as

$$\Theta_e \approx \Theta + \frac{\Theta}{T_{LCL}} \frac{\ell_v w}{c_p}. \tag{6.132}$$

Potential temperature and the temperature at the lifting condensation level are absolute temperatures; so their ratio is close to 1, and we can further approximate Θ_e as

$$\Theta_e \approx \Theta + \frac{\ell_v w}{c_p} \simeq T + \frac{gz}{c_p} + \frac{\ell_v w}{c_p}. \tag{6.133}$$

Each g/kg of water vapor elevates the equivalent potential temperature about 2.5 K above the potential temperature. Note the similarity between Eqs. (6.133) and (6.79), the defining equation for equivalent temperature. Whereas this isobaric equivalent potential temperature is defined for a nonexistent (thermodynamically forbidden) adiabatic, isobaric process, equivalent potential temperature is defined for an allowed adiabatic process that is far from being isobaric. From Eq. (6.133) it is evident that moist static energy is also related to equivalent potential temperature as

$$h \simeq c_p \Theta_e. \tag{6.134}$$

Because moist static energy is based on hydrostatic equilibrium, it is not strictly conserved in the adiabatic ascent or descent of a parcel, although equivalent potential temperature is conserved.

For the moist reversible processes considered to this point we have assumed that parcels are closed systems, which retain all their liquid water condensed during moist ascent. In reality, however, parcels in which condensation occurs can lose liquid water as precipitation, and thus are open systems. In what is called a *pseudoadiabatic* process it is assumed that all condensed water is immediately removed from a parcel. Such a process is adiabatic in the sense that $Q = 0$, but the mass of water in the parcel is not constant. Equation (6.118) can be modified to apply to such a process by replacing w_t with w_s so that

$$(w_s c_w + c_{pd}) \frac{1}{T} \frac{dT}{dt} + \frac{d}{dt} \left(\frac{\ell_v w_s}{T} \right) - \frac{R_d}{p_d} \frac{dp_d}{dt} = 0. \tag{6.135}$$

With this equation, unlike Eq. (6.118), it is not necessary to know the total water mixing ratio w_t to obtain a relationship between temperature and pressure during saturated isentropic ascent because $w_s = w_s(T, p)$. Equations (6.118) and (6.134) (for $dS/dt = 0$) have the form

$$\frac{1}{T} \frac{dT}{dt} = \frac{1}{c_p} \left[\frac{R_d}{p_d} \frac{dp_d}{dt} - \frac{d}{dt} \left(\frac{\ell_v w_s}{T} \right) \right], \tag{6.136}$$

where c_p is $c_{pd} + w_t c_w$ (adiabatic) or $c_{pd} + w_s c_w$ (pseudoadiabatic). For an ascending parcel (decreasing p_d) the right side of Eq. (6.136) is negative, and hence, temperature decreases. But the rate of decrease is higher in pseudoadiabatic ascent because the quantity c_p is less than for adiabatic ascent, in which all water is carried with the parcel (closed system). To illustrate this, the temperature associated with saturated ascent from 900 to 200 hPa of a parcel that is initially at 25 °C is shown in Fig. 6.4 for both adiabatic and pseudoadiabatic (isentropic) processes. These curves were obtained by numerically integrating Eq. (6.135). For an adiabatic process the temperature can be as much as 3 °C higher than for a pseudoadiabatic process.

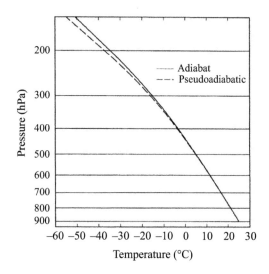

Fig. 6.4: Temperature as a function of pressure for adiabatic and pseudoadiabatic ascent from initial conditions of 900 hPa at 25 °C. For adiabatic ascent the total water mixing ratio w_t is fixed at the saturation value for 900 hPa and 25 °C.

Natural atmospheric processes lie somewhere between saturated adiabatic and pseudoadiabatic, and hence, we must consider both extremes unless we can accurately predict the liquid water content of a parcel upon ascent. For some shallow clouds, precipitation may be negligible, and the adiabatic (closed system) assumption may apply in regions of the cloud not affected by entrainment. In deep clouds, however, precipitation can substantially deplete liquid water. For such clouds the temperature difference between the saturated adiabatic and pseudoadiabatic processes can result in substantial differences in density between a rising parcel and its surroundings. These differences, however, are partially offset by the increase in the parcel density due to liquid water. But even for deep cumulus clouds one would not expect all liquid to be converted to precipitation. The liquid water content of clouds can be estimated as $w_b - w_s(T, p)$, where w_s is the saturation mixing ratio for adiabatic ascent and w_b is the mixing ratio at cloud base. This liquid water content, called the *adiabatic liquid water content*, is an upper limit to the water content of a cloud.

Before proceeding, we clarify our assertion about the liquid water contribution to parcel density ρ:

$$\rho = \frac{M_d + M_v + M_w}{V}. \tag{6.137}$$

The density of air in a parcel is

$$\rho_{air} = \frac{M_d + M_v}{V} = \frac{p}{R_d T_v} \tag{6.138}$$

and hence, Eq. (6.132) can be written

$$\rho = \rho_{air}\left(1 + \frac{M_w}{M_d + M_v}\right) = \rho_{air}(1 + q_w), \tag{6.139}$$

where q_w is the liquid water specific humidity. For fixed air density, parcel density increases with liquid water, whereas for fixed liquid water, parcel density decreases with water vapor (increased virtual temperature T_v). Both vapor and liquid can affect buoyancy forces on cloud parcels because these forces depend on density differences rather than on absolute values. The quantity $1+q_w$ in Eq. (6.139) sometimes is called the *liquid water loading term*.

Equivalent potential temperature can be defined for a pseudoadiabatic ($dS/dt = 0$) process, where $c_p = c_{pd} + w_s c_w$. Unfortunately, the resulting equation cannot be integrated analytically to obtain a simple expression for the pseudoadiabatic potential temperature because c_p is not constant. One advantage of the pseudoadiabatic process, however, is that it can be represented graphically because of the unique relation between temperature and pressure in this process. In an adiabatic reversible process undergone by a closed parcel, the relation between temperature and pressure depends on the total water content w_t.

A prime motive for determining the temperature of a parcel during saturated ascent (or descent) is to estimate the difference between its temperature and that of its surroundings. Such differences are needed to estimate density differences, and hence, buoyancy forces on parcels. Although the temperature of a parcel can be obtained by numerically integrating Eq. (6.135), calculations can be simplified by using Θ_e, which is conserved in an isentropic process both before and after saturation occurs.

If we retain only the first two terms in an expansion of the exponential function in Eq. (6.124) and replace Θ_d by Θ, we obtain

$$\Theta_e \approx \Theta + \frac{\Theta}{T}\frac{\ell_v}{c_p}w_s. \tag{6.140}$$

Θ_{es} is approximated by a similar equation. Now subtract Θ_e of the parcel from Θ_{es} of its (possibly unsaturated) surroundings:

$$\Theta_e - \Theta_{es} = \Theta_p - \Theta_s + \frac{\ell_v}{c_p}\left(\frac{\Theta_p}{T_p}w_s(T_p) - \frac{\Theta_s}{T_s}w_s(T_s)\right), \tag{6.141}$$

where the subscripts p and s denote parcel and surroundings, respectively. Because both are at the same pressure, $\Theta_p/T_p = \Theta_s/T_s$, and hence, Eq. (6.141) can be written

$$\Theta_e - \Theta_{es} = \Theta_p - \Theta_s + \frac{\ell_v}{c_p}\frac{\Theta_s}{T_s}\left[w_s(T_p) - w_s(T_s)\right]. \qquad (6.142)$$

The difference between the temperatures of parcel and surroundings is sufficiently small compared with their magnitudes that we can expand w_s in a Taylor series and neglect all terms but the first:

$$w_s(T_p) - w_s(T_s) \approx (T_p - T_s)\frac{\partial w_s}{\partial T}. \qquad (6.143)$$

With this equation and $(T_p - T_s)/T_s = (\Theta_p - \Theta_s)/\Theta_s$, we can write Eq. (6.137) as

$$\Theta_e - \Theta_{es} \approx \Theta_p - \Theta_s + \frac{\ell_v}{c_p}\frac{\partial w_s}{\partial T}(\Theta - \Theta_s). \qquad (6.144)$$

Rearranging terms yields

$$\Theta_p - \Theta_s \approx \frac{\Theta_e - \Theta_{es}}{1 + \beta}, \qquad (6.145)$$

where

$$\beta = \frac{\ell_v}{c_p}\frac{\partial w_s}{\partial T} = \frac{\epsilon \ell_v^2 e_s p}{c_p R_v (p - e_s)^2 T^2}. \qquad (6.146)$$

Thus, the potential temperature (and hence, temperature) difference between a parcel that has reached saturation by isentropic ascent and its surroundings is proportional to the difference between the equivalent potential temperature of the parcel and the saturation equivalent potential temperature of its surroundings. Consequently, the equivalent potential temperature that feeds convection affects the buoyancy of cloud parcels. A higher Θ_e of the air below the base of cumulus clouds will result in greater buoyancy of cloud parcels. Because Θ_e depends on both the temperature and moisture, at higher temperature and moisture content the intensity of moist convection can increase. Thus, Θ_e of air near the surface that feeds cumulus clouds is critical to cloud buoyancy for deep convection or convective circulations like hurricanes.

If Θ_e is conserved, the difference in potential temperature of parcel and surroundings can be estimated using Eq. (6.145) with β calculated at the temperature and pressure of the surroundings. For applications to shallow atmospheric layers in which temperature does not vary greatly, β may be taken to be constant. At 900 hPa and 25 °C, $1 + \beta = 4.6$; at 500 hPa and -10 °C, $1 + \beta = 1.7$. Thus, the equivalent potential temperature difference is greater than the potential temperature difference.

Although the treatment here assumes that Θ_e of the parcel is conserved, this restriction can be relaxed. For example, if the parcel mixes with surrounding air, the equivalent potential temperature of the parcel can be obtained as a mass-weighted average of Θ_e of parcel and surroundings (see Section 6.7).

The potential temperature difference Eq. (6.145) can be used to estimate the density difference between a parcel and its surroundings. Refinements in this estimate can be made by including the effects of water vapor and liquid water on the parcel density. If the parcel temperature is known, the corresponding saturation vapor pressure can be used to calculate the virtual temperature of the parcel. If no liquid water leaves the parcel during ascent, the liquid water content can be estimated, and hence, its contribution to the parcel density. If, however, precipitation removes water from the parcel, estimates of its liquid water content are more difficult to make.

In the derivation of equivalent potential temperature we considered moist ascent of a parcel to a height at which all its water vapor condenses (humidity mixing ratio of zero). Another process that can be considered is the descent of a saturated parcel moist adiabatically to a pressure p_o (usually taken to be 1000 hPa). This may require that water be added to the parcel as it descends in order to maintain saturation. The temperature upon descent to p_o is defined as the *wet-bulb potential temperature* Θ_w. If the parcel under consideration is initially unsaturated, it first would have to be lifted to the lifting-condensation level and then taken moist adiabatically to a pressure p_o with water vapor added to maintain saturation. To obtain an approximation for Θ_w, we assume c_p in Eq. (6.122) is constant. We take the initial state of the parcel to be that at the lifting-condensation level. By definition, $\Theta_{df} = T_f = \Theta_w$. Thus, Eq. (6.122) can be written as

$$c_p \ln \left(\frac{\Theta_w}{\Theta_d} \right) = \frac{\ell_v(T_{LCL})w}{T_{LCL}} \frac{\ell_v(\Theta_w)w_s(\Theta_w)}{\Theta_w}, \tag{6.147}$$

where w is the saturation mixing ratio evaluated at the lifting-condensation level. We can go further by approximating Θ_d as Θ and ignoring the difference between ℓ_v evaluated at T_{LCL} and Θ_w;

$$\Theta_w = \Theta \exp \left[\frac{p_v}{c_p} \left(\frac{w}{T_{LCL}} - \frac{w_s(\Theta_w)}{\Theta_w} \right) \right]. \tag{6.148}$$

This transcendental equation can be solved numerically. Because the argument of the exponential is less than about 0.2, we can further approximate Eq. (6.148) as

$$\Theta_w \approx \Theta + \left[\frac{p_v}{c_p} \left(\frac{\Theta}{T_{LCL}} w - \frac{\Theta}{\Theta_w} w_s(\Theta_w) \right) \right]. \tag{6.149}$$

Again, the ratios Θ/T_{LCL} and Θ/Θ_w are close to unity; so further simplification is possible:

$$\Theta_w \approx \Theta + \frac{\ell_v}{c_p}[w - w_s(\Theta_w)] = \Theta_e - \frac{\ell_v}{c_p} w_s(\Theta_w). \tag{6.150}$$

This equation has the same form as the expression for wet-bulb temperature given by Eq. (6.75). The wet-bulb potential temperature is the potential temperature that a parcel would have if it were cooled by evaporation during saturated descent to p_o. Although the relation between wet-bulb potential temperature and entropy is not as obvious as it is for equivalent potential temperature, Θ_w is also conserved.

Calculating temperature and pressure relations for moist adiabatic ascent becomes more complicated when temperatures drop below $0\,°C$, and hence, ice can form. It is difficult to estimate when freezing occurs and to what extent because subcooled water can exist at temperatures as low as $-40\,°C$. For a pseudoadiabatic process in which all liquid water freezes and is immediately removed from a parcel when its temperature drops below $0\,°C$, Eq. (6.135) is applicable with c_w replaced with c_i, ℓ_v replaced with ℓ_s, and w_s is the mixing ratio for water vapor in equilibrium with ice. An example of how freezing affects the temperature of an ascending saturated parcel is shown in Fig. 6.5 for the same conditions used to produce Fig. 6.4. Because the enthalpy of sublimation is greater than the enthalpy of vaporization, ice formation results in a temperature as much as $2\,°C$ higher than that of a parcel in which liquid subcools, rather than freezes. Temperature increases of this magnitude can result in relatively large increases in buoyancy forces and consequent greater upward acceleration of rising parcels. Thus, the intensity of updrafts in deep cumulus clouds is sometimes observed to increase rapidly as ice begins to form.

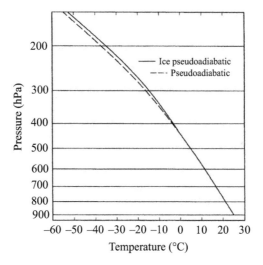

Fig. 6.5: Temperature as a function of pressure for pseudoadiabatic ascents calculated assuming no freezing of liquid water (dashed line) and freezing of all liquid water at temperatures below $0\,°C$ (solid line). The initial state of the parcel is the same as that for Fig. 6.4.

An Overview of Temperatures, Real and Fictitious

We introduced several new temperatures in this chapter. Some of them are not real and cannot be measured. Virtual temperature T_v, for example, is a fictitious temperature that provides a convenient way to calculate the density of moist air from the ideal gas law using the specific gas constant for dry air. An important application of T_v is calculation of the buoyancy force on a moist parcel of air because of the difference between its density and that of its environment. The virtual temperature differences due to a mixing ratio difference of 1 g/kg between a parcel and its environment is equivalent to a temperature difference of about $0.2\,^\circ$C. In a warm, moist environment changes in virtual temperature can affect density as much as temperature can.

The wet-bulb temperature T_w is a real temperature, which, like the dew point, can be measured, and is a useful measure of human comfort. Further, T_w is conserved for isobaric evaporation or condensation process but is not conserved for isobaric warming or cooling. The equivalent temperature T_e, which is not a real temperature, shares the same conservation properties as T_w. This conservation, for example, can be used to estimate the cooling of air below precipitating clouds as the precipitation evaporates where one gram of water evaporating in a kilogram of air cools the air by $2.5\,^\circ$C

Potential temperature Θ, which is not a real temperature, was introduced in Chapter 3 and is conserved for non-saturated adiabatic compression or expansion, but not for isobaric warming or cooling. In this chapter we introduce equivalent potential temperature Θ_e, which is conserved for both unsaturated and saturated adiabatic expansions or compressions and isobaric evaporation or condensation. Thus, it is a useful parameter when considering moist processes in the atmosphere. The wet-bulb potential temperature Θ_w shares the same conserved parameters as Θ_e. These conserved potential temperatures can be used to tag and track air masses and also to quantify mixing processes. For example, Θ_e can be used to treat mixing of dry air that is entrained into the sides and the tops of clouds. In its simplest form, when an unsaturated air mass surrounding a cloud is mixed with a saturated cloud air mass, the mean Θ_e of the mixture is the mass-weighted average of Θ_e in the environment and the cloud. The Θ_e in the lowest layers of the atmosphere is a key factor for deep convection and surface Θ_e can be an effective indicator of global warming. Although several of the temperatures introduced in this chapter are not real, their conserved properties facilitate the analyses of observations and the study of processes in the atmosphere.

6.6   Thermodynamic Diagrams

Thermodynamic processes can be visualized clearly and efficiently with diagrams. For example, in Section 4.4, we used a *pv* diagram to depict an idealized thermodynamic engine; on this diagram, total work done is proportional to the area enclosed by a curve describing a cyclic process. In Section 5.5 we followed the vapor-to-water phase transition along isotherms on a *pv* diagram. In atmospheric applications we often consider processes that involve both energy transformations and phase changes, but the associated thermodynamic diagrams used by meteorologists are more than just a means of depicting such processes. These diagrams also provide a convenient way to estimate various thermodynamic variables, to display vertical thermodynamic structure of the atmosphere obtained from soundings, and to determine stability. Since the advent of computers, the use of diagrams for computations has declined, but they still are used widely for depicting atmospheric structure, visualizing processes, and determining stability.

A Smattering of History

The origins of atmospheric thermodynamic diagrams in their present form can be traced back to the late 1800s. Around this time progress in understanding the thermodynamics of atmospheric processes and unraveling the mysteries of cloud formation were accompanied by a greater need for estimates of temperature changes in saturated ascent. The task at hand was, in essence, solving Eq. (6.136). A quantitative saturated adiabatic theory originated with the work of William Thomson (Lord Kelvin), described in a paper read in 1862 (not published until 1865). But the distinction between saturated and unsaturated ascent was made some 20 years before Thomson's work. James Espy, an American meteorologist, used experimental data to estimate both moist and dry adiabatic lapse rates. He constructed and used a device he called a *nephelescope* (an expansion cloud chamber) to make these estimates. His pioneering contributions to atmospheric thermodynamics can be found in his 1841 treatise *Philosophy of Storms*. Espy is best known for his disagreement with William Redfield about the way winds are distributed in storms. James McDonald, in the second of two excellent historical papers, argues that this controversy may have obscured Espy's important contributions to atmospheric thermodynamics.

Although Thomson established the theoretical basis for describing saturated adiabatic processes, McDonald avers that Joule may have provided the conceptual impetus for Thomson's work. McDonald also notes that Thomson's paper is difficult to follow because it is poorly written and because volumes, rather than masses, are used to formulate the problem. A year before publication of this paper, a Swiss meteorologist, Reye, submitted a paper that elegantly treated saturation theory, including a formulation of and exact solution to the differential equation describing the moist adiabatic process. Reye used his results to determine the stability of a saturated layer and compared it with that of a dry layer. McDonald argues that Reye gave the definitive theory with little room for improvement and that his work was independent of Thomson's.

Despite these early treatments, an 1874 paper by an Austrian meteorologist, Hann, often is cited as the origin of saturation theory, even though his purpose was to summarize previous work of Thomson and Reye and an 1868 paper by a French engineer Peslin. Hann also gave a table of temperatures and pressures for saturated adiabatic ascent.

In 1884 Heinrich Hertz, a physicist best known for his experimental verification of the electromagnetic theory of light and immortalized by the frequency unit, the hertz (Hz), developed a thermodynamic diagram that eliminated the need for tedious calculations of thermodynamic variables. The overall features of this diagram do not differ substantially from those used today.

Although it is not clear why Hertz became interested in the saturated adiabatic process, this may have been a consequence of the interest of his mentor, Helmholtz, in hydrodynamics. McDonald unearthed one reference suggesting that Hertz's work on thermodynamic diagrams was a recreational activity. Hertz included on his diagram moist adiabats calculated as pseudo-adiabats, although he did not refer to them as such.

In 1888 von Bezold, a Bavarian meteorologist, carefully derived a mathematical description of the pseudoadiabatic process. In addition to calculating temperature differences between this process and the adiabatic process, he considered a complete cloud cycle involving pseudoadiabatic ascent followed by dry adiabatic descent. By this process condensational warming (enthalpy of vaporization transformed into a temperature increase) followed by the removal of water through precipitation can result in net warming of the atmosphere. Although global net evaporation of water from the surface is balanced by precipitation over long times, pseudoadiabatic processes yielding precipitation result in warming of the atmosphere, which helps balance net radiative cooling. The work of von Bezold provided the framework for describing atmospheric energetics associated with precipitation.

Although several thermodynamic diagrams were developed following the pioneering work of Hertz, only the skew $T - \log p$ (often referred to briefly as skew-T) and the tephigram are in common use today. We discuss the details of construction for the skew-T diagram; the general approach can be applied to any diagram. Isopleths included on standard thermodynamic diagrams are pressure (isobars), temperature (isotherms), potential temperature (dry adiabats), equivalent potential temperature or wet-bulb potential temperature (moist adiabats), and saturation mixing ratio. Although computers and pocket calculators have diminished the need for diagrams to calculate thermodynamic variables, these tools now make it easier to construct diagrams and display atmospheric data on them.

Skew-T–log p Diagram

The basic coordinates of the skew-$T - \log p$ diagram (also known as the skew emagram) are temperature T and $- \ln p$ (see Fig. 6.6). Isotherms are rotated $45°$ relative to isobars to increase the angle between isotherms and adiabats, thereby facilitating the analysis of stability. Because of the nearly exponential variation of pressure with height, $- \ln p$ is approximately proportional to height. Many skew-T diagrams include a height scale based on the U.S. Standard Atmosphere, which is convenient for estimating heights for data obtained as a function of pressure.

Dry adiabats are curves defined by

$$T = \Theta \left(\frac{p}{p_0} \right)^{R_d/c_{pd}} \tag{6.151}$$

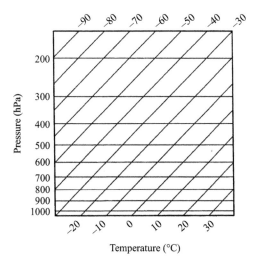

Fig. 6.6: Basic coordinates of the skew-T diagram.

for constant Θ. Because the difference between T and Θ is small compared with either of them, we can rewrite this equation so that the (approximate) shape of dry adiabats is evident:

$$-\ln\left(\frac{p}{p_0}\right) \approx \frac{c_{pd}}{R_d}\left(1 - \frac{T}{\Theta}\right) \tag{6.152}$$

For constant Θ, the curve of $-\ln(p/p_o)$ versus T is approximately a straight line with negative slope (see Fig. 6.7). The $45°$ skew of isotherms results in adiabats about $90°$ to isotherms.

Even with only dry adiabats on a thermodynamic diagram, we can illustrate its use. To estimate potential temperature, for example, plot temperature and pressure; the corresponding potential temperature is obtained directly from the dry adiabat. If a sounding is used to display atmospheric temperature as a function of pressure, stability at each height is evident from the slope of the sounding compared with that of the dry adiabat it intersects.

We must carefully distinguish between lines on a diagram that display atmospheric structure and those that depict processes. Figure 6.7 shows temperature versus pressure in a layer AB, the lapse rate for which is less than dry adiabatic. Air in this layer is therefore stable. This application illustrates atmospheric structure. A process can be illustrated by imagining air to be taken from some pressure within the layer to another pressure; the temperature variation (provided there is no condensation) follows a dry adiabat on the diagram. Figure 6.7 shows adiabatic ascent of air originating at C and terminating at D; the temperature of a parcel along this adiabat can be read directly from the diagram.

Moist adiabats (adiabats for processes undergone by air at saturation) can be included on a thermodynamic diagram by numerically integrating Eq. (6.136) for a

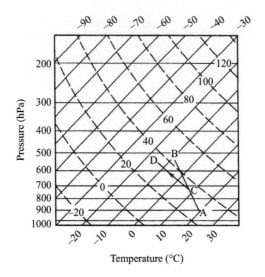

Fig. 6.7: Skew-T diagram of Fig. 6.6 with dry adiabats (dashed) added. The line segment AB shows the temperature structure for a hypothetical sounding between about 500 and 900 hPa. Dry adiabatic ascent from 725 hPa is illustrated by the line segment CD.

pseudoadiabatic process $(c_p = c_{pd} + w_s c_w)$. These adiabats often are labeled with Θ_w values because the temperature at 1000 hPa is the wet-bulb potential temperature. Moist adiabats as they appear on a skew-T diagram are illustrated in Fig. 6.8. At low pressures and temperatures moist adiabats become parallel to dry adiabats. Each moist adiabat could be labeled with a Θ_e value corresponding to the dry adiabat the moist adiabat converges to at low pressure.

From the definition of mixing ratio at saturation $(w = w_s)$, it follows that

$$p = e_s(T) + \frac{\varepsilon e_s(T)}{w_s} . \tag{6.153}$$

For constant w_s Eq. (6.153) defines a p–T curve, which can be depicted on a skew-T diagram. Lines of constant saturation mixing ratio, which are approximately straight, are shown in Fig. 6.9. The dew point of unsaturated air at any pressure p determines the mixing ratio w because $w = w_s(T_d, p)$. The moisture structure of a sounding can be illustrated by plotting dew point or w as a function of pressure. In unsaturated ascent or descent of a parcel w is conserved. The dew point variation in such a process can be determined from the skew-T diagram by following the constant-mixing-ratio line and finding the temperature at each pressure. This is shown in Fig. 6.9 for a layer with mixing ratio profile AB. If air at the middle of this layer (C) is displaced adiabatically to a lower pressure, the variation in dew point can be determined along CD because w is constant (in the absence of condensation).

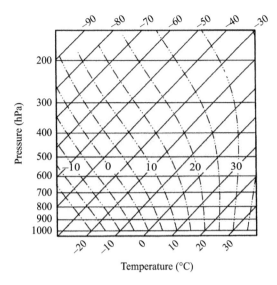

Fig. 6.8: Skew-T diagram of Fig. 6.6 with moist adiabats (chain-dashed) added; the isopleths are labeled with the value of the wet-bulb potential temperature (temperature at 1000 hPa).

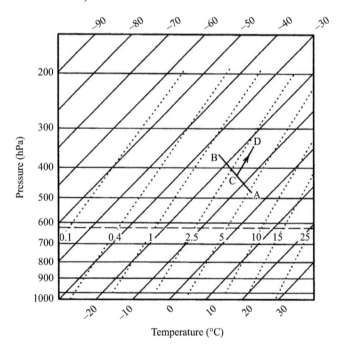

Fig. 6.9: Skew-T diagram of Fig. 6.6 with constant mixing ratio (g/kg) isopleths (dotted) added. The horizontal dashed line is at 622 hPa. At this level the saturation mixing ratio (expressed in g/kg) is approximately numerically equal to the saturation vapor pressure (expressed in hPa).

Mixing ratio is approximately $\epsilon e/p$, where $\epsilon = 0.622$. To express w in g/kg instead of kg/kg multiply by 1000:

$$w \approx 622\frac{e}{p}. \tag{6.154}$$

Thus, at $p = 622$ hPa, the value of w (in g/kg) is numerically equal to e (in hPa), which provides a quick way to estimate saturation vapor pressure from a skew-T diagram. For example, in Fig. 6.9 follow the $0\,°C$ isotherm until it intersects the 622 hPa isobar. By interpolation, the corresponding (saturation) mixing ratio is about 6 g/kg, and thus the saturation vapor pressure at $0\,°C$ is about 6 hPa (6.11 hPa is the correct value).

A skew-T diagram with all isopleths previously described is shown in Fig. 6.10. To provide more resolution, actual diagrams have isopleths at more frequent intervals. Consequently, these diagrams may appear more cluttered than that in Fig. 6.10.

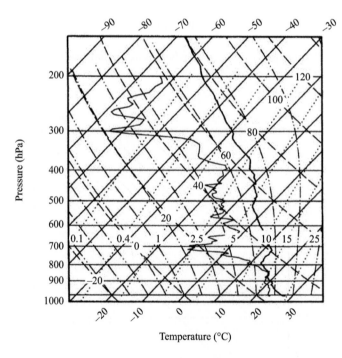

Fig. 6.10: A complete skew-T diagram obtained by combining Figs. 6.7–6.9. The thick solid line shows a sounding of temperature; the thin solid line is the corresponding dew point.

With both moist and dry adiabats on the same diagram, the asymptotic approach of a moist to a dry adiabat at low pressure is evident. To illustrate how atmospheric temperature and moisture structure can be depicted on a skew-T diagram, we show a sounding of temperature and dew point as a function of pressure obtained from a ship in the central equatorial Pacific. Two variables are required to specify the thermodynamic state of unsaturated air at a given pressure. For saturated air only one variable is required (liquid water is not represented on a skew-T diagram). A detailed stability analysis of the sounding shown in Fig. 6.10 is deferred to the following section, but for the moment we note the stable inversion layer at 850 hPa. Above this inversion the mixing ratio drops markedly.

Relations among many of the thermodynamic variables discussed in this and previous chapters can be illustrated on thermodynamic diagrams, in particular, a skew-T diagram (Fig. 6.11). Isotherms and most isobars have been removed for clarity.

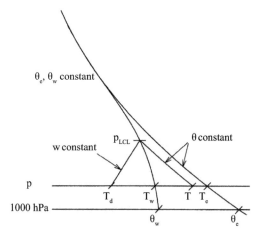

Fig. 6.11: Depiction of thermodynamic variations on a skew-T diagram from which all but the essential isopleths are omitted.

Consider air with pressure p, temperature T and dew point T_d, which uniquely determine the mixing ratio and potential temperature. If a parcel characterized by these two thermodynamic variables is lifted adiabatically, during ascent they follow the appropriate w and Θ isopleths. Temperature and dew point decrease until the two are equal at the lifting condensation level (LCL). From Eq. (6.13) and the definition of dew point it follows that $w =$ constant is a curve of dew point versus pressure. Thus, the LCL is where the Θ and w isopleths extending from the original temperature and dew point intersect. As ascent continues past the LCL, temperature and mixing ratio follow a moist adiabat.

The equivalent potential temperature of a parcel can be estimated by following the moist adiabat to low pressure and obtaining the Θ value for the dry adiabat to which the moist adiabat converges. If this adiabat is not labeled with a value (on the diagram), follow it down to 1000 hPa and estimate the temperature at this level (see Fig. 6.11). If, after ascent to high level, the parcel descends along the dry adiabat to the original pressure rather than 1000 hPa, the resulting temperature is the equivalent temperature T_e. This (adiabatic or, more correctly, pseudoadiabatic) equivalent temperature differs from the (isobaric) equivalent temperature discussed in Section 6.4. Unlike that equivalent temperature, the pseudoadiabatic equivalent temperature is defined by a real process. A parcel is lifted from some initial level dry adiabatically until it reaches saturation. Above this level, temperature decreases at the moist adiabatic rate and any condensed water is removed. The parcel is lifted to low pressure until all its water vapor condenses and then is returned dry adiabatically to the original pressure. The final temperature is the (pseudoadiabatic) equivalent temperature. Although this process is not isobaric, the initial and final pressures are the same.

Wet-bulb temperature and wet-bulb potential temperature also can be estimated on a skew-T diagram (see Fig. 6.11). Imagine the parcel to be lifted dry adiabatically to the LCL, then taken moist adiabatically back to its original pressure. Water must be added to the parcel to maintain saturation. The temperature at this level is sometimes called the adiabatic wet-bulb temperature and differs from the (isobaric) wet-bulb temperature defined in Section 6.4. The term *adiabatic wet-bulb tempera-ture* is confusing given that the process defining the (isobaric) wet-bulb temperature is also adiabatic. To add to the confusion, the process defining Θ_w is really pseudoadiabatic because water vapor is added to the parcel (it is not a closed system). If the moist adiabat is followed down to 1000 hPa, the corresponding temperature is the wet-bulb potential temperature (see Fig. 6.11). Representing wet-bulb temperature on a skew-T diagram again illustrates that $T_d \leq T_w \leq T$.

The work done in a cyclic process represented by a closed curve on a skew-T diagram is proportional to the area enclosed by this curve. To show this, note that the work done in such a process can be written as [see Eq. (4.146)]

$$\oint p \, dv = \oint d(pv) - \oint v \, dp = -\oint v \, dp. \qquad (6.155)$$

By using the ideal gas law Eq. (6.155) becomes

$$\oint p \, dv = -\oint \frac{RT}{p} dp = R \oint T(-d \ln p). \qquad (6.156)$$

The skew-T diagram has coordinates T and $-\ln p$, and hence Eq. (6.156) shows that work done is proportional to the area enclosed on this diagram; the skew of the axes merely changes the proportionality factor.

Tephigram

The tephigram (also known as the $T-s$ diagram) derives its name from the coordinates temperature T and entropy ($c_p \ln \Theta$), sometimes denoted by ϕ. The basic coordinates for this diagram are shown in Fig. 6.12; the abscissa is temperature, and the ordinate is $\ln \Theta$. Isobars are generated from the Poisson relation; moist adiabats and saturation mixing lines are calculated as for the skew-T diagram but transformed to $T-\ln \Theta$ coordinates. A complete diagram is shown in Fig. 6.13. If this diagram is rotated $45°$ clockwise, it appears similar to the skew-T diagram, but isobars are slightly curved and dry adiabats are straight lines.

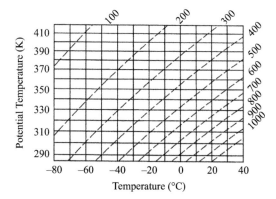

Fig. 6.12: Basic coordinates of the tephigram. Isobars are dashed lines.

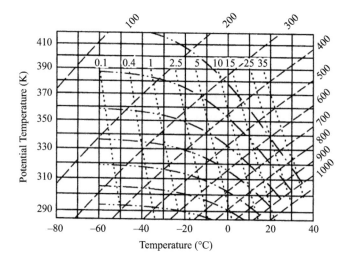

Fig. 6.13: Tephigram. Dotted lines are isopleths of mixing ratio (g/kg). Chain-dashed lines are moist adiabats. Although these adiabats are not labeled, the Θ_e values for them can be obtained from the Θ values they converge to at low pressure, and their Θ_w values are the temperatures at 1000 hPa. This diagram was designed to give good resolution of tropical soundings. As a result, temperatures at 1000 hPa range 12–40 °C, which is not adequate for many mid-latitude applications.

The area on a tephigram enclosed by a closed curve is proportional to the work done in the cyclic process represented by the curve. To show this, use the definition of potential temperature

$$c_p \ln \Theta = c_p \ln T - R \ln p + \text{ const.} \qquad (6.157)$$

in Eq. (6.154)

$$\oint p\, dv = R \oint T(-d\ln p) = c_p \oint T\, d\ln\Theta - c_p \oint dT = c_p \oint T\, d\ln\Theta. \qquad (6.158)$$

Other Diagrams

Although skew-T and $T-\phi$ diagrams are used extensively for weather forecasting and analysis, depicting thermodynamic structure and processes can be aided by graphs of potential temperature, mixing ratio, equivalent potential temperature, and other derived parameters. For example, Fig. 6.14 shows potential temperature, virtual potential temperature, equivalent potential temperature, saturation equivalent potential temperature, and mixing ratio for the tropical sounding on the skew-T diagram in Fig. 6.10. This example demonstrates the usefulness of the parameters defined in Section 6.5 for displaying vertical structure. The inversion layer evident on the skew-T diagram lies between 2.2 and 2.8 km, where slightly stable layers are evident at 0.8 and 1.4 km. The layer extending from 0.2 to 0.6 km is well mixed, as evidenced by nearly constant Θ_v. The mixing ratio structure is fairly complicated, decreasing by about 3 g/kg above the mixed layer, approximately constant from 1.0 to 1.5 km, and decreasing to a minimum $\partial\Theta/\partial z > 0$ at about 3 km. Although the Θ_e profile reflects features in both the mixing ratio and potential temperature, variability in the vertical structure is dominated by moisture. The saturation equivalent potential temperature, which is a function of temperature (not moisture), mirrors and amplifies the vertical structure in the potential temperature profile. The stable layer centered at 2.5 km is characterized by Θ_{es} increasing with height. A nearly saturated layer is evident at 1.5 km, where Θ_e approaches Θ_{es}.

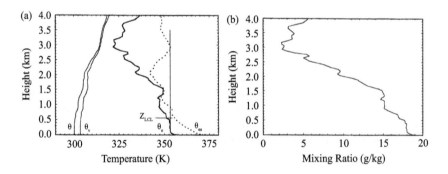

Fig. 6.14: Potential temperature, virtual potential temperature, equivalent potential temperature, saturation equivalent potential temperature (a) and mixing ratio (b) versus height calculated using a sounding over the central equatorial Pacific (shown in Fig. 6.10). The straight vertical line indicates the equivalent potential temperature for a parcel ascending from 0.2 km. The lifting condensation level is indicated by z_{LCL}.

Because equivalent potential temperature is conserved during adiabatic ascent or descent, such processes can be depicted with a diagram on which Θ_e is plotted. For example, if a parcel is lifted adiabatically from 0.2 km (see Fig. 6.14), its equivalent potential temperature is conserved and thus follows the path shown by the vertical line. Above the LCL, the difference between Θ_e of the parcel and Θ_{es} of its surroundings is proportional to their temperature difference (see previous section). If the effects of vapor and liquid on the density of the parcel are neglected, this temperature difference can be used to estimate the buoyancy force on the parcel. For this example a parcel would experience negative buoyancy from the LCL to about 1 km and then positive buoyancy to about 2.7 km. The level at which buoyancy becomes positive is called the *level of free convection* (LFC). The buoyancy force experienced by a parcel increases where $\partial \Theta_{es}/\partial z < 0$, and decreases where $\partial \Theta_{es}/\partial z > 0$.

Although derived thermodynamic parameters are useful for displaying vertical structure and processes, *conserved parameter diagrams* are useful for depicting processes in which mixing may occur. A conserved parameter diagram in common use has equivalent potential temperature Θ_e as abscissa and total water mixing ratio w_t as ordinate (see Fig. 6.15). Because mixing ratio usually decreases with height, the ordinate is labeled with decreasing mixing ratio to display the vertical structure. The usefulness of this diagram is in the treatment of mixing processes. Because w_t and Θ_e are conserved during adiabatic ascent or descent, parcels from different levels can be mixed, and the thermodynamic coordinates of the resulting mixture are mass-weighted averages of those of the parcels. To illustrate this, consider two levels in the atmosphere with thermodynamic coordinates (Θ_{e1}, w_{t1}) and (Θ_{e2}, w_{t2}). If mass M_1 from one level is mixed with mass M_2 from the other level, the conserved parameter coordinates of the mixture are

$$\Theta_e = \chi \Theta_{e1} + (1 - \chi)\Theta_{e2} \qquad (6.159)$$

and

$$w_t = \chi w_{t1} + (1 - \chi)w_{t2}. \qquad (6.160)$$

where the mixing fraction $\chi = M_1/(M_1 + M_2)$. Note the similarity between these equations and Eq. (3.174), obtained for isobaric, adiabatic mixing of parcels with different temperature and vapor pressure. On a $\Theta_e - w_t$ diagram, mixtures with χ varying from 0 to 1 fall on a line extending between the points representing the thermodynamic coordinates of the two layers (see Fig. 6.15). The slope of this line is

$$\frac{\partial w_t}{\partial \Theta_e} = \frac{w_{t2} - w_{t1}}{\Theta_{e2} - \Theta_{e1}}. \qquad (6.161)$$

Although the thermodynamic coordinates of a mixture resulting from adiabatic mixing of parcels from the two layers will fall on this line, other processes can be represented on a $\Theta_e - w_t$ diagram. For example, precipitation can remove water from a saturated layer and decrease w_t, whereas evaporation of rain falling through an unsaturated layer will increase w_t. But both processes leave Θ_e unchanged. If, however, a layer is warmed or cooled by nonadiabatic processes, Θ_e will increase (decrease) while w_t remains constant. These processes are depicted in Fig. 6.15. For a sounding obtained from a rawinsonde or an instrumented aircraft that does not penetrate a cloud, the total water mixing ratio needed for the $\Theta_e - w_t$ diagram is simply the water vapor mixing ratio. If the sounding includes cloudy layers, total water can be estimated if liquid water measurements are available.

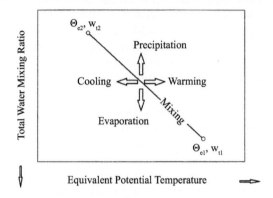

Fig. 6.15: Schematic conserved parameter diagram showing a mixing line and direction of changes associated with nonadiabatic cooling and heating and the addition or removal of water by precipitation and evaporation.

The use of a $\Theta_e - w_t$ diagram is illustrated in Fig. 6.16, where w_t and Θ_e are for the sounding in Figs. 6.13 and 6.14. The open circles represent a level near the surface (A) and at the top of the inversion layer (B). Connecting these two points is the mixing line generated from mixtures of air at these two levels. Despite the

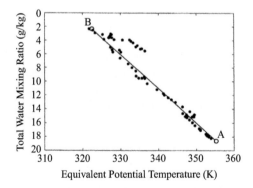

Fig. 6.16: Conserved parameter diagram for the sounding shown in Fig. 6.14. The data points are at 5 hPa intervals. A is near the surface at 1005 hPa (~ 10 m), and B is at about 700 hPa (~ 3 km), near the top of the inversion. Points to the right of the mixing line near B are for the 695–600 hPa level (3–4 km) and indicate a different air mass than in the lowest 3 km.

complicated structure in the mixing ratio and derived temperature plots, the data points closely follow the mixing line. Furthermore, because air above 3 km does not follow the mixing line for the lower levels, processes maintaining the thermodynamic structure of the upper layer are clearly separate from those that maintain the lower layer. This separation was not evident in either Fig. 6.14 or 6.15. A disadvantage of the $\Theta_e - w_t$ diagram is that neither variable is readily interpreted in terms of density.

6.7   Stability and Cloud Formation

The stability criteria discussed in Sections 3.5 and 6.3 were established by imagining a parcel to be displaced slightly from equilibrium and determining if buoyancy would force it back to (stable) or away from (unstable) equilibrium, or neither (neutral). Although density is the relevant thermodynamic quantity for determining stability [Eq. (3.106)], temperature often is more convenient [Eq. (3.108)]. For example, our introduction of virtual temperature in Section 6.3 was motivated in part by the need to account for density variations associated with water vapor. Although stability criteria are defined by temperature gradients at a given height, these criteria often are applied to layers of finite thickness. A *layer* is a region between two heights in which the lapse rate is approximately constant. Inversion layers often are fairly thin, whereas mixing can result in much thicker layers.

The key to static stability is how the density (or temperature) of a parcel changes with height compared with that of its surroundings. In previous treatments, the displaced parcel was assumed to be unsaturated, and hence its temperature changed at the dry adiabatic lapse rate. But if a parcel is saturated, its temperature changes at the moist adiabatic lapse rate. According to the approach described previously for establishing stability, a layer with a lapse rate less than dry adiabatic but greater than moist adiabatic (at a given temperature and pressure) would be stable for displacement of an unsaturated parcel and unstable for a saturated parcel. Such a layer is said to be *conditionally unstable*. The *condition* for the instability is that the parcel be saturated, although an unsaturated layer still can be classified as conditionally unstable depending on its lapse rate. A layer with a lapse rate less than moist adiabatic is classified as *absolutely stable*. Likewise, a layer with a lapse rate greater than dry adiabatic is classified as *absolutely unstable. Absolute* indicates that these criteria apply to both saturated and unsaturated parcels. Keep in mind that lapse rates are rates of decrease, and hence, a greater lapse rate means a more negative temperature gradient.

Sometimes it is stated that the test for conditional instability should be restricted to upward displacements of parcels because the pseudoadiabatic assumption does not allow for saturated downward motions. But this is an unnecessary restriction because water vapor can be imagined to be added to a parcel to maintain saturation during descent in a pseudoadiabatic process. This is not entirely a fiction: Evaporation of cloud droplets or precipitation can maintain saturation, although sometimes precipitation may not evaporate quickly enough to do so.

Although a layer can be tested for conditional instability by calculating the moist adiabatic lapse rate using Eq. (6.116) and comparing this lapse rate with that of the layer, this test also can be made using saturation equivalent potential temperature Θ_{es}, which can be calculated from Eq. (6.130). If the temperature in a layer decreases at the moist adiabatic rate, Θ_{es} is constant with height. On a thermodynamic diagram the temperature in such a layer would follow a moist adiabat. If, however, Θ_{es} in the layer increases with height, the lapse rate is less than moist adiabatic, and the layer is absolutely stable. If Θ_{es} decreases with height, the layer is conditionally unstable. Thus, the stability criteria can be summarized as:

$$\Gamma < \Gamma_w \text{ or } \frac{\partial \Theta_{es}}{\partial z} > 0, \qquad \text{absolutely stable} \qquad (6.162)$$

$$\Gamma_d > \Gamma > \Gamma_w \text{ or } \frac{\partial \Theta_{es}}{\partial z} < 0, \qquad \text{conditionally unstable} \qquad (6.163)$$

$$\Gamma > \Gamma_d \text{ or } \frac{\partial \Theta}{\partial z} < 0 \qquad \text{absolutely unstable} \qquad (6.164)$$

where Γ is the lapse rate in the layer, Γ_d is the dry adiabatic lapse rate, and Γ_w is the lapse rate of the moist adiabat that intersects the layer temperature sounding.

The criteria for conditional instability sometimes are incorrectly specified using Θ_e instead of Θ_{es}. The test for conditional instability involves comparing the sounding lapse rate with the moist adiabatic lapse rate. But Θ_e depends on both temperature and moisture. However, the vertical gradient of Θ_e is used to specify *potential* (sometimes called *convective*) *instability*. This instability can arise if an entire layer (not just a parcel) is lifted by flow over a mountain or in association with fronts and other atmospheric flows. When a layer is lifted, its bottom can become saturated before its top. The layer then can be destabilized as its bottom cools at the moist adiabatic lapse rate while its top cools at the dry adiabatic lapse rate. Thus, even an absolutely stable layer can become conditionally unstable through lifting. This is illustrated in Fig. 6.17, which shows an initially absolutely stable layer AB. As evidenced by its smaller dew point depression, the bottom of this layer is moister than its top, and hence, would reach saturation before the top if the entire layer were lifted. As lifting continues, the layer is destabilized and at its final level is conditionally unstable. As a layer is lifted until it becomes saturated, it is unstable if the lapse rate is greater than the moist adiabatic lapse rate. The equivalent potential temperature in the layer is conserved during lifting both before and after saturation. Once saturation occurs, however, Θ_e of the lifted air is equal to Θ_{es}. When both the top and bottom of the layer are saturated, it is conditionally unstable if Θ_e at the top of the layer is less than at the bottom. Thus, the criterion for potential instability is $\partial \Theta_e / \partial z < 0$.

The discussion of stability given here does not consider the effects of liquid and water vapor on buoyancy. A more exact formulation of stability criteria would require using density. Virtual temperature accounts for water vapor, but the contribution of liquid water to buoyancy can be included only if the liquid water content

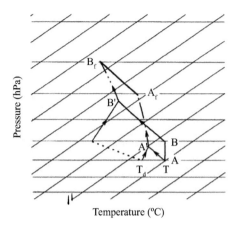

Fig. 6.17: Schematic diagram showing how a layer that is potentially unstable can be destabilized as it is lifted. Solid lines are temperature soundings, the dashed line is a dew point sounding. Lines with arrows indicate processes. Initially the layer AB is absolutely stable. As this layer is lifted, its bottom becomes saturated before its top. The LCL for a parcel beginning at $A(B)$ is A' (B'). Along AA' (BB') a parcel follows a dry adiabat; along $A'A_f(B'B_f)$ a parcel follows a moist adiabat. The final layer $A_f B_f$ is conditionally unstable.

is known. If pseudoadiabatic processes are considered, however, all the liquid water is assumed to be removed, and hence, virtual temperature is sufficient to account for density variations due to water vapor.

Although the static stability tests discussed previously were made by considering vanishingly small displacements, stability associated with finite displacements is more relevant to real convection. Not only the immediate environment from which a parcel originates determines the vertical extent of convection, but also the temperature and moisture structure above that level. To investigate further stability for finite vertical displacements, consider the acceleration due to buoyancy experienced by a parcel (Eq. 3.106):

$$\frac{d^2z}{dt^2} = g\left(\frac{\rho_s - \rho_p}{\rho_p}\right).$$
(6.165)

We can use this equation to determine the work done on the parcel (per unit mass) by the buoyancy force acting over some vertical distance. First, we rewrite Eq. (6.165) in terms of specific volume v and integrate from an initial height z_i to a final height z_f:

$$\int_{z_i}^{z_f} \frac{d^2z}{dt^2}dz = \int_{z_i}^{z_f} g\left(\frac{v_p - v_s}{v_s}\right)dz.$$
(6.166)

This integral depends on the characteristics of both the parcel and its surroundings. With the help of the hydrostatic equation, we can write Eq. (6.166) as

$$\int_{z_i}^{z_f} \frac{d^2 z}{dt^2} dz = - \int_{p_i}^{p_f} (v_p - v_s) dp \, . \tag{6.167}$$

By using the ideal gas law and neglecting the contribution of liquid water to specific volume, we can write Eq. (6.167) as

$$\int_{z_i}^{z_f} \frac{d^2 z}{dt^2} dz = -R_d \int_{p_i}^{p_f} (T_{vp} - T_{vs}) d \ln p \tag{6.168}$$

where T_{vp} is the virtual temperature of the parcel and T_{vs} is that of its surroundings. According to Eq. (6.168), the work done on the parcel by the buoyancy force is proportional to the area between two curves on a skew-T diagram: one curve describing the virtual temperature of the parcel and the other that of its surroundings. Usually, virtual temperature is not displayed on a sounding (although it could be calculated from temperature and specific humidity). Moreover, adiabats on standard skew-T diagrams are not virtual temperature adiabats. Thus, as a practical matter, the area defined by Eq. (6.168) often is estimated from temperature profiles on a skew-T diagram rather than from virtual temperature profiles. If this area is positive, the parcel is (on average) positively buoyant. Keep in mind when you interpret Eq. (6.168) that on a skew-T diagram, pressure decreases along the vertical axis, and hence, the variable of integration in Eq. (6.168) is $-\ln p$.

An example in which a parcel subject to finite displacement can result in a positive value for Eq. (6.168) is shown by a sounding taken in late spring at Fort Worth, Texas in the vicinity of a severe storm that spawned a tornado (Fig. 6.18). A parcel originating from near the surface and lifted to its lifting condensation level (LCL), then along a moist adiabat, is negatively buoyant above its LCL and positively buoyant from its level of free convection (LFC) to the *level of neutral buoyancy* (LNB) at 200 hPa, where the parcel path intersects the temperature sounding (the LFC is the lower and the LNB the upper level of zero buoyancy). This results in a large positive area between the temperature profile of the parcel and that of its surroundings. The temperature difference between parcel and sounding increases as the parcel rises through conditionally unstable layers (e.g. 700–500 hPa) and decreases in absolutely stable layers (above 250 hPa).

Despite the large positive area between the LCL and the LNB for the sounding shown in Fig. 6.18, no deep cumulus clouds were observed when it was taken. The negative area between the LCL and LFC was a potential well that inhibited convection. This sounding was taken in early morning. A mid-afternoon sounding at the same site is shown in Fig. 6.19. Although the temperature profile above 700 hPa is not substantially different from that for the early morning sounding, subsequent warming of air near the surface and from 850 to 750 hPa removed the negative area above the LCL. This sounding would therefore support deep convection, and the tops of resulting clouds would be expected to be near the LNB, although overshooting is likely because the positive area is large. Shortly after this

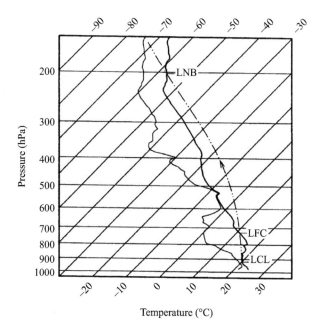

Fig. 6.18: Early morning sounding at Fort Worth, Texas, obtained about nine hours before a severe thunderstorm produced a tornado 150 km to the north. The thick solid line is the temperature sounding; the thin solid line is the dew point sounding. Air near the surface has an LCL as indicated. The path of a parcel following a moist adiabat from the LCL is indicated by the chain-dashed line. The level of free convection (LFC) is where the parcel becomes positively buoyant. The area between the path of the parcel along the moist adiabat and the temperature sounding is positive from the LFC to the level of neutral buoyancy (LNB) at about 200 hPa.

sounding was taken, a severe storm formed about 150 km to the north and produced a tornado. Astute forecasters would have recognized the potential for severe weather indicated in the early morning sounding. Their challenge would have been to predict the changes in atmospheric structure near the surface that allowed convection to ignite later in the afternoon. During the summer in many parts of the United States there is frequently the potential for deep convection. But parcels originating from the surface during the morning may not be able to overcome the region of negative buoyancy often present just above the LCL. As the surface warms during the day, however, mixing can warm and destabilize the lower layers of the atmosphere and trigger afternoon thunderstorms, following which the surface cools at night and the low levels are stabilized. Convection ceases and the cycle begins anew the following day. Anyone who backpacks in the mountains of the western United States learns from experience that blue skies in the morning are likely to give way to intense afternoon thunderstorms, followed, mercifully, by clearing at evening.

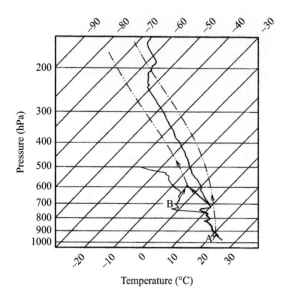

Fig. 6.19: Another sounding from Fort Worth, Texas, taken about eight hours after the sounding shown in Fig. 6.18, near the time when severe thunderstorms developed in the area. A parcel lifted from A experiences no negative buoyancy and a large amount of CAPE. Although there is an inversion layer at 730 hPa, it is insufficient to inhibit deep convection. In contrast, parcels lifted from 700 hPa (B) have an LCL of about 600 hPa and would experience no positive buoyancy past this level.

The positive area obtained by applying Eq. (6.168) to soundings can be used to estimate the potential severity of thunderstorms by providing an estimate of updraft velocities. The integral on the left side of Eq. (6.168) is the change in kinetic energy (per unit mass) of a parcel between the two levels, which can be shown by changing the variable of integration from z to t, where dz/dt is the parcel's vertical velocity V_z:

$$\int_{z_i}^{z_f} \frac{d^2 z}{dt^2} dz = \int_{t_i}^{t_f} \frac{dV_z}{dt} V_z \, dt = \frac{1}{2}V_{z_f}^2 - \frac{1}{2}V_{z_i}^2 . \tag{6.169}$$

If we take the lower limit of integration in Eq. (6.168) as the LFC and the upper limit as the LNB, this integral is the maximum work done by buoyancy, which is given a fancy name, *convective available potential energy*, and its own acronym (CAPE):

$$\text{CAPE} = -R_d \int_{p_{LFC}}^{p_{MB}} (T_{vp} - T_{vs}) \, d\ln p . \tag{6.170}$$

If $V_{zi} = 0$, the maximum vertical velocity the parcel might attain is

$$V_{zf} = \sqrt{2 \, \text{CAPE}} . \tag{6.171}$$

For the soundings shown in Figs. 6.17 and 6.18, the average temperature difference between the sounding and the moist adiabat is approximately $5\,^\circ$C in the layer from 700 to 200 hPa. This gives a CAPE of approximately 1800 J kg^{-1} and a maximum updraft velocity of 60 m s^{-1}, not the conditions most pilots (nor their passengers) would care to fly through. This example, however, is still comparatively mild. CAPE as large as 7000 J kg^{-1} has been reported.

The temperature difference between the parcel and the surroundings in the integral for CAPE in Eq. (6.170) can be expressed as the difference between Θ_e of the parcel and Θ_{es} of the surroundings if virtual effects are neglected. Because Θ_e of the parcel is conserved for both dry and moist ascent, CAPE depends directly on Θ_e in the boundary layer (subcloud layer) since this will be the Θ_e at the LFC. Thus, high Θ_e associated with warm, moist air in the boundary layer will contribute to large CAPE and subcloud layer Θ_e can be a critical indicator for intense convection.

Although CAPE is one indicator of the severity a convective storm might attain, the negative area or the *convective inhibition energy* (CINE) controls whether convection will occur or not. The magnitude of this negative area is

$$\text{CINE} = \text{R}_d \int_{p_i}^{p_{LFC}} (T_{vp} - T_{vs})\mathrm{d}\ln p . \tag{6.172}$$

The initial pressure p_i often is taken to be that near the surface. For a sounding preceding a severe storm (Fig. 6.18), the parcel temperature is about $1\,^\circ$C lower than that of its surroundings between 950 and 850 hPa, and hence, CINE is about 30 J kg^{-1}. A parcel would need an initial upward velocity of about 8 m s^{-1} to penetrate this layer and reach the LFC.

As stated previously, assessing stability for finite displacements involves parcel temperature and that of the surroundings. Consequently, the level from which a parcel is imagined to be lifted has a substantial effect on the values of CAPE and CINE. For example, if a parcel at 700 hPa for the sounding shown in Fig. 6.18 is lifted to its LCL and then up a moist adiabat, the parcel temperature would always be lower than that of its surroundings, and there would be no CAPE, even though the layers at pressures less than 600 hPa are conditionally unstable. In fact, any parcels originating from levels with pressures less than 850 hPa would have no CAPE. A *sounding* (not just a layer) is often said to be *conditionally unstable* or in the conditional state if a parcel from any level of the sounding has positive CAPE. This convention, however, allows for the possibility that a sounding can be conditionally stable even if it has layers that are conditionally unstable.

Thermodynamic stability is treated in many atmospheric science textbooks, from books on general meteorology to atmospheric dynamics. Although the treatment for unsaturated conditions is similar, when phase changes are involved, terminology sometimes diverges. Moreover, one can find a host of ambiguous and confusing terms that use every possible combination of the adjectives "potential," "latent," and "pseudo" to describe different types of stability. We do not attempt to clarify all these terms. Instead, we content ourselves with warning readers to beware and, when in doubt, to adhere to the *Glossary of Meteorology*, as we have done, to resolve any conflicts.

Entrainment

In assessing stability using finite displacements and in evaluating CAPE, we assumed that a parcel does not mix with its surroundings. In reality, as a parcel rises, it can entrain and mix with surrounding air, and as a consequence the maximum possible vertical velocity usually is not realized. For a cloudy parcel surrounded by unsaturated air, entrainment can result in cooling of the parcel by evaporation. Temperature changes associated with entrainment can appreciably change the buoyancy of a cloudy parcel. Furthermore, changes in the water vapor and liquid water content due to entrainment also can affect density.

To quantify the effect of entrainment on the temperature of a parcel, we take advantage of the conservative property of Θ_e discussed in Section 6.5. We begin with a parcel of mass M and equivalent potential temperature Θ_{ep} and consider how the parcel changes as it mixes with surrounding air during ascent. As the parcel ascends a small distance Δz, it mixes with a mass ΔM of air from its surroundings. If no other processes modify the parcel, the equivalent potential temperature of the mixture with mass $M + \Delta M$ is obtained by applying Eq. (6.159) to give

$$\Theta_{ep} + \Delta\Theta_{ep} = \frac{M\Theta_{ep} + \Delta M\Theta_e}{M + \Delta M}, \tag{6.173}$$

where $\Delta\Theta_{ep}$ is the change in the parcel's equivalent potential temperature upon mixing and Θ_e is the equivalent potential temperature of the surroundings. This equation is valid even if water is removed from the parcel by precipitation or if there is evaporation of water in the parcel as drier air is entrained. Divide Eq. (6.168) by Δz, rearrange terms, and take the limit as $\Delta z \to 0$:

$$\frac{d\Theta_{ep}}{dz} = -\frac{1}{M}\frac{dM}{dz}(\Theta_{ep} - \Theta_e). \tag{6.174}$$

The quantity $(1/M)(dM/dz)$ is an entrainment parameter, positive because the mass of an entraining parcel increases as it rises. This entrainment parameter has dimensions of inverse length, and hence, can be written as $1/H_e$, where H_e is a length scale for the entrainment process. Laboratory experiments indicate that H_e is proportional to the radius of a rising plume. If the entrainment equation is applied to clouds, H_e often is assumed to be proportional to cloud radius. This conforms to our expectation that the larger a cloud is, the more it is protected from buoyancy reduction by entrainment of air at its boundary.

As an entraining parcel rises, its equivalent potential temperature is not conserved but becomes closer to that of the surroundings. If, for example, Θ_e of the surroundings decreases with height, so does that of the parcel. The effect of this decrease on the buoyancy of a cloudy parcel can be quantified by applying Eq. (6.145), according to which buoyancy is proportional to the difference between Θ_e and Θ_{ep}. The effect of entrainment is illustrated in Fig. 6.20 using the tropical sounding of Fig. 6.14a, which showed a constant Θ_e path for a nonentraining parcel originating from 200 m and rising to 3 km. The path followed by an entraining parcel is included on Fig. 6.20 by applying Eq. (6.172) with $H_e = 2$ km. As expected,

Θ_e of the parcel decreases with height, along with that of its surroundings. This reduction of Θ_e results in less buoyancy. Entrainment tends to diminish CAPE as well as lower the level of neutral buoyancy and cloud top height. Because of entrainment, the smaller the cloud, the lower the height of its top (in a given environment). Small, shallow clouds often are seen in the vicinity of large, deep clouds, but skinny cumulus clouds rarely extend to the tropopause.

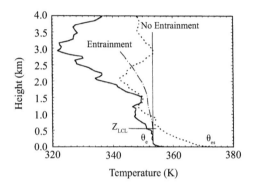

Fig. 6.20: Θ_e and Θ_{es} for the tropical sounding in Fig. 6.14a with Θ_e paths for nonentraining and entraining parcels originating from 200 m. The difference between the parcel Θ_e and Θ_{es} of the surroundings is proportional to the buoyancy force on a parcel. An entrainment length H_e of 2 km was used to calculate Θ_e for the entraining parcel.

In the preceding discussion we showed that the density of a rising cloud parcel can be reduced by entrainment. When we used the parcel method for this analysis, we implicitly assumed horizontally uniform thermodynamic and dynamic properties at any level in the cloud. This approach may push the limits of the simple parcel approach, however, because in reality entrainment can result in horizontal inhomogeneities within clouds. The cauliflower appearance of cumulus clouds suggests that entrainment at their tops and sides is anything but a smooth, uniform process. Furthermore, pilots (and especially their queasy passengers) can attest to the presence of substantial fluctuations in vertical motion (turbulence) within cumulus clouds. Entrainment can contribute to these fluctuations.

As unsaturated air is entrained into the top of a cumulus cloud, evaporative cooling can result in parcels with a higher density than the surrounding cloud (see Fig. 6.21). These parcels, mixtures of entrained and cloudy air, can penetrate through some depth of the cloud as downdrafts. Parcels can remain colder than their environment (the cloud) because if they remain saturated, their temperature increase upon descent (following a moist adiabat) is the same as that of the cloud. But as with positively buoyant parcels, negatively buoyant parcels within a cloud can be further diluted, and their negative buoyancy reduced as they mix with their environment upon descent. This effect of dilution will be greater the smaller the parcel.

Negatively buoyant parcels can be generated by entrainment into clouds even though the air surrounding the cloud is usually at a higher temperature. To show this we compare the potential temperature of a parcel composed of a mixture of entrained and cloud air with the potential temperature of the cloud at that level (Fig. 6.21). In this treatment we ignore the density difference between the parcel and the cloud due to vapor and liquid water differences, although sometimes it may be important to account for them. We can use the conserved properties of Θ_e to determine the equivalent potential temperature of a mixture as a mass-weighted average of Θ_e of the cloud and surroundings. From Eq. (6.159) we have

$$\Theta_{em} = \chi\Theta_e + (1-\chi)\Theta_{ec}, \tag{6.175}$$

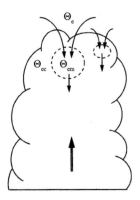

Fig. 6.21: Schematic diagram showing entrainment of air into a cumulus cloud and the formation of mixtures of entrained and cloudy air that can be negatively buoyant relative to the cloud. The dimensions of the mixtures vary depending on the processes involved in the entrainment. The equivalent potential temperature of the environment is denoted by Θ_e, that of the cloud by Θ_{ec}, and that of the mixture by Θ_{em}.

where Θ_e (no subscript) is the equivalent potential temperature of the cloud surroundings, the subscript m denotes the mixture produced by entrainment, the subscript c denotes the cloud, and χ, the mixing fraction, is defined here as the ratio of the mass entrained to the mass of the resulting mixture. For no entrainment χ is zero. Although sufficiently large values of χ may result in unsaturated mixtures, we restrict our treatment to mixtures (parcels) that remain saturated, so that for both mixtures and surrounding cloud air $\Theta_e = \Theta_{es}$. We thus can use Eq. (6.140) to write the potential temperature difference between the mixture (entrainment parcel) and the cloud as

$$\Theta_m - \Theta_c = \frac{\Theta_{em} - \Theta_{ec}}{1+\beta}. \tag{6.176}$$

By combining Eqs. (6.175) and (6.176), we can write

$$\Theta_m - \Theta_c = \frac{\chi(\Theta_e - \Theta_{ec})}{1 + \beta} . \tag{6.177}$$

Because χ and β are positive, it follows from this equation that the potential temperature (and hence, temperature) of the mixture will be less than that of the cloud if the equivalent potential temperature of the unsaturated air surrounding the cloud is less than that of the cloud. This condition can be met even if the temperature of the air surrounding the cloud is higher than that of the cloud. Because Θ_e usually decreases with height in the lower troposphere, as evidenced by the tropical sounding shown in Fig. 6.14a, the conditions required for penetrative downdrafts in cumulus clouds by entrainment often are satisfied. Entrainment of dry air locally can produce negatively buoyant parcels relative to the cloud as a whole. On average, the cloud may be rising, but within the cloud some parcels are going up and others down. This is analogous to an ascending elevator filled with people who are jumping up and down.

Although our previous treatment of entrainment focused on cumulus clouds, stratiform clouds also can entrain unsaturated air at the cloud top. Stratocumulus clouds, which are maintained by turbulent mixing, are capped by inversions. Thus, the cloudy air below such an inversion is denser than the air above, and hence, energy is required (e.g. from a shear flow) to entrain unsaturated air from above the inversion into the cloud. Once air is entrained, evaporative cooling can produce penetrative downdrafts.

To illustrate conditions that may support the production of negatively buoyant parcels due to entrainment in stratocumulus clouds, we show observations made off the coast of California, where stratocumulus clouds are observed frequently during the summer. The cloud-top structure is shown in a high-resolution satellite image (Fig. 6.22). In addition to the variations in cloud top structure over length scales of 5–10 km shown in this image, there is substantial structure at smaller scales (features smaller than 1 km are present), indicating the turbulent nature of the cloud and the possibility for entrainment at various scales of motion.

The thermodynamic structure associated with this cloud (Fig. 6.23) was obtained from an instrumented aircraft flying near the area shown in the satellite image. The stable layer at the top of the cloud is indicated by the sharp 10 K increase in potential temperature. From the approximate relation obtained from Eq. (6.134)

$$\Delta\Theta_e \approx \Delta\Theta + \frac{\ell_v \Delta w}{c_p} \tag{6.178}$$

it follows that although the *increase* $\Delta\Theta$ in potential temperature contributes to an increase in Θ_e at the cloud top, the *decrease* Δw in mixing ratio is sufficient to counter this increase and results in a net decrease in Θ_e across the inversion. For this example, entrainment can produce negatively buoyant parcels even though air above the inversion is 10 K warmer than in the cloud.

Fig. 6.22: High-resolution satellite (LANDSAT) image of stratocumulus clouds off the coast of California. This image is for an area 60 km × 60 km.

Although Eq. (6.177) indicates that the negative buoyancy of an entraining parcel increases linearly with mixing fraction when $\Delta\Theta_e$ across cloud top is negative, this result holds only if the mixture remains saturated. But large mixing fractions can result in unsaturated mixtures. The largest negative buoyancy possible is for a mixing fraction that gives a just saturated parcel (no liquid remaining). Any increase in the mixing fraction above this value will result in a decrease in the negative buoyancy of the parcels because unsaturated air above the inversion has a lower density than the cloudy air and the increase in density with mixing is due to evaporative cooling. Consequently, the mixing fraction that gives a just-saturated mixture increases as the cloud liquid water content increases. Thin stratus clouds with little liquid water or clouds depleted of liquid water by precipitation are less likely to generate penetrative downdrafts by entrainment than are thick, juicy clouds.

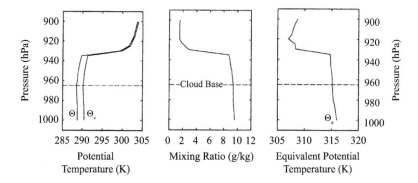

Fig. 6.23: Potential temperature, virtual potential temperature, mixing ratio, and equivalent potential temperature obtained from a sounding made with an instrumented aircraft in a stratocumulus deck off the coast of California. Measurements were made in an area close to that corresponding to the satellite image in Fig. 6.22.

6.8   Mixing Clouds

A common winter experience for those who live in cold climates is to see clouds form on our breath. These clouds, and many others like them, often are called steam, a misnomer given that steam is water vapor at a temperature above the boiling point. We can't see steam, but we can see light scattered by water droplets. That is, we can see water only when it has ceased to be vapor. The clouds formed on our breath are sometimes called steam clouds or, better yet, *mixing clouds*. Although this term is not in everyday use, unlike steam and steam clouds, the term *mixing clouds* provides a clue to their physical origin.

In our experience, it is virtually impossible to restrain students, even by resorting to electroshock and other diabolical forms of torture, from explaining the mixing clouds they exhale by saying that their warm breath is cooled, and, as everyone knows, cold air can't hold as much water vapor, etc., etc., *ad nauseam*. This is a consequence of students having been told countless times that clouds result from cooling. Mixing clouds, however, are an exception. Although it is not incorrect to say that your breath cools upon coming into contact with surrounding colder air, it is equally correct to say that this air is warmed by your breath. But neither assertion comes to grips with why mixing clouds form. Heating and cooling per se are irrelevant. As their name implies, mixing clouds are formed by mixing different parcels of air. Because of the shape of the saturation vapor pressure curve for water, two parcels with different temperature and vapor pressure, both subsaturated, can mix to form a supersaturated parcel.

All the words and equations in the world won't serve you nearly so well as invoking a simple diagram to explain why mixing clouds form in some circumstances, but not in others. If you find yourself trying to concoct an explanation by waving your arms and shouting but not making a crude sketch, you have taken the wrong path to enlightenment about mixing clouds.

We can show the moisture states of air by means of an $e-T$ diagram (Fig. 6.24) on which the saturation (equilibrium) curve divides all the subsaturated from all the supersaturated states. Remember that supersaturation is not impossible; it just is not something we commonly experience. Denote by A and B the states of two parcels of air, both of which are unsaturated. We showed at the end of Section 3.7 that in isobaric, adiabatic mixing of two parcels (without condensation), the final state of the mixture lies on a straight line joining the states of its two component parcels. The precise point on this line depends on the relative masses of the two parcels. Regardless of these masses, the line joining A and B lies below the saturation curve. But suppose that moist air represented by point A were to be mixed with moist air represented by point C (Fig. 6.24). The straight line joining these two points lies partly above the saturation curve. For the moment, assume that the air is free of condensation nuclei. Because AC lies above the saturation curve, the door is open to possible supersaturation of the mixture. Now imagine dropping a few condensation nuclei into such a supersaturated mixture: The result will be a mixing cloud. Should we say that A was cooled or that C was heated? It is a matter of taste, not of physical necessity. What is necessary is that the set of possible intermediate states lying between A and C fall on a straight line, whereas the saturation curve be concave. If the saturation curve were a straight line, mixing clouds could not form, which is why we said that heating and cooling per se are irrelevant to their formation.

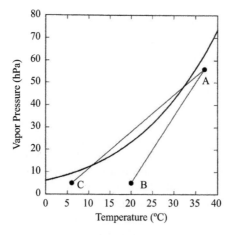

Fig. 6.24: Vapor pressure and temperature states of moist air. The thick curve is the equilibrium (saturation) curve. When two parcels the states of which are represented by A and B mix adiabatically and isobarically, the state of the mixture lies on the line between A and B, below the saturation curve. But when air at A mixes with air at C, supersaturation is possible, and a cloud may form.

Once a mixing cloud forms, the assumption of no condensation underlying the linearity of curves like AB is no longer valid. Condensation warms air and removes water vapor from it, thereby changing the state of the mixed parcel. But this in no way affects our ability to predict the existence of mixing clouds. By drawing lines on an $e - T$ diagram between points representing two different parcels of air, all we can say is that cloud formation upon mixing of these two parcels is possible or impossible. We cannot say how fast the cloud will form or dissipate, how large its droplets will be, and so on. By simple thermodynamics, we can establish only whether or not a mixing cloud is possible.

We noted at the beginning of this section that our attempts to disabuse students of the notion that mixing clouds form because moist air is cooled have mostly failed. Here, as elsewhere, it is very difficult to unteach what has been taught students almost from their infancy, namely, that air must be cooled for clouds to form. An amusing footnote to this is that at one time the conventional wisdom was quite the opposite. In 1784, the eminent Scottish geologist James Hutton (b. Edinburgh, Scotland, 1726; d. 1797) presented a paper on his theory of rain to the Royal Society of Edinburgh. Solely on the basis of the existence of clouds formed on breath in cold weather, Hutton deduced the shape of the curve of saturation vapor pressure (which he called the "dissolving power" of air) versus temperature. Whereas we explained mixing clouds as resulting from the shape of the Clausius–Clapeyron equation, Hutton reasoned in the opposite direction. This was a remarkable achievement. But he went too far by invoking mixing as the fundamental cause of rain and snow. Such was his reputation that his theory of rain prevailed for nearly 50 years until it was demolished by Luke Howard in 1833. And, as so often happens, one monolithic theory was replaced by another. Nature, however, often takes more than one path to the same end.

6.9   Cloud Formation on Ascent and Descent

Among the many advantages of abandoning childish notions about the alleged holding power of warm air is that this opens up a world of possibilities beyond the wildest dreams of sponge theorists. The thoughts you are able to entertain about atmospheric phenomena are limited by your mental models of how the atmosphere works. If these models are faulty or incomplete, so will be your thinking.

When a parcel of moist air rises, we are told, it expands, cools, and can't hold as much water vapor; so the excess condenses as a cloud. A better explanation is to point out that saturation vapor pressure decreases with decreasing temperature, so that all else being equal, the relative humidity of an ascending parcel

$$r = \frac{e}{e_s} \tag{6.179}$$

increases because e_s decreases. But this is not the entire story because all else is *not* equal, which is rarely noted. In an expansion the vapor pressure e in Eq. (6.179) also decreases. For r to increase therefore requires e to decrease more slowly in an expansion than e_s. If this condition is not met, a cloud cannot form in ascent, but *can* form in descent. Cloud formation on ascent is what we have come to consider the norm, but it is not inevitable. Whether clouds form on ascent or descent is determined by the slopes of the $e(T)$ and the $e_s(T)$ curves, which in turn depend on thermodynamic variables:

$$\frac{de}{dT} = \frac{\gamma}{\gamma - 1} \frac{e}{T}, \qquad \frac{de_s}{dT} = \frac{\ell_v}{R_v T} \frac{e_s}{T}. \tag{6.180}$$

What we have come to consider the normal state of affairs is shown in Fig. 6.25a. A subsaturated parcel of air at an initial temperature of 28 °C cools upon adiabatic expansion (ascent) to a temperature of about 15 °C, at which point the vapor pressure is equal to the saturation vapor pressure. That is, the two curves $e(T)$ and $e_s(T)$ intersect. But this intersection occurs only because of the particular values of the enthalpy of vaporization and gas constant of water vapor and the ratio of specific heats of air. None of these values is dictated by law, and hence, could be otherwise, with profound consequences for cloud formation. For example, suppose that γ for air were 1.04 instead of 1.4. With this assumption, an unsaturated parcel at an initial temperature of 20 °C warms upon adiabatic compression (descent) to a temperature of about 25 °C, at which point the vapor pressure is equal to the saturation vapor pressure (Fig. 6.25b). For this example a cloud could form on descent but not on ascent. To go further we need the criteria, specified by thermodynamic variables, that determine whether clouds form on ascent or descent.

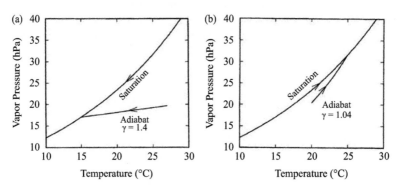

Fig. 6.25: Clouds form in ascending air in Earth's atmosphere because the ratio of specific heats of air (1.4) is such that the curve of adiabatic cooling can intersect the saturation curve (a). But if the ratio of the specific heats of air were 1.04, clouds would form in descending air because the curve of adiabatic heating can intersect the saturation curve (b).

From Eq. (6.13) it follows that the vapor pressure e in a parcel undergoing adiabatic expansion (or compression) at constant mixing ratio satisfies

$$\frac{e}{e_o} = \frac{p}{p_o}, \tag{6.181}$$

where p is the total pressure and the subscript o indicates some initial state. In this process total pressure and temperature are related by the Poisson equation

$$\left(\frac{p}{p_o}\right)^{(1-\gamma)/\gamma} = \frac{T_o}{T},$$ (6.182)

where γ is the ratio of specific heats at constant pressure and constant volume. By combining Eqs. (6.179)–(6.182) with the integrated Clausius–Clapeyron equation we obtain the relative humidity of the parcel

$$r = r_o\xi \exp\left[\frac{\ell_v}{R_v T_o}\left(\xi^{(1-\gamma)/\gamma} - 1\right)\right],$$ (6.183)

where r_o is the relative humidity at the beginning of the process, T_o is the initial temperature, and $\xi = p/p_0$. Underlying Eq. (6.183) is the assumption that the enthalpy of vaporization is constant with temperature. For a cloud to form in the process described by Eq. (6.182), relative humidity must increase. In an expansion (ascent), ξ decreases, and hence, for a cloud to form on ascent requires that $dr/d\xi < 0$ at the beginning of the process ($\xi = 1$). By differentiating Eq. (6.183) at $\xi = 1$, we obtain the criterion for cloud formation on ascent:

$$\frac{\ell_v}{R_v T_o} > \frac{\gamma}{\gamma - 1}.$$ (6.184)

The criterion for cloud formation on descent is obtained by reversing the inequality:

$$\frac{\ell_v}{R_v T_o} < \frac{\gamma}{\gamma - 1}.$$ (6.185)

Thus, if the enthalpy of vaporization of water and the ratio of the specific heat capacities of air were lower than their actual values, clouds would form on descent, rather than ascent.

The criteria of Eqs. (6.184) and (6.185) can be written more compactly by noting that

$$\frac{\ell_v}{R_v} = \frac{\ell_{vm}}{R^*},$$ (6.186)

where ℓ_{vm} is the *molar* enthalpy of vaporization and R^* is the universal gas constant. With Eq. (6.186) in Eqs. (6.184) and (6.185), only two variables, the molar enthalpy of vaporization and the ratio of specific heat capacities determine whether clouds form on ascent or descent.

If your head is swimming after being shown that clouds (but not of water droplets in an atmosphere containing mostly nitrogen and oxygen) could form on descent, prepare yourself for an even stranger revelation: clouds could form in the same atmosphere on either ascent or descent. The criteria of Eqs. (6.184) and (6.185) were obtained by examining the sign of $dr/d\xi$ when the pressure ratio is unity. But we said nothing about the third possibility:

$$\frac{\ell_v}{R_v T_o} = \frac{\gamma}{\gamma - 1}. \tag{6.187}$$

When Eq. (6.187) is satisfied, r has an extremum at $\xi = 1$:

$$\frac{dr}{d\xi} = 0, \qquad \xi = 1. \tag{6.188}$$

To determine if this extremum is a minimum or a maximum, we take the second derivative

$$\frac{d^2 r}{d\xi^2} = r_o \exp(\ell_v/R_v T_o) \frac{\ell_v}{R_v T_o} \left(\frac{\gamma - 1}{\gamma} \right)^2, \qquad \xi = 1 \tag{6.189}$$

which cannot be negative. Therefore, r has a minimum at $\xi = 1$ when Eq. (6.187) is satisfied. For this special case, r increases when the pressure is increased or decreased, which allows for the possibility of cloud formation on descent or ascent.

It is not difficult to find vapors that can condense on compression. As a rough rule of thumb, γ decreases with increasing molecular weight: the more atoms in a molecule, the more ways for partitioning energy (see Section 3.7). For example, the ratio of specific heats of diethyl ether ($C_4H_{10}O$), also called ethyl ether, is about 1.08, and its molar enthalpy of vaporization is sufficiently low that the criterion for cloud formation on descent is satisfied. The saturation vapor pressure of diethyl ether is 1 atm at a temperature of 308 K. So on a planet having an atmosphere with about the same temperature and pressure as ours, but composed of diethyl ether, if clouds of ether droplets formed, they would do so in descending motion. If this atmosphere were seasoned with low-molecular-weight gases, γ of the mixture (see Section 3.7) would be increased to the point where clouds of ether droplets could form on both ascent and descent. Diethyl ether is not a singularity. Many organic vapors possess a sufficiently low γ to satisfy Eq. (6.185).

Annotated References and Suggestions for Further Reading

Moist thermodynamics is treated by Julio Victor Iribarne and W. L. Godson (1981) *Atmospheric thermodynamics*. 2nd edn. Dordrecht: D. Reidel.

For atmospheric thermodynamics integrated with atmospheric dynamics see, for example, John A. Dutton (1976) *The ceaseless wind*. New York: McGraw-Hill; James R. Holton (2004) *An introduction to dynamic meteorology*. 4th edn. London: Elsevier/Academic Press.

For more details on how to calculate equivalent potential temperature, see David Bolton (1980) "The computation of equivalent potential temperature," *Monthly Weather Review*, 108, pp. 1046–53.

Two excellent historical papers are James E. McDonald (1963) "Early developments in the theory of the saturated adiabatic process," *Bulletin of the American Meteorological Society*, 44, pp. 203–11; James E. McDonald (1963) "James Espy and the beginnings of cloud thermodynamics," *Bulletin of the American Meteorological Society*, 44, pp. 634–41.

Another good historical paper, which outlines the contributions of Hertz to atmospheric thermodynamics, is by Elizabeth Garber (1976) "Thermodynamics and meteorology (1850–1900)," *Annals of Science*, 33, pp. 51–65.

For more about Hertz and geophysical science, see Joseph F. Mulligan and H. Gerhard Hertz (1997) "An unpublished lecture by Heinrich Hertz: 'On the energy balance of the Earth'," *American Journal of Physics*. 65, pp. 36–45.

In Russell McCormmach's outline in *Dictionary of scientific biography* of the scientific contributions of Hertz, no mention is made of Hertz's interest in meteorology.

For a comprehensive treatment of conserved parameter diagrams and the effects of vapor and liquid on cloud density, see Kerry A. Emanuel (1994) *Atmospheric convection*. New York: Oxford University Press.

The use of conserved parameter diagrams to identify air masses and track their evolution was first described by C. G. Rossby (1932) "Thermodynamics applied to air mass analysis," *Meteorological Papers*, Vol. 1, No. 3. Cambridge, MA: MIT.

The use of conserved parameter diagrams and mixing line analysis for nonprecipitating cloud systems is developed and described by Alan K. Betts (1985) "Mixing line analysis of clouds and cloudy boundary layers," *Journal of the Atmospheric Sciences*, 42, pp. 2751–63.

For an illustration of how conserved parameter diagrams can be used for understanding atmospheric thermodynamic processes, see Alan K. Betts and Bruce A. Albrecht (1987) "Conserved variable analysis of the convective boundary layer thermodynamic structure over the tropical oceans," *Journal of the Atmospheric Sciences*, 44, pp. 83–99.

Procedures for calculating CAPE and its use in forecasting severe weather can be found in Howard B. Bluestein (1993) *Synoptic-dynamic meteorology in midlatitudes*, Vol. 2. Oxford: Oxford University Press.

For a discussion of convective inhibition energy (CINE) and its implications for tropical convection, see Earle Williams and Nilton Renno (1993) "An analysis of the conditional instability of the tropical atmosphere," *Monthly Weather Review*, 121, pp. 21–36.

Discussions of entrainment can be found in Robert A. Houze, Jr. (1993) *Cloud dynamics*. London: Academic; and William R. Cotton and Richard A. Anthes (1989) *Storm and cloud dynamics*. London: Academic, 1989).

For a discussion of cloud-top entrainment processes, see Q. Wang and Bruce A. Albrecht (1994) "Observations of cloud-top entrainment in marine stratocumulus," *Journal of the Atmospheric Sciences*, 51, pp. 1530–47.

The use of observations, laboratory experiments, and models to explore the nature of cloud-top entrainment and buoyancy reversal is given in Steven T. Siems, Christopher S. Bretherton, Marcia B. Baker, Shenqyang Shy, and Robert E. Breidenthal (1990) "Buoyancy reversal and cloud-top entrainment instability," *Quarterly Journal of the Royal Meteorological Society*, 116, pp. 705–39.

A general description of the thermodynamic structure of stratocumulus clouds and a treatment of the thermodynamics of cloud-top mixing is given by Bruce A. Albrecht, Richard S. Penc, and William H. Schubert (1985) "An observational study of cloud-topped mixed layers," *Journal of the Atmospheric Sciences*, 42, pp. 800–22. An excellent overview of stratocumulus clouds is provided by Robert Wood (2012) "Stratocumulus clouds," *Monthly Weather Review*, 140, pp. 2373–2423.

For an illustration of the use of conserved diagrams for studying entrainment in cumulus clouds, see Ilga R. Paluch (1979) "The entrainment mechanism in Colorado cumuli," *Journal of the Atmospheric Sciences*, 36, pp. 2467–78.

For one of the first analyses of penetrating downdrafts in cumulus clouds, see Patrick Squires (1958) "Penetrative downdraughts in cumuli," *Tellus*, 10, pp. 381–9.

A good general discussion of cloud processes and applications of atmospheric thermodynamics (with a superb collection of cloud photographs) is Frank Ludlum (1980) *Clouds and storms*. University Park, PA: Pennsylvania State University Press.

P. M. Saunders (1957) "The thermodynamics of saturated air: a contribution to the classical theory," *Quarterly Journal of the Royal Meteorological Society*, 83, 342–50, compares temperature–pressure relations for adiabatic and pseudoadiabatic processes, including the effects of ice.

A laboratory determination of the entrainment factor is given in J. S. Turner (1962) "The starting plume in neutral surroundings," *Journal of Fluid Mechanics*, 13, pp. 356–8. See also J. S. Turner (1966) "Jets and plumes with negative or reversing buoyancy," *Journal of Fluid Mechanics*, 26, pp. 779–92.

Mixing clouds, including contrails, are discussed at an elementary level by Craig F. Bohren (1987) *Clouds in a glass of beer*. Hoboken: John Wiley & Sons, ch. 5.

Hutton's theory of rain is discussed by W. E. Knowles Middleton (1965) *A history of the theories of rain*. New York: Franklin Watts, pp. 106–10. For a biographical sketch of Hutton see the entry by V. A. Eyles in *Dictionary of scientific biography*.

Cloud formation on descent is discussed by David Brunt (1934) "The possibility of condensation by descent of air," *Quarterly Journal of the Royal Meteorological Society*, 60, pp. 279–84; James E. McDonald (1964) "On a criterion governing the mode of cloud formation in planetary atmospheres," *Journal of the Atmospheric Sciences*, 21, pp. 76–82; Craig F. Bohren (1986) "Cloud formation on descent revisited," *Journal of the Atmospheric Sciences*, 43, pp. 3035–7.

In a footnote (p. 107) of an early twentieth-century book, Thomas Preston (1904)
The theory of heat, 2nd edn, in its day a popular textbook, we find that

> vapour is usually and erroneously represented by clouds and this perhaps fos-
> ters the idea commonly prevalent ... that steam is visible like a cloud ...
> What we now term steam or vapour is an invisible substance, but when this
> condenses into small globules it becomes visible and is then called cloud or
> mist.

More than a century later conflating water droplets and water vapor is still
"commonly prevalent." Despite the title of this book it is filled with detailed
descriptions of experiments. If you like instruments and measurements and
their history, this is the book for you. The third edition is also a gem. Both
can be found online.

A discussion of the differences between potential temperatures and static energies is given by Alan K. Betts (1974) "Further comments on 'A Comparison
of the Equivalent Potential Temperature and the Static Energy'," *Journal of
the Atmospheric Sciences*, 31, pp. 1713–15.

Archived upper-air soundings from anywhere in the world can be retrieved
from the Wyoming Weather Web site (http://weather.uwyo.edu/upperair/)
supported by the University of Wyoming. Thermodynamic and wind data or
skew-T plots from soundings at selected upper-air stations and dates can be
downloaded.

Approximate expressions for wet-bulb temperatures are evaluated by Knox,
J. A., Nevius, D. S., and P. N. Knox (2017) "Two simple and accurate
approximations for wet-bulb temperature in moist conditions, with forecasting applications," *Bulletin of the American Meteorological Society*, 98, pp.
1897–1906.

The geographical distribution of observed surface wet-bulb temperatures from
28 °C to greater than 35 °C and the increased occurrence of these wet-bulb temperatures from 1979–2017 are provided by Collin Raymond, Tom Matthews,
and Radley M. Horton (2020) "The emergence of heat and humidity too severe
for human tolerance," *Science Advances*, 6, Number 19, eaaw1838, pp. 1–8. Climate model projections of wet bulb temperatures in the geographical region
from 20°N to 20°S indicating a global temperature increase greater than 1.5 °C
would lead to wet-bulb temperatures greater than 35 °C in this area are given
by Yi Zhang, Isaac Held, and Stephan Fueglistaler (2021) "Projections of tropical heat stress constrained by atmospheric dynamics," *Nature Geoscience*, 14,
pp. 133–7. A global warming of 7 °C could make some areas of Earth uninhabitable for wet-bulb temperatures exceeding 35 °C is argued by Steven C.
Sherwood and Matthew Huber (2010) "An adaptability limit to climate change
due to heat stress," *Proceedings National Academy of Sciences of the United
States of America*, 107, pp. 9552–5.

For satellite measurements of precipitable water, see A. N. Maurellis, R. Lang, W. J. van der Zande, W. J. Aben, and W. M. G. Ubachs (2000) "Precipitable water column retrieval from GOME data," *Geophysical Research Letters*, 27, pp. 903–6.

Because equivalent potential temperature depends on both temperature and moisture it can be used as an effective indicator of global warming as described by Fengfei Song, Guang J. Zhang, V. Ramanathan, and L. Ruby Leung (2022) "Trends in surface equivalent potential temperature: a more comprehensive metric for global warming and weather extremes," *Proceedings of the National Academy of Sciences of the United States of America*, 119(6), e2117832119.

Problems

1. Wet-bulb temperatures are often measured by wrapping the bulb of a thermometer with an absorbent material (e.g. cotton cloth) thoroughly moistened with clean water. The lowest temperature recorded while the cloth is still wet is the wet-bulb temperature. Suppose that you were to use very dirty water. What would happen to the measured wet-bulb temperature (for the same air)? Explain your answer.

2. In Section 5.2 we defined the dew point and mentioned the frost point. For temperatures below $0\,°C$, the frost point T_f of air with vapor pressure e is defined by $e_{si}(T_f) = e$, where e_{si} is the equilibrium vapor pressure of ice. Show that the frost point is always greater than the dew point (provided, of course, that the dew point exists). *Hint*: A simple sketch will be of great help.

3. If there were no condensation nuclei in the atmosphere, this would not mean that clouds could not form, merely that they would form at much higher relative humidities than we have come to accept as normal. Under the assumption that without condensation nuclei in the atmosphere clouds won't form by lifting moist air until the relative humidity reaches 400%, *estimate* the increase of the height of cloud base (i.e. the lifting condensation level). *Hint*: The easiest way to do this problem is to redefine the dew point as the temperature to which moist air must be cooled isobarically such that the relative humidity becomes 400%.

4. We usually think of mixing clouds forming on our breaths in cold weather. Yet they are *possible* when air temperatures are greater than breath temperature. Show this.

   There are limits to the greatest possible air temperature. Let us take this limit (near the surface) to be $40\,°C$. Suppose that breath temperature is $30\,°C$ and that exhaled air is saturated at this temperature. What constraint does this place on the relative humidity if a mixing cloud is to form on our breaths for an air temperature of $40\,°C$?

   Discuss why mixing clouds on your breath are less likely when air temperatures are above breath temperature than when below.

Hint: It would be madness to do this problem without appealing to an $e - T$ diagram. You may use the approximate expression for the saturation vapor pressure (in hPa) versus temperature, $e_s = \exp(21.653 - 5420/T)$, where T is in degrees K.

5. Estimate the *lowest* temperature to which *very* dry and hot (40 °C) Arizona air can be cooled by evaporation.

6. On an uncomfortable summer day in Washington, D.C., the maximum air temperature may reach 95 °F while the dew point may be as high as 75 °F. For these conditions, estimate the lowest temperature to which air can be cooled by evaporation.

7. Observant dog owners know that mixing clouds can form on the breaths of dogs when no such clouds form on the breaths of their owners (dog and owner in the same environment). There are (at least) two differences between humans and dogs: (1) The normal deep body temperature of humans is around 37 °C, whereas that of dogs is slightly higher, between 37.8 °C and 39.3 °C. (2) Dogs have few sweat glands, and hence, evaporative cooling of dogs occurs mostly from their moist tongues and nasal passages. Is the higher temperature of dogs *by itself* sufficient to account for the difference between the ability of dogs and humans to produce mixing clouds on their breaths? After you have answered this question, explain why a mixing cloud forms on the breath of a dog but not on its owner's. *Hint*: A diagram is essential.

8. Show that when specific humidity decreases with height, as it often does, determining static stability by means of the gradient of potential temperature (instead of virtual potential temperature) errs on the side of predicting a more stable atmosphere than really exists. *Hint*: To begin, you must express virtual potential temperature as a function of potential temperature and specific humidity q.

9. During a period of thawing, the ground in and around State College, Pennsylvania, was a mosaic of bare and snow-covered patches. On the south slope of Mt. Nittany the following was observed late one afternoon. Within less than a meter or so of the ground, thin clouds would form, last for as long as a minute, and then dissipate. Previously, under more or less the same conditions but when the ground was completely snow-covered, these thin clouds were not observed. Please explain.

10. An elementary textbook asserts that when air over snow is "both cold and dry," solar radiation absorbed by the snow "tends to evaporate [sublimate] the snow rather than melt it." Please give the missing physical reasons to support this assertion. Then design an experiment to determine how much of the ablation (reduction in mass) of a snowpack is a result of net sublimation and how much is the result of melting. Give careful thought to the design of your experiment. State the experimental problems you might face. Do not discuss trivial mechanical problems. Concentrate on the inherent limitations of your design and the consequent errors.

11. Throughout this book we made extensive use of the hydrostatic approximation, according to which the vertical pressure gradient in the atmosphere is exactly balanced everywhere and at all times by the weight (per unit volume) of air. What common observation is evidence that the atmosphere cannot be in strict hydrostatic equilibrium?

12. To avoid interrupting an already lengthy derivation of the moist adiabatic lapse rate, we gave Eq. (6.101) without proof. Provide the missing proof. *Hint*: Everything you need is in Chapter 2, especially Section 2.8.

13. Clouds sometimes form over the wings of airplanes, especially on takeoff or landing in humid air. These clouds, unlike contrails, are not mixing clouds. Airplanes fly because of a pressure difference between the upper and lower surfaces of their wings. Thus, clouds formed over wings are a consequence of (adiabatic) pressure changes, just as are ordinary clouds (other than mixing clouds). Estimate the upper limit for the dew point depression (difference between air temperature and dew point) such that a cloud is possible over the wing of a large airplane flying at constant altitude. That is, what is the dew point depression above which a wing cloud cannot form? After you have obtained this estimate, discuss the flying conditions under which clouds might form over wings at even higher dew point depressions. *Hints*: Data for the Boeing 747 are given in Chapter 2. This problem can be done quickly if you use results in Section 6.2.

14. In Section 6.4 we defined the equivalent temperature as the final temperature of moist air in an adiabatic, isobaric process in which all the water vapor in the air condenses to liquid water. This hypothetical process is allowed by the first law of thermodynamics. Show that it is not allowed by the second law.

15. Give your best estimate of the *greatest* possible, yet still defensible ($1000\,°C$ is not defensible), difference between temperature and virtual temperature (in the troposphere). You would be willing to bet a large sum of money that although this difference is not physically impossible, any greater differences are. After you have obtained this maximum temperature difference, convert it to a maximum (relative) density difference between a parcel and its environment. Then determine the greatest possible updraft velocity this density difference could produce. You can imagine that a juicy parcel near the surface breaks through an inversion into a dry layer where relative humidity drops to zero. Discuss your result.

16. In Section 6.9 we showed curves to illustrate how clouds could form on ascent (Fig. 6.25a) or on descent (Fig. 6.25b), depending on the value of the ratio of specific heats. We also gave a criterion for when clouds could form on both ascent and descent, but without an accompanying figure. Provide the missing figure. That is, sketch a plausible saturation vapor pressure curve and adiabat that correspond to cloud formation on both ascent and descent.

17. At air temperatures below $0\,°C$ it is sometimes possible to measure, using an ordinary sling psychrometer, a wet-bulb temperature slightly higher than the dry-bulb temperature. Explain why and under what conditions. *Hint*: This problem is related to Problem 2.

18. Oceanographers never seem to tire of reminding us that the bulk of Earth's surface is covered by oceans, and hence, oceanographic research deserves financial support in proportion to its areal importance. Two can play the same game. Show that the surface area of Earth's oceans pales in comparison with the total surface area of all the cloud droplets in Earth's atmosphere. *Hints*: You don't need anything more than what can be found in this chapter, together with an estimate of the average size of cloud droplets and a few facts that every earth scientist can be expected to know (Earth's radius, approximate planetary fractional cloud coverage, etc.).

19. What is the temperature of the hottest possible air (at sea level) such that its wet-bulb temperature is $0\,^\circ$C?

20. When we derived the wet-bulb temperature by way of arguments that led to Eqs. (6.68)–(6.70) and Fig. 6.2, we implicitly assumed that the air was unsaturated. Suppose that it had been supersaturated (which is certainly possible). How would you now define the wet-bulb temperature, and how would it stand in relation to the air temperature and dew point? *Hint*: Do this problem graphically in a manner similar to what was done to obtain the wet-bulb temperature of unsaturated air.

21. If you live where winters are cold, you may have observed the following. Immediately after a car is started on a cold winter's morning, a thick cloud billows from its tail pipe. With the passage of time, this cloud disappears. Please explain.

22. A chinook is a down-slope wind observed along the east side of the Rocky Mountains, usually in winter and early spring. Often a chinook is associated with moist adiabatic ascent and precipitation on the windward side of the mountains and dry adiabatic descent on the leeward side (see Fig. 6.26). Suppose that a parcel originating at A (850 hPa or about 1.5 km) has a temperature of $10\,^\circ$C and relative humidity of 100%. What is the temperature of the parcel when it reaches B (600 hPa or about 4.5 km)? If all the moisture that condenses on ascent is removed by precipitation, what is the temperature and relative humidity of the parcel after it descends to C? Your results should tell you why the word *chinook* is reputed to be an American Indian word meaning snow eater.

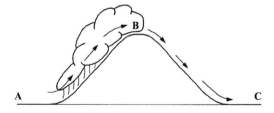

Fig. 6.26: Schematic diagram of a chinook.

23. In winter in northern latitudes, precipitation (at the ground) sometimes begins as rain but can turn to snow or another frozen form even though colder air is not advected into the region. To predict this occurrence, forecasters consider

not only the temperature but also the expected below-cloud humidity at the onset of precipitation. Why? Which single thermodynamic parameter would be the most useful predictor of a switch from rain to snow?

24. Meteorologists sometimes are asked how it is possible that air near the surface can have a relative humidity less than 100% when it is raining. Please explain in such a way that someone who is not an expert can understand.

25. In Section 6.4 we derive an expression for the equivalent temperature T_e, defined for a nonexistent isobaric, adiabatic process undergone by a closed system [see Eq. (6.74)]. In Section 6.6 we describe a real process (adiabatic, nonisobaric, but same initial and final pressures) undergone by a parcel from which liquid water is removed as soon as it is formed (pseudoadiabatic process). Obtain an approximate expression for the difference between the equivalent temperature for the real process and the (isobaric) equivalent temperature and estimate its magnitude. *Hints*: Use Eq. (6.126) and expand the exponential term. Figure 6.10 also is helpful.

26. We noted that isopleths of saturation mixing ratio on a skew-T diagram are approximately straight lines. Please show this without plotting these isopleths. *Hints*: Use the (approximate) expression for the saturation mixing ratio, the integrated Clausius–Clapeyron equation, and the definition of the coordinates on a skew-T diagram.

27. Early one late August morning in Central Pennsylvania, a thin, white, turbulent "smoke" was observed gently rising above a large pile of horse manure. Conditions were calm under a clear sky and an air temperature of about $10\,°C$. Was this really smoke? If not, what was it and why? What additional observations would support your explanation? This is a multi-faceted problem that requires pulling together several ideas, none of which are especially complicated. If you carefully think about this observation, you will have learned a lot. If you just give a knee-jerk response of the kind expected in multiple-choice exams, you will have learned nothing.

28. People who live in arid western areas of the United States often describe their summertime weather as hot, but add that "it's a dry heat." Using what you learned in this chapter, discuss what they really mean by this characterization. To help focus your discussion consider typical summertime conditions in Las Vegas, Nevada where the high temperature is about $40\,°C$ and the relative humidity is 20%. Compare these conditions with those in Miami, Florida where the summertime high temperature is about $35\,°C$ and the relative humidity is 75%.

29. Rain falling from a cloud can cool the air below it by evaporative cooling. Consider a layer at 900 hPa below a precipitating cloud that is initially at a wet-bulb temperature of $-2\,°C$ and a temperature of $2\,°C$. What is the relative humidity in this layer? If rain falling through this layer evaporates at a rate of 0.5 g/kg per hour, what will the temperature, wet-bulb temperature and relative humidity be after one hour? How long would this evaporation rate need to persist to cool the layer to the freezing point? Explain why forecasters consider the wet-bulb temperature below precipitating clouds when trying to predict a rain to snow changeover.

7

Energy, Momentum, and Mass Transfer

Although the province of thermodynamics is uniform systems, it would be exces-sively provincial to not peer over the low fence separating thermodynamics from neighboring fields. We therefore devote a chapter to the transfer of energy, momen-tum, and mass driven by spatial gradients. Meteorologists should know about the relevance of these transfers, which depend on thermodynamic variables and prop-erties, to everyday phenomena, and enough to be able to converse intelligently with engineers about problems of common interest.

7.1 Energy Transfer by Thermal Conduction

By this we mean energy transfer *within* a medium in which there is *negligible mass motion* (motion at the macroscopic scale) over times of interest. Most often it is solid, possibly liquid, a gas only in stagnant thin layers near boundaries. A *fixed* volume V within this medium is enclosed by a surface A that could coincide with a physical boundary or be a mathematical surface. U is the total internal energy of everything within V. For a process in which no work is done, the first law of thermodynamics is

$$\frac{dU}{dt} = \dot{Q}.$$ (7.1)

The heating (cooling) rate $\dot{Q}$ is, in general, nonzero because of a temperature dif-ference between V and its surroundings (see Section 1.7 for the ambiguity of this difference). Temperature can vary over A, and hence, $\dot{Q}$ can be a *surface functional* (Section 3.1). An example is a surface integral, a rule for assigning a number to a surface. We assume that $\dot{Q}$ is a surface integral

$$\dot{Q} = \int_A f \, dA,$$ (7.2)

Atmospheric Thermodynamics. Second Edition. Craig F. Bohren and Bruce A. Albrecht, Oxford University Press.

where f is a function defined on A. We can go further by making a judicious guess about the form of f.

Surfaces of constant temperature within a medium are called *isotherms*, defined by

$$T(\mathbf{x},t) = \text{const.} \tag{7.3}$$

We assume parallel planes for simplicity and take A to be a cylinder with its axis perpendicular to the isotherms, the union of three surfaces: two flat, circular ends parallel to the isotherms and a curved cylindrical surface perpendicular to them. Q is therefore a sum of three integrals, one for each surface. Temperature does not change in directions perpendicular to the curved surface, and hence the part of the integral Eq. (7.2) contributed by it is zero. If we tilt the cylinder, however, there is a contribution to Q from the curved surface. The isotherms are unchanged, as are the shape and area of the surface, yet Q should be different. A way to account for this is to write Eq. (7.2) as

$$Q = \int_A -\mathbf{q} \cdot \mathbf{n} \, dA, \tag{7.4}$$

where $\mathbf{n}$ is the outwardly directed unit vector normal to A at each point and $\mathbf{q}$ is the (conductive) *energy flux density vector*. The minus sign ensures that where $\mathbf{q} \cdot \mathbf{n} < 0$ the integrand is positive. The dimensions of $\mathbf{q}$ are energy per unit time and *area*, thus the designation energy flux *density*.

We assume local thermodynamic equilibrium (Section 2.5). If T varies in space and time, the local specific internal energy is $u[T(\mathbf{x},t)]$, where $u(T)$ is the equilibrium specific internal energy of the medium. We further assume that $u = \rho c_v T$ (adding a constant to u does not change $\partial u/\partial t$), where ρ is the local density of the medium and c_v its specific heat capacity at constant volume. With these assumptions

$$\frac{dU}{dt} = \frac{d}{dt} \int_V \rho c_v T \, dV = \int_V \frac{\partial}{\partial t}(\rho c_v T) \, dV. \tag{7.5}$$

If we combine Eqs. (7.1), (7.4), and (7.5), we obtain

$$\int_V \frac{\partial}{\partial t}(\rho c_v T) dV = \int_A -\mathbf{q} \cdot \mathbf{n} \, dA. \tag{7.6}$$

From the divergence theorem, the surface integral is equal to a volume integral:

$$\int_A -\mathbf{q} \cdot \mathbf{n} \, dA = \int_V -\nabla \cdot \mathbf{q} \, dV. \tag{7.7}$$

Equations (7.6) and (7.7) yield

$$\int_V \left(\frac{\partial}{\partial t}(\rho c_v T) + \nabla \cdot \mathbf{q} \right) dV = 0. \tag{7.8}$$

Because V is arbitrary, Eq. (7.8) can be satisfied only if the integrand vanishes identically:

$$\frac{\partial}{\partial t}(\rho c_v T) + \nabla \cdot \mathbf{q} = 0.$$ (7.9)

Thus, the local time rate of change of the specific internal energy is equal to the negative divergence of the energy flux density vector. This is similar to the continuity equation of fluid mechanics

$$\frac{\partial \rho}{\partial t} + \nabla \cdot \rho \mathbf{v} = 0,$$ (7.10)

where ρ is the mass density, $\mathbf{v}$ is the velocity field, and $\rho \mathbf{v}$ is the mass flux density vector. This similarity is not an accident: both energy and mass are conserved for isolated systems. Equation (7.10) expresses local mass transfer, whereas Eq. (7.9) expresses local energy transfer. By *local*, in contrast with *global*, is meant point by point. Equation (7.1) is a global equation of energy transfer.

Fourier Thermal Conduction Law

To go further requires assumptions. $\mathbf{q} \cdot \mathbf{n}$ must vanish on surfaces perpendicular to isotherms, and the normal to an isotherm at any point is parallel to the gradient of T. The simplest physically plausible form of $\mathbf{q}$ satisfying this condition is

$$\mathbf{q} = -k_t\, \nabla T.$$ (7.11)

This is often called the *Fourier heat conduction law*, and k_t is variously called the *coefficient of thermal conductivity, thermal conductivity*, or simply *conductivity* if there is no danger of confusing it with electrical conductivity. Although $\mathbf{q}$ is sometimes called the heat flux density vector, we prefer energy flux density vector because it determines a transfer of *energy*. The negative sign indicates that energy is transferred from regions of higher to regions of lower temperature. Although called a law, Eq. (7.11) does not have the same generality as thermodynamic laws. If k_t is called the *thermal* conductivity, logic would dictate that Eq. (7.11) describes *thermal conduction*. Books like *Conduction of Heat in Solids* could have titles that do not perpetuate the notion of "heat" as a kind of ethereal fluid but clearly convey their contents. *Thermal Conduction in Solids* is short and to the point but retains just a tinge of ambiguity. "Solids" includes stationary air. A better title would be *Thermal Conduction in Stationary Media*, unless the media are restricted to solids.

The essential physics underlying Eq. (7.11) is embodied in Eq. (1.55). If molecules with different energies interact, those with lower energies gain energy whereas those with higher energies lose energy *on average*. Conduction is a random process driven by internal energy density gradients, and hence, temperature gradients. This is called energy transfer even though nothing palpable is transferred, and transfer connotes carrying a thing rather than a quality from one place to another. Transfer of energy by conduction more closely resembles electronic transfer of money than trucking artichokes from California to Pennsylvania.

The Fourier law combined with Eq. (7.9) yields a partial differential equation for T:

$$\frac{\partial}{\partial t}(\rho c_v T) = \nabla \cdot (k_t \nabla T). \tag{7.12}$$

If we further *assume* that ρc_v is independent of time and k_t is independent of position, Eq. (7.12) becomes

$$\frac{1}{\kappa}\frac{\partial T}{\partial t} = \nabla^2 T, \tag{7.13}$$

where the *thermal diffusivity* κ (sometimes denoted by α) is

$$\kappa = \frac{k_t}{\rho c_v}, \tag{7.14}$$

and could depend on T. Equation (7.13) determines the distribution in time and space of temperature, *not* "heat," in a body in which there is no mass motion. But this does not rule out the possibility that it is surrounded by a fluid in motion (e.g. air, water), the state and properties of which determine solutions to Eq. (7.13) satisfying *boundary conditions*. An implicit assumption underlying this equation is that no chemical reactions or phase changes occur within V.

A single ball with speed v *directly* transfers kinetic energy at this speed. But if a collection of many colliding balls with different speeds are moving *randomly* in all directions, the *net* rate of energy transfer (possibly zero) is much less than their average speed. Because of the *dimensions* of the diffusivity (and its name), length squared over time, we hazard a guess that the speed of energy transfer in a solid for a given temperature gradient is specified not by the time to be transferred *directly* a distance r but the much longer time to be transferred a distance r^2. Detailed analysis shows that this guess is correct. If a tiny sphere in an infinite medium is suddenly given an increase of temperature and hence internal energy, energy diffuses radially outward. At any distance r from the sphere, the temperature increases to a maximum at $t = r^2/2\kappa$ and then decreases.

If ρc_v is independent of position we can write Eq. (7.11) as $\mathbf{q} = -\kappa \nabla u$, which reveals the physics underlying Eq. (7.13), the random *diffusion* of specific internal energy, not a "flow" of anything palpable like the directed flow of water in pipes impelled by pumps or gravity. By replacing T in Eq. (7.13) with u, we obtain an equation for the diffusion of energy. Easier to visualize is the diffusion of mass (Section 7.3) because of *concentration gradients*. Because its governing equation has the same form as Eq. (7.13), this lends credence to diffusion of energy as a random process. Temperature changes with time at a given point are a *measurable* manifestation of this diffusion caused by *internal energy gradients*. Equation (7.13) puts the effect before its cause. Unlike energy, temperature is *not* a conserved quantity. The internal energy of an isolated system is conserved, but not necessarily its temperature.

Given a characteristic length L of a finite body, we can roughly estimate, without having to solve a differential equation, a characteristic time L^2/κ over which its temperature changes. A good habit to develop is to use physical reasoning before dragging out heavy mathematical artillery. Another approach is to write Eq. (7.13) as $\nabla_\xi^2 T = \partial T/\partial \tau$ in nondimensional space coordinates $\xi = \mathbf{x}/L$ and nondimensional time $\tau = t\kappa/L^2$.

The time-dependence of an arbitrary initial temperature distribution in isolated solid bodies, especially simple shapes such as spheres and slabs, of any size and composition follows from solutions to Eq. (7.13). The final equilibrium temperature is uniform, as predicted by entropy maximization arguments (Section 4.3), which cannot predict how long this takes.

We cannot pretend that air, even in an isolated cylinder, could ever be absolutely still, like a solid except with a different thermal diffusivity. That of air at 20 °C and one atmosphere is approximately that of lead, $\approx 2 \times 10^{-5}\,\mathrm{m^2 s^{-1}}$. Inevitable temperature fluctuations in the stillest air result in density fluctuations, and hence, buoyancy, and hence, motion (Section 3.5). In lead this does not happen (at least over hundreds of thousands of years). The thermal expansion coefficient of air at 300 K is about 300 times that of lead, the atoms in which are more tightly bound than molecules in air. If we were to naively apply Eq. (7.13) to a 500-meter column of isolated air, we would predict that an initially nonuniform vertical temperature profile would take around 250 years to evolve to isothermal ($t = L^2/\kappa$). Convection is much more rapid, even in perceptibly still air (Section 7.3), but the equilibrium profile is *not* isothermal (Section 4.3).

If the temperature dependence of k_t and ρc_v could not have been neglected for solids, Eq. (7.12) would have been nonlinear, analytical solutions difficult, and Fourier might have been lost in the mists of history.

The least ambiguous and most distinctive term we could come up with for Eq. (7.13), if it needs one, is thermal conduction equation. Even better would be to call Eq. (7.11) Fourier's *first law*, and Eq. (7.13) his *second law* by analogy with the first and second laws of Fick (Section 7.3), which are identical in form but describe diffusion of matter, rather than energy. Equation (7.13) is in Fourier's *The Analytical Theory of Heat* (1822), but not in this form because ∇ had not yet come into use, nor did he use the "curly *d*" symbol ∂ for partial derivatives.

Equation (7.13) is not time-reversal symmetric, and hence, describes irreversible processes. A body initially at a temperature higher (or lower) than ambient will spontaneously evolve to that temperature but not spontaneously return to its initial temperature even though permitted by the first law of thermodynamics (Section 4.3).

Only the *product* of ρ and c_v, the specific heat capacity per unit *volume*, determines the space and time dependence of T. We wrote this as a product because often "specific heats" are invoked without saying which is meant. There is no correlation between k_t and ρc_v. For example, ρc_v for aluminum and ice are about the same, but the thermal conductivity of aluminum is about 100 times greater.

Fourier begins his 1822 treatise with "Primary causes are unknown to us ... Heat, like gravity, penetrates every substance of the universe, its rays occupy all parts of space." He further writes "Of the nature of heat only uncertain hypotheses could be formed." He understood temperature but was confused about "heat." At one point he calls it an "element." He variously refers to the "diffusion," "propagation," "movement," "transmission," "conduction," "communication," and "distribution" of heat. A plethora of terms for the same thing indicates confusion that persists to this day. His treatise is an account of a first-class scientist exploring new territory; but it must be read critically rather than as a collection of immutable truths. What has stood the test of time is his mathematics, which has found applications he could not have dreamed of.

If temperature is independent of time,

$$\nabla^2 T = 0. \tag{7.15}$$

This is *Laplace's equation*, which pops up in many seemingly disparate physical problems: electrostatics, magnetostatics, steady-state diffusion, steady-state flow of fluids through porous media (e.g. soils), and irrotational flow of fluids. Whatever you learn about Laplace's equation in one physical context is transferrable to another by changing notation and interpreting symbols differently.

Two material properties appear in space- and time-dependent thermal conduction problems, k_t and ρc_v of solid objects, *as well as* the properties and state of any fluid surrounding them. Slipshod treatments often invoke only conductivity or heat capacity in inadequate or downright incorrect explanations. We give a few examples in this section. Don't be awed by formidable mathematical solutions to Eq. (7.13) for different geometries, boundary, and initial conditions. These are solutions to contrived or idealized problems. This equation is a kind of playground for applied mathematicians who, despite their label, sometimes have little or no interest in applications. Time and space variations of temperature can be measured in the laboratory but not as easily as solving Eq. (7.13). Solutions to idealized problems, however, serve to sharpen physical intuition, guide the design of experiments, and suggest observations.

Equations (7.12) and (7.13) are not of much use without boundary and initial conditions for finite bodies. Percy W. Bridgman, who appears in Chapters 1 and 3, emphasized that

> no physical significance can be directly given to a flow of heat, and there are no operations for measuring it. All we can measure are temperature distributions and rates of rise [or fall] of temperature. As at present defined, a heat current [q] is a pure invention, without physical reality, for any determined heat flow may always be modified by the addition of a solenoidal vector, with change in no measurable quantity.

That is, if **A** is any continuous vector function, we can add $\nabla \times \mathbf{A}$ to the right side of Eq. (7.11) without changing Eq. (7.12) or boundary conditions, and hence the temperature distribution.

Thermal Resistance

The simplest steady-state problems are those in which temperature varies with only one space coordinate, x, say. For such one-dimensional problems the isotherms are planes, and Eq. (7.15) becomes

$$\frac{d^2T}{dx^2} = 0. \tag{7.16}$$

We apply this equation to a slab of area A with thickness $d \ll \sqrt{A}$, a large windowpane, for example. Because it is relatively large in lateral extent, T depends (approximately) only on x, and the steady-state temperature distribution satisfies Eq. (7.16), the general solution to which is

$$T = Bx + D, \tag{7.17}$$

where B and D are constants determined by boundary conditions, which we can take as $T(0) = T_0, T(d) = T_d$. With these boundary conditions, Eq. (7.17) becomes

$$T = \frac{T_d - T_0}{d}x + T_0. \tag{7.18}$$

The energy flux density vector is $\mathbf{q} = q\mathbf{e}_x$, where $\mathbf{e}_x$ is a unit vector in the positive x direction, and

$$q = k_t \frac{T_0 - T_d}{d}. \tag{7.19}$$

If $T_0 > T_d$, $\mathbf{q}$ is in the positive x direction. Multiply both sides of Eq. (7.19) by the cross-sectional area A of the slab

$$qA = \frac{(T_0 - T_d)Ak_t}{d} = \frac{T_0 - T_d}{R} = \frac{\Delta T}{R}, \tag{7.20}$$

where the *thermal resistance* R (or R-value) of the slab is defined as

$$R = \frac{d}{k_t A}. \tag{7.21}$$

The term *resistance* is from electric circuit theory. qA may be interpreted as an *energy current* analogous to the electrical current in a resistor; ΔT is analogous to the potential difference across the resistor; R has the same form as the resistance of an electrical resistor of length d, cross-sectional area A, and electrical conductivity k_t. And the analogy goes further. Suppose that N slabs with the same cross-sectional area but different thickness and composition are stacked together like pancakes. Under steady conditions, qA is the same for each slab:

$$qA = \frac{\Delta T_j}{R_j}, \tag{7.22}$$

where ΔT_j is the temperature difference across the jth slab with thermal resistance R_j. From Eq. (7.22) it follows that

$$\Sigma \Delta T_j = \Delta T = \Sigma R_j qA = qAR, \tag{7.23}$$

where ΔT is the temperature difference across the entire stack of slabs and R is its total resistance. These thermal resistances are said to be in series because the energy current is the same for each. Thermal resistances in series, like electrical resistances, are additive:

$$R = \Sigma R_j. \tag{7.24}$$

What about thermal resistances in parallel?

Consider N slabs with the same thickness but different cross-sectional areas and thermal conductivities butted together like tiles on a kitchen floor. For the jth slab

$$q_j A_j = \frac{\Delta T}{R_j}, \tag{7.25}$$

where ΔT is the common temperature difference across the slabs. The total energy current is

$$\Sigma q_j A_j = \sum \frac{\Delta T}{R_j} = \frac{\Delta T}{R}, \tag{7.26}$$

where

$$\frac{1}{R} = \sum \frac{1}{R_j}. \tag{7.27}$$

Thermal resistances in parallel add like electrical resistances in parallel. Sometimes R is called the *absolute thermal resistance* to distinguish it from

$$RA = \frac{d}{k_t}, \tag{7.28}$$

the *R-value*, with SI units m^2KW^{-1}, familiar to homeowners and builders. The inverse of the R-value is the *U-value*. The primary function of most thermal insulation is to *restrict* the movement of air. What do wool, cork, felt, down, and foams have in common? They all are fluffy or porous with many small air-filled spaces in a lightweight solid network. The *specific surface* of a pore with surface A and volume V is A/V, which increases with decreasing size. The *porosity* of a material with n_p pores per unit volume is $n_p V$. The maximum insulating value of such a material requires specific surface as large as possible to suppress convection within pores and porosity as large as possible (by increasing n_p) to decrease the effective conductivity [see Eq. (7.46)]. But increased porosity is paid for by decreased strength. To better insulate buildings the only variable under much control is d, and the brute force way to increase insulation is to pile it on. Space and bank accounts are finite, so inexpensive, nontoxic, and durable superinsulation has the potential to reduce the cost of heating, air conditioning, and refrigeration.

The function of insulation sometimes is misunderstood. A thick wool sweater is not "warm" except figuratively. Outside on a bitterly cold day measure the temperature of your "warm" sweater with an infrared thermometer. A sweater can only passively reduce q, and hence, the burden on your body to keep your core temperature within a narrow range. Sweaters are warmth *preserving* rather than inherently warm.

Convective Transfer of Energy

Thermal conductivities are different for different materials. Those of metals are considerably higher than those of electrical insulators such as glass, which are considerably higher than that of air. But it moves readily when subjected to temperature gradients. The Fourier law in the form Eq. (7.11), where k_t is solely a property of the medium and not its state of motion, is *not* valid in fluids except very near solid surfaces. By motion we mean that the separate parts of the medium move *relative* to each other. If you kick a football, it moves, but all its parts move together.

In a solid with a nonuniform temperature distribution, the (specific) energy of the colder parts increases at the expense of a decrease of the energy of the hotter parts without the transfer of anything palpable, just as money is transferred from one bank account to another without the transfer of wads of bills and bags of coins. In a fluid with a nonuniform temperature distribution, however, temperature differences give rise to density differences, which acted upon by a mass-dependent force such as gravity cause mass motion of the fluid. This process irons out temperature gradients more rapidly, in general, than does energy transfer without mass motion (usually called conduction). The discovery of thermally driven mass motion in fluids and its role in transferring energy was made in 1797 by Benjamin Thompson, Count of Rumford, one of the strangest characters in the history of science. Many years had to elapse before a name was coined for what he discovered. In 1834 William Prout averred that "There is at present no single term in our language employed to denote this mode of propagation of heat; but we venture to propose for that purpose the term *convection* (*convectio*, a carrying or conveying) ...". Within 20 years the term *convection* had become firmly implanted in the scientific literature, where it resides undisturbed today.

Energy transfer by convection between a solid object and a surrounding fluid is more accurately called conduction *modulated* by convection. The thermal resistance of a stagnant thermal boundary layer adjacent to an object is determined by mass motion (convection) beyond it. Air would be a superb thermal insulator if it did not move under the slightest perturbation. Near solid surfaces, however, air and other fluids are assumed to be immobilized, often called the *no-slip boundary condition*: the tangential component of the velocity of a fluid at a solid boundary vanishes. If the surface is impermeable, the perpendicular component vanishes. To demonstrate the no-slip boundary condition, sprinkle some baby powder on a clean sheet of black matte paper. Blow on it with all your might. Even with our geriatric lungs we could produce a wind speed of over 18 ms^{-1}, and yet powder was still detectable on the paper. But a puff of powder released at chest height into a wind $\approx 1 \text{ ms}^{-1}$ disappeared in seconds. This demonstration is complicated because small particles on surfaces are acted on by adhesive forces, unlike particles released into the atmosphere. All we can say with certainty is that the aerodynamic force on *some* of the particles was insufficient to overcome adhesion and that we measured wind speeds of 2 ms^{-1} with a hand-held anemometer at 2 m above asphalt, 0.3 ms^{-1} at 2 cm.

Convection without qualification often means *external* convection: energy transfer between an object and an unbounded surrounding fluid. But there is also *internal* convection, for example, within porous media, which depends on pore size, or within double-pane windows, which depends on the separation between panes. Internal convection is characterized by more than one boundary layer. At the same temperature and pressure, the thermal conductivity of argon is about 0.68 that of air, its density about 1.38 that of air. This is why some double-pane windows are filled with argon. The claim is that the *R-value* increases, although over the years argon will gradually leak out and be replaced by air. Estimates are 1% a year, which corresponds to a loss of 20% in about 25 years.

Conductivity of Gases: A Few Fallacies Dispelled

Before reading about thermal conductivity or any of the transport coefficients of gases, read the subsection on the nonexistence of still air in Section 7.3.

Consider a *stationary* gas of identical molecules in which a temperature gradient in the x direction is maintained (how need not concern us). For simplicity, we take the molecules to be monatomic so that their only energy is translational kinetic energy. If n is the number density of molecules, $n/2$ molecules are moving in the positive x direction, and $n/2$ are moving in the negative x direction. The flux density E of (kinetic) energy in either direction at any point x is therefore

$$E = \frac{n}{2} \langle v_x \frac{1}{2} m v^2 \rangle, \qquad (7.29)$$

where $v_x = v \cos \theta$ is the x component of the velocity and θ is the angle between the velocity vector and the x axis. If we assume no correlation between the speed of a molecule and its direction,

$$\langle v_x v^2 \rangle = \langle v^3 \cos \theta \rangle = \langle v^3 \rangle \langle \cos \theta \rangle = \frac{\langle v^3 \rangle}{2}, \qquad (7.30)$$

where we used the result that the average cosine (over a hemisphere of directions) is $1/2$. To evaluate the mean cubed speed in Eq. (7.30), we need the Maxwell–Boltzmann distribution function $f(v)$ for molecular speeds (Section 2.4):

$$\langle v^3 \rangle = 4\pi \left(\frac{m}{2\pi kT} \right)^{3/2} \int_0^\infty v^5 \exp\left(\frac{-mv^2}{2kT} \right) dv. \tag{7.31}$$

To evaluate this integral transform the variable of integration from v to $\sqrt{u}$ and integrate repeatedly by parts (or look up the integral in a table). With this integral in hand, we obtain

$$E = \frac{n\langle v \rangle kT}{2}, \tag{7.32}$$

where the mean speed $\langle v \rangle$ (Section 2.4) is

$$\langle v \rangle = \sqrt{\frac{8kT}{m\pi}}. \tag{7.33}$$

Except for the factor $1/2$ we could have guessed the form of Eq. (7.32). Energy flux density is a product of number density, speed, and a quantity proportional to mean kinetic energy. But E is *not* the energy flux density q in the Fourier law. E is the flux density in *one* direction, whereas q is the *net* energy flux density. We can estimate q from E by the following reasoning.

Consider a constant x plane in a gas (Fig. 7.1). All the molecules at $x - \lambda$, where λ is the mean free path (Section 2.5), travel an (average) distance λ in the positive x

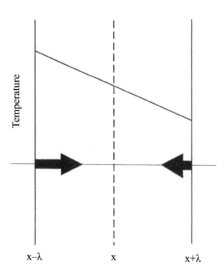

Fig. 7.1 Because of a temperature gradient in a gas, there is a net transfer of energy across the plane $x = $ const. Molecules at $x - \lambda$ have a greater average kinetic energy than molecules at $x + \lambda$.

direction without suffering a collision, and hence, without change in kinetic energy. Similarly, all the molecules at $x + \lambda$ travel a distance λ in the negative x direction without change in kinetic energy. Thus, the *net* kinetic energy flux density (in the positive x direction) across the plane is

$$q = E(x - \lambda) - E(x + \lambda). \tag{7.34}$$

This equation describes an unequal two-way *exchange* of energy across a plane. If we expand $E(x \pm \lambda)$ in a Taylor series about x and retain only the first two terms, Eq. (7.34) becomes

$$q = -2\lambda \frac{dE}{dx}. \tag{7.35}$$

Both n and T (and hence, $\langle v \rangle$) in Eq. (7.32) depend on x. For the gas to be stationary, the pressure gradient must vanish, and hence,

$$\frac{dp}{dx} = \frac{d}{dx}(nkT) = 0. \tag{7.36}$$

By combining Eqs. (7.32), (7.33), (7.35), and (7.36), we obtain

$$q = -\frac{1}{3}\lambda \rho \langle v \rangle c_v \frac{dT}{dx}, \tag{7.37}$$

where we used

$$\frac{k}{m} = \frac{2}{3}c_v \tag{7.38}$$

for the specific heat capacity at constant volume c_v of a monatomic gas (Section 3.6). Could we have guessed at the form of Eq. (7.37)? Perhaps. The quantity $\rho \langle v \rangle c_v \, \Delta T$ is an energy flux density; so to obtain Eq. (7.37) we need a guess for ΔT. Because energy fluxes are driven by temperature gradients, a plausible guess for ΔT is the temperature gradient times a characteristic length, and the only one at hand is the mean free path. Except for a multiplicative constant, we could have made an intelligent guess at the form of Eq. (7.37) without any mathematical fireworks. We cannot write q solely as the product of a number density, speed, and a property of individual molecules (k/m) because the temperature gradient is *not* such a property.

From Eq. (7.37) and the definition of thermal conductivity in the Fourier law it follows that the conductivity of a gas is

$$k_t = \frac{1}{3}\lambda \rho \langle v \rangle c_v = \frac{m\langle v \rangle c_v}{3\sigma} = \frac{c_v}{\sigma}\sqrt{\frac{8kT}{9m\pi}}, \tag{7.39}$$

where $\lambda = 1/n\sigma$ [Eq. (2.79)] and $\rho = nm$. We are left with the disturbing result that the thermal conductivity of a gas is independent of pressure but not of temperature. This is disturbing because we previously mentioned vacuum flasks (thermos bottles). Why is the space between their walls evacuated if thermal conductivity does not depend on pressure?

An implicit assumption in the preceding analysis is that the relevant character-istic length is the mean free path. This is indeed true for gases at pressures where it is much smaller than the linear dimensions of the gas volume. The separation between the walls of a vacuum flask is of order several millimeters. From Fig. 2.12, the height at which the mean free path drops to 1 mm is about 60 km. Thus, if the gap between the walls of a vacuum flask is evacuated to pressures lower than those above this height, the mean free path becomes greater than the gap width. A molecule that begins its journey on the hotter side of the gap travels more or less unimpeded to the colder side, and vice versa. For pressures low enough that the mean free path is much greater than the gap width, the characteristic length λ in Eq. (7.37) is replaced by the gap width, and the energy flux density becomes

$$q \propto \rho c_v \, \Delta T, \qquad (7.40)$$

where ΔT is the temperature difference across the gap. Because density is propor-tional to pressure (for fixed temperature), the net energy flux density decreases linearly with pressure at sufficiently low pressures. At a pressure of ≈ 1 hPa, the mean free path is comparable to the gap between the walls of a vacuum flask. Although this may seem low, in high-vacuum technology it is rather high, easily obtainable with mechanical pumps.

According to Eq. (7.39), $k_t \propto \sqrt{T}$. Figure 7.2 shows measured thermal conduc-tivity of the two gases of most interest to us—dry air and water vapor—as a function of $\sqrt{T}$ for temperatures between about 200 K and 350 K. According to the preced-ing simple theory of gas conductivity, this relation should be a straight line (see Section 7.2 for more about this) and it apparently is. What about its magnitude? At sea level the density of air is ≈ 1.2 kgm^{-3} and its mean free path is $\approx 0.8 \times 10^{-7}$m. For air $c_v \approx 700$J kg^{-1}K^{-1} and the mean speed of an air molecule (at 0 °C) is ≈ 450 ms^{-1}. With these values in Eq. (7.39) we obtain a thermal conductivity of

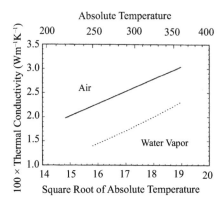

Fig. 7.2 Thermal conductivity of dry air and of water vapor.
Data are taken from *Thermal Conductivity: Nonmetallic Liquids and Gases*, Volume 3 of *Thermophysical Properties of Matter*, by Y. S. Touloukian, P. E. Liley, and S. C. Saxena. IFI/Plenum, New York, 1970.

$\approx 10^{-2}$ Wm^{-1}K^{-1}, comparable to measurements (Fig. 7.2). Simple theory gives the approximate magnitude and temperature dependence of the thermal conductivities of air and water vapor over the range of normal terrestrial temperatures, but it does not accurately predict their ratio.

Because the molar specific heat capacity of the noble (monatomic) gases is independent of molecular weight, we expect the thermal conductivities of these gases to be inversely proportional to $m^{3/2}$. Although simple theory does give the correct ordering of conductivities, the error in their ratios is as much as a factor of two.

Figure 7.2 dispels two fallacies we encounter from time to time. One is that we feel cold on high mountains because of the "thin air," by which we assume is meant low pressure (or perhaps we misheard "thin hair" as "thin air," and it is only bald people who are cold on mountains). This fallacy may result from a false syllogism: pressure is lower at high elevations than at sea level; I feel colder at high elevations; therefore, I feel colder at high elevations because the pressure is lower there. That is, thin air is *inherently* colder. But, as we have seen, the thermal conductivity of air is independent of pressure (at least in the troposphere), although it does decrease as the square root of absolute temperature. Thus, the thermal conductivity of mountain air is *lower* than that of sea-level air because of lower temperatures at higher elevations. Shouldn't the (slightly) lower thermal conductivity of the air give a bit *more* insulation from the colder air on mountains? Both the decrease of pressure and temperature with elevation have a *common* cause: Earth's gravitational field (g); see Eqs. (2.49) and (3.94). We defer discussion of the insulating value of "thin" air to the subsection Chilliness at High Altitudes: Forced Convection after erecting the necessary theoretical framework.

The other fallacy is that people feel colder in wet climates than in dry, presumably all else being equal, because the thermal conductivity of moist air is so much greater than that of dry air. The determinants of human comfort are so many that we are not going to settle once and for all why some people say they feel colder in winter in, say, Minnesota, than in, say, Arizona (at the same air temperature). Suffice it to say that this has nothing to do with a supposed great difference between the thermal conductivities of dry and moist air. Water vapor is only a few percent of air. Although the theory of the thermal conductivity of mixtures of gases is complicated, especially if one of the components is a polar molecule (such as water), it is plausible that, like specific heat capacities of mixtures, the thermal conductivity of a mixture is approximately a weighted average of the conductivities of its components. Water vapor could greatly increase the thermal conductivity of air only if the conductivity of water vapor were much greater than that of dry air—but it is *less* (Fig. 7.2). The best evidence from measurements and theory is that for normal atmospheric temperatures and relative humidities up to 100%, the thermal conductivity of moist air is almost that of dry air, if anything slightly *less*, as expected. Nonsense about the thermal conductivity of moist air may reflect a failure to distinguish between water *vapor* and *liquid* water with a thermal conductivity appreciably higher than that of air. Or this could be yet another example of water vapor mysticism (Section 3.7).

Although the thermal conductivity of air is independent of humidity, the notion persists that "A cold damp day feels colder than a 'dry' one because moist air conducts heat away from the body better than a 'dry' air." This gem is from the third edition of a textbook. What is meant by "damp"? Can humans sense the difference between "dry" and "damp" air at the same temperature and in the same environment? Does atmospheric water *vapor* ever condense onto skin? Dew point is lower than or equal to air temperature (Section 5.2), and hence, for water vapor to condense onto skin requires that its temperature drop below air temperature, which we never have observed. If you are rain soaked, you may feel cold. But "feel" gets us into the murky subject of *subjective* sensations of hot and cold. If you say that you feel colder, there is nothing to argue about. But physical explanations are open to criticism.

We measured cooling times for a 250 ml flask of water initially at 100 °C in a room at 20 °C for relative humidities of 50% and 100%. There was no significant difference in the cooling times. Why, then, do people think that there is a difference between "dry" cold and "moist" cold, *all else being equal* (air temperature, wind, cloud cover, etc.)? Only *one* variable can be different, namely, humidity. Perceptions of hot and cold are not entirely objective: humans are not thermometers. We offer a tentative explanation based on observations. Sitting at a desk inside on a cold, overcast day in winter we look out a window at a gloomy sky and feel a bit chilly. But if the clouds suddenly break and the surroundings are bathed in sunlight, we feel warmer even though our immediate environment is the same. If we are outside and feel colder when cloudiness is accompanied by higher humidity, as it often is, then of course we feel colder because of less sunshine. The senior author monitors 27 solar panels on his roof. This has taught him that he is not a good judge of downward solar radiation on different days on which the cloudy sky looks about the same, but the panels indicate otherwise.

The Effective Conductivity of Porous Materials

To explore further the function of common insulating materials, consider a porous medium composed of N parallel solid slabs with the same thickness Δx_s and thermal conductivity k_s separated by $N-1$ air gaps, each of thickness Δx_a. We assume that these gaps are sufficiently narrow that the air within them is more or less stagnant. The thermal resistance of a slab is $\Delta x_s/k_s A$, where A is its cross-sectional area, and the thermal resistance of a gap is $\Delta x_a/k_a A$. Thus, the total resistance R of this medium is obtained by adding all resistances:

$$R = (N-1)\frac{\Delta x_a}{k_a A} + N\frac{\Delta x_s}{k_s A}. \tag{7.41}$$

The porosity f, defined as the ratio of the air volume to the total volume, is

$$f = \frac{(N-1)\Delta x_a}{d}, \tag{7.42}$$

where the total thickness is

$$d = N \Delta x_s + (N-1)\Delta x_a . \tag{7.43}$$

By using Eq. (7.42), we can write Eq. (7.41) as

$$R = \left(\frac{f}{k_a} + \frac{1-f}{k_s} \right) \frac{d}{A}. \tag{7.44}$$

The quantity in parentheses multiplying d/A may be interpreted as the reciprocal of an effective conductivity

$$\frac{1}{k_{eff}} = \frac{f}{k_a} + \frac{1-f}{k_s}. \tag{7.45}$$

This is correct at the end points $f = 0$ and $f = 1$, but cannot capture the complicated geometry of real porous media. We simplified a porous medium as resistors in series, but real three-dimensional "thermal circuits" are more like a network of resistors in series *and* in parallel.

If $(1-f)(k_a/k_s) \ll f$,

$$k_{eff} \approx \frac{k_a}{f}. \tag{7.46}$$

The porosity of fresh powder snow with a density 10% that of water is 0.9, which, from Eq. (7.46), yields $k_{eff} \approx 1.1 k_a$. An empirical fit to many measurements, $k_{eff} = 2.22[0.917(1-f)]^{1.88}$, gives $k_{eff} \approx 1.02 k_a$. But at $f = 0.8$, $k_{eff} \approx 3.7 k_a$, three times greater than Eq. (7.46), because of the labyrinthine structure of snow and a thermal conductivity of ice 90 times that of air. The simple formula becomes worse with decreasing porosity. Only the lowest-density snow is a "good insulator." The R-value of a layer of snow, d/k_{eff}, decreases because k_{eff} increases *and* d decreases as the snow settles. For adequate structural strength the porosity of compacted snow for an igloo can't be much greater than about 0.5, corresponding to an effective thermal conductivity comparable to that of common building brick.

A good rule of thumb based on measurements is that $k_{eff} \approx 2 k_a$ for common porous insulation. For example, on a plot of R versus pelt thickness of Arctic mammals ranging from shrews to polar bears, the data points fall close to a line with slope $1/2k_a$. Shrews and polar bears are insulated in the same way; polar bears do so with more rather than with markedly different hair. Humans have shed most of their body hair, invented adjustable pelts (clothing), and created comfortable artificial microclimates.

We don't want to leave you with the impression that convection does not take place within insulation. Only ideal insulation completely suppresses convection. It still takes place within real insulation, especially when buffeted by wind. This is why a common form of protection against cold is a tightly woven, thin windbreaker worn over a fluffy, thick sweater. The sweater, a porous medium, provides insulation by trapping air (although not perfectly), and the windbreaker is a barrier to air currents within the sweater.

The thermal conductivity of air even *at standard pressure*, but confined within nanometer-size pores, can be appreciably smaller than the usual value for still air [see Eq.(7.40)]. From Fig. 2.12, the mean free path in air at sea level is about 100 nm. Aerogel, a porous material, has the lowest density of any solid (100 kgm^{-3}, mostly from the silica skeleton) and an average pore diameter about 20 nm. It is the world's best thermal insulator, superinsulation—but fragile.

The Skin Diver's Fallacy

Some skin divers seem to believe that their wetsuits keep them warm in cold water because it leaks into the space between suit and skin, is warmed, and therefore, warms the diver. To convince them that this is nonsense would probably be a fruitless (possibly even dangerous) task; they have heard the warm-water argument so many times that it has become an immutable truth, beyond criticism. Skin divers who also backpack in cold weather should have the courage of their convictions. If they really believe that they are warmed by water that infiltrates their wetsuits, they should be willing to saturate their sleeping bags with water before settling into them on a cold night. This is an easy experiment to perform and should settle the issue once and for all. Any takers?

Until word arrives from the next of kin of skin divers who have done the definitive experiment, we have to settle for a bit of simple theory. The total thermal resistance of the foam of a wetsuit and the air gap between it and the diver's skin is

$$R = \frac{\Delta x_{foam}}{Ak_{foam}} + \frac{\Delta x_{gap}}{Ak_{gap}}. \tag{7.47}$$

We can rewrite Eq. (7.47) as

$$R = \left(\frac{\xi}{k_{foam}} + \frac{1-\xi}{k_{gap}} \right) \frac{\Delta x}{A}, \tag{7.48}$$

where Δx is the sum of foam and gap thicknesses and ξ is the ratio of foam thickness to total thickness. A good estimate for the thickness of the foam of a wet suit is 8 mm; the gap thickness is around 1 mm or less; so we take it to be 0.5 mm. This corresponds to $\xi = 0.94$. Based on results in the previous subsection, we estimate that k_{foam} is about $2k_a$. Most of the pores in the foam are not connected to the surroundings, which you can verify for yourself by trying to blow through a wetsuit pressed against your lips. With the assumption that the gap is filled with air having an effective thermal conductivity twice that of still air, the total resistance is

$$R = 0.5 \frac{\Delta x}{k_a A}. \tag{7.49}$$

The thermal conductivity of liquid water is about 20 times that of air. Thus, if the gap is filled with water, the total thermal resistance of foam and gap is about

$$R = 0.47 \frac{\Delta x}{k_a A}. \tag{7.50}$$

We conclude from this crude analysis that the thermal resistance *decreases*, although not by much, when water displaces air from the gap between wetsuit and skin. Foam provides the bulk of the thermal resistance, both because of its greater thickness and its lower effective conductivity. Skin divers use wetsuits (that leak) because drysuits (that do not leak) cost much more, not because the water layer provides insulation. The warming of this layer indicates that the foam insulation is doing its job. But it takes energy to warm the water, and this must be supplied by the skin diver's metabolism. The area of the human body is about 1 m^2. For a gap thickness of 0.5 mm the corresponding mass of water is about 0.5 kg. Suppose that the water temperature entering the wetsuit is 10 °C and is heated to 34 °C. The specific heat capacity of liquid water is about 4200 J kg^{-1}K^{-1}. Thus, about 5×10^4 J is required to heat water in the gap between skin and wetsuit. How long, on average, does a molecule from the surrounding water spend in the gap before returning to this water? One second seems too small, and one hour seems too large. For ease of computation, we assume that the residence time for water in the gap is 1000 s. This corresponds to a heating rate of about 50 W, roughly half the basal metabolic rate of an adult male. By admittedly crude calculations, we conclude that skin divers pay a price for heating the water that infiltrates their wetsuits. This water not only does not insulate divers, but also it presents them with a heating bill they must pay by drawing on their own reserves of stored potential energy.

Before leaving the subject of wetsuits, we note that the residence time we crudely estimated in the previous paragraph is not the same as the filling time. When a freshly suited skin diver jumps into cold water, the gap fills quickly with cold water, resulting in an initial shock. The transient energy flux density from the diver to the water trapped in the suit is many times the steady-state value when the rates of water flowing into and out of the suit are equal. Humans are sensitive to energy fluxes, not just temperatures, which you can verify for yourself on a cold morning in the outhouse of a cabin equipped with wooden and metal toilet seats. Both are at the same temperature, but your backside, which is not a good thermometer, is nevertheless very effective at telling you which is which.

The skin diver's fallacy takes different forms and pops up in different places. We stumbled upon the following gem in a *National Geographic* article (October 1996, p. 133) about a traverse of Baffin Island, some of it on skis across shallow water: "Despite being soaked for hours, his feet stay tolerably warm, thanks to the insulating effect of water trapped in his boots." Set aside that water is not "trapped" in the boots of someone sloshing through it, water flows in and out of them. If water in boots increases their insulating value, cold-weather boots should be equipped with nozzles so that they could be filled with water under conditions of extreme cold. Believing this twaddle about the "insulating effect of water" could be dangerous to your health, even fatal. Suppose that in winter you find yourself on the edge of hypothermia, your teeth chattering. A lake is nearby. Can you save yourself by jumping into the water, thereby availing yourself of its alleged marvelous warming power? Not advisable unless you are eager to hasten your departure from this globe. In a following subsection we treat in detail the cooling of bodies in air and in water, a fertile source of fallacies.

Metabolism is an umbrella term for all the chemical reactions going on in a human (or animal) body, not "heat generation" unless by "heat" is meant temperature, which perpetuates confusion supposedly clarified more than a hundred years ago. These chemical reactions (see Section 3.9) result in a temperature increase that counteracts what otherwise would be a steady decrease of *core* body temperature. The net result is a nearly constant temperature, an example of *homeostasis*.

Newton's Law of Cooling: A Study in Historical Error Propagation

We considered previously the steady-state temperature distribution in a slab with the temperatures of its two faces maintained constant at T_0 and T_d, a contrived problem abstracted from the real world. When we tackle concrete problems, we have to recognize that slabs (e.g. walls) of interest are likely to be surrounded by air or water. For example, the slab of interest could be a window or the wall of a house. The temperature T_0 could be that of the inside surface, T_d that of the outside surface. These two temperatures are not, however, the inside and outside air temperatures. Large temperature gradients can occur over short distances in fluids near solid surfaces. Although the Fourier thermal conduction law is not generally valid within a fluid, it is approximately valid at the boundary between a fluid and a solid because of the no-slip boundary condition. For example, if the air immediately adjacent to a wall is stagnant, we can apply the Fourier thermal conduction law there

$$q = -k_a \left(\frac{dT}{dx}\right)_s,$$ (7.51)

where k_a is the thermal conductivity of (still) air and the subscript s indicates that the gradient is evaluated in the air at the wall (solid interface). From Eq. (7.51) we can *define* the quantity h_c, called the *film coefficient, heat transfer coefficient, convective heat transfer coefficient,* or *surface conductance* (its inverse, *surface resistance*),

$$q = -k_a \left(\frac{dT}{dx}\right)_s = -h_c(T_s - T_a),$$ (7.52)

where T_s is the temperature at the wall and $T_a < T_s$ is the temperature in the air at a distance from the wall where temperature is not markedly changing. For the moment we assume that radiant energy transfer is negligible (not always valid) and do not distinguish between free and forced convection. Equation (7.52) is deceptive. Hidden in the *symbol h_c* are the thermophysical properties and state of the fluid, as well as the size, shape, and orientation of the cooling body (see subsections at the end of this section). Moreover, each of the three conventional modes of energy exchange—radiation, convection, conduction—rarely occurs by itself nor are they strictly independent, as evidenced by the difficulty of measuring the thermal conductivity of air and other gases. Invoking this trinity is a tactic for reducing a complex whole into simpler parts, a gain not without losses.

We can write Eq.(7.52) as $Q = Q_h + Q_c$, where $Q_h = Ah_cT_a$ is a constant heating rate and $Q_c = -Ah_cT_s$ is a variable cooling rate (see Section 1.8).

Equation (7.52) is almost universally referred to as *Newton's law of cooling* despite the fact that Newton never annunciated it. The spread of this historically incorrect attribution is an example of error propagation by textbook writers. Instead of taking the trouble to read Newton's words (available in many libraries), they merely repeat what someone else said Newton said. This error persists and cannot be dislodged with dynamite. What has appeared in a textbook three times, no matter how easily it can be demonstrated to be false, is almost universally accepted as true. This is a law of error propagation with an applicability and scope comparable to the laws of thermodynamics.

For many years, one of the most widely used textbooks on engineering heat transfer was that by the late William McAdams. On page 5 of the third edition of his book we find the following assertion: "In 1701 Newton defined the heat-transfer rate ... from a surface of a solid to a fluid by the equation [Eq. 7.52] ..." The citation here is to Newton's paper in the *Philosophical Transactions of the Royal Society* for March–April 1701. This assertion in an authoritative textbook, accompanied by citation of an ancient and musty paper by Newton, creates the impression of scholarly veracity. Nevertheless, McAdams's assertion is a fabrication, which you can verify for yourself by comparing his words with what Newton actually wrote in 1701: "if equal times of cooling be taken, the degrees of heat will be in geometrical proportion, and therefore easily found by the tables of logarithms." By "degrees of heat" Newton meant what is now called temperature. The modern version of Newton's law of cooling is that the temperature of a cooling body decreases exponentially with time. Even by mangling words and their meanings, we cannot translate Newton's statement of his law of cooling into McAdams's assertion made more than 200 years later. At issue here is more than just correct attribution of a physical law to its proper author. It's also the difference between a definition and a law. Equation (7.52) *defines* the film coefficient as the ratio of a temperature gradient at a wall to the difference between the wall temperature and that of the surrounding fluid:

$$h_c = k_a \left(\frac{dT}{dx} \right)_s \bigg/ (T_s - T_a) = \frac{q}{T_a - T_s}. \tag{7.53}$$

This equation is impossible to refute or prove. Definitions are either useful or not. The historically incorrect Newton's law of cooling isn't a law until it is supplemented in such a way that it becomes testable. For example, by asserting that h_c is independent of the temperature difference or specifying how it depends on the fluid motion and its thermophysical properties we have testable propositions.

We can, at least in principle, measure h_c and determine if it exhibits, within experimental error, the behavior attributed to it. But by making several additional assumptions, we can transform Eq., (7.53) into a testable proposition and also square it with Newton's form of his law. We defer this task.

The temperature gradient divided by a temperature difference in Eq. (7.53) has the dimensions of an inverse length. We may call the inverse of this quantity a length and give it a name, *thermal boundary layer thickness*, and symbol:

$$\frac{1}{\delta_t} = \left(\frac{dT}{dx}\right)_s \Big/ (T_a - T_s). \tag{7.54}$$

With this definition, the film coefficient becomes

$$h_c = \frac{k_a}{\delta_t}. \tag{7.55}$$

Because of the form of Eq. (7.55), h_c is sometimes called the *film coefficient* or *film conductance* (its inverse, *film resistance*), more physically descriptive terms because of the qualifier "film." We may interpret δ_t as the thickness of a region near a wall over which temperature varies appreciably. To make this clearer, consider the integral

$$\int_0^\delta \frac{dT}{dx}\,dx = T(\delta) - T(0), \tag{7.56}$$

where x is measured outward from the wall as origin. The mean-value theorem of integral calculus yields

$$\int_0^\delta \frac{dT}{dx}\,dx = \left(\frac{dT}{dx}\right)_{av}\delta, \tag{7.57}$$

where, if dT/dx is monotone, our best estimate for the average value of the temperature gradient is

$$\left(\frac{dT}{dx}\right)_{av} = \frac{1}{2}\left[\left(\frac{dT}{dx}\right)_0 + \left(\frac{dT}{dx}\right)_\delta\right]. \tag{7.58}$$

We choose $x = \delta$ to be such that the temperature gradient there is negligible compared with that at the wall:

$$\left(\frac{dT}{dx}\right)_{av} \approx \frac{1}{2}\left(\frac{dT}{dx}\right)_0. \tag{7.59}$$

If we combine Eqs. (7.56), (7.57), and (7.58), we obtain

$$\frac{1}{\delta} = \frac{1}{2}\left(\frac{dT}{dx}\right)_0 \Big/ [T(\delta) - T(0)], \tag{7.60}$$

which, is twice the thermal boundary layer thickness. This confirms our physical interpretation of δ_t. In air not adjacent to solid and liquid surfaces, energy transfer is convective (mass motion), and the notion of a stagnant thin boundary layer is meaningless. We can estimate δ_t for surfaces in air by crude reasoning. The thermal conductivity of air is $\approx 2 \times 10^{-2}\,\mathrm{Wm^{-1}K^{-1}}$ at room temperature. We need an estimate of h_c. What is the magnitude of energy flux densities in our everyday lives? The primary source of energy for Earth is solar radiation, which bathes every square meter pointed toward the Sun at the top of the atmosphere with about 1360 W. Thus, we might guess that q because of energy exchange with air is of order 100 Wm^{-2}, although this depends on its motion. With this estimate for q and $\Delta T = 10\,°\mathrm{C}$, the difference between the surface temperature and that of air well away from it, we obtain $\delta_t \approx 2\,\mathrm{mm}$ from Eq. (7.55). Our clumsy measurements of thermal boundary layers (Section 1.7) support this estimate, as do more skillful noninvasive (interferometric) measurements near heated cylinders by Ralph Kennard.

Because δ_t is inversely proportional to h_c, we expect δ_t to decrease with increasing wind speed, given the common experience of feeling more uncomfortable on a cold day if the wind suddenly picks up. Clinging to every surface is a layer of stagnant air of order millimeters thick, which adds a surface resistance to the thermal resistance discussed previously. To determine how much, consider a slab of thickness d; for concreteness, take it to be the wall of a house. The inner and outer wall temperatures are T_i and T_o, respectively, with corresponding inner and outer air temperatures T_{ai} and T_{ao}. Under steady-state conditions

$$q = h_i(T_{ai} - T_i) = \frac{k_w(T_i - T_o)}{d} = h_o(T_o - T_{ao}), \tag{7.61}$$

where h_i and h_o are the film coefficients at the inner and outer surfaces of the wall, d is its thickness, and k_w is its thermal conductivity. These three equations can be rewritten

$$\frac{q}{h_{ci}} = T_{ai} - T_o, \quad \frac{qd}{k_a} = T_i - T_o, \quad \frac{q}{h_{co}} = T_o - T_{ao} \tag{7.62}$$

and added

$$\frac{q}{h_i} + \frac{qd}{k_w} + \frac{q}{h_o} = T_{ai} - T_{ao}. \tag{7.63}$$

If we multiply and divide the left side of Eq. (7.63) by the cross-sectional area A of the wall

$$qA = \frac{T_{ai} - T_{ao}}{R_t}, \tag{7.64}$$

where the *total thermal resistance* is the sum of three resistances in series, one for the wall and one each for the inside and outside surfaces:

$$R_t = R_i + R_w + R_o = \frac{1}{h_i A} + \frac{d}{k_w A} + \frac{1}{h_o A} = \frac{\delta_{ti}}{k_a A} + \frac{d}{k_w A} + \frac{\delta_{to}}{k_a A}. \tag{7.65}$$

Although the thicknesses of thermal boundary layers may be small (compared with wall thicknesses), their contribution to the total thermal resistance is not necessarily small given that the thermal conductivity of air is likely to be considerably less than that of the wall.

Authors are sometimes ambiguous about just what kind of energy transfer Eq. (7.51) describes. Sometimes it is said to be convection, followed by the assertion that it is "really" conduction. Or to cover all bases, the term conduction–convection is sometimes used. Equation (7.51) contains a quantity that is a property of a stagnant fluid (its thermal conductivity) and the temperature gradient at the wall. The thermal conductivity depends only on the thermodynamic state of the fluid, whereas the temperature gradient also depends on the state of *motion* of the fluid beyond the wall. Thus, Eq. (7.51) embodies both conduction and convection, but is deceptively simple or even misleading because it does not display the explicit dependence of q on this motion. At a surface, energy is transferred molecule by molecule (no mass motion), whereas in the body of a fluid energy is transferred by many molecules moving together as groups (mass motion). Equation (7.52) is sometimes said to describe "sensible heat." Is not cold also "sensible"? Sensible heat is an outmoded term long overdue for retirement. The noun convection or the adjective convective conveys the essential physics: energy transfer in fluids, unlike in solids, entails mass motion.

Engineers distinguish between *free* (or *natural*) and *forced convection*. Forced convection occurs in fluids moved by pumps, blowers, and fans. The fan of an automobile idling at curbside cools its radiator by forced convection. When the engine is turned off, the radiator continues to cool by free convection currents around the radiator driven by temperature (and hence, density) differences between it and *nominally* (see Section 7.3) still air. To a meteorologist, the distinction between free and forced might seem a bit strained because winds on all scales are driven by temperature differences, and therefore, an example of free convection—unless you believe that Aeolus is the divine keeper of the winds who releases them at the command of the gods.

We now can better understand something we alluded to in Section 3.4: near surfaces, temperature gradients in air often greatly exceed the dry adiabatic lapse rate. Equation (7.51) is strictly applicable only in the vicinity of surfaces where adjacent air can be considered stagnant (no-slip boundary condition). Well away from surfaces, where the air may be turbulent, Eq. (7.51) is not valid—at least not with k_a equal to its value for still air. But a plausible approximation sometimes made is that the same form of Eq. (7.51) applies even in freely moving air:

$$q = -K\frac{dT}{dz}. \tag{7.66}$$

The *eddy conductivity* K, which unlike the molecular conductivity k_a depends on the details of the air motion (and hence, on height above the surface), is associated with the transfer of energy by groups of molecules moving collectively (eddies) rather than individually. This view of energy transport by eddies implies that in general K is considerably larger than k_a. For steady-state energy transfer, q must be independent of height, and hence, according to Eq. (7.66), the temperature gradient must increase if the conductivity decreases. Near the ground, where the conductivity is least, we therefore expect the highest temperature gradients, in accord with what is observed.

Freezing of Lakes

The concept of thermal resistance helps us understand the freezing of lakes. Suppose that the temperature at the surface of an initially unfrozen lake drops and stays below 0 °C. How does the thickness of the resulting ice increase with time?

Enthalpy is conserved in an adiabatic, isobaric process. When water freezes, its specific enthalpy decreases (Section 5.3). Thus, freezing of water cannot be an adiabatic process, and the ice must exchange energy with its surroundings. Denote by ξ the ice thickness at any time t after the surface drops to a subfreezing temperature T_0 (i.e. the lake is entirely liquid at $t = 0$). The temperature at the interface between ice and water is the melting point of ice T_w. Determining the rate of advance of this interface is an example of a *Stefan problem*. Such problems are difficult because they require solving a moving boundary-value problem. That is, the temperature must satisfy Eq. (7.13) within a medium subject to certain conditions on its boundary. When the medium is fixed, this boundary is fixed, whereas in a Stefan problem the boundary moves, and hence, is a part of the solution. To determine the temperature distribution we need to know how the boundary moves, but to know this we need to know the temperature distribution. If we are willing to settle for an approximation, we can sidestep these difficulties and approximate the energy flux through the ice as

$$qA = \frac{k_i A(T_w - T_0)}{\xi},$$ (7.67)

where k_i is the thermal conductivity of the ice and A is its cross-sectional area. The source of qA is freezing of water, and thus must be equal to the rate of change of ice volume times the enthalpy of fusion (per unit volume) of ice:

$$qA = A\frac{d\xi}{dt}\ell_f \rho_i,$$ (7.68)

where ρ_i is the density of ice. If we combine Eqs. (7.67) and (7.68), we obtain the differential equation

$$\xi\frac{d\xi}{dt} = \frac{k_i(T_w - T_0)}{\rho_i \ell_f},$$ (7.69)

which has the solution

$$\xi = \sqrt{\frac{2k_i(T_w - T_0)t}{\rho_i \ell_f}} = \sqrt{\frac{2k_i}{\rho_i \ell_f}}\sqrt{\Delta T t},$$ (7.70)

subject to the condition that $\xi = 0$ at $t = 0$, where ΔT is the number of degrees air temperature is below 0 °C. A square-root dependence of ice thickness on time goes a long way toward explaining why ice on lakes does not reach great thicknesses even in the coldest regions of the world. The interpretation of this square-root law is that as ice thickens, its thermal resistance increases, hence, the energy flux decreases, and the water freezes more slowly. All this is well and good, but it is not quite right. Although Eq. (7.70) is the correct (subject to our approximations) solution to a mathematical problem, it is not the correct solution to a physical problem because we ignored the thermal resistance of the boundary layer of air above the ice. We should have written Eq. (7.67) as

$$qA = \frac{T_w - T_a}{R_t},$$
(7.71)

where T_a is the air temperature (above the thermal boundary layer) and the total thermal resistance is

$$R_t = \frac{\xi}{k_i A} + \frac{1}{h_c A}.$$
(7.72)

From Eqs. (7.64), (7.68), (7.71) and (7.72) we obtain

$$\left(\xi + \frac{k_i}{k_a}\delta_t\right)\frac{d\xi}{dt} = \frac{k_i(T_w - T_a)}{\rho_i \ell_f}.$$
(7.73)

This equation is no more difficult to solve than Eq. (7.69), but instead we consider two limiting cases. According to Eq. (7.73), when the ice thickness $\xi \ll k_i\delta_t/k_a$, we can ignore ξ in the function multiplying its time derivative, which, therefore, is approximately constant. This inequality is always satisfied for sufficiently small ξ. Thus, in the early stages of growth, ice thickness increases *linearly* with time. When the ice thickness $\xi \gg k_i\delta_t/k_a$, we retain the (approximate) square-root dependence of ice thickness on time. The thermal conductivity of ice is about 90 times that of air, and, as we have seen, the thickness of thermal boundary layers is of order a millimeter. We therefore conclude that ice grows linearly with time initially but decreases its rate of growth when its thickness has reached a few centimeters. In the initial stages of growth the total thermal resistance is dominated by the boundary layer of air above the ice. But this resistance is fixed (under fixed external conditions), whereas the thermal resistance of the ice increases and eventually dominates the total resistance.

The form of Eq. (7.70) suggests that the time-integrated depression of air temperature below 0 °C would be a good indicator of cumulative ice thickness once freezing has begun. And, indeed, the seasonal increase of ice thickness on lakes sometimes correlates well with *cumulative freezing degree days*, the daily average $\Delta T > 0$ summed over the number of days with subfreezing temperatures, raised to a power close to 0.5.

Radiative Energy Transfer

Radiation contributes to energy transfer because all objects at all temperatures at all times emit electromagnetic radiation and absorb it from their surroundings to varying degrees, sometimes immeasurably. This is incessant even after the Sun has dipped below the horizon. During the day, we can see the consequences of sunlight illuminating terrestrial objects, but there always is unseen radiation, which we have in mind here, its spectrum extending over a broad region centered at wavelengths in the range 8–12 μm.

The term "blackbody" is one of the most misleading scientific terms ever coined. Clean, fresh, fine-grained snow is the *whitest* natural substance on Earth and also the *blackest*. How can this be? Our eyes sense only a tiny part of the electromagnetic spectrum, which greatly colors how we use the terms white and black. We perceive snow to be white and soot black, both illuminated by sunlight, but are blind to the infrared radiation emitted by snow, which is close to a "blackbody" spectrum. Painted surfaces *visibly white* and *visibly black* are about equally "black" in the infrared, which humans without rattlesnake ancestry cannot see. Extrapolation from what we see to what we cannot is done all the time, sometimes valid, other times hopelessly wrong. For example, although we may call aluminum foil shiny, this should be qualified by *to our eyes*. As it happens, it *is* shiny (highly specularly reflecting) from infrared down to radio wavelengths, but to show this requires measurements or calculations based on theory, not visual observations. Calling a blackbody a "perfect" emitter is meaningless. Perfect for what purpose? Aluminum foil is a highly "imperfect" radiator, which makes it "perfect" for insulation.

Consider a cavity large relative to relevant wavelengths and composed of *any* material. If its walls are sufficiently thick, the spectral distribution of radiation in equilibrium with them is determined solely by the wall temperature T, is *isotropic* (the same in all directions), and unpolarized. A term better than blackbody spectrum for this radiation is the historically correct Planck spectrum or distribution. We may consider cavity radiation to be a gas of photons with an equilibrium distribution of energies. The analogy goes deeper. The pressure of an ideal gas is *two-thirds* its kinetic energy density [Eq. (2.9)], whereas the pressure of a photon gas is *one-third* its energy density. And like a molecular gas, a photon gas has entropy, rarely mentioned even in advanced textbooks. The Planck distribution is a maximum entropy distribution.

By luck, over a limited range of infrared wavelengths emission by many objects—skin, soil, plants, wood, fabrics, painted surfaces—but not all, *approximates* Planckian emission at normal terrestrial temperatures. If this were not so, the radiative thermometers touted in this book would be of much less use.

To proceed requires a bit of janitorial work. The mantra "a good absorber is a good emitter" has been passed on to many generations of students. Set aside that "good" and "bad" are meaningless without definite criteria. Gasoline is a "good" fuel for automobile engines but a "bad" fuel for human engines. This mantra is devoid of any explanatory power and (worse) is a source of befuddlement. For example, we are told that Earth's protective ozone layer is a "good" absorber of solar ultraviolet radiation. It logically follows that it is a "good" emitter, thereby undoing the "good" it is alleged to do.

The spectrum of Planckian radiation is determined solely by temperature. Emission integrated over all wavelengths and in all directions in a hemisphere is given by the Stefan–Boltzmann law, $\bar{\sigma}T^4$, where the Stefan–Boltzmann constant $\bar{\sigma}$ is $5.67 \times 10^{-8} \mathrm{Wm^{-2}K^{-4}}$; we write $\bar{\sigma}$ to avoid confusion with the collision cross section

σ (Section 2.5). To express emission by a real body at the same temperature we introduce the concept of *emissivity*, a dimensionless quantity ≤ 1, often denoted as ε. The product of spectral emissivity and the Planck spectrum is the emission spectrum of the real body. Rarely mentioned is that ε depends on the direction of emission and state of polarization, as well as the wavelength. Normal emissivity is that in the direction normal to the emitter surface, whereas hemispherical emissivity is an average over all directions: they are not the same. Also, emissivities are difficult to measure, highly dependent on surface conditions, and difficult to calculate except for highly idealized bodies. Emissivity is a somewhat squishy concept. We can *define* an average emissivity $\langle \varepsilon \rangle$ such that $\langle \varepsilon \rangle \bar{\sigma} T^4$ is the integrated emission over all wavelengths and in all directions by a real body with temperature T. But $\langle \varepsilon \rangle$ is an average over the Planck spectrum, which depends on temperature and hence so does $\langle \varepsilon \rangle$.

The absorptivity, often denoted by α, is the *fraction* of incident radiation that is absorbed. One form of *Kirchhoff's law* is $\alpha = \varepsilon$, which should be qualified. Because ε depends on wavelength, direction, and polarization, so does α. Emission and absorption are inverse processes and the fundamental laws of physics do not distinguish between past and future. If time is imagined to be reversed, an emitted photon becomes an absorbed photon. Imagine a movie of someone diving into water to be run backwards: an "absorbed" diver becomes a never-observed "emitted" diver.

Absorption of solar ultraviolet radiation by the ozone layer is the product of its ultraviolet absorptivity *and* the amount of solar ultraviolet radiation incident on it. The solar spectrum is approximately Planckian at 6000 K; the temperature of the ozone layer is, say, 250 K. The ratio of the ultraviolet radiation emitted by the ozone layer to absorption of solar ultraviolet radiation by it is about 10^{-63}, even though its ultraviolet emissivity is equal to its absorptivity. Confusion arises from failing to distinguish between dimensionless emissivity (≤ 1) and *emission* (Wm^{-2}) and between dimensionless absorptivity (≤ 1) and *absorption* (Wm^{-2}). The ozone layer is a "good" absorber of (solar) ultraviolet radiation and a "very bad" emitter of ultraviolet radiation in the same wavelength interval. A body at 250 K emits no measurable ultraviolet radiation.

The average emissivity $\langle \varepsilon \rangle$ is *not*, in general, equal to the average absorptivity $\langle \alpha \rangle$ even if $\varepsilon = \alpha$. The average of α weighted by a (normalized) function f is not the average of ε weighted by a function g unless $f = g$; the greater the difference between f and g, the greater the difference in the averages. For example, the average absorptivity of, say, bare soil illuminated by infrared radiation from an overcast sky is not equal to the average emissivity of the soil. Nevertheless, $\langle \alpha \rangle \approx \langle \varepsilon \rangle$ is often an acceptable approximation because spectra and angular distributions of radiation emitted by sky and soil are not greatly different. And this is true for many objects (walls, clothing, skin, etc.) at normal terrestrial temperatures.

A simple example of radiative exchange is that between two parallel plates separated by a nonparticipating medium. The temperatures of these plates are T_1 and T_2, with average emissivities $\langle \varepsilon_1 \rangle$ and $\langle \varepsilon_2 \rangle$, respectively. Plate 2 emits radiation and also reflects radiation from plate 1, and vice versa. We assume that both plates are opaque (no transmission) to the radiation they emit. The upward $(1 \to 2)$ and downward $(2 \to 1)$ radiative flux densities are solutions to

$$F_\uparrow = \langle \varepsilon_1 \rangle \bar{\sigma} T_1^4 + F_\downarrow (1 - \langle \varepsilon_1 \rangle), \quad F_\downarrow = \langle \varepsilon_2 \rangle \bar{\sigma} T_2^4 + F_\uparrow (1 - \langle \varepsilon_2 \rangle), \tag{7.74}$$

where $1 - \langle \alpha \rangle = 1 - \langle \varepsilon \rangle$ is the reflectivity of a plate. Upward and downward denote two opposite directions and have nothing to do with gravity. The net radiant flux density is therefore

$$q_r = F_\uparrow - F_\downarrow = \frac{\langle \varepsilon_1 \rangle \langle \varepsilon_2 \rangle \bar{\sigma} (T_1^4 - T_2^4)}{\langle \varepsilon_1 \rangle + \langle \varepsilon_2 \rangle - \langle \varepsilon_1 \rangle \langle \varepsilon_2 \rangle} = \langle \varepsilon_{12} \rangle \bar{\sigma} (T_1^4 - T_2^4). \tag{7.75}$$

As it must, $q_r \to 0$ when $T_1 \to T_2$ and also when $\langle \varepsilon_1 \rangle$ or $\langle \varepsilon_2 \rangle \to 0$; its sign depends on the sign of the temperature difference. The infrared emissivity of metals, especially if highly polished, most notably aluminum foil, is low. The limiting case is 100% reflection (unattainable) when $\langle \varepsilon \rangle = 0$, and hence, there is no net radiative energy transfer. If one of the emissivities is 1 (Planckian emitter), $\langle \varepsilon_{12} \rangle \to \langle \varepsilon_1 \rangle$ or $\langle \varepsilon_2 \rangle$.

We sometimes wrap hot sandwiches in aluminum foil to keep them warm. The "obvious" reason for this, the foil reflects radiation emitted by the sandwich, is wrong. This can be shown by measuring cooling times of hot water in a glass flask, bare and wrapped with aluminum foil. The cooling time for the latter (in still air) is measurably greater. Paint one side of the foil black and repeat the experiment with this side facing inward. The difference in cooling times is negligible. What is the function of the foil? The emissivity on the side facing the sandwich is largely irrelevant because they are at about the same temperature. This is *not* true of the side facing the room at a lower temperature. If we take plate 1 to be aluminum foil and plate 2 to be its surroundings, there is net radiation from the foil *if* it is hotter than its surroundings. The low emissivity of the foil minimizes q_r. Wrapping sandwiches with aluminum foil does indeed substantially slow the rate of *radiative* cooling, all else being equal. This is one of many examples of an observation being unassailable, but not its explanation. Do this experiment in wind (produced indoors with a variable-speed fan) to observe the relative contributions of radiative and convective cooling.

The temperatures in Eq. (7.75) are absolute temperatures, and for most meteorological applications relative differences in absolute temperature are not large. Because of this we can *linearize* Eq. (7.75) as follows. Write $T_2 = T_1 - \Delta T$, which, when substituted into Eq. (7.75), yields

$$q_r = \langle \varepsilon_{12} \rangle \bar{\sigma} T_1^4 \left[1 - \left(1 - \frac{\Delta T}{T_1} \right)^4 \right]. \tag{7.76}$$

With the assumption that $|\Delta T / T_1| \ll 1$, we can approximate the quartic term in Eq. (7.76) as

$$\left(1 - \frac{\Delta T}{T_1} \right)^4 \approx 1 - \frac{4\Delta T}{T_1}. \tag{7.77}$$

This approximation in Eq. (7.76) yields the linearized net radiative flux density

$$q_r \approx h_r(T_2 - T_1), \tag{7.78}$$

where the *radiative film coefficient* is

$$h_r = 4\langle \varepsilon_{12} \rangle \bar{\sigma} T_1^3. \tag{7.79}$$

To the same degree of approximation $h_r = 4\langle \varepsilon_{12} \rangle \bar{\sigma} T_2^3$. At room temperature, $h_r \approx 5\,\mathrm{Wm}^{-2}\mathrm{K}^{-1}$ if $\langle \varepsilon_{12} \rangle \approx 1$. Even if Eq. (7.78) is valid, h_r cannot be readily disentangled from h_c *by experiment*. The definition Eq. (7.53) of δ_t based on a measured temperature difference between a surface and surrounding fluid and a measured temperature gradient at the surface includes energy exchange by net radiation even if it is not described adequately by Eq. (7.79). If the temperature difference between surface and surroundings is sufficiently large, the contribution of radiation to total energy exchange becomes nonlinear, at which point we turn tail and run. A good rule of thumb is that when the temperature of the surface or object of interest is 50 °C higher (or lower) than that of its surroundings, both temperatures in the range of normal terrestrial temperatures, the linearization error climbs above 25%.

From Eq. (7.11), the limit of **q** as $k_t \to 0$ is zero if $|\nabla T|$ is finite. If such a material existed (it does not), we could wrap it around a system of interest so that any processes it undergoes is adiabatic. Would this be sufficient? No, unless the system is also wrapped with yet another nonexistent material, one with zero emissivity over the range of relevant wavelengths. This underscores that absolutely adiabatic processes are unrealizable idealizations.

Radiation and Convection Combined: Dew and Frost Formation

Both radiation and convection contribute to energy flux densities. An example is energy exchange between ground and atmosphere. A radiative thermometer pointed at the overhead sky, clear or cloudy, will record a *radiative temperature* T_r, which is not necessarily air temperature. T_r is an *inferred* temperature based on the assumption that the radiating object is a Planckian emitter in the infrared. A typical value for a mid-latitude clear sky is about 250 K, that of a cloudy sky is appreciably higher, perhaps 20–30 °C, even if the air temperature is the same. You can verify this on a day with broken clouds. Point a radiation thermometer at clear and adjacent cloudy sky. On a cold early December afternoon in Central Pennsylvania we measured an air temperature of 274 K and radiative temperatures of 232 K for a patch of blue sky, 261 K for a nearby not-very-thick cloud. The lowest value we found for downward infrared radiation, a four-year average from an automated weather station on the Antarctic plateau, is 125 Wm^{-2}, which from the Stefan-Boltzmann law corresponds to a radiative temperature 217 K. The seasonal variation in this radiation is much less than that of downward solar radiation. The atmosphere radiates 24 hours a day, 365 days a year, at all latitudes, Consequently, the total downward infrared radiation over a year usually appreciably *exceeds* the total downward solar radiation. And yet this infrared radiant energy is ultimately transformed solar radiant energy mediated by Earth. This does *not* violate conservation of energy. The globally averaged annual radiant energy budget of Earth is balanced to within a small but uncertain amount, extremely difficult to measure.

Because radiative thermometers cannot distinguish between emitted and reflected radiation, it is possible to "trick" them. How do we know that the difference between radiative temperatures is not simply a difference between reflection by clear and cloudy skies of radiation emitted by the ground? Tough question. Calculated reflection of infrared radiation by clouds is small, even if they are only tens of meters thick. But why should you believe calculations? One reason is that the kinds of theories that predict high reflection of visible solar radiation by clouds also predict very low infrared reflection by them, even less by clear skies. Perhaps the best indirect experimental evidence is measured reflection of snow (on the ground), the optical properties of which are similar to those of clouds. Reflection by snow plummets at wavelengths greater than about 2 μm.

We simplify the sky as a nonreflecting (in the infrared) plate with a single temperature T_r and downward radiative flux density $\bar{\sigma}T_r^4$. Terrestrial surfaces are approximately Planckian emitters and hence upward emission is $\bar{\sigma}T_s^4$, where T_s is the surface temperature. Because T_r and T_s are not greatly different, Eq. (7.78) is a good approximation for the net radiative flux density. If we ignore conduction from below ground and assume that the Sun is well below the horizon, the total net energy flux density at the surface is

$$q = q_c + q_r = h_c(T_s - T_a) + h_r(T_s - T_r), \qquad (7.80)$$

where T_a is the air temperature above the surface. By rearranging, Eq. (7.80) can be written

$$q = h_t(T_s - T_e), \qquad (7.81)$$

where the overall film coefficient is the sum of the radiative and convective coefficients,

$$h_t = h_c + h_r, \qquad (7.82)$$

and the *effective temperature* of the environment is a weighted average of the air temperature and the radiative temperature of the sky

$$T_e = \frac{h_c}{h_c + h_r}T_a + \frac{h_r}{h_c + h_r}T_r . \qquad (7.83)$$

From Eqs. (7.81) and (7.83), we gain insight into the conditions under which dew and frost form overnight. If the ground temperature is initially greater than T_e, as it is likely to be at sundown, the ground temperature will decrease until $q = 0$ (again, we ignore conduction from below ground), at which point $T_s = T_e$. Thus, the lowest temperature to which the ground can fall is the effective temperature, which in general is *not* air temperature. For dew or frost to form, the ground temperature must drop below air temperature (Section 5.2). This is possible if $T_e < T_a$, which is more likely to occur the smaller the value of T_r and the greater the ratio h_r/h_c.

Underlying this result and everything in this chapter is the implicit assumption that net evaporation from or net condensation onto the ground is negligible. This is not always a valid assumption. Appreciable net evaporation from a surface can result in its temperature dropping well below air temperature (Section 5.3).

All else being equal, morning temperatures are higher after a cloudy night because *emission* by clouds is greater than by clear air at the same temperature, *not* because clouds reflect radiation emitted by the ground. The same amount of water substance emits differently depending on its phase. Although reflection of infrared radiation by clouds is not zero, it is usually negligible. Our eyes deceive us. We cannot see infrared radiation. Merely because thick clouds (or lakes) highly reflect radiation we *can* see does not mean that they necessarily highly reflect radiation we *cannot* see. Early morning frost on lawns may be punctuated by narrow bare spots around the bases of trees or other objects, evidence that they are radiatively warmer than the clear night sky but not as extensive.

Because T_r is generally higher on cloudy nights, clear nights favor the formation of dew or frost. But a clear night is not sufficient. Winds must be light. You don't have to know much about convective energy transfer to know that h_c increases with wind speed. You learn this the first time you stand shivering in a cold, howling wind. If $h_c \gg h_r$, then $T_e \approx T_a$ according to Eq. (7.83). Thus, if convection dominates the energy flux density, ground temperature cannot fall (much) below air temperature, and hence, neither dew nor frost is likely to form.

Equation (7.83) might not pass muster with boundary-layer meteorologists (who live in Earth's boundary layer and study it) because they tend not to think about surface energy flux densities in the heat transfer language of engineers. This simple equation is intended mostly to spur readers to make and interpret observations. On mornings after a night of light wind and a clear sky, measure ground temperature in an open area with an infrared thermometer and also air temperature. Do the same after cloudy nights or windy nights or nights both cloudy and windy. Compare what is observed in light of this equation. A morning ground temperature ten or more degrees *below* air temperature is no more mysterious than an afternoon ground temperature ten or more degrees *above* air temperature. The only difference is that we can see visible solar radiation but not infrared radiation.

We may rewrite Eq. (7.81) as

$$qA = \frac{T_s - T_e}{R_t}, \tag{7.84}$$

where the total thermal resistance R_t is

$$\frac{1}{R_t} = \frac{1}{R_c} + \frac{1}{R_r} = \frac{1}{h_c A} + \frac{1}{h_r A} = \frac{1}{h_t A}. \tag{7.85}$$

Resistances for radiation R_r and convection R_c combine as resistors in parallel because radiation is an additional mechanism by which a system can exchange energy with its surroundings. The resistance of resistors in parallel is always less than the same resistors in series.

To Insulate or Not to Insulate?

Suppose that you were faced with deciding whether or not to wrap insulation around hot water pipes with the aim of lowering your heating bill. Would this decision be clear cut? This question is sometimes used to trip up engineering students because the knee-jerk answer may be wrong.

For simplicity, consider a slab made of a metal, for example, copper or aluminum. Will tacking insulation onto the slab increase or decrease its thermal resistance? The answer seems so obvious. The function of insulation is to insulate, the more the better. This would indeed be true if radiation were negligible, but if not, we have to exercise a bit of thought. Radiation sometimes is the forgotten mode of energy transfer because unless a radiating object is so hot that it emits appreciable amounts of visible radiation (e.g. the heating element of a stove), we cannot see it by its emitted radiation.

As rough approximations the infrared emissivity of polished aluminum and copper is about 0.05, that of insulators such as asbestos paper and rubber foam are 0.90–94; Styrofoam is an outlier with an emissivity of 0.60. We continue to write emissivities in brackets to emphasize that they are imprecise. Emissivity tables usually are short on details such as temperature, wavelength range, and emission directions—normal to surfaces or a hemisphere of directions—and rarely cite original sources. Moreover, emissivities for a given material vary from sample to sample and depend on surface conditions.

Consider first a bare metal slab with emissivity $\langle \varepsilon \rangle$, thickness d_m, and thermal conductivity k_m, in surroundings with an emissivity ≈ 1 and at an actual temperature T_s. The thermal resistance of the bare metal slab is compounded of convective and radiative resistances in parallel with a conductive resistance in series,

$$R = \frac{1}{h_t A} + \frac{d_m}{k_m A}, \tag{7.86}$$

where h_t is Eq. (7.85). If the bare slab is insulated with a material having thermal conductivity k_i and thickness d_i, the total thermal resistance increases by $d_i/k_i A$. But this is not all that happens. The insulation is now the outermost layer, and its emissivity is likely to be appreciably greater than that of the bare metal. Although the conductive resistance is increased, the radiative resistance is decreased.

Now suppose that $h_c \ll h_r$ and neglect the thermal resistance of the slab, a good assumption for a thin, metallic slab. With these assumptions, the thermal resistance of the bare slab is

$$R_{ba} \approx \frac{1}{4\bar{\sigma}\langle \varepsilon \rangle T_s^3 A}, \tag{7.87}$$

whereas that of the insulated slab $\langle \varepsilon \rangle \approx 1$ is

$$R_{in} \approx \frac{1}{4\bar{\sigma}T_s^3 A} + \frac{d_i}{k_i A}. \tag{7.88}$$

The thermal resistance of the insulated slab is *less* than that of the bare slab if

$$\frac{1 - \langle \varepsilon \rangle}{\langle \varepsilon \rangle} > 4\bar{\sigma}T_s^3 \frac{d_i}{k_i}, \tag{7.89}$$

which is possible because $\langle \varepsilon \rangle$ can be $\ll 1$. This is not a theoretician's pipe dream because measurements show that the thermal resistance of a bright tin pipe *decreases* substantially by covering it with a layer of asbestos paper. If d_i is large enough the conductive thermal resistance eventually dominates, but this doesn't come for free.

If $h_c \gg h_r$, the thermal resistance of the bare slab is

$$R_{ba} \approx \frac{1}{h_c A}, \tag{7.90}$$

whereas that of the insulated slab is

$$R_{in} \approx \frac{1}{h_c A} + \frac{d_i}{k_i A}. \tag{7.91}$$

For this example, the resistance of the insulated slab is *greater* than that of the uninsulated slab. The convective thermal resistance is smaller the higher the wind speed. Thus, for pipes to be used outdoors in a windy environment, insulating bare pipes may increase their thermal resistance. But for pipes to be used indoors, sheltered from wind, insulation may decrease thermal resistance. You already may have hit on how to have the best of both worlds. Wrap the pipe with insulation to increase the conductive resistance and then wrap this insulation with metallic foil to increase radiative resistance. This is why vacuum flasks are silvered. If you watch houses being built, you'll notice that walls are often insulated with fiberglass batting sandwiched between sheets of aluminum foil.

We did a crude kitchen experiment. We filled a thin-walled steel cylinder, its surface polished but not to a mirror finish, 6 cm in diameter, 12 cm long, with 350 ml of water at 59.7 °C. In 25 min the water temperature decreased by 12.3 °C in the bare cylinder and by 12.8 °C in the cylinder wrapped with a cotton athletic sock. The difference is not significant, but we conclude that the sock increased the conductive resistance at the expense of the radiative resistance.

Doubling the thickness of insulation for a wall doubles its thermal resistance. Although doubling the insulation around long cylindrical pipes carrying fluids hotter than room temperature is similar, there are differences. For a fixed inner diameter the insulation thickness increases with outer diameter but so does the *surface area per unit length*. If the thickness is small compared with the inner diameter, insulating a pipe is no different from insulating a house. As insulation is added, however, eventually the rate of increase of thermal resistance per unit length changes from linear to logarithmic.

Radiation in Porous Media

As noted in the previous paragraph, the radiative resistance of a porous insulating medium can be increased by encasing it in low-emissivity metal foil. This radiative resistance might be called the *external* radiative resistance of the medium. What about its *internal* radiative resistance?

To answer this we appeal to the simplest conceptual model of a porous medium, two identical parallel slabs with emissivity $\langle\varepsilon\rangle$ separated by a stagnant air gap of thickness d_a. We assume that the slabs are opaque and that absorption and emission by the air between them is negligible. But we do not neglect reflection by the slabs. The net radiative flux density follows from Eq. (7.75) with $\langle\varepsilon_1\rangle = \langle\varepsilon_2\rangle = \langle\varepsilon\rangle$

$$q_r = \frac{\bar{\sigma}\langle\varepsilon\rangle(T_1^4 - T_2^4)}{2 - \langle\varepsilon\rangle}, \tag{7.92}$$

where T_1 is the temperature of the surface (facing the gap) of the upper slab and T_2 is the temperature of the lower slab. As previously, we set $\Delta T = T_1 - T_2$ and expand Eq. (7.92) in a power series in $\Delta T/T_1$ to obtain

$$q_r \approx h_r(T_1 - T_2), \tag{7.93}$$

where

$$h_r = \frac{4\bar{\sigma}\langle\varepsilon\rangle T_1^3}{2 - \langle\varepsilon\rangle}. \tag{7.94}$$

The corresponding radiative resistance is

$$R_r = \frac{1}{h_r A}. \tag{7.95}$$

The total thermal resistance R_t of the two slabs separated by an air gap is

$$\frac{1}{R_t} = \frac{1}{2R_s + R_a} + \frac{1}{R_r}, \tag{7.96}$$

where R_a is the thermal resistance of the gap and R_s is the thermal resistance of a slab. The thermal resistances of slabs and gap are in series, and this combined resistance is in parallel with the radiative resistance. With the assumption that $R_s \ll R_a$,

$$R_t \approx \frac{R_r R_a}{R_r + R_a}. \tag{7.97}$$

In the limit of zero emissivity the radiative resistance is infinite, and $R_t = R_a = d_a/k_a A$. The ratio of the gap resistance to the total resistance is

$$\frac{R_a}{R_t} = 1 + \frac{R_a}{R_r} = 1 + \frac{4\bar{\sigma}\langle\varepsilon\rangle T_1^3 d_a}{k_a(2 - \langle\varepsilon\rangle)}. \tag{7.98}$$

We may write this more compactly as

$$\frac{R_a}{R_t} = 1 + \frac{k_r}{k_a},$$

(7.99)

where the *radiative conductivity* k_r is defined by

$$k_r = \frac{4\bar{\sigma}\langle\varepsilon\rangle T_1^3 d_a}{2 - \langle\varepsilon\rangle}.$$

(7.100)

This equation resembles the expression for gas conductivity, Eq. (7.39), with d_a playing the same role for radiation that the mean free path does for conduction. As it must, radiative conductivity vanishes in the limit of zero gap width.

The maximum value of $\langle\varepsilon\rangle/(2 - \langle\varepsilon\rangle)$ is 1, which we use to estimate the radiative conductivity. At $298\,\mathrm{K}$ the thermal conductivity of air is about $0.025\,\mathrm{Wm^{-1}K^{-1}}$. With these values in Eq. (7.100)

$$\frac{R_a}{R_t} = 1 + 240 d_a.$$

(7.101)

For a gap width of 1 mm, the radiative resistance contributes about 25% to the total; for a gap width of 100 μm, the radiative contribution is about 2.5%. As temperature increases, the radiative resistance decreases, and hence, q_r increases for a given ΔT. The radiative conductivity at 373K is about twice that at 298 K, all else being equal. On the basis of this admittedly simple analysis, we conclude that although radiation contributes to the energy flux density in porous media, it is not the major contributor. Real porous media have a complicated three-dimensional structure a slab model cannot capture. And a more thorough analysis would include the thermal resistance of the slabs.

Newton's Law of Cooling According to Newton

We have laid the groundwork for the simplest form of Newton's law of exponentially decreasing temperature with time. A solid body with thermal conductivity k_b and linear dimension L may be characterized *approximately* by a single time-dependent temperature if the *Biot number*, $\mathrm{Bi} = h_t L / k_b$, is $\ll 1$, which makes physical sense, especially if we call k_b/L the thermal conductance of the body and h_t its surface conductance. The inverse of a conductance is a resistance. The smaller the body and the larger its thermal conductivity, the more that temperature gradients within it are ironed out quickly, whereas the greater the film coefficient, the greater these gradients. For iron with $L = 10$ cm and $h_t \approx 10\,\mathrm{Wm^{-2}K^{-1}}$, $\mathrm{Bi} \approx 0.01$.

For simplicity take the body to be a sphere of radius a surrounded by an unbounded fluid at constant pressure. Given the assumption of a spatially uniform but time-varying temperature T, the time rate of change of the enthalpy of a sphere of volume V with density ρ and specific heat capacity per unit mass c_p, all assumed independent of temperature,

$$\frac{dH}{dt} = \rho c_p V \frac{dT}{dt} \tag{7.102}$$

is equal to the cooling rate

$$Q = qA = -h_t(T - T_\infty)A, \tag{7.103}$$

where A is the area of the sphere, $T_\infty \leq T$ is the temperature of the surrounding fluid at distances beyond which it is not sensibly changing, and h_t is the total film coefficient, assumed to be independent of the temperature difference (not true for free convection). For solids, $c_p \approx c_v$ and we can drop the subscript. These two equations can be combined to yield

$$\frac{dT}{dt} = -\frac{T - T_\infty}{\tau}, \tag{7.104}$$

where the *cooling time* τ is

$$\tau = \frac{\rho c}{h_t}\left(\frac{A}{V}\right)^{-1} = \frac{\rho c}{h_t}\frac{a}{3}. \tag{7.105}$$

Regardless of the shape of the body, we expect the cooling time to decrease the greater the surface to volume ratio. For a fixed ratio, τ decreases with increasing h_t and with decreasing specific heat capacity per unit volume ρc. Thus, τ passes a physical plausibility test, which doesn't guarantee that it is correct, only that it is not *obviously* incorrect. Although not explicit in Eq, (7.105), k_b is hidden in the assumption that it is sufficiently large that temperature gradients *within* the body are negligible.

If T_∞ is independent of time, Eq. (7.104) can be written

$$\frac{d}{dt}(T - T_\infty) = -\frac{T - T_\infty}{\tau}, \tag{7.106}$$

the solution to which is

$$T - T_\infty = (T - T_\infty)_0 e^{-t/\tau}, \tag{7.107}$$

where the subscript 0 indicates the temperature difference at $t = 0$. If Newton were alive, he might recognize this form of his cooling law (which is a *heating* law if $T_0 < T_\infty$), especially if we took its logarithm. More important, Eq. (7.107) is testable: measure the difference between the temperature of the sphere and that of the surrounding fluid over time and plot the results on a semilogarithmic scale. If the data points fall on a straight line to within experimental error, Newton's law of exponential cooling is obeyed. It may be obeyed, although not with the cooling time Eq. (7.105), for a range of temperature differences and times, even if Bi is *not* $\ll 1$ (Fig. 7.3). This is only a *sufficient* condition. It may not always be *necessary*, especially if by temperature is meant that in the center of a cooling body with a simple symmetric shape (e.g. sphere, cylinder) and an initially uniform temperature. The solution to Eq. (7.13) for a sphere of any size, thermal conductivity, and initial temperature distribution, subject to the *boundary condition* $-k_b \partial T / \partial r = h_t(T - T_\infty)$

at $r = a$, is an infinite series of functions of the form $f_n(r) \exp(-t/\tau_n)$, where $\tau_1 > \tau_2 > \ldots$, which we might call Newton's extended law of cooling. Eventually the time dependence will be a *single* exponential determined by τ_1. If $h_t = 0$, $a \neq 0$, and k_b is finite, the sphere is isolated and any initial temperature distribution will evolve to a uniform temperature (not necessarily ambient) in a time $t \gg \tau_1$.

The cooling time τ in Eq. (7.105) does not depend on the thermal conductivity because it is assumed to be so large that its exact value is irrelevant. We now face another apparent puzzle. The cooling times $\tau_n = a^2/\kappa\bar{\alpha}_n^2$, where $\kappa = k_b/\rho c$, appear not to depend on h_t. But this dependence is hidden in the $\bar{\alpha}_n$, which are roots of a transcendental equation that includes the Biot number $h_t a/k_b$.

Equation (7.107) may look familiar. We gave it without proof in Section 2.1 in a discussion of the temporal response of thermometers. With Eq. (7.105) we have an expression telling us how the response time of a thermometer depends on its physical characteristics.

How well does Eq. (7.105) predict the response time of a thermometer (in air)? To answer this question we first rewrite this equation as

$$\tau = \frac{\rho c \delta_t}{k_a} \frac{V}{A}. \tag{7.108}$$

A mercury-in-glass thermometer is, as its name implies, a composite object, but we ignore the glass sheath of the bulb and assume that it is entirely mercury. The density of mercury is 1.36×10^4 kgm^{-3} and its specific heat capacity is 140 J kg^{-1}K^{-1}. Take the thermal boundary layer thickness δ_t to be 2×10^{-4} m and the ratio V/A to be 5×10^{-3} m. With these values and a thermal conductivity for air of 2×10^{-2}Wm^{-1} K^{-1}, we obtain a response time τ of about 100 s. This is close to the response time we gave in Section 2.1. In fact, it is so close that we could be accused of cheating. But we didn't consciously cheat; we just made plausible guesses for the values of all the quantities in Eq. (7.108), and out popped a plausible response time. A thermometer must be wispy to respond to fluctuations over time scales much smaller than one second. A response time this small requires a thermometer 50 times smaller than the bulb of an ordinary thermometer.

*Thermometers and Soils as Low-pass Filters

We assumed in deriving Eq. (7.106) that air temperature T_a is independent of time. But air temperature continuously varies on many time scales: seasonally, diurnally, hourly, and even shorter time intervals (although with a smaller amplitude). We can express this by writing

$$T_a = \Sigma A_n \cos \omega_n t, \tag{7.109}$$

where A_n is the amplitude of the nth component of the time variation of air temperature and ω_n is its angular frequency. With T_a thus expressed, Eq. (7.104) becomes

$$\frac{dT}{dt} + \frac{T}{\tau} = \frac{1}{\tau}\Sigma A_n \cos \omega_n t. \tag{7.110}$$

Multiply both sides of Eq. (7.110) by the integrating factor $\exp(t/\tau)$ to obtain

$$\frac{d}{dt}\left(e^{r/\tau}T\right) = \frac{1}{\tau}\Sigma A_n e^{r/\tau}\cos\omega_n t. \tag{7.111}$$

Perhaps the easiest way to find the antiderivative of $e^{r/\tau}\cos\omega_n t$ is to recognize that it must have the form $e^{t/\tau}(a\cos\omega_n t + b\sin\omega_n t)$ and determine a and b accordingly. The result is

$$\frac{1}{\tau}e^{r/\tau}\cos\nu_n t = \frac{d}{dt}\left(\frac{e^{t/\tau}(\omega_n\tau\sin\nu_n t + \cos\omega_n t)}{1 + \omega_n^2\tau^2}\right), \tag{7.112}$$

and hence, the solution to Eq. (7.111) is

$$T = Ce^{-t/\tau} + \Sigma\frac{A_n(\omega_n\tau\sin\nu_n t + \cos\omega_n t)}{1 + \omega_n^2\tau^2}, \tag{7.113}$$

where C is a constant. For times large compared with τ, $e^{-t/\tau}$ is negligible. If we neglect this term and do a bit of rearranging,

$$T = \Sigma\frac{A_n}{\sqrt{1 + \omega_n^2\tau^2}}\cos(\omega_n t - \phi_n), \tag{7.114}$$

where

$$\sin\phi_n = \frac{\omega_n\tau}{\sqrt{1 + \omega_n^2\tau^2}}, \quad \cos\phi_n = \frac{1}{\sqrt{1 + \omega_n^2\tau^2}}, \quad \tan\phi_n = \omega_n\tau. \tag{7.115}$$

T is the instantaneous temperature recorded by the sensor. The amplitude and phase of each component of this temperature are, in general, not the same as those of each component of air temperature. The nth component of air temperature is

$$T_{an} = A_n \cos\omega_n t, \tag{7.116}$$

whereas the nth component of the sensor temperature is

$$T_n = \frac{A_n}{\sqrt{1 + \omega_n^2\tau^2}}\cos(\omega_n t - \phi_n). \tag{7.117}$$

The low-frequency components of T and T_a are those for which $\omega_n \ll 1/\tau$, and hence, from Eqs. (7.115)–(7.117),

$$T_n \approx T_{an}. \tag{7.118}$$

The high-frequency components are those for which $\omega_n \gg 1/\tau$, and hence,

$$T_n \approx \frac{A_n}{\omega_n\tau}\sin\omega_n t. \tag{7.119}$$

The amplitude and phase of the low-frequency components of T and T_a are about the same, whereas the phase of the high-frequency components of T is shifted by

90° and, more importantly, their amplitude is considerably less than that of the corresponding component of T_a. A filter is a device such that what is fed into it is not what comes out. Thus, a thermometer is a *low-pass filter*: all components of air temperature are fed into the thermometer, but it passes the low-frequency components and suppresses the high-frequency components. What is meant by low and high depends on the response time τ, which, in turn, depends on the size of the sensor. The larger the sensor, the greater its response time.

This kind of filtering occurs in soils. At the surface, soil temperature is more or less in step with the diurnal and seasonal variations of air temperature. But with increasing depth, the high-frequency components of air temperature are increasingly shifted in phase and reduced in amplitude. Sufficiently deep in the soil, there is neither night nor day, nor spring, summer, winter, nor fall. Joseph Fourier knew 200 years ago that "No diurnal variation can be detected at the depth of about three metres; and the annual variations cease to be appreciable at a depth much less than sixty metres." It is fitting that Eq. (7.109) is today called a Fourier series.

Chilliness at High Altitudes: Forced Convection

Air pressure and temperature decrease with elevation because the relentless downward pull of gravity causes ceaselessly moving molecules to be concentrated more near the surface, and for some to gain gravitational potential energy they must lose kinetic energy. If we are uncomfortably cold at high elevations (outdoors, not in chalets sitting in front of fires) it is partly because of lower temperatures. But does the "thinness" of mountain air contribute in any way?

To this point, h_c is only a symbol masking a complicated dependence on the thermophysical properties and states of fluids and their motions adjacent to finite objects of different size, shape, and even orientation. In contrast, c_p for a given pure substance depends mostly on temperature. h_c is a can of worms rarely opened in physics courses; students are unlikely even to learn that it exists. But if they don't, they will not be able to understand everyday phenomena such as cooling of humans in wind, still air, and still water. h_c does not make the hearts of physicists beat faster. The eminent physicist Paul Dirac opined that "it is more important to have beauty in one's equations than to have them fit experiment." By his criterion the equations for h_c are unsightly, cabalistic power-law fits to experimental data. But to do calculations in order to answer questions we have to open the can and let out a few worms, ugly though they may be. In this and the following section we try to make h_c sufficiently presentable to appear in public.

Our starting point is the dimensionless *Nusselt number* Nu defined as

$$h_c = \frac{k_t}{D}\text{Nu}, \tag{7.120}$$

where D is a characteristic linear dimension of a finite object and k_t is the thermal conductivity of the surrounding fluid (air, water). For a sphere, D could be its diameter. From Eq. (7.21) $D/k_t = RA$ is the thermal resistance of a slab of thickness

D composed of a stationary material with the thermal conductivity of the fluid. $h_c A$ is a conductance, as is $1/RA$, so we can interpret Nu as a ratio of convection to conduction. From Eq. (7.55) we can write

$$\text{Nu} = \frac{D}{\delta_t} \tag{7.121}$$

and interpret Nu as the size of the object measured in units of thermal boundary layer thickness.

High mountains engender high winds, for example, the 110 ms^{-1} (231 mph) winds recorded on Mt. Washington, New Hampshire, considered the world's most dangerous small mountain (1916 m) on which around 50 hikers have died of hypothermia, and not just in winter. To grasp the reason for this requires Nu for forced convection (wind). For convenience we pretend that forced and free convection are separable, even though a fluid in motion does not "know" how it got that way.

We model a human as a long circular cylinder. The dominant determinant of Nu, and hence, h_c, is the Reynolds number, discussed in Section 7.2:

$$\text{Re} = \frac{\rho D u_\infty}{\mu}, \tag{7.122}$$

where D is the cylinder diameter, u_∞ is the undisturbed wind speed perpendicular to the cylinder, μ is the dynamic viscosity of air, and ρ is its density. Its thermal conductivity is independent of pressure and proportional to $\sqrt{T}$ [Eq. (7.39)]. We removed the warts from a *correlation* for Nu to obtain

$$\text{Nu} \approx 0.63 \text{Re}^{1/2} \tag{7.123}$$

for a cylinder in crossflow. This is rightly called a correlation, not a theory. Adding frills to it does not change our conclusions. We show in Section 7.2 that μ [Eq. (7.141] is independent of pressure and proportional to $\sqrt{T}$. Thus, h_c is proportional to $T^{1/4}$, and a 3% decrease from, say, 293 K to 263 K is insignificant. But ρ is proportional to the number density n. At 4000 m, close to the greatest height of any mountain in the continental United States and Europe excluding Russia, we estimate a number density 60% of that at sea level, the square root of which is 78%. Thus, there is a potential maximum *decrease* in h_c of 20% at this elevation, all else being equal. We conclude that there is nothing special about h_c for thin air. Whenever possible it is instructive to not only debunk faulty physical explanations, the supply of which is limitless, but trace their origins. Often, they are giants who, in the course of distinguished careers, inevitably leave behind detritus mixed with gems. But the pygmies who follow in their wake are incapable of distinguishing between them and errors are faithfully passed on uncorrected in textbooks for hundreds of years (*not* an exaggeration). The inherent coldness of "thin" air is an example, and we give a problem on this at the end of the chapter.

Take D for an erect adult to be 0.3 m. At a wind speed of 30 ms^{-1} (63 mph) and $\mu = 1.7 \times 10^{-5}$Nsm^{-2} at 0 °C and 1 atm, Re $= 700,000$ and $h_c = 43$Wm^{-2}K^{-1}.

The product $h_c \Delta T$ is the quantity that counts, the outward energy flux density, which for $\Delta T = 20\,°C$ is a bone-chilling 840 Wm^{-2}, well within the capability of Himalayan and Andean peaks and Mt. Washington to deliver. In the Prairie Provinces of Canada, q (Wm^{-2}) once was reported in winter alerts. Many years ago someone sent us one that made us shiver: 1000 Wm^{-2}. But by 2001, Wm^{-2} was abandoned as being "too technical" for Canadians, as it always had been for their southern neighbors.

Given the inverse relation, Eq. (7.121), between Nu and the thickness of the boundary layer, increasing wind speed thins this layer, the limit being zero, and hence, zero external resistance. But for an increase in wind speed from 1 m s^{-1} to 100 m s^{-1}, near the limit of what is possible, the decrease is a factor of 10.

Another contribution to the chilliness of mountaintops is decreased downward infrared radiation from less atmosphere overhead. Our exponential fit to calculations by Bruce Kindel (University of Colorado) of downward clear-sky infrared radiation for the mid-latitude summer standard atmosphere from sea level to the top of Mt. Everest yielded a scale height of 5.3 km with a coefficient of determination of 0.99. We therefore predict that downward infrared radiation at the summit of Mt. Washington is diminished by 30%, which would be compensated for somewhat by increased solar radiation. Clouds decrease solar radiation but increase infrared radiation. These are secondary factors. What kills unprepared or unlucky hikers and climbers in mountains is high winds combined with low temperatures, not the decreased density ("thinness") of mountain air. According to the Centers for Disease Control and Prevention, from 1999 to 2011 a total of 16,911 deaths between 1999 and 2011 in the United States were "associated with exposure to excessive natural cold,"—twice as many men as women (draw your own conclusions).

From Eqs. (7.120), (7.122), and (7.123) $q \propto 1/\sqrt{D}$, which *appears* to conflict with the observation that dogs and other animals, as well as humans, instinctively try to curl up into a ball in cold weather. For simplicity, take the characteristic length D to be the inverse of the surface to volume ratio. Q/V is $qA/V \propto h/D \propto 1/D^{3/2} \propto (A/V)^{3/2}$. Volume is fixed but D is not, and a sphere has the least surface to volume ratio of any regular solid, so curling up in cold, windy weather to decrease A is a good strategy. Assertions about Q/V being proportional to surface to volume ratio fail to account for the dependence of h on D in forced convection. Children have a *higher* surface to volume ratio than adults, so for this reason alone (there are others) are more vulnerable in cold weather.

Wind chill, a controversial and misunderstood concept, is at heart no more than explicit recognition that q depends on ΔT *and* h, which itself depends on wind speed. And a high value of q is what causes hypothermia and frostbite. But if someone presents at a hospital with symptoms of hypothermia, core temperature is measured, *not* "heat loss." Wind chill originated from measurements made in Antarctica (1941) by Paul Siple and Charles Passel of freezing rates of water in a 6.5-cm diameter plastic cylinder, from which they inferred a heat transfer coefficient similar to Eqs. (7.120) and (7.123), which they called the wind chill factor. Its product with degrees below 33 °C they called the wind chill index. Their intent was *not* to model cooling of the human body, but rather to combine ambient temperature and wind speed into a single index to indicate quantitatively how frostbite risk depends on more than just air temperature—a laudable goal. This work was subsequently criticized and spawned a succession of competing wind chill temperatures aimed at the general public, which hasn't the foggiest idea what exactly they mean. But q for a human depends on air temperature and wind speed as well as cloud cover, solar radiation, bare or covered skin, body size, where on the body, clothing, etc. Wind chill temperature is the meteorological equivalent of the body mass index.

Cooling in Air and Water: Free Convection

We can be moderately comfortable lightly dressed in *still air* at 15 °C, whereas a dip in *still water* at this temperature would be chilling. Why? The widespread mantra is that the thermal conductivity of water is about 24 times that of air (true), and so we "lose heat" 24 times faster in water (rubbish). We once thoughtlessly attributed greater cooling in water mostly to the greater conductivity of water but dug deeper and discovered that reality is more interesting. Although it takes more effort to unearth, it requires only basic algebra and experiments anyone can do using a bathtub and a plastic bucket.

First, we must define what we mean by cooling of a human body. We are *homeotherms*. Our bodies marshal various feedback mechanisms outside our conscious control (autonomic) to maintain our *core* temperature at about 37 °C. If our fingers and toes become very cold, we may be uncomfortable but would not likely die. We have measured (with a thermocouple) temperatures of our fingers as low as 15 °C, which we obviously survived. A core temperature this low, or even 25 °C, would likely be fatal. The importance of a nearly constant brain temperature is evidenced by altered mental states accompanying high fevers (hyperthermia) and hypothermia.

By cooling we mean the rate of decrease of core temperature, which is slowed by metabolism (Section 3.9) and thermoregulation, most notably cutaneous vasoconstriction, the shrinking of blood vessels to the skin to prevent warm blood in the essential core from being wasted on the expendable periphery. Our brain senses that frostbite is preferable to death and responds accordingly.

Because of the complexity of the human body we cannot rigorously calculate the rate of decrease of core temperature in living humans immersed in air and in water at the same temperature. Nor is it acceptable to toss people into water at various temperatures and observe how long they live. We can, however, do calculations for simple models of an inert human body and measure cooling curves in air and water for simulated humans lacking metabolism and thermoregulation. This will tell us what our bodies have to combat, sometimes a losing battle, as noted in the previous subsection.

As a kind of warmup for the main bout, we first address the following question. The faster you walk, the higher your metabolic rate, which helps in cold weather to keep your core temperature up to standard. But the faster you walk, the greater the convective energy flux from your body. How fast do you have to walk before forced convection (self-induced wind) swamps free convection?

Free convection is buoyancy driven. The vertical acceleration of a parcel is proportional to $g\Delta T/T$, where ΔT is the difference between the parcel temperature and that of its surroundings T (Section 3.5). Any correlation for free convection must incorporate this. Nu is often written as a constant times the *Rayleigh number* Ra $=$ Gr Pr raised to a fractional power. The *Grashof number* is

$$\text{Gr} = \frac{g\alpha\Delta T L^3}{\nu^2}, \tag{7.124}$$

where α is the volume coefficient of thermal expansion and L is a characteristic (vertical) dimension. For a gas, $\alpha = 1/T$. The appearance of L is expected from experience with bonfires. To get one roaring quickly, stack kindling vertically. The *Prandtl number* Pr (Section 7.2) for air is about 0.71. A simplified expression for Nu adequate for our purposes is

$$\text{Nu} = 0.15\,\text{Ra}^{1/3}. \tag{7.125}$$

Thus, the ratio of forced to free convection is

$$4.2\frac{L}{D}\frac{\text{Re}^{1/2}}{\text{Ra}^{1/3}}. \tag{7.126}$$

For a cylindrical human 1.75 m tall, $L/D = 5.8$. For estimating Gr we can take $\Delta T/T \approx 1/30$ to obtain Gr $= 10^{10}$ and Ra $= 0.71 \times 10^{10}$. In a previous example Re was 700,000 for a wind speed of 30 ms^{-1}, so for 2 ms^{-1}, Re $=$ 47,000. Thus, the ratio Eq. (7.126) is about 2.6. The average pace for a male marathon runner is about 2.5 ms^{-1}; a brisk walking pace is 1.5 ms^{-1}.

Because h_c is proportional to ΔT, q is proportional to $(\Delta T)^{4/3}$, and we cannot expect free convection cooling rates to be exactly exponential although they may be approximately if the range of temperatures is sufficiently small (see Fig. 7.3).

Now to the main event: the ratio of free convection h for water and air *all else being equal*. The Prandtl number can be written in different ways but for our purposes we take it to be $\rho c_p \nu/k$. From Eqs. (7.120), (7.124), and (7.125)

$$\frac{h_w}{h_a} = \left(\frac{k_w}{k_a}\right)^{2/3}\left(\frac{\rho_w}{\rho_a}\right)^{1/3}\left(\frac{\alpha_w}{\alpha_a}\right)^{1/3}\left(\frac{c_{pw}}{c_{pa}}\right)^{1/3}\left(\frac{\nu_a}{\nu_w}\right)^{1/3}, \tag{7.127}$$

where w and a denote water and air. We implicitly assumed that our model cylindrical human is completely immersed in water or air. The 24-fold greater thermal conductivity of water results in only a factor of 8.3. At sea level the density of water is about 800 times that of air, and so the cube root of the ratio of densities is 9.3; the cube root of the ratio of specific heat capacities is 1.6.

The other two factors require a bit more thought because ν and α in the thermal boundary layer (film) depend on temperature. It is customary to take the film temperature to be the average of the surface temperature and that beyond the film.

The kinematic viscosity of air increases with temperature, that of water decreases. For a film temperature of 20 °C, the cube root of the ratio of viscosities is 2.5, whereas the cube root of the ratio of coefficients of thermal expansion is 0.4. These two factors exactly cancel but not by design: we picked 20 °C at random.

The final result for h_w/h_a is a whopping 123. Does this mean that we 'lose heat" more than a hundred times faster in water? We hope not. This is five times more ridiculous than 24. There must be more to the story.

Radiative energy transfer is quite different in air and in water. Suppose that you are in a room with a wall temperature less than your surface temperature. For example, on a cold December morning we measured a wall temperature of 13.5 °C while wearing a shirt with surface temperature of 22.5 °C. Transmission in *air* of infrared radiation emitted by the wall to the shirt is $\approx$ 100%. But transmission in *water* over distances of order 10 μm is $\approx$ 0%, and thermal boundary layers are of order millimeters thick. Because only water from within a small fraction of a millimeter from the surface contributes to net radiation, and the temperature of this water is close to surface temperature, net radiative exchange is negligible. From Eq. (7.82), the total film coefficient in air is $h_t = h_r + h_a$ and now the relevant ratio is

$$\frac{h_w}{h_r + h_a} = \frac{h_w/h_a}{1 + h_r/h_a}, \tag{7.128}$$

with h_w/h_a from Eq. (7.127). As a rough rule of thumb, $h_r \approx h_a$ for free convection and the ratio 123 drops to about 60, which does not mean that you "lose heat" 60 times faster in water. What is missing is that the energy flux from the core is proportional to the difference between core and environmental temperatures and inversely proportional to the *total thermal resistance* [Eq. (7.64)], which includes that of muscle, fat, and bone. The thickness of thermal boundary layers in air and water is of order a millimeter, whereas the thickness of a human hand is about 25 mm. Although the thermal conductivity of muscle is about that of water, that of fat is about half. Our natural insulation greatly reduces cooling rates in air and water. The fraction of body fat depends on sex and age but is in the range 30–40%, greater for women. Before invoking measurements of relative cooling rates, we turn over one more stone.

The major contributor to h_w/h_a is the greater density of water, followed closely by its greater thermal conductivity; its greater specific heat per unit mass is only a minor factor. But why is the thermal conductivity of water greater? This is a bit difficult to answer because we have to compare two different chemical substances in different phases. To get around this we compared the thermal conductivity of water vapor (close to that of air) with that of liquid water, a factor of about 27 (at 300 K), far less than the ratio of densities. In fact, the thermal conductivity of an ideal gas is independent of density. From this, we conclude that density is the single factor, explicit and implicit, that makes h_w appreciably greater than h_a. But if we reasoned *linearly* solely on the basis of density, we would have to conclude that we "lose heat" 1000 times faster in water. The relative rate is much smaller. To support this we turn to a simple experiment.

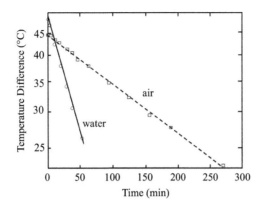

Fig. 7.3 Cooling curves of water, initially at 60 °C in a 15.6-liter plastic bucket, in water and in air. The solid lines are least-squares fit to an exponential function. The difference between water and ambient temperature is on a logarithmic scale. From Bohren, C. F. (2012).

Figure 7.3 shows cooling curves in air and in water for a simulated human the size of a 4–6 year-old child. A 15.6-liter rectangular plastic bucket (uninsulated) was stuffed with polyester fiberfill, to suppress internal convection, and saturated with water at an initial temperature of 60 °C. The temperature T_c in the center (core) of the bucket immersed in a water-filled bathtub (12 °C) and in air (16 °C). was measured versus time. The difference between core temperature and ambient temperature T_∞ was fitted to $(T_c - T_\infty)_0 \exp(-t/\tau)$; the fit is excellent even though the Biot number is *not* $\ll 1$ and $q \propto (\Delta T)^{4/3}$. The cooling time τ was about 6.67 hours in air, and 1.35 hours in water, a ratio of 4.9. We do not know for certain how this ratio scales with size but do know that the ratio of cooling times of a one-liter water-filled flask in air and in water is about 13.5. If we take the ratio of volume to surface to be the cube root of volume, this ratio for the bucket relative to that for the flask is $(15.6/1)1/3 = 2.5$, its product with 4.9 is 12.2. This agreement suggests that the ratio of cooling times scales with surface to volume ratio. And indeed this is consistent with Eqs. (7.120), (7.124), and (7.125), according to which h is independent of L. Because the thermal conductivity of fats is appreciably less than that of water, it would be interesting to do the same experiment using fiberfill saturated with canola oil, its thermal conductivity one-third that of water.

Bogus explanations about relative cooling rates in air and in water arise from single-factor, linear reasoning combined with a failure to distinguish between q, k, h, dT_c/dt (as well as its time integral), and incomplete because the natural insulation of a cooling human is ignored. We now can write

$$T_c - T_\infty = (T_c - T_\infty)_0 \exp(-t/\tau). \qquad (7.129)$$

The *initial* rate of cooling $(dT_c/dt)_0$ is inversely proportional to τ, and the ratio of these rates for water and air is τ_w/τ_a. If the ambient temperature is 15 °C and the initial core temperature is 37 °C, after one hour in water core temperature drops to about 26 °C, likely fatal for a human. Estimates of survival times in water based on experiments with humans (who were not "sacrificed") predict 4.3 hours. Our estimate is too short because it does not account for metabolism and thermoregulation, but it is not ridiculous and tells us the extent to which our bodies ward off hypothermia.

Our criticism of the 24-fold ratio is not new. In a 1946 paper in a medical journal, G. W. Molnar begins with "the usual explanation that water conducts heat from the body twenty to twenty-five times more rapidly than air, is inadequate." Toward the end of his paper he concludes, on the basis of analysis of all the measurements he could find, that "heat loss from the body in water is only twice that in air, or only one-tenth that expected on the basis of the ratio of water/air conductivity." This was written around the time of the Second World War when merchant ships were being torpedoed in the North Atlantic and dumping seamen into cold water to drown or die of hypothermia.

By unqualified hypothermia is meant accidental hypothermia, an involuntary decrease of core temperature to below about 35 °C. Surviving core temperatures below about 25 °C would be unlikely, but not impossible, even with immediate and expert medical intervention. In Norway—the best country in which to be hypothermic because of its unexcelled resources for and expertise in hypothermic resuscitation—a 29-year-old women whose core temperature had dropped to 13.7 °C was resuscitated, but this took heroic efforts and a recovery time of many months. This is why we get angry when we read in newspapers or hear on the radio that someone "froze to death." No one dies by turning into an icicle. Death comes long before solidification. But "freeze to death" may lead some people to think that if air temperature is above 0 °C they can't freeze to death. Many hypothermia deaths occur in late fall and early spring when temperatures are 5–10 °C, or even higher. On May 23, 2021, 21 lightly clad ultramarathon runners died of hypothermia in China even though the minimum temperature was 6 °C. Wind, rain, and clouds tipped the balance. Believe it or not, people die of hypothermia in Florida.

At the other extreme, hyperthermia, the highest core temperature a human has survived (after 24 days in hospital) is 46.5 °C. The range of survivable temperature differences from "normal," is appreciably greater for hypothermia (23.3 °C) than for hyperthermia (9.5 °C). This is no surprise, given that the structure of many proteins begins to change (denature) above 41 °C. An everyday example of denaturation is hard-boiled eggs. Normal is in scare quotes because the measurements on which 37 °C was based may have been inaccurate and average temperatures in the United States appear to have been steadily decreasing. Although often used synonymously, "normal" and "average" have different connotations. Deviations from average are not necessarily abnormal. Core temperatures vary diurnally, and older people have lower temperatures. That of one of the authors is often 35 °C. Normal core temperature is a misnomer for the average of a statistical distribution of temperatures measured for a specified sample of people not obviously sick.

Human bodies are fine-tuned chemical retorts. Not only are absolute rates of chemical reactions highly dependent on temperature, but also so are *relative* rates. A rule of thumb among chemists is that rates of chemical reactions increase (or decrease) by roughly a factor of two for every 10 °C increase (decrease) of temperature. This rule follows from the Arrhenius equation, according to which rates of chemical reactions are proportional to

$$A \exp(-E_a/R^*T), \qquad (7.130)$$

where $R^* = 8.341\,\mathrm{J\,mol^{-1}K^{-1}}$ is the universal gas constant (Section 2.8) and E_a is (misleadingly) called the activation energy. Although it has the dimensions of energy and looks superficially like proper energies in the denominators of the exponents in Eqs. (2.49), (2.56) and (5.50), it is not always an energy barrier nor even has an approximate physical interpretation. Unique constants A and E_a for each reaction do not exist and themselves may depend on temperature. Equation (7.130) is best looked upon as an *empirical*, rather than a theoretical, relation obtained by fits to measurements. If we take $E_a = 50\,\mathrm{kJ\,mol^{-1}}$ as a typical value, a 10 °C decrease (increase) from core temperature does indeed yield a factor of two decrease (increase) in reaction rate. Although "normal" core temperature is different for different people and varies diurnally, a core temperature of only 40 °C is considered extreme *hyperthermia* (heatstroke). A core temperature change of only ±5 °C disrupts *absolute* and *relative* reaction rates sufficiently to require a quick trip to a hospital emergency room to stave off a slow trip to the morgue.

7.2 Momentum Transfer: Viscosity

We interpret gas pressure as a molecular momentum flux density (Section 2.3) for a gas at rest (or, equivalently, one in which mass motion is everywhere the same). The pressure on solids or liquids immersed in such a gas acts normal to their surfaces. How could it do otherwise? If there were a tangential component of force, it would have to act in a particular direction, yet all directions in a plane parallel to a surface in a gas at rest are equivalent. Symmetry considerations rule out the possibility of a *shear stress* (tangential force per unit area) on a surface in a gas at rest. Shear stresses require some kind of asymmetry, one source of which is a *velocity gradient*.

The intrinsic property (independent of state of motion) of a fluid that determines shear stresses is its *coefficient of viscosity*, or simply *viscosity*, a term that may evoke images of oil flowing slowly through a funnel into the crankcase of an automobile. Viscosity is defined by the following *thought* experiment. Imagine that a thin, flat plate of area A is towed through a fluid at constant velocity u parallel to a surface at a uniform distance d from the plate. The linear dimensions of the plate are large compared with d, and u is sufficiently small that the flow is smooth (or laminar), rather than turbulent. A force F in the direction of u is required to tow the plate. For *some* fluids the shear stress τ is given by

$$\tau = \frac{F}{A} = \mu\frac{u}{d}, \tag{7.131}$$

where μ is the coefficient of viscosity or, to be precise (many books are not), the *dynamic viscosity*. When in doubt, look at the units: the SI unit for dynamic viscosity is Pa s^{-1}; the cgs unit is the *poise* (P); 1 P = 0.1 Pa s^{-1}. Because of the no-slip boundary condition, the fluid velocity at the surface is zero, that at the plate is u, and hence, u/d is the velocity gradient. Thus, we can write Eq. (7.131) more generally as

$$\tau = \mu\frac{du}{dz}, \tag{7.132}$$

where u is the component of velocity in the x direction, and the z direction is perpendicular to the surface. Gravity is irrelevant, so don't interpret z as necessarily being the direction of gravity.

Equation (7.132) is not an inviolable physical law; it is an example of a *constitutive relation*, similar to the Fourier law (which, despite its name, is also not a law). Equation (7.132) is the simplest assumption one can make about shear stresses in fluids. It satisfies the requirement that shear stresses vanish if velocity gradients do. Fluids for which Eq. (7.132) is a good approximation are called *Newtonian*, all others being denoted, naturally enough, as *non-Newtonian*. By a stroke of good luck, the two fluids of greatest interest to us—air and water—are properly Newtonian. But things could have been otherwise, and if they had, it is likely that progress in fluid mechanics would have been slower. Paints and multi-grade motor oils are non-Newtonian.

What would happen if things were different? For example, suppose that the relation between shear stress and velocity gradient for a particular fluid happened to be described more accurately by

$$\tau = \mu'\left(\frac{du}{dz}\right)^2. \tag{7.133}$$

This also satisfies the requirement that the shear stress be zero if the velocity gradient is zero. Suppose we didn't know that Eq. (7.133) is a better approximation to reality. We still could *define* the viscosity of this fluid by Eq. (7.132), which from Eq. (7.135) would imply that

$$\mu = \mu'\frac{du}{dz}. \tag{7.134}$$

Now we would be in a pickle. If μ' were independent of the flow, μ would not be, and the viscosity of this fluid could not be considered one of its intrinsic properties. Viscosity would be less useful if it depended on the details of flow rather than solely on thermodynamic state variables such as pressure and temperature.

Constitutive relations of the form Eq. (7.132) (as well as the Fourier law) are useful because for some fluids under some conditions, viscosity (or thermal conductivity) depends only on the thermodynamic state of the fluid, not on its motion. Constitutive relations supplement the fundamental (macroscopic) laws of conservation of energy, mass, and momentum. Although all fluid motion is constrained by the same conservation laws, all fluids are not described by the same constitutive relations. Moreover, they cannot be derived without appealing to a higher order theory. If you want to know how the viscosity of a fluid depends on thermodynamic variables, or why the viscosity of one fluid is different from that of another, the macroscopic equations of fluid flow are of no help. Constitutive relations are independent of these equations, and in the absence of experimental evidence or a microscopic theory Eqs. (7.132) and (7.133) and many others would be equally possible constitutive relations.

Before we tackle a microscopic theory of Newtonian viscosity, consider a simple analogy that sheds light on the mechanism for viscosity in gases. Two very long trains are traveling parallel to each other in the same direction but at different speeds. The trains are sufficiently far apart that they never touch. Passengers on the two trains continually throw identical balls back and forth to each other at the same speed in the direction perpendicular to the tracks. Although there is no net transfer of balls from one train to the other, there is a net transfer of momentum. Passengers on the faster train throw balls having greater momentum (along the direction of travel) than that of the balls they receive from passengers on the slower train. As a consequence, the faster train loses momentum in its direction of travel, whereas the slower train gains momentum, and in the absence of other forces the faster train decelerates while the slower train accelerates until both are traveling at the same speed.

Given this train analogy, let's turn to momentum transfer in a gas in which the x component of velocity varies in the z direction. This velocity component is the sum of the mass motion (center-of-mass) velocity u and the component about the center of mass v_y. At any point $n/2$ molecules per unit volume are moving in the positive z direction (and also in the negative z direction). The flux density of x momentum in the z direction is therefore

$$P_{zx} = \frac{n}{2} \langle v_z m (u + v_x) \rangle, \tag{7.135}$$

where v_z is the z component (either positive or negative) of the velocity of a molecule. If the z and x components of velocity are completely uncorrelated

$$\langle v_z v_x \rangle = \langle v_z \rangle \langle v_x \rangle = 0 \tag{7.136}$$

because $\langle v_x \rangle = 0$. The average of the z component of velocity does not vanish because it is taken over all positive (upward) or all negative (downward) directions:

$$\langle v_z \rangle = \langle v \cos \theta \rangle = \frac{1}{2} \langle v \rangle. \tag{7.137}$$

By combining Eqs. (7.135)–(7.137), we obtain

$$P_{zx} = \frac{1}{4}nm\langle v\rangle u. \tag{7.138}$$

Although this is a momentum flux density, it is not the *net* momentum transfer flux density. To obtain this we have to make the same kind of arguments as in the derivation of the thermal conductivity of a gas. Molecules at $z - \lambda$ carry their x momentum a distance λ (on average) upward without a collision. Similarly, molecules at $z + \lambda$ carry their x momentum a distance λ downward. Thus, the net transfer of x momentum upward across the plane $z = $ const is

$$\tau_{zx} = P_{zx}(z - \lambda) - P_{zx}(z + \lambda). \tag{7.139}$$

Assume that temperature and number density are independent of z; only u, the mass motion x component of velocity, varies with z. If we expand $P_{zx}(z \pm \lambda)$ in a Taylor series about z and retain only the first two terms,

$$\tau_{zx} = -\frac{1}{2}mn\langle v\rangle\frac{du}{dz}. \tag{7.140}$$

For a positive velocity gradient du/dz, x momentum is transported downward.

Now return to the plate towed through a fluid. For the velocity of the plate to be constant, no net force must act on it. Thus, the fluid must exert a force $-F$ on the plate (F is the external towing force). This force is the net momentum flux density Eq. (7.140), which, together with the definition of viscosity Eq. (7.132), implies that

$$\mu = \frac{1}{2}nm\langle v\rangle\lambda = \frac{1}{2}\frac{m\langle v\rangle}{\sigma}. \tag{7.141}$$

About this simple equation Sir James Jeans asserted that it "needs innumerable adjustments." He said a mouthful. We could have labored mightily to obtain a possibly more accurate expression for viscosity, but instead we gave a simple derivation in the same spirit as that for the thermal conductivity of a gas (Section 7.1). It seemed pointless to go to greater efforts because the gas of interest to us (air) is a mixture. If we had done a more rigorous derivation of the viscosity of a single-component gas, we then would have had to tackle the knotty problem of how to determine the viscosity of a mixture knowing the viscosities of its components. Don't be surprised if you find expressions for viscosity somewhat different from Eq. (7.141), especially the numerical constant. Even viscosities obtained more rigorously don't agree much better with measured values.

How accurate is Eq. (7.141) for air? At sea level the density of air is about $1.2\,\mathrm{kgm^{-3}}$, its mean free path is about $0.8 \times 10^{-7}\mathrm{m}$, and $\langle v\rangle = 450\,\mathrm{ms^{-1}}$(at 0 °C). With these values in Eq. (7.141) $\mu = 2.2 \times 10^{-5}\mathrm{kgm^{-1}s^{-1}}$. The measured value is about $1.7 \times 10^{-5}\mathrm{kgm^{-1}s^{-1}}$. An expression for μ similar to Eq. (7.141) but with $1/2$ replaced by $1/3$ yields a viscosity closer to the measured value for air, but too low instead of too high.

The mean free path λ is inversely proportional to number density n, and hence, to mass density ρ. Thus, the product $\rho\lambda$ in Eq. (7.141) is independent of density (and temperature). A viscosity independent of density (equivalently, pressure for fixed temperature) may run counter to notions of viscosity shaped by everyday terms such as heavy oil (high viscosity) and light oil (low viscosity). We are on shaky ground when we compare the viscosities of liquids to those of gases. To drive home this point, note that, according to Eq. (7.141), the viscosity of a gas *increases* as the square root of absolute temperature (because of the dependence of $\langle v \rangle$ on T). A gas becomes *more* viscous with increasing temperature, which makes little sense to anyone whose only experience with the temperature dependence of viscosity was acquired by heating cold maple syrup (a Newtonian fluid) so that it spreads over and seeps into pancakes.

Gases become more viscous with increasing temperature, whereas liquids usually become less viscous because the mechanisms for viscosity are quite different in gases and liquids. This should come as no surprise because the ideal gas equation is an excellent approximation for air but fails miserably for liquid water. In gases viscosity is a manifestation mostly of momentum transfer (which is why viscosity increases with temperature), whereas in liquids viscosity is affected both by momentum transfer and by the nonnegligible forces exerted by the closely packed molecules on each other. The viscosity of a gas is independent of density (or pressure) for the same reason that its thermal conductivity is independent of pressure. The greater the gas density, the greater the number density of carriers of momentum. But for a given velocity gradient, the momentum carried between collisions decreases with density, and hence, density drops out of the picture. To good approximation the viscosity of a gas is independent of density (pressure) and increases approximately as the square root of absolute temperature.

The striking difference in the viscous behavior of gases and liquids is illustrated by water in its two fluid forms. Over the temperature range 0–100 °C the viscosity of water vapor *increases* by about 17%, whereas over this same range the viscosity of liquid water *decreases* by more than a factor of six. For dry air at a pressure of 0.1 MPa, measured viscosities in multiples of 10^{-6} Pa s^{-1} are 16.06 at 250 K and 23.1 at 400 K, their ratio 1.438. The ratio of the square roots of the absolute temperatures is 1.265, about 12% too *low*, whereas the ratio of absolute temperatures is 1.6, about 11% too *high*. In 1866, Maxwell reported measurements of viscosity at 185 F and 51 F, the ratio of absolute temperatures almost equal to the ratio of viscosities, sometimes adduced as evidence for a linear dependence on absolute temperature. But his ratio of viscosities is too high by about 8%. Neither a linear nor a square root dependence fully captures the temperature dependence. If you squint, either seems to fit. The pattern for thermal conductivities is similar. For the same two temperatures the linear dependence is 8% too high, the square root dependence 15% too low. This similarity is no surprise because viscosity and thermal conductivity are transport coefficients for momentum and energy, respectively.

Air is sometimes said to be a fluid with a "small" viscosity. Unless properly qualified, this is meaningless. Small and large are relative, not absolute: small relative to what? Merely because the viscosity of air (in SI units) is a number much smaller than unity means nothing because we could choose the units of mass, length, and time to give the viscosity of air any magnitude we wish. We have to probe a bit deeper to understand what is meant by saying that the viscosity of air is small.

Shear stress is proportional to a velocity gradient. What is a good upper limit (although still realistic) for a velocity gradient? The fluid velocity at a surface is zero because of the no-slip condition. What is the characteristic distance from the surface over which the velocity rises from zero to that of the surrounding fluid? In the previous section we discussed the thermal boundary layer near a surface, the region over which temperature changes markedly. By analogy, we expect a viscous boundary layer, the region over which velocity changes markedly. Because viscosity and thermal conductivity are both transport coefficients (note the similarity in the expressions for the two quantities), we expect the thickness of the viscous boundary layer to be comparable to that of the thermal boundary layer. Take the viscous boundary layer thickness to be 10^{-3} m. Suppose that the free-stream velocity of the air in which the surface is immersed is 50 ms^{-1}, a very stiff wind. With these assumptions the velocity gradient near a surface is 5×10^4 s^{-1}. For a viscosity of 2×10^{-5} kgm^{-1}s^{-1} this gradient corresponds to a shear stress of 1 Pa, which is indeed small compared with sea-level atmospheric pressure (100 kPa). For this reason the viscosity of air is assumed to be negligible in many atmospheric applications. One application for which viscosity cannot be neglected is determining the drag on falling cloud droplets and rain drops.

For a sufficiently small and slow droplet (we'll clarify what is meant by "sufficiently"), the drag force is dominated by viscosity. The area of a sphere of radius a is proportional to a^2 (we don't fuss over the constant of proportionality because all such constants are folded together). The total shear force acting on the sphere is therefore proportional to τa^2, where the shear stress τ is proportional to the viscosity μ and a velocity gradient. If u is the velocity of the sphere (relative to the surrounding fluid), it is reasonable to take the velocity gradient to be proportional to u/a. With these assumptions we obtain the drag force

$$F = C\mu a u, \tag{7.142}$$

where the constant C is unobtainable by our crude analysis. A careful derivation yields $C = 6\pi$, and the drag force is called *Stokes's law*. When a cloud droplet falls at its terminal velocity u_∞, the drag force and gravitational force (weight) balance:

$$\frac{4}{3}\pi a^3 \rho_w g = 6\pi a \mu u_\infty, \tag{7.143}$$

and hence,

$$u_\infty = \frac{2a^2 \rho_w g}{9\mu}. \tag{7.144}$$

A typical cloud droplet radius is 10^{-5}m. For this radius, a water density ρ_w of 10^3 kgm^{-3}, and a viscosity for air of 2×10^{-5} kgm^{-1}s^{-1}, the terminal velocity is about 0.01 ms^{-1}. In a distance of about twice its diameter the droplet falls from rest to more than 99% of its terminal velocity. The lifetime of a typical cloud is about half an hour, say 2000 s. Even if we ignore updrafts, a cloud droplet falls only about 20 m during the lifetime of its parent cloud, usually a tiny distance relative to the height of cloud base. When we are asked why clouds don't fall out of the sky, we respond that they *do*—but not very far. And they also evaporate below cloud base. Cloud droplets, a thousand times denser than air, moist or dry, are *negatively* buoyant. Clouds are composed of cloud droplets. Ergo, clouds are negatively buoyant. Despite this, we found on a U.S. government agency website: "Just as oil floats on water because it is less dense, clouds float on air because the moist air in clouds is less dense than dry air." If u_∞ for a cloud droplet is less (in magnitude) than the updraft velocity, it will have a positive upward velocity *relative to the ground*. But this has nothing to do with the tiny density difference between moist and dry air (for the same p/T). *All* natural air is moist, so if cloud droplets were positively buoyant in moist air, humans could float in it. The simplest way to demolish the wrong explanation of floating clouds, which we have seen often, is to note that rain is by definition *falling* water drops. Why aren't they positively buoyant given that they are different from cloud droplets only in size, not density? Buoyancy is determined by density *differences*. According to Eq. (3.104), the apparent weight of a cloud droplet or rain drop in air is 99.9% of its weight in a vacuum.

A terminal velocity 0.01 m s^{-1} of typical cloud droplets is appreciably smaller than typical updraft velocities in clouds, and much smaller than wind speeds aloft. Thus, the trajectories of such droplets are more horizontal than vertical. For particles ten times larger, the terminal velocity increases to ≈ 1 m s^{-1}, and hence, they could fall vertically as much as 1000 meters in 20 minutes, although they move appreciably greater distances horizontally. Their trajectories are complicated by *wind shear*, an increase of wind speed with height. Below high altitude (> 5000 m) wispy, isolated cirrus clouds one can sometimes see *fall streaks*. These are *not* the trajectories of *individual* ice particles, but rather the instantaneous locus of *many*. This is an example in which the consequences of cloud particles falling can be observed. But they are not precipitation particles in that they are unlikely to reach the ground.

Before we can accept Eq. (7.142), we have to address what is meant by "sufficiently small and slow." An object moving in a fluid has inertia and is acted upon by a viscous force. One or the other of these two forces can dominate. For example, a large ship steaming across a lake does not know that viscosity exists. The ship's captain can order it to be stopped, but that doesn't mean it will stop. It has so much inertia relative to the viscous force that it keeps moving for a long distance (relative to its length) after the engines are cut. A small swimming organism, in contrast, is almost completely under the control of viscosity. When the organism ceases swimming, it stops in a distance comparable to its length. The quantity that determines the relative magnitudes of inertia and viscous forces is the Reynolds number Eq. (7.122) written somewhat differently

$$\text{Re} = \frac{Du}{\nu}, \tag{7.145}$$

where D is a characteristic length and u is a characteristic velocity. The characteristic length of a cloud droplet is its radius (or diameter) 10^{-5} m. For a cloud droplet in air the appropriate density in Eq. (7.145) is that of air, about $1\,\mathrm{kg\,m^{-3}}$. We estimated a droplet terminal velocity of 10^{-2} ms^{-1}. With these values in Eq. (7.145), Re < 0.01. Stokes's law is valid only for Reynolds numbers appreciably less than 1, which is satisfied by cloud droplets. Raindrops, however, are about 100 times larger and fall 10–100 times faster, and hence, the Reynolds number of a falling raindrop is too large for the drag on it to be described accurately by Stokes's law. We could have guessed this just by applying Eq. (7.144) to a raindrop, which yields the unrealistically high terminal velocity of about $100\,\mathrm{ms^{-1}}$.

The ratio of viscosity to density in Eq. (7.145) crops up so often that it gets a name, *kinematic viscosity*, and its own symbol:

$$\nu = \frac{\mu}{\rho}. \tag{7.146}$$

The SI unit for kinematic viscosity is m^2 s^{-1}. The distinction between dynamic and kinematic lies in the absence of mass in the latter. Kinematic parameters in mechanics are velocity and acceleration. We can study motions independent of how they come about. Mass times acceleration is dynamic because it is equal to the force that causes the acceleration. The dimensions of ν are the same as those of the thermal diffusivity κ, and hence, ν is often called the *diffusivity for momentum*. Now we can define the Prandtl number that appeared in Section 7.1:

$$\mathrm{Pr} = \frac{\nu}{\kappa}. \tag{7.147}$$

The simple physical interpretation of Pr is that $\mathrm{Pr}^{1/2}$ is of order the ratio of the velocity (or viscous) boundary layer to the thermal boundary layer. For air, Pr $\approx$ 0.71, for water $\approx$ 8. For both fluids the two boundary layers are about the same thickness.

Now for a bit of controversy. We have read many times that the Grashof number [Eq, (7.124)] is the ratio of the buoyancy force to the viscous force. How can this be correct, given that ν *squared* appears in Gr whereas the viscous force is proportional to ν? Gr/Re is the ratio of the buoyancy force to the viscous force. Only if Re *happens* to be of order 1 does this physical interpretation also apply to Gr. The measured velocities of thermal plumes around humans (Section 1.6) are ≈ 0.1 ms^{-1}. The kinematic viscosity of air is $\approx 10^{-5}\mathrm{m^2s^{-1}}$, and hence, for Re to be of order 1 would require a characteristic dimension D of a human to be about 0.1 mm. Even if the velocity is decreased by a factor of 10, D still corresponds to a ridiculously small human. And we showed that Re for a cloud droplet is less than 0.01. Thus, we reject the unqualified physical interpretation of Gr. The *product* of the ratio of buoyant to viscous force with the Reynolds number is the Grashof number.

Another viscosity you may encounter is *eddy viscosity*. This viscosity may be invoked when the flow is turbulent, momentum being transported in groups of molecules moving together, called eddies, rather than molecule by molecule. But eddy viscosity is not a nice viscosity because its value depends on the flow, not simply on thermodynamic variables.

As noted previously, and as is evident from Eqs. (7.48) and (7.141), thermal conductivity and viscosity are transport coefficients, the first for energy, the second for momentum. To complete the triumvirate of transport coefficients, we turn to the transport coefficient for mass.

7.3   Mass Transfer: Diffusion

We have seen in the previous two sections that energy is transferred in a gas because of temperature gradients, and momentum is transferred because of velocity gradients. We therefore should hardly be surprised to learn that mass is transferred because of concentration (number density) gradients.

As the gaseous mixture of interest let us take dry air (mostly oxygen and nitrogen) mixed with water vapor. The concentration of each component varies with position. Purely by random processes, molecules flow from regions of high to regions of low concentration. No force impels molecules in a preferential direction other than the force of numbers. We assume that (total) pressure and temperature are uniform, which requires that the total number density n of molecules

$$n = n_a + n_v \tag{7.148}$$

be uniform, where n_a is the number density of dry air molecules and n_v is that of water vapor molecules. For the moment, assume that these two number densities vary only in the x direction. Uniformity of n requires that

$$\frac{\partial n_a}{\partial x} = -\frac{\partial n_v}{\partial x}. \tag{7.149}$$

Denote by J_v the net flux density (in the x direction) of water vapor: the net number of molecules crossing unit area of the yz plane in unit time; J_a denotes the net flux density of dry air. For the total number density of molecules in any region to be constant, the flux density of one of these components must balance that of the other:

$$J_v + J_a = 0. \tag{7.150}$$

A sufficient condition that Eq. (7.150) be satisfied, given Eq. (7.149), is

$$J_v = -D\frac{\partial n_v}{\partial x}, \quad J_a = -D\frac{\partial n_a}{\partial x}, \tag{7.151}$$

where D is called the *diffusion coefficient*. The minus sign ensures that a net flow of molecules occurs from regions of higher toward regions of lower concentration. Mathematically, it makes no difference which sign we choose, but physically it does. Equation (7.151) is called *Fick's law* or, to be more precise, Fick's first law, obtained solely by analogy with Fourier's law. His second law follows from a mass balance.

Consider two planes with area A perpendicular to the x axis, one at x and one at $x + \Delta x$. The net flux density of water molecules into this region is equal to the time rate of change of the total number of water molecules between these planes:

$$A J_v(x) - A J_v(x + \Delta x) = A \, \Delta x \frac{\partial n_v}{\partial t} . \tag{7.152}$$

Divide both sides of this equation by Δx and take the limit as Δx goes to zero

$$-\frac{\partial J_v}{\partial x} = \frac{\partial n_v}{\partial t} , \tag{7.153}$$

which from Eq. (7.151) becomes

$$\frac{\partial}{\partial x} \left(D \frac{\partial n_v}{\partial x} \right) = \frac{\partial n_v}{\partial t} . \tag{7.154}$$

If D is independent of position, Eq. (7.154) reduces to

$$D \frac{\partial^2 n_v}{\partial x^2} = \frac{\partial n_v}{\partial t} . \tag{7.155}$$

Equation (7.154) or (7.155) is Fick's second law.

Now we generalize these results to three dimensions. Again, we assume that the total number density of gas molecules is uniform:

$$\nabla (n_a + n_v) = 0 \tag{7.156}$$

Denote by $\mathbf{J}$ (with appropriate subscripts) the vector flux density of diffusing molecules: $\mathbf{J} \cdot \mathbf{n}$ is the rate at which molecules cross a unit area of an arbitrary plane the normal to which is $\mathbf{n}$. No net flow of molecules (air and water vapor) into a region occurs if

$$\mathbf{J}_a + \mathbf{J}_v = 0. \tag{7.157}$$

Again, Eq. (7.157) is satisfied, in light of Eq. (7.156), if

$$\mathbf{J}_v = -D \, \nabla n_v, \quad \mathbf{J}_a = -D \nabla n_a. \tag{7.158}$$

This is the three-dimensional version of Fick's first law. To obtain the corresponding second law for, say, water vapor diffusing in air, we integrate the vector flux density over an arbitrary surface A enclosing a volume V.

$$-\int_A \mathbf{J}_v \cdot \mathbf{n} \, dA = \frac{d}{dt} \int_V n_v dV = \int_V \frac{\partial n_v}{\partial t} dV . \tag{7.159}$$

By using the divergence theorem, we can rewrite Eq. (7.159) as

$$\int_V \left(\nabla \cdot \mathbf{J}_v + \frac{\partial n_v}{\partial t} \right) dV = 0 . \tag{7.160}$$

For Eq. (7.160) to be satisfied for all arbitrary volumes V

$$\nabla \cdot \mathbf{J}_v + \frac{\partial n_v}{\partial t} = 0. \tag{7.161}$$

This is the equation of continuity (Eq. 7.10), which from Fick's first law for water vapor yields

$$\nabla \cdot (D \nabla n_v) = \frac{\partial n_v}{\partial t}. \tag{7.162}$$

If D is independent of position, Eq. (7.162) reduces to

$$D \nabla^2 n_v = \frac{\partial n_v}{\partial t}, \tag{7.163}$$

the three-dimensional form of Eq. (7.155). Note the similarity between Eqs. (7.163) and (7.13). The differential equations governing mass transfer in a (stationary) diffusing medium are the same as those governing the transfer of energy. The mathematics is identical but interpreted differently.

The linear relation between energy flux density and temperature gradient is attributed to Jean Baptiste Joseph Fourier (b. Auxerre, France, 1768; d. Paris, 1830). Although Fourier is not quite a household name unless we restrict ourselves to the households of scientists, especially physical scientists, mathematicians, and electrical engineers. The linear relation between momentum flux density and velocity gradient is attributed to Newton. If you were to ask people at random on the street to name a famous physicist, the response would likely be Newton or perhaps Einstein. The third in the trio of linear relations between a flux density and a gradient is attributed to Fick. Who was Fick, and why is he so obscure? We can't recall ever having seen mention of Fick the man. When did he live, what was his nationality, what did he do other than get his name attached to two laws of diffusion? Curious, we decided to investigate.

Adolf Eugen Fick (b. Kassel, Germany, 1829; d. Blankenberge, Belgium, 1901) began his university studies in mathematics and physics but soon switched to medicine. This explains his neglect by the authors of physics and chemistry textbooks. Fick had the misfortune to belong to the wrong tribe. According to his biographical sketch by K. E. Rothschuh, Fick's lifework was "concerned primarily with problems on the borderline between medicine, physiology, and physics." The fate of those who live in border regions is to be ignored by inhabitants of the interior. Most of Fick's career was spent as professor of physiology in the University of Würzburg. He made many contributions to his field, including what is called the *Fick principle* for determining cardiac output, the volumetric flow rate of blood pumped by the heart.

Fick measured diffusion driven by concentration gradients in *aqueous salt solutions.* How applicable are his laws to diffusion in gases? We return to this question after digressing on the diffusion coefficient.

Diffusion Coefficient

The diffusion coefficient is an inert mathematical symbol until we breathe life into it. Determining the diffusion coefficient for a mixture of air and water vapor is a bit tricky because the mean free paths of water vapor and air are different, as are their mean speeds (for the same absolute temperature). We can save ourselves labor without losing physical understanding if we consider *self-diffusion* within a single gas. We can imagine that some of its molecules are distinguished from others solely by virtue of being radioactive. Thus, we have like diffusing in like. Consider a gas with uniform temperature and total number density (radioactive plus nonradioactive molecules). A concentration gradient of both kinds of molecules exists (in the x direction). Consider the plane $x = $ const. Molecules at $x - \lambda$, where λ is the mean free path, travel an unimpeded distance λ (on average) in the positive x direction. Similarly, molecules at $x + \lambda$ travel an unimpeded distance λ in the negative x direction. Thus, the net flow of molecules across the plane $x = $ const is

$$\frac{1}{4}n(x - \lambda)\langle v \rangle - \frac{1}{4}n(x + \lambda)\langle v \rangle . \tag{7.164}$$

Expand $n(x \pm \lambda)$ in a Taylor series about x and neglect all but the first two terms. The result for the net flux density is

$$J = -\frac{1}{2}\lambda\langle v \rangle \frac{\partial n}{\partial x} . \tag{7.165}$$

It therefore follows from Fick's law that the diffusion coefficient is

$$D = \frac{1}{2}\lambda\langle v \rangle = \frac{\langle v \rangle}{2n\sigma} . \tag{7.166}$$

At sea level, the mean free path in air is about 0.8×10^{-7} m, and the mean speed is about $450\,\mathrm{m s^{-1}}$. This yields a diffusion coefficient (self-diffusion) for air of about $1.8 \times 10^{-5}\,\mathrm{m^2 s^{-1}}$. At 15 °C and a pressure of 1 atm, the coefficient of self-diffusion of oxygen and nitrogen is about $1.8 \times 10^{-5}\,\mathrm{m^2 s^{-1}}$. This agreement between theory and measurement is partly luck. You'll find expressions for D differing from Eq. (7.166) by a multiplicative constant. We chose a simple derivation and didn't worry about fine details. A more rigorous derivation of D may make you feel better, but the result may not agree as well with the quick-and-dirty result. Such is the injustice of life: do more and get less. The coefficient of diffusion for water vapor in air, about $2.5 \times 10^{-5}\,\mathrm{m^2 s^{-1}}$, is comparable in magnitude to the coefficients of self-diffusion of nitrogen and oxygen.

We can interpret molecular diffusion as a random walk. Suppose that a molecule initially at the origin of a line moves a distance λ along it at a constant speed v in either direction with equal probability, then either continues in its original direction or reverses it with equal probability. Call this event a collision. After N collisions, each independent of all the others, a molecule has traveled a total distance $N\lambda$. Suppose that this process is repeated many times, each with a different total displacement from the origin (positive and negative). Although the average displacement is zero, the root mean square displacement is $L = \sqrt{N}\lambda$, a result derived in many places (see, e.g. Reif, 1967, sects. 1.1–1.4). The time t to make N collisions is $N\lambda/v$. Thus, $L^2/t = \lambda v$, the form of the diffusion coefficient Eq. (7.166). $L/t = (\lambda/L)v = v/\sqrt{N}$ as a kind of speed $\ll v$. Diffusion is a random process in which molecules moving rapidly between collisions don't go anywhere on average; however, a few beat the average and slowly iron out concentration gradients.

The Nonexistence of Still Air

Century-old debates about the equilibrium temperature profile of Earth's atmosphere (Section 4.3), with no resolution in sight, are, we believe, a consequence of the *solid air fallacy*. Suppose that an anemometer indicates 0.0 inside a house. If the air were *absolutely* still this would violate building codes, which typically require air turnover times of 15 minutes. Zero more likely reflects a lower limit to what the anemometer can measure. From our treatment of molecular diffusion, we can show that absolutely still air exists only in textbooks and in the minds of theoreticians who have yet to wake up and smell the roses. The notion that buildings are "hermetically sealed," which we have seen in print, is nonsense.

The diffusion coefficient D has the dimensions of length-squared over time, and so by dimensional arguments we can estimate how fast molecules in air move strictly by diffusion. They move at high speeds (of order 100 ms^{-1}) *between* collisions but follow a zigzag trajectory, because of which odorant molecules released at a point in a room make much slower outward progress. Determining the time- and space-dependent concentration of molecules from an instantaneous localized source (plane or point) in hypothetical still air by solving the diffusion equation does not change our estimate. Dimensional arguments often give good enough estimates without having to solve differential equations.

We did two experiments using a cotton swab infused with a few drops of pentanethiol (pentyl mercaptan), which is similar to the odorants put in natural gas in parts per million. The odor threshold of pentanethiol is probably a few parts per billion. When the measured outdoor wind speed was zero, we pulled the swab out of a bottle briefly, then put it back. We smelled an unpleasant odor only for about five minutes. We repeated the experiment in an unventilated, unheated, and uncooled room ($\approx 150 \text{ m}^3$). Again, our anemometer did not budge from zero. But in only a few minutes we smelled the same odor at a distance of about a meter. The molecular weight of pentanethiol is 104, that of ethyl ether is 74, and its diffusion coefficient in air is about $8 \times 10^{-4} \text{ m}^2\text{s}^{-1}$. Diffusion coefficients of different molecules in air are not greatly different, so we used this value for pentanethiol. At a distance d from the point of release, the time after release at which the concentration is greatest is approximately d^2/D, which for ethyl ether and $d = 1$ m, yields a time of about thirty hours, much longer than observed. After about five hours with no one in it the entire room reeked. By purely molecular diffusion this would have taken weeks. We conclude from this that our anemometer is not up to the task of detecting imperceptible winds with detectable consequences.

Fick's law is valid only for a stationary gas, which we argue does not exist. Our experiment shows that Fick's law is *not* valid even within an *apparently* stationary gas. But there is an exception: stagnant *thin* boundary layers near surfaces (the no-slip boundary condition). Beyond these layers mass motion of fluids causes Fick's law, like Fourier's law of thermal conduction and Newton's law of viscosity, to break down. Unlike equilibrium properties such as compressibility and thermal expansion, the transport coefficients k_t, ν, and D explicitly depend on the *molecular* scattering cross section σ and in the same way. For a fixed number density, the collision rate [Eq. (2.80)] increases with increasing σ, as does the rate at which a local nonequilibrium distribution of molecular speeds evolves to the equilibrium distribution, whereas gradients are ironed out more slowly.

Growth of Cloud Droplets

Let us apply Fick's and Fourier's laws to a problem of considerable meteorological interest: the growth of cloud droplets. As a droplet grows by diffusion of water vapor to it, its radius changes, but the rate of diffusion depends on droplet radius. This is another example of a Stefan problem (like the freezing of lakes discussed in Section 7.1), a moving boundary-value problem difficult to solve exactly. We assume that we can use the steady-state solutions to the temperature and diffusion equations at any instant. That is, at each stage of a droplet's growth, it quickly adjusts to the steady-state vapor and temperature distribution appropriate to its instantaneous radius.

The steady-state temperature distribution around a sphere of radius a satisfies Laplace's equation

$$\nabla^2 T = 0 . \tag{7.167}$$

We assume that the temperature T depends only on the radial distance r from the center of the sphere, and hence, Eq. (7.167) becomes

$$\frac{1}{r^2}\frac{\partial}{\partial r}\left(\gamma^2\frac{\partial T}{\partial r}\right) = 0, \quad r > a. \tag{7.168}$$

The solution to this equation is

$$T = -\frac{\bar{B}}{r} + \bar{C}, \tag{7.169}$$

where the integration constants $\bar{B}$ and $\bar{C}$ are determined by boundary conditions. One such condition is that the temperature at the surface ($r = a$) be T_o:

$$T_o = -\frac{\bar{B}}{a} + \bar{C}. \tag{7.170}$$

A second boundary condition is obtained by specifying the temperature T_∞ at the edge of the thermal boundary layer:

$$T_\infty = -\frac{\bar{B}}{a + \delta_t} + \bar{C}. \tag{7.171}$$

Equations (7.170) and (7.171) can be solved for $\bar{B}$ (and $\bar{C}$, which we don't need for our purposes here):

$$\bar{B} = (T_\infty - T_o)a\left(1 + \frac{a}{\delta_t}\right). \tag{7.172}$$

In Section 7.1 we estimated the thickness of thermal boundary layers to be of order 0.1 mm. Because the diameter of a typical cloud droplet is about 10 μm, and hence, $a \ll \delta_t$, we may approximate $\bar{B}$ by

$$\bar{B} \approx a(T_\infty - T_o). \tag{7.173}$$

The total energy flux at the surface of the droplet is

$$4\pi a^2 q = -4\pi a^2 k_a \left(\frac{\partial T}{\partial r}\right)_{r=a} = 4\pi a k_a (T_o - T_\infty). \tag{7.174}$$

The steady-state concentration of water vapor n around a droplet satisfies the same equation as the temperature. Subject to the same approximations that led to Eq. (7.174), we can write down the total flux density of water vapor to the droplet as

$$4\pi a^2 J = 4\pi a D(n_o - n_\infty). \tag{7.175}$$

Multiply both sides of Eq. (7.175) by m, the mass of a water molecule, to obtain the mass flux density:

$$4\pi a^2 J m = 4\pi a D(\rho_0 - \rho_\infty), \tag{7.176}$$

which is equal to the time rate of change of a droplet mass with M:

$$\frac{dM}{dt} = \rho_w 4\pi a^2 \frac{da}{dt} = -4\pi a^2 J m. \tag{7.177}$$

We assume that at the surface of the droplet, the vapor density is equal to the equilibrium (saturation) vapor density at temperature T_o:

$$\rho_o = \frac{e_s(T_o)}{R_v T_o}. \tag{7.178}$$

The *saturation ratio* S is defined as

$$S = \frac{e_\infty}{e_s(T_\infty)}, \tag{7.179}$$

from which follows

$$\rho_\infty = \frac{S e_s(T_\infty)}{R_v T_\infty}. \tag{7.180}$$

By combining Eqs. (7.177), (7.179), and (7.180), we obtain

$$4\pi a^2 Jm = \frac{4\pi a D}{R_v}\left(\frac{e_s(T_o)}{T_o} - \frac{S e_s(T_\infty)}{T_\infty}\right). \tag{7.181}$$

We may take S and T_∞ to be properties of the environment in which a droplet finds itself. The temperature T_o at the droplet surface is not arbitrary but is determined by the energy flux density. We assume that this is given by Eq. (7.174). Consider a system composed of water vapor surrounding a droplet. If some of this water vapor condenses, the result is a decrease in enthalpy. To balance the enthalpy budget would require the temperature of the droplet to increase. But if this temperature is constant (steady state), the decreased enthalpy associated with condensation is instead balanced by an enthalpy flux density to the surroundings:

$$\ell_v\frac{dM}{dt} = 4\pi a k_a(T_o - T_\infty). \tag{7.182}$$

By combining Eqs. (7.177), (7.181), and (7.182), we obtain

$$\frac{\ell_v D}{R_v}\left(\frac{e_s(T_o)}{T_o} - \frac{S e_s(T_\infty)}{T_\infty}\right) = k_a(T_\infty - T_o). \tag{7.183}$$

We could, in principle, solve Eq. (7.183) for T_o given S and T_∞. Instead, we omit this step and proceed directly to the quantity of interest, the mass flux density. We assume a droplet sufficiently large that we need not worry about the dependence of equilibrium vapor pressure on size (see Section 5.10). First, we expand e_s/T in a Taylor series to obtain

$$\frac{e_s(T_o)}{T_o} = \frac{e_s(T_\infty)}{T_\infty} + \frac{d}{dT}\left(\frac{e_s}{T}\right)_\infty(T_o - T_\infty) + \cdots . \tag{7.184}$$

By using the Clausius–Clapeyron equation to determine the derivative of e_s/T, truncating the series Eq. (7.184) after the second term, and assuming that $\ell_v/R_v T_\infty \gg 1$ (a good assumption at normal temperatures), we obtain

$$T_\infty - T_o \approx \frac{R_v T_\infty^3}{\ell_v e_s(T_\infty)}\left(\frac{e_s(T_\infty)}{T_\infty} - \frac{e_s(T_o)}{T_o}\right). \tag{7.185}$$

Now substitute Eq. (7.185) in Eq. (7.183) and do a lot of rearranging to obtain

$$\frac{e_s(T_o)}{T_o} = \frac{e_s(T_\infty)}{T_\infty}\left(\frac{A + BS}{A + B}\right), \tag{7.186}$$

where

$$A = \frac{k_a R_v T_\infty^3}{\ell_v e_s(T_\infty)}, B = \frac{\ell_v D}{R_v}. \tag{7.187}$$

This equation provides us with what we need to obtain the mass flux density by way of Eq. (7.181):

$$4\pi a^2 Jm = \frac{4\pi a(1 - S)}{C}, \tag{7.188}$$

where

$$C = \frac{R_v T_\infty}{D e_s(T_\infty)} + \frac{\ell_v^2}{R_v k_a T_\infty^2} . \tag{7.189}$$

The equation for droplet growth follows from Eqs. (7.177) and (7.188):

$$a\frac{da}{dt} = \frac{S-1}{C\rho_w} . \tag{7.190}$$

A droplet cannot grow unless it is in a supersaturated environment $(S > 1)$. If S is constant, Eq. (7.190) is readily solved:

$$a = \sqrt{a_o^2 + \frac{2(S-1)t}{C\rho_w}}, \tag{7.191}$$

where a_o is the droplet radius at $t = 0$. For times t such that

$$\frac{2(S-1)t}{C\rho_w} \gg a_o^2, \tag{7.192}$$

the rate of growth is independent of initial droplet radius:

$$a \approx \sqrt{\frac{2(S-1)t}{C\rho_w}} . \tag{7.193}$$

According to Eq. (7.193), all droplets in the same environment (same S) should eventually reach the *same* size regardless of their size when they began growing. The interpretation of this becomes clearer by writing Eq. (7.190) as

$$\frac{da}{dt} = \frac{S-1}{aC\rho_w} . \tag{7.194}$$

The smaller the droplet, the greater its rate of growth. If a droplet begins its life small, it compensates by growing at a faster rate.

Equation (7.193) implies that cloud droplets should all be about the same size. This is observed in some, but *not* in all, clouds. Droplet size distributions in wave clouds often are fairly narrow, as evidenced by the coronas and iridescence seen in such clouds. But droplet size distributions in cumulus clouds are *not* narrow; so there is more to cloud droplet growth than simple diffusion in a uniform, constant environment.

Annotated References and Suggestions for Further Reading

The transport coefficients for energy, momentum, and mass are derived from a kinetic theory point of view in, for example, the books by Jeans, Loeb, and Kennard cited at the end of Chapter 2.

Because Joseph Fourier's 1822 classic was translated into English and reprinted many years later, it is not difficult to get your hands on his *The Analytical Theory of Heat* (Dover, 1955).

For a good article on the discovery of convection, see Sanborn C. Brown (1947) "The discovery of convection currents by Benjamin Thompson, Count of Rumford," *American Journal of Physics*. 15, pp. 273–74.

For a simple experiment demonstrating that water vapor does not affect the thermal conductivity of air to any substantial degree, see Craig F. Bohren (1991b) *What light through yonder window breaks?* Hoboken: John Wiley & Sons, ch. 12.

Our assertion in Section 7.1 about the effective thermal conductivity of the pelts of Arctic mammals is based on measurements by P. F. Scholander, Vladimir Walters, Raymond Hock, and Laurence Irving (1950) "Body insulation of some Arctic and tropical mammals and birds," *Biological Bulletin*, 99, pp. 225–36, fig. 3, although the adaptation of this figure by Richard C. Birkebak (1966) "Heat transfer in biological systems," *International Review of General and Experimental Zoology*, 2, pp. 269–344, fig. 19, is more suitable for quantitative purposes.

For a criticism of misconceptions about Newton's law of cooling, see Craig F. Bohren (1991) "Comments on 'Newton's law of cooling—A critical assessment' by Colm T. O'Sullivan (1990, *American Journal of Physics*, 58, pp. 956–60)," *American Journal of Physics*, 59, pp. 1044–6.

For more on boundary layers, see Hermann Schlichting (1968) *Boundary-layer theory*. 6th edn. London: McGraw-Hill. For a physical interpretation of the Prandtl number see pp. 263–5.

The question of whether wrapping metal pipes with insulation increases or decreases thermal resistance is an old one, as evidenced by the brief note by William Schriever (1933) "Increased heat emissivity caused by asbestos 'insulation'," *American Journal of Physics*, 1, pp. 48–9, who cites even earlier (1920) experimental measurements made at the University of Illinois.

A detailed derivation of Stokes flow (low Reynolds number) of a sphere in a fluid is given by G. K. Batchelor (1967) *An introduction to fluid mechanics*. Cambridge: Cambridge University Press, pp. 230–4.

For a delightful discussion of swimming organisms, for which "inertia is totally irrelevant," see Edward M. Purcell (1977) "Life at low Renolds number," *American Journal of Physics*, 45, pp. 3–11.

The only person we have ever seen address the issue of using a diffusive vapor flux density for a water droplet of fixed radius to determine the rate of growth of a droplet with a time-dependent radius is Sean Twomey (1977) *Atmospheric aerosols*. London: Elsevier. Twomey concludes (pp. 71–4) that one can use the steady-state flux density without introducing appreciable error provided that the density of the vapor phase is appreciably less than that of the condensed phase, a condition likely to be satisfied in Earth's atmosphere.

For more on the growth of cloud droplets, see Horace Robert Byers (1965) *Elements of cloud physics*. Chicago: University of Chicago Press; R. R. Rogers and M. K. Yau (1989) *A short course in cloud physics*. 3rd. edn. Oxford: Pergamon; Hans R. Pruppacher and James D. Klett (1980) *Microphysics of clouds and precipitation*. Dordrecht: D. Reidel.

See *Dictionary of scientific biography* for biographical sketches of Fourier (Jerome R. Ravetz and I. Grattan-Guinness) and Fick (K. E. Rothschuh).

For measurements of the viscosity and thermal conductivity of air, see K. Kadoya, N. Matsunaga, and A. Nagashima (1986) "Viscosity and thermal conductivity of dry air in the gaseous phase," *Journal of Physical and Chemical Reference Data*, 14, pp. 947–70. For Maxwell's much earlier measurements, see J. C. Maxwell (1867) "The Bakerian Lecture: on the viscosity or internal friction of air and other gases," *Proceeding of the Royal Society of London*, 15, pp. 14–17.

For properties of moist air, see P. T. Tsilingiris (2008) "Thermophysical and transport properties of humid air at temperature range between 0 and 100 °C," *Energy Conversion and Management*, 49, pp. 1098–1100, figs. 2 and 3.

For diffusion coefficients in air, see G. A. Lugg (1968) "Diffusion coefficients of some organic and other vapors in air," *Analytical Chemistry*, 40, pp. 1072–6.

For a more rigorous solution to the problem of diffusion of an odorant in still air treated by simple dimensional arguments at the end of Section 7.3, see J. Crank (1975) *The mathematics of diffusion*. 2nd edn. Oxford: Oxford University Press, secs. 2.2.1 and 6.1.

For measured radiative and free-convective heat transfer coefficients for the human body see Richard J. de Dear, Edward Arens, Zhang Hui, and Masayuki Oguro (1997) "Convective and radiative heat transfer coefficients for human body segments," *International Journal of Biometeorology*, 40, pp. 141–56. Their measurement of h_r is close to what we estimate; h_c is comparable to h_r but slightly less. See also Yoshihito Kurazumi, Tadahiro Tsuchikawa, Jin Ishii, Kenta Fukagawa, Yoshiaki Yamato, and Naoki Matsubara (2008) "Radiative and convective heat transfer coefficients of the human body in natural convection," *Building and Environment*, 43, pp. 2142–53.

For a good treatment of Newton's law of cooling, see Michael Vollmer (2009) "Newton's law of cooling," *European Journal of Physics*, 30, pp. 1063–84. The solution to the temperature distribution in an arbitrary sphere and initial temperature distribution is given in H. S. Carslaw and J. C., Jaeger (1959) *Conduction of heat in solids*. 2nd edn. Oxford: Oxford University Press, para. 9.4. With a bit of mathematical labor one can show that this solution reduces to Eqs. (7.104) and (7.105) in the limit of small Biot number.

Thermal boundary layers are so thin that it is difficult to measure undistorted temperature distributions in them with ordinary probes. This problem was gotten around nicely by Ralph B. Kennard (1941) "Temperature distributions and heat flux in air by interferometry," in *Temperature and its measurement and control in science and industry*, Vol. 1. New York: Reinhold, pp. 685–706. The temperatures of his hot surfaces were sufficiently above ambient (100+ °C) that radiation was not negligible, the linear approximation was not valid, and yet he makes no mention of radiation.

One monograph on micrometeorology does use, briefly, the language of engineering heat transfer: O. G. Sutton (1953) *Micrometeorology: a study of physical processes in the lowest layers of Earth's atmosphere*. New York: McGraw-Hill, sec. 4.5.

For thermograms of men and women see Eduardo Borba Neves, Ana Carla Chierighini Salamunes, and Rafael Melo de Oliveira (2017) "Effect of body fat and gender on body temperature distribution," *Journal of Thermal Biology*, 70, pp. 1–8.

In the subsection Temperature Scales and Thermometers in Section 2.1 we mention the experience of stepping on a coin on a carpet with a bare foot and the coin feeling colder even though carpet and coin are at the same temperature. The knee-jerk explanation for this is the greater thermal conductivity of the coin. There is more to the story. For a more nuanced explanation see Craig F. Bohren (2015) "What my dogs forced me to learn about thermal energy transfer," *American Journal of Physics*, 83, pp. 443–6. This also is relevant to the toilet seat observation mentioned in Section 7.1.

Stephen G. Brush (1970) "Interatomic forces and gas theory from Newton to Lennard–Jones," *Archive for Rational Mechanics and Analysis*, 39, pp. 1–29. Reprinted in an anthology of classic paper in the kinetic theory of gases by the same author, (2003) *The Kinetic Theory of Gases*. London: Imperial College Press.

For more about the Fick principle in physiology see Leroy D. Vandam and John A. Fox (1998) "Adolf Fick (1829–1901), physiologist," *Anesthesiology*, 88, pp. 514–18.

Bridgman's assertion about "heat current" is in P. W. Bridgman (1927) *The logic of modern physics*. New York: Macmillan, p. 130.

For more about "superinsulation" see Matthias Koebel, Arnaud Rigacci, and Patrick Achard (2012) "Aerogel-based superinsulation: an overview," *Journal of Sol-Gel Science and Technology*, 63, pp. 315–39.

See Kennard, Chapter IV B for a detailed derivation of the thermal conductivity of a gas and Chapter VIII for a discussion of slip at boundaries. For a philosophical and historical examination of the no-slip condition see Michael A. Day (1990) "The no-slip boundary of fluid mechanics," *Erkenninis*, 33, pp. 285–96.

For measurements of ice thickness on a lake over three months in winter, see Justin C. Murfitt, Laura C. Brown, and Stephen E. L. Howell (2018) "Estimating ice thickness in Central Ontario," *PloS ONE*, 13(12), 0208519. The maximum observed thickness was about 50 cm, and the steady increase of ice thickness approximately followed cumulative freezing degree days to the power 0.5.

A remarkably simple experiment shows that emission of infrared radiation by a can filled with water near the boiling point and painted black is no different from that emitted by the same can painted white. Both cans are radiatively "black" although perceptually quite different. See Richard A. Bartels (1990) "Do darker objects really cool faster?", *American Journal of Physics*, 58, 244–8, fig. 3 and chapter 7 in Craig F. Bohren (1991b).

For the whiteness and blackness of snow see Tsutomu Nakamura, Osamu Abe, Tomohiro Hasegawa, Reina Tamura, and Takeshi Ohta (2001) "Spectral reflectance of snow with a known grain-size distribution in successive metamorphism," *Cold Regions Science and Technology*, 32, pp. 13–26, fig. 3; and Junsei Kondo and Hiromi Yamazawa (1986) "Measurement of snow surface emissivity," *Boundary Layer Meteorology*, 34, pp. 415–16. The measured reflectivity of new snow over the visible (400–700 nm) is close to 100%; the measured infrared emissivity is 0.97.

For measurements of shortwave and longwave radiation budgets in Antarctica over four years at five different stations, see Michiel van den Broeke, Carleen Reijmer, and Roderick van de Wal (2004) "Surface radiation balance in Antarctica as measured with automatic weather stations," *Journal of Geophysical Research*, D09193. The average ratio of downward longwave to downward shortwave was 1.44.

Equation (7.123) is a simplified version of eq. (9) in Stuart W. Churchill and M. Bernstein (1977) "A correlation equation for forced convection from gases and liquids to a circular cylinder in cross flow," *Journal of Heat Transfer*, 99, pp. 300–5.

Equation (7.125) is a simplified version of eq. (9) in Stuart W. Churchill and Humbert H. S. Chu (1975) "Correlating equations for laminar and turbulent free convection from a vertical plate," *International Journal of Heat and Mass Transfer*, 18, pp. 1323–9.

An example how the term "blackbody" is confusing is the assertion that black people are more suited to life in hot climates because they are more efficient radiators. All skin is "black" at terrestrial infrared wavelengths. See fig. 7 in R. Bowling Barnes (1963) "Thermography of the human body," *Science*, 140, pp. 870–7. Imaging of *emitted* infrared radiation (thermography) is used in building diagnosis and firefighting, among other applications. Many infrared cameras are frightfully expensive, but ones to be used with smart phones now cost around $200, and the price will likely drop with time.

For a review of thermography see Angeliki Kylili, Paris A. Fokaides, Petrous Christou, and Soteris A. Kalogirou (2014) "Infrared thermography (IRT) applications for building diagnostics: a review," *Applied Energy*, 134, pp. 531–49.

For an extensive review of the thermal properties of snow and ice, which includes the empirical fit to snow conductivity in Section 7.1, see Yin-Chao Yen (1981) *Review of thermal properties of snow, ice, and sea ice: CRREL Report 81–10*. Hanover, NH: US Army Corps of Engineers.

For criticism of the notion that you "lose 40% of your heat from your head," see Rachel C. Vreeman and Aaron E. Carroll (2008) "Festive medical myths," *British Medical Journal*, 337, a2769 (December 17). If you are wearing a thick down-filled hooded parka but barelegged, then what?

For a good history of wind chill, see Randall J. Osczevski (1995) "The basis of wind chill," *Arctic*, 48, pp. 372–82. He measured h versus wind speed in a wind tunnel for a simulated human head. He concludes that "For clothed individuals, wind chill is primarily caused by the local cooling effect of wind on the bare face." The key qualifier here is "clothed." Skin on cheeks is likely to be the only part of the body exposed to cold wind, and hence, the only basis for a single wind chill index.

The measurements of cooling times for a water flask in air at the same temperature but relative humidities of 50% and 100% is described in chapter 12 in Craig F. Bohren (1991b).

For an example of a strong correlation between relative humidity and cloud cover see Chris J. Walcek (1994) "Cloud cover and its relationship to relative humidity during a springtime midlatitude cyclone," *Monthly Weather Review*, 122, pp. 1021–35, fig. 1.

For more details about the fall streaks mentioned briefly in Section 7.2, see Craig F. Bohren and Alistair B. Fraser (1992) "Fall streaks: parabolic trajectories with a twist," *American Journal of Physics*, 60, pp. 1030–3.

For a simulation of global vertical velocity distributions within cirrus see Donifan Barahoma, Andrea Molod, and Heike Kalesse (2017) "Direct estimation of the global distribution of vertical velocity within cirrus clouds," *Scientific Reports*, 7, 6840.

For measurements of updraft and downdraft velocities in fair-weather cumulous clouds see Pavlos Kollias, Bruce A. Albrecht, Roger Lhermitte, and Andrey Savtchenko (2001) "Radar observations of updrafts, downdrafts, and turbulence in fair-weather cumuli," *Journal of the Atmospheric Sciences*, 58, pp. 1750–66.

For an account of the lowest core temperature a human has survived see Mads Gilbert, Rolf Busund, Arne Skagseth, Paul Åge, and Jan P. Solbø (2000) "Resuscitation from accidental hypothermia of 13.7 °C with circulatory arrest," *The Lancet*, 355, pp. 375–6. For the highest temperature see Corey M. Slovis, Gail F. Anderson, and Anthony Casolaro (1982) "Survival in a heat stroke victim with a core temperature in excess of 46.5 °C," *Annals of Emergency Medicine*, 11, pp. 269–71.

Simple theory of the growth of cloud droplets by condensation (Section 7.3) predicts that they all should grow to about the same size, whereas measured size distributions in cumulus clouds are broad. But such clouds are inherently turbulent and hence droplets do not grow in a uniform supersaturated environment (an implicit assumption in the simple theory). For more about this see Raymond A. Shaw, Walter C. Reade,

Lance R. Collins, and Johannes Verlinde (1998) "Preferential concentration of cloud droplets by turbulence: effects on the early evolution of cumulus cloud droplet spectra," *Journal of the Atmospheric Sciences*, 55, pp. 1965–76. They argue that coherent turbulent structures in clouds are the source of inhomogeneous supersaturation and vorticity that broadens size distributions.

Problems

1. Suppose that your house were to be perfectly insulated: R-infinity for all walls, windows, and the roof. This would not mean that you could disconnect your furnace. All livable houses leak. Air circulates between outside and inside. What would it cost (in power) to keep room temperature 25 °C when the outside temperature is 0 °C? You may take the turnover time for air in a house to be about one hour. By turnover time is meant the average time for all inside air to be replaced by outside air. When you obtain the required power, put it in perspective by comparing it with the power output of small space heaters. You will need to make a judicious estimate of the volume of a house.

2. Estimate a plausible upper limit (1000 m is not plausible) for the thickness of lake ice (or sea ice). This thickness is such that if anyone were to tell you traveler's tales about thicker ice you would laugh, whereas you would be highly impressed by reports of somewhat thinner ice but would not consider them outrageous lies. You may take 2.3 $Wm^{-1}K^{-1}$ for the thermal conductivity of ice, 920 kgm^{-3} for its density, and $2.4 \times 10^5 Jkg^{-1}$ for its enthalpy of fusion (at 20 °C). Don't bother about the initial stages of ice growth during which thickness increases linearly with time. Try to verify your guess by digging into the literature on Arctic and Antarctic exploration.

3. In Chapter 2 we noted that humans are not very reliable thermometers. For example, we usually sense a slab of metal to be colder than a slab of wood even though both are at the same temperature. At what temperature would metal and wood be perceived by touch to have the same temperature? The answer is not absolute zero or some other equally absurd temperature. Our sense of warmth gives way to heat pain at skin temperatures above about 44 °C, and our sense of cold gives way to cold pain at temperatures below about 17 °C. For more on thermoreception, see *Fundamentals of Sensory Physiology*, edited by Robert F. Schmidt.

4. In the depths of winter we have been asked by anxious people the following question: "I heard on the radio that the minimum temperature forecast for tonight is 50 °F and the wind-chill temperature is supposed to drop to 30 °F. But the antifreeze in the radiator of my car protects only to 34 °F. Do I run the risk of a frozen radiator tonight?" Please answer this question.

5. One of our students brought to class a large, black, metallic plate, 0.5 cm thick, with a mass of about 1 kg. This "magic plate," for which the student paid $20, had the amazing ability to melt frozen objects very quickly. It was reputedly made of a "space-age" material and, indeed, an ice cube placed on the plate melted at a phenomenal rate. Yet another triumph of modem science? A spinoff of the space program? Please discuss. You can go a long way toward unraveling the mystery of the "magic plate" by doing some simple experiments in your kitchen.

6. What distance does a typical cloud droplet fall from rest before reaching 99% of its terminal velocity? Express this distance relative to the droplet diameter.

7. The viscosity of air at normal terrestrial temperatures is about $2 \times 10^{-5} \mathrm{kgm}^{-1}\mathrm{s}^{-1}$. To drive home the message that this viscosity cannot be said to be small simply because 2×10^{-5} is small (compared with 1), choose a set of units such that the numerical value of viscosity is huge (compared with 1).

8. The ground hog or woodchuck (*Marmota monax*) is a large rodent whose range extends throughout much of the Eastern United States and sub-Arctic Canada. Ground Hog Day is February 2nd. The story goes that if a ground hog emerges from its burrow on this day, sees its shadow, and is frightened by it, wintry weather will return. Although this story may seem to be pure poppycock, it embodies some sound biology and physics. Try to give a physical explanation for this story about ground hogs. *Hints*: The ground hog is a true hibernant, going into a deep sleep in its burrow in winter. Nature has thoughtfully provided the ground hog with an internal mechanism that wakens the animal when its temperature drops to near freezing. Its burrow may be four to five feet deep. To unravel the physics of ground hogs, the starred subsection in Section 7.1 on the thermometer as a low-pass filter should be helpful.

9. One of the authors was driving with a friend, in her car, on a cold but sunny day. When the windshield became dirty, she considered turning on the windshield washer, but hesitated for fear that the washing fluid would freeze because of the low air temperature (about 0 °F) and the even lower "wind-chill temperature" as a result of the fairly high speed of the car (55 mph). The washing fluid, which contained antifreeze (presumably an alcohol), was rated at about −10 °F. She was assured that the fluid would not freeze on contact with the windshield. But, alas, it did. Please explain. What is the difference between this problem and Problem 5? *Hint*: Section 6.4 may help you. Another way of phrasing this question is to ask why neither air temperature nor wind-chill temperature is an infallible guide to determining whether a windshield washing fluid, containing antifreeze, will freeze on contact with the windshield of a moving car?

10. Misconceptions about energy transfer by radiation are so widespread and entrenched that it is difficult to know the direction in which to swing a wrecking ball. One example is black dogs. Owners of such dogs sometimes fret that they risk hyperthermia if exposed for long to intense sunlight. "Obvious" evidence for this is that a black dog's coat may be noticeably quite hot to the touch. But dogs don't die of hyperthermia because of hot hair. Devise a simple

experiment to show that hair hot to the touch is not necessarily a good indicator of the temperature of underlying skin. An infrared thermometer makes the experiment easier to perform, but other than this no special equipment is necessary. What you need can be found in your bedroom. You might also measure the air temperature and skin temperature of a dog (try the belly where there is not much hair). This is *not* solely a radiation problem. Convection and conduction also play roles and solely from temperature measurements it is not possible to assign relative weights to the three modes of energy transfer. In the real world they interact and are not completely independent. How do you expect the thickness of a dog's coat to affect your results? How could you test your expectations? Keep in mind that there are no right or wrong answers here. Experiments provide different answers for different experimental conditions.

11. If the previous problem has whetted your appetite for dog physics, try the following. We have been told that dogs should not be misted (sprayed with water droplets) in sunshine to cool them because the droplets will focus sunlight and increase the incident radiation. How would you test this? Based on the previous problem, if misting is not harmful would it be effective? Misting a dog's coat is not the same as soaking it.

12. In his *Lectures on the elements of chemistry*, Vol. 1, p. 23, Joseph Black gives measurements made by Benjamin Franklin and collaborators of the temperature of a cooling object versus time. At 17, 54, 85, and 167 minutes the temperatures of the object suspended in an evacuated space were $50°, 37°, 30°,$ and $20°$, respectively. For the same object in air and the same temperatures, the times were 7, 22, 29, and 63 minutes. The initial temperature was $60°$. Temperatures are in Réaumur degrees, which are proportional to Celsius degrees. From this data *estimate* the ratio of the radiation film coefficient and the free-convection film coefficient. It is not necessary to use all the measurements.

13. Long (≈ 1 m) electrical resistance heaters with a reflective backing hang from the rafters of a building one of the authors uses in winter. The heating elements glow red. He has heard the criticism, "How stupid to put these heaters so high! Everyone knows that heat rises." Discuss.

14. A famous experiment proposed by Bishop Berkeley (eighteenth century), an Irish philosopher, was the following. Take three pans of water, one with cold water, one with lukewarm water, and one with hot water (temperatures unspecified). Place one hand in the cold water and one in the hot water long enough for the sensation of hot and cold to be constant. Then place both hands in the lukewarm water. The initially colder hand will (probably) feel warmer than the initially hotter hand. Without knowing anything about the neurophysiology of thermal skin sensations or thermoregulation, and setting aside Berkeley's philosophical aims, interpret this experiment physically.

15. One of the authors exposed the back of his hand to direct solar radiation on a clear, dry (50% relative humidity), calm summer day with the Sun high in the sky. Ambient temperature was 24.5 °C Although he did not sense warming

or cooling, after ten minutes his skin temperature, measured with an infrared thermometer and a thermocouple was 27 °C six degrees *lower* than the *initial* temperature of his indoor-adapted skin. At the same time and under the same conditions the temperature of a reddish-brown wooden railing reached 53 °C. No sweat was *perceptible* on skin dry to the touch. A six-degree *drop* in skin temperature because of fairly intense solar radiation seems physically impossible. Upon returning to an unairconditioned room at 24 °C skin temperature increased to its previous adaptation value within a few minutes. In another experiment he immersed his hand in hot water ($\approx$ 40 °C) and his skin temperature increased. In both experiments skin was subjected to heating, yet its temperature decreased in the first, increased in the second. Explain.

16. Estimate the maximum possible surface temperature on Earth. By surface temperature is not meant air temperature near the surface but the temperature of a palpable surface. The solar irradiance (at the top of the atmosphere) is about 1365 Wm^{-2}. State all your assumptions. You can check your results against measured and modeled surface temperatures in John R. Garratt (1992) "Extreme maximum land surface temperatures," *Journal of Applied Meteorology*, 31, pp. 1096–1105.

17. For free convection the film coefficient is not independent of the temperature difference. Suppose that the energy flux density is proportional to the $p = 1 + \eta$ power, where $\eta > 0$. Under what conditions could we expect the cooling rate to be approximately an exponential decrease? If you put your mind to it, you can answer this question using only physical reasoning and simple mathematics. Or you can attack it with hammer and tongs. Try the easy way first.

18. Downward infrared radiation from clear air would be quite small absent infrared-active water vapor and carbon dioxide. But these same molecules are in the air in a room. Can you do a simple experiment to show that they have a negligible effect on infrared radiation incident on someone in the room? *Hint*: An infrared thermometer and air thermometer are essential. First do the experiment with a wall and air at the same temperature. Explain what you observe. What if the wall temperature were appreciably lower than air temperature? Why is the field of view of the thermometer relevant? And what has this to do with the temperature distribution on the wall? What is the difference (radiatively) between air in a room and a vertical column of atmospheric air?

19. Knowing that the thermal conductivity of water is about 24 times that of air, you can estimate the thickness of the thermal boundary layer above your bare skin in room-temperature air. Make as many reasonable approximations as possible. The aim is a rough magnitude not a number to five decimal places—or even two.

20. A colleague wrote to us that he often reads in books that snow and ice are good thermal insulators. He therefore expected the thermal conductivity of ice to be small, but instead found it to be appreciably greater than that of liquid water, 2.2 $Wm^{-1}K^{-1}$ compared with 0.55 $Wm^{-1}K^{-1}$ (both at 0 °C). Explain. *Hint*:

The density of new snow can be 10% that of solid ice. Do a simple calculation to support your explanation.

21. The simplest model of a human body in order to understand energy transfer is a uniform slab of thickness L. T_c is the constant temperature of one face. This represents the constant core temperature of a human body. The ambient temperature is $T_a < T_c$. The thermal conductivity of the slab is k_t. Include a convective film coefficient h in your analysis. For steady state determine the surface temperature T_s and the energy flux density q. For simplicity write the thermal resistance as $R_t = L/k_t$ and the convective resistance as $R_f = 1/h$. What is the surface temperature in the limit of infinite R/R_f? Does the result make sense? How does q vary with R and R_f for fixed $T_c - T_a$?

22. The preceding problem was inspired by a paper showing thermograms of a lightly clad man and woman. These are infrared images of surface (skin) temperature converted to false colors. The skin temperature of the woman is appreciably lower than that of the man. And this pattern was seen for a sample of 47 men and 47 women. The difference in skin temperatures was attributed to greater body fat percentage for the women. We have sometimes heard people puzzling about why women have more body fat, and hence, natural insulation, but seem to complain more about "feeling cold." Explain.

    You can observe the role of insulation. In a cool room measure the bare skin temperature on your belly with an infrared thermometer. Then put on a cotton T-shirt and wait until the surface temperature of the shirt in the same area is more or less constant. Then put on a wool or flannel shirt and make the same measurement. Finally, put on a wool sweater. Note your level of comfort with each successive layer.

23. This is related to the previous two problems. A man and a woman put on the same sweater in the same environment. The natural insulation of the woman is thicker than that of the man. In the steady-state, which has the lower skin temperature? Before doing any analysis, make an educated guess.

24. "Normal" body temperature supposedly is 37 °C. But without making any measurements you should realize that skin temperature must be appreciably lower. Why?

25. We sometimes have read that the human body (presumably naked) radiates as much "heat" as a 60 W light bulb. To our knowledge this is never qualified and has been repeated so often that it has become the scientific equivalent of an urban legend. This wattage makes no sense unless qualified. Why? The fact that part of the radiation from a light bulb is in the visible has nothing to do with this. *Hints*: This problem is related to the previous one. Measure skin temperatures on different parts of your body. Measure inside wall temperatures, including floor and ceiling. Skin and walls are to good approximation black bodies. You will need to find total skin area of humans and make some approximate corrections to estimate the radiative area.

26. Can you think of a simple way to trick a radiation thermometer into displaying an obviously incorrect temperature?

27. For slab geometries, equal thicknesses give equal thermal resistances. Is this also true for insulation wrapped around pipes? Try to give a simple physical explanation. If you are ambitious, you can solve Laplace's equation for the temperature in cylindrical geometry and obtain an expression for the thermal resistance. Try to include cost–benefit arguments.

28. We have seen the following assertion. If you hold the palms of your hands sufficiently close to each other for sufficiently long, their temperatures will ultimately approach 98.6 °F. Before doing a measurement to support or refute this assertion, criticize it.

29. In Section 2.3 we argued that if the upper 5 km of the atmosphere were to be removed suddenly, its weight per unit area would decrease as would surface pressure because the average number density would decrease by a random net upward diffusion of molecules. Make a rough estimate of how long it might take for the surface pressure to be measurably less. *Hint*: The self-diffusion coefficient of oxygen and nitrogen is $1.8 \times 10^{-5}\,\mathrm{m^2 s^{-1}}$. How does this differ from lifting a 10-kg plate off the ground?

30. In Section 1.8 we suggest that you measure skin temperature on various parts of your body to critically examine the notion that you "lose 40% of your heat from your head." Based on this chapter you can do a bit more analysis. Assume that head area is 10% of total body area. If the energy flux from your head is 40% of the total, what does this imply? Make as many simplifying assumptions as necessary for a first approximation. With an infrared thermometer you can test your analysis. Make some measurements indoors. If you live where winters are cold, try the following experiment. Bundle up and go outside on an overcast day with temperatures near freezing. Wear a heavy shirt and thick sweater but nothing on your head. Measure your head temperature (crown, forehead, cheek), the surface temperature of your sweater at various spots, and air temperature. Unfortunately, you will have to sit in the cold until all these temperatures stabilize. We did this experiment in Pennsylvania in early January.

31. Problem 91 in Chapter 2 will be of help. What is your best guess for the effective thermal conductivity of a cloud considered as a porous medium? The thermal conductivity of water is about 24 times that of air. This is a simple problem solely to provide ammunition to demolish notions about the conductivity of "damp" air.

32. Very roughly estimate the thickness of a porous insulator such that its thermal resistance is about the same as that of the film resistance of air. Then measure the thickness of a wool shirt or sweater. What do you conclude from this? How do your conclusions depend on wind speed?

33. We must eat if for no other reason than to compensate for $Q < 0$ from our bodies in order to keep our core temperature constant. Consider only the radiative component of Q because of a difference between our surface temperature and the surrounding temperature (indoors). Take the surrounding (radiative) temperature to be 293 K. What is the difference between Q integrated over 24 hours between two people with the same effective area of 1 m^2 if their average

surface temperatures differ by 1 K? Express your results in kilocalories ("food" calories). What do you conclude from this? By surface temperature we do not necessarily mean skin temperature. Your surface temperature will vary in time and over your body depending on what you are wearing.

34. We put a shiny steel can, probably tin-coated, about 6 cm in diameter in the freezing compartment of a refrigerator at about −20 °C. About an hour later we removed the can and quickly measured its temperature with an infrared thermometer, about 17 °C. This was measured from the side (holding the thermometer parallel to our body). We rotated to face the can and the indicated temperature increased by 2 °C for the *same* spot. Explain.

35. The temperature of the inner core of Earth is believed (no one has been there to make the measurement) to be around 6000 K, the core radius around 1000 km. Despite this high temperature it appears to have little consequences for Earth's climate. Why? Only a rough *quantitative* answer is wanted (i.e. *one* significant digit). Don't agonize over the geometry or exact details. You need to find some readily available data.

36. Joseph Fourier may be the source of the notion that "thin" air is inherently cold. He wrote (1822) that in "highest regions of the atmosphere the air is very rare and transparent, and retains but a minute part of the heat of the solar rays; this is the cause of the excessive cold of elevated places" even though he adds that they "are cooled by expansion." Examine his first assertion about the "excessive cold" by considering the rate of temperature increase per unit volume by absorption of solar radiation and its dependence on elevation. *Hint:* Use Eq. (7.9) but *not* Eq. (7.11) because q in this problem is the flux density of solar radiation. You will have to make some plausible physical arguments. Detailed calculations are not necessary. You will also have to learn just a bit about absorption of solar radiation by the different molecular constituents of air.

37. For a sedentary adult male to maintain a constant weight, the recommended daily calorie intake is around 2500 kcal. Exceeding this may result in accumulated fat, which is stored potential energy. If all these calories resulted *solely* in a temperature increase, what would it be? Make any reasonable assumptions. Then try to balance the energy budget. A few simple measurements using an infrared thermometer will help. Keep in mind that the thermal environment under a blanket during eight hours of sleep is different from the daytime environment. Make any reasonable assumptions and approximations. Exact calculations are neither wanted, necessary, nor possible.

38. Some websites assert that mercury is (was?) used as a liquid in thermometers because it "has a much higher coefficient of thermal expansion than water," a consequence of not bothering to spend a few minutes searching for measurements. Thermal expansion of mercury is *less*, about a factor of 2.5 at core body temperature. Based on this chapter and other chapters, what advantages does mercury have over water as a thermometric fluid? For what temperature range

would pure water be useless? *Hints*: See Chapter 3. You will need to search for measured thermophysical properties.

39. Estimate the *relative* spatial variation of *absolute* temperature of a human. What do you conclude?

40. Accounts of travel in hot deserts, for example, the Empty Quarter and the Sahara, almost invariably express surprise at how low morning temperatures can be despite high daytime temperatures. Explain. There are at least two components to a plausible explanation.

41. The convective film coefficient h_c may become less mysterious if you do a simple experiment to estimate it. We filled an uninsulated thin-walled polished metal cylinder with about 0.4 kg of water. The diameter was 3.25 cm, its length 11.5 cm. The initial still-air temperature was 24.2 °C; the initial water temperature was 25.7 °C. We put a lid on the cylinder to suppress (net) evaporation and took it off only briefly to measure the water temperature and to stir the water from time to time to try to keep its temperature uniform. After 82 minutes the temperature difference had dropped from 1.5 °C to 0.5 °C. From this, estimate the value of h_c. State all assumptions. Is the value you calculate comparable to values in this chapter for the free-convection coefficient? Why did we not wait much longer before measuring the temperature difference?

42. We filled a shiny can (see Problem 34) with boiling water. Air temperature was about 21 °C, whereas the temperature indicated by an infrared thermometer pointed at the can was 29 °C. Explain. We wrapped the can with shiny aluminum foil and repeated the experiment. The measured can temperature dropped to ≈ 22 °C. What does this tell you about the difference between the bare can and wrapped with foil? In Section 1.8 we state that emission of radiation by snow at 0 °C is about seven times *greater* than by highly polished aluminum at 100 °C. What is true for snow is also true for liquid water. We brought the palm of a hand close to, but not touching, the foil-covered can filled with hot water. There was a definite feeling of increased warmth. The temperature of cold water in which small ice cubes were floating was close to 0 °C (measured with the infrared thermometer) and the palm of a hand brought close to the water cooled slightly. Why was the sensation of warming stronger than of cooling? Resolve the apparent contradiction with what is in Section 1.8. This is not a complicated problem but if you think carefully about it you will have gone a long way to making sense of mantras about "radiation" of heat.

Selected Physical Constants

Avogadro's constant N_a	6.0231×10^{23} mol^{-1}
Boltzmann's constant k	1.3806×10^{-23} J K^{-1}
Gravitation acceleration at sea level g	9.807 ms^{-2}
Molecular weight of dry air M_d	28.966 g mol^{-1}
Molecular weight of water M_v	18.016 g mol^{-1}
Universal gas constant R^*	8.3143 J mol^{-1} K^{-1}
Gas constant for dry air R_d	287.04 J kg^{-1} K^{-1}
Gas constant for water vapor R_v	461.50 J kg^{-1} K^{-1}
Specific heat capacity of dry air at 0 °C	
c_{vd}	718 J kg^{-1} K^{-1}
c_{pd}	1005 J kg^{-1} K^{-1}
Specific heat capacity of water vapor at 0 °C	
c_{vv}	1390 J kg^{-1} K^{-1}
c_{pv}	1850 J kg^{-1} K^{-1}
Specific heat capacity of water at 0 °C	
c_{vv}	4218 J kg^{-1} K^{-1}
Specific heat capacity of ice at 0 °C	
c_i	2106 J kg^{-1} K^{-1}
Enthalpy of fusion of water at 0 °C	
ℓ_f	0.334×10^6 J kg^{-1}
Enthalpy of vaporization of water at 0 °C	
ℓ_v	2.501×10^6 J kg^{-1}
Enthalpy of vaporization of water at 100 °C	
ℓ_v	2.257×10^6 J kg^{-1}
Enthalpy of sublimation of water at 0 °C	
ℓ_s	2.834×10^6 J kg^{-1}
Conversions	

Thermochemical calorie = 4.184 J
Standard atmosphere = 101.325 kPa
760 m of mercury
29.213 inches of mercury
14.696 lb in^2

Sources: List, R. J. (1949) *Smithsonian meteorological tables*. 6th edn. Washington, DC: Smithsonian. Grigull, U., Straub, J., and Scheiber, P. (eds.) (1990) *Steam tables in SI-units*. New York: Springer. Each year the best values of fundamental physical constants such as the mass of the electron are tabulated in *Physics Today*.

Saturation Vapor Pressure over Water

$e_s(hPa)$ for 0 to 40 °C

Temp (° C)	0.0	0.1	0.2	0.3	0.4	0.5	0.6	0.7	0.8	0.9
0	6.108	6.152	6.197	6.242	6.288	6.333	6.379	6.426	6.472	6.519
1	6.566	6.614	6.661	6.709	6.758	6.807	6.856	6.905	6.955	7.004
2	7.055	7.105	7.156	7.207	7.259	7.311	7.363	7.416	7.469	7.522
3	7.575	7.629	7.683	7.738	7.793	7.848	7.904	7.960	8.016	8.072
4	8.129	8.187	8.245	8.303	8.361	8.420	8.479	8.538	8.598	8.659
5	8.719	8.780	8.842	8.903	8.966	9.028	9.091	9.154	9.218	9.282
6	9.346	9.411	9.477	9.542	9.608	9.675	9.742	9.809	9.877	9.945
7	10.01	10.08	10.15	10.22	10.29	10.36	10.43	10.50	10.58	10.65
8	10.72	10.79	10.87	10.94	11.02	11.09	11.17	11.24	11.32	11.40
9	11.47	11.55	11.63	11.71	11.79	11.87	11.95	12.03	12.11	12.19
10	12.27	12.35	12.44	12.52	12.61	12.69	12.77	12.86	12.95	13.03
11	13.12	13.21	13.29	13.38	13.47	13.56	13.65	13.74	13.83	13.93
12	14.02	14.11	14.20	14.30	14.39	14.49	14.58	14.68	14.77	14.87
13	14.97	15.07	15.17	15.27	15.37	15.47	15.57	15.67	15.77	15.87
14	15.98	16.08	16.19	16.29	16.40	16.50	16.61	16.72	16.83	16.93
15	17.04	17.15	17.26	17.38	17.49	17.60	17.71	17.83	17.94	18.06
16	18.17	18.29	18.41	18.52	18.64	18.76	18.88	19.00	19.12	19.24
17	19.37	19.49	19.61	19.74	19.86	19.99	20.12	20.24	20.37	20.50
18	20.63	20.76	20.89	21.02	21.15	21.29	21.42	21.56	21.69	21.83
19	21.96	22.10	22.24	22.38	22.52	22.66	22.80	22.94	23.08	23.23
20	23.37	23.52	23.66	23.81	23.96	24.11	24.26	24.41	24.56	24.71
21	24.86	25.01	25.17	25.32	25.48	25.63	25.79	25.95	26.11	26.27
22	26.43	26.59	26.75	26.92	27.08	27.25	27.41	27.58	27.75	27.92
23	28.09	28.26	28.43	28.60	28.77	28.95	29.12	29.30	29.47	29.65
24	29.83	30.01	30.19	30.37	30.56	30.74	30.92	31.11	31.30	31.48
25	31.67	31.86	32.05	32.24	32.43	32.63	32.82	33.02	33.21	33.41
26	33.61	33.81	34.01	34.21	34.41	34.62	34.82	35.03	35.23	35.44
26	33.61	33.81	34.01	34.21	34.41	34.62	34.82	35.03	35.23	35.44
28	37.80	38.02	38.24	38.46	38.69	38.91	39.14	39.37	39.59	39.82

29	40.05	40.29	40.52	40.76	40.99	41.23	41.47	41.71	41.95	42.19
30	42.43	42.67	42.92	43.17	43.41	43.66	43.91	44.17	44.42	44.67
31	44.93	45.18	45.44	45.70	45.96	46.22	46.49	46.75	47.02	47.28
32	47.55	47.82	48.09	48.36	48.64	48.91	49.19	49.47	49.74	50.03
33	50.31	50.59	50.87	51.16	51.45	51.74	52.03	52.32	52.61	52.90
34	53.20	53.50	53.80	54.10	54.40	54.70	55.00	55.31	55.62	55.93
35	56.24	56.55	56.86	57.18	57.49	57.81	58.13	58.45	58.77	59.10
36	59.42	59.75	60.08	60.41	60.74	61.07	61.41	61.74	62.08	62.42
37	62.76	63.11	63.45	63.80	64.14	64.49	64.84	65.20	65.55	65.91
38	66.26	66.62	66.98	67.35	67.71	68.08	68.45	68.81	69.19	69.56
39	69.93	70.31	70.69	71.07	71.45	71.83	72.22	72.61	72.99	73.39
40	73.78	74.17	74.57	74.97	75.37	75.77	76.17	76.58	76.98	77.39

$e_s(hPa)$ for 0 to -40 °C

Temp (° C)	0.0	−0.1	−0.2	−0.3	−0.4	−0.5	−0.6	−0.7	−0.8	−0.9
0	6.108	6.064	6.020	5.976	5.933	5.889	5.847	5.804	5.762	5.720
−1	5.678	5.637	5.595	5.554	5.514	5.473	5.433	5.393	5.354	5.314
−2	5.275	5.236	5.198	5.160	5.121	5.084	5.046	5.009	4.972	4.935
−3	4.898	4.862	4.826	4.790	4.754	4.719	4.683	4.649	4.614	4.579
−4	4.545	4.511	4.477	4.444	4.410	4.377	4.344	4.312	4.279	4.247
−5	4.215	4.183	4.151	4.120	4.089	4.058	4.027	3.997	3.966	3.936
−6	3.906	3.876	3.847	3.818	3.788	3.759	3.731	3.702	3.674	3.646
−7	3.618	3.590	3.562	3.535	3.508	3.481	3.454	3.427	3.401	3.374
−8	3.348	3.322	3.297	3.271	3.246	3.221	3.195	3.171	3.146	3.121
−9	3.097	3.073	3.049	3.025	3.001	2.978	2.954	2.931	2.908	2.885
−10	2.863	2.840	2.818	2.796	2.773	2.752	2.730	2.708	2.687	2.665
−11	2.644	2.623	2.602	2.582	2.561	2.541	2.520	2.500	2.480	2.461
−12	2.441	2.421	2.402	2.383	2.363	2.345	2.326	2.307	2.288	2.270
−13	2.252	2.233	2.215	2.197	2.180	2.162	2.144	2.127	2.110	2.092
−14	2.075	2.059	2.042	2.025	2.009	1.992	1.976	1.960	1.944	1.928
−15	1.912	1.896	1.880	1.865	1.850	1.834	1.819	1.804	1.789	1.774
−16	1.760	1.745	1.731	1.716	1.702	1.688	1.674	1.660	1.646	1.632
−17	1.619	1.605	1.592	1.578	1.565	1.552	1.539	1.526	1.513	1.500
−18	1.488	1.475	1.463	1.450	1.438	1.426	1.414	1.402	1.390	1.378
−19	1.366	1.355	1.343	1.332	1.320	1.309	1.298	1.287	1.276	1.265
−20	1.254	1.243	1.233	1.222	1.211	1.201	1.191	1.180	1.170	1.160

−21	1.150	1.140	1.130	1.120	1.111	1.101	1.091	1.082	1.072	1.063
−22	1.054	1.045	1.035	1.026	1.017	1.008	1.000	0.991	0.982	0.973
−23	0.965	0.956	0.948	0.940	0.931	0.923	0.915	0.907	0.899	0.891
−24	0.883	0.875	0.867	0.859	0.852	0.844	0.837	0.829	0.822	0.814
−25	0.807	0.800	0.793	0.785	0.778	0.771	0.764	0.757	0.751	0.744
−26	0.737	0.730	0.724	0.717	0.711	0.704	0.698	0.691	0.685	0.679
−27	0.673	0.667	0.660	0.654	0.648	0.642	0.637	0.631	0.625	0.619
−28	0.613	0.608	0.602	0.597	0.591	0.586	0.580	0.575	0.569	0.564
−29	0.559	0.554	0.549	0.543	0.538	0.533	0.528	0.523	0.518	0.514
−30	0.509	0.504	0.499	0.495	0.490	0.485	0.481	0.476	0.472	0.467
−31	0.463	0.458	0.454	0.450	0.445	0.441	0.437	0.433	0.429	0.425
−32	0.421	0.417	0.413	0.409	0.405	0.401	0.397	0.393	0.389	0.386
−33	0.382	0.378	0.374	0.371	0.367	0.364	0.360	0.357	0.353	0.350
−34	0.346	0.343	0.340	0.336	0.333	0.330	0.327	0.323	0.320	0.317
−35	0.314	0.311	0.308	0.305	0.302	0.299	0.296	0.293	0.290	0.287
−36	0.284	0.281	0.279	0.276	0.273	0.270	0.268	0.265	0.262	0.260
−37	0.257	0.254	0.252	0.249	0.247	0.244	0.242	0.239	0.237	0.235
−38	0.232	0.230	0.228	0.225	0.223	0.221	0.218	0.216	0.214	0.212
−39	0.210	0.208	0.205	0.203	0.201	0.199	0.197	0.195	0.193	0.191
−40	0.189	0.187	0.185	0.183	0.181	0.180	0.178	0.176	0.174	0.172

Reference

The saturation vapor pressures were calculated using the formulation of F. W. Murray (1967) "On the computation of saturation vapor pressure," *Journal of Applied Meteorology*, 6, pp. 203–4, which is equivalent to the internationally accepted formulation of saturation vapor pressure by J. A. Goff and S. Gratch(1946) "Low-pressure properties of water from −160 to 212 F," *Transactions of the American Society of Heating and Ventilating Engineering*, 52, pp. 95–121. The values calculated are identical to those in R. J. List (1949) *Smithsonian meteorological tables*. 6th edn. Washington, DC: Smithsonian.

Bibliography

Adair, R. K. (1994) *The physics of baseball.* 2nd edn. New York: Harper.

Agrawal, D. C. (2009) "A simplified version of the Curzon–Ahlborn engine," *European Journal of Physics*, 30, pp. 173–9.

Aitken, J. (1923) *Collected scientific papers.* Cambridge: Cambridge University Press.

Albrecht, B. A. (1989) "Aerosols, cloud microphysics, and fractional cloudiness," *Science*, 245, pp. 1227–30.

Albrecht, B. A., Penc, R. S., and Schubert, W. H. (1985) "An observational study of cloud-topped mixed layers," *Journal of the Atmospheric Sciences*, 42, pp. 800–22.

Allan, R. P. (2006) "Variability in clear-sky longwave radiative cooling of the atmosphere," *Journal of Geophysical Research* [online], 111(D22). https://doi.org/10.1029/2006JD007304

Allan, R. P. (2009) "Examination of relationships between clear-sky longwave radiation and aspects of the atmospheric hydrological cycle in climate models, reanalyses, and observations," *Journal of Climate*, 22, pp. 3127–45.

Allendoerfer, C. B. (1952) "Editorial [differentials]," *American Mathematical Monthly*, 59, pp. 403–6.

American Institute of Physics. (1941) *Temperature, its measurement and control in science and industry.* New York: Reinhold.

Apostol, T. M., *et al.* (eds.) (1969) *Selected papers on calculus.* Providence, RI: American Mathematical Association.

Atkins, P. W. (1978) *Physical chemistry.* New York: W. H. Freeman.

Baird, D. (2004) *Thing knowledge: a philosophy of scientific instruments.* Berkeley: University of California Press.

Ball, F. K. (1956) "Energy changes involved in disturbing a dry atmosphere," *Quarterly Journal of the Royal Meteorological Society*, 82, pp. 15–29.

Barahoma, D., Molod, A., and Kalesse, H. (2017) "Direct estimation of the global distribution of vertical velocity within cirrus clouds," *Scientific Reports*, 7, 6840.

Barnes, R. B. (1963) "Thermography of the human body," *Science*, 140, pp. 870–7.

Bartels, R. A. (1990) "Do darker objects really cool faster?" *American Journal of Physics*, 58, 244–8.

Bartlett, A. A. (1975) "Introductory experiment to determine the thermodynamic efficiency of a household refrigerator," *American Journal of Physics*, 44, pp. 555–8.

Batchelor, G. K. (1967) *An introduction to fluid mechanics*. Cambridge: Cambridge University Press.

Bauer, L. (1908) "On the relationship between 'potential temperature' and 'entropy'," *Physical Review*, 26, pp. 177–83.

Becker, P. (2001) "History and progress in the accurate determination of the Avogadro constant," *Reports on Progress in Physics*, 64, pp. 1945–2008.

Beckman, O. (1997) "Anders Celsius and the fixed points of the Celsius scale," *European Journal of Physics*, 18, pp. 169–75.

Benedict, F. G. (1912) *The composition of the atmosphere with special reference to its oxygen content*. Publication No. 166. Washington, DC: Carnegie Institution of Washington.

Bent, H. A. (1965) *The second law*. Oxford: Oxford University Press.

Bereberan-Santos, M. N., Bodunov, E. N., and Pogliani, L. (1997) "On the barometric formula," *American Journal of Physics*, 65, pp. 404–11.

Berry, M. V. (1971) "The molecular mechanism of surface tension," *Physics Education*, 6, pp. 79–84.

Betts, A. K. (1974) "Further comments on 'A comparison of the equivalent potential temperature and the static energy'," *Journal of the Atmospheric Sciences*, 31, pp. 1713–15.

Betts, A. K. (1985) "Mixing line analysis of clouds and cloudy boundary layers," *Journal of the Atmospheric Sciences*, 42, pp. 2751–63.

Betts, A. K. and Albrecht, B. A. (1987) "Conserved variable analysis of the convective boundary layer thermodynamic structure over the tropical oceans," *Journal of the Atmospheric Sciences*, 44, pp. 83–99.

Birkebak, R. C. (1966) "Heat transfer in biological systems," *International Review of General and Experimental Zoology*, 2, pp. 269–344.

Black, J. (1807) *Lectures on the elements of chemistry*, vol. 1. Philadelphia: Mathew Carey.

Blanchard, D. C. (1979) "Science, success, and serendipity," *Weatherwise*, 32, pp. 236–41.

Blanchard, D. C. (1996) "Serendipity, scientific discovery, and Project Cirrus," *Bulletin of the American Meteorological Society*, 77, pp. 1279–85.

Bluestein, H. B. (1993) *Synoptic–dynamic meteorology in midlatitudes*, vol. 2. New York: Oxford University Press.

Boer, J. D. (1974) "van der Waals in his time and the present revival—opening address," in Prins, C. (ed.) *Proceedings of the van der Waals centennial conference on statistical mechanics*. North Holland: Elsevier, pp. 1–27.

Bohren, C. F. (1986) "Cloud formation on descent revisited," *Journal of the Atmospheric Sciences*, 43, pp. 3035–7.

Bohren, C. F. (1987) *Clouds in a glass of beer*. New York: John Wiley & Sons.

Bohren, C. F. (1990a) "You never miss the water until the well runs dry," *Weatherwise*, 43, pp. 342–7.

Bohren, C. F. (1990b) "All that glistens is not dew," *Weatherwise*, 43, pp. 284–7.

Bohren, C. F. (1991a) "Comments on 'Newton's law of cooling—a critical assessment' by Colm T. O'Sullivan [1990: *American Journal of Physics*, 58, pp. 956–60]," *American Journal of Physics*, 59, pp. 1044–6.

Bohren, C. F. (1991b) *What light through yonder window breaks?* New York: John Wiley & Sons.

Bohren, C. F. (1993) "Melting with salt and heating with ice," *Weatherwise*, 46 (6), pp. 46–8.

Bohren, C. F. (1995) "Rain, snow, and spring runoff revisited," *The Physics Teacher*, 33, pp. 79–81.

Bohren, C. F. (2015) "What my dogs taught me about thermal energy transfer," *American Journal of Physics*, 83, pp. 443–6.

Bohren, C. F. and Fraser, A. B. (1992) "Fall streaks: parabolic trajectories with a twist," *American Journal of Physics*, 60, pp. 1030–3.

Bolton, D. (1980) "The computation of equivalent potential temperature," *Monthly Weather Review*, 108, pp. 1046–53.

Bolton, H. C. (1900) *Evolution of the thermometer: 1592–1743*. New York: Chemical Publishing Co.

Boorse, H. A., and Motz, L. (eds.) (1966) *The world of the atom*. New York: Basic Books.

Booth, J. (1977) "A short history of blood pressure measurement," *Proceedings of the Royal Society of Medicine*, 70, pp. 793–9.

Bos, H. J. M. (1974) "Differentials, higher-order differentials, and the derivative in the Leibnizian calculus," *Archive for History of the Exact Sciences*, 14, pp. 1–90.

Bridgman, P. W. (1959) "*The Logic of Modern Physics* after thirty years," *Daedalus*, 88, pp. 518–26.

Bridgman, P. W. (1960) *The logic of modern physics*. London: MacMillan.

Bridgman, P. W. (1961) *The nature of thermodynamics*. London: Harper.

Brown, S. C. (1947) "The discovery of convection currents by Benjamin Thompson, Count of Rumford," *American Journal of Physics*. 15, pp. 273–4.

Brown, S. C. (1950) "The caloric theory of heat," *American Journal of Physics*, 18, pp. 367–73.

Brunt, D. (1927) "The period of simple vertical oscillations in the atmosphere," *Quarterly Journal of the Royal Meteorological Society*, 53, pp. 30–2.

Brunt, D. (1934) "The possibility of condensation by descent of air," *Quarterly Journal of the Royal Meteorological Society*, 60, pp. 279–84.

Brush, S. G. (1961) "Development of the kinetic theory of gases. V. The equation of state," *American Journal of Physics*, 29, pp. 593–605.

Brush, S. G. (1965) *Kinetic theory, vol. 1 : The nature of gases and of heat*. Oxford: Pergamon.

Brush, S. G. (1970) "Interatomic forces and gas theory from Newton to Lennard-Jones," *Archive for Rational Mechanics and Analysis*, 39, pp. 1–29. [Reprinted in an anthology of classic papers in the kinetic theory of gases by the same author (2003) *The kinetic theory of gases*. London: Imperial College Press.]

Brush, S. G. (1983) *Statistical physics and the atomic theory of matter, from Boyle and Newton to Landau and Onsager*. Princeton: Princeton University Press.

Buck, R. C. (1956) *Advanced calculus*. New York: McGraw Hill.

Burko, L. M. and Price, R. H. (2005) "Ballistic trajectory: parabola, ellipse, or what?" *American Journal of Physics*, 73, pp. 516–20.

Byers, H. R. (1965) *Elements of cloud physics*. Chicago: University of Chicago Press.

Caccomo, M. T., *et al.* (2019) "Rüchardt's experiment treated by Fourier transforms," *European Journal of Physics*, 40, 025703.

Cajori, F. (1952) *A history of mathematical notations*, Vol. 2. Chicago: Open Court.

Callen, H. B. (1985) *Thermodynamics and an introduction to thermostatistics*. 2nd edn. New York: John Wiley & Sons

Cardwell, D. S. L. (ed.) (1968) *John Dalton & the progress of science*. Manchester: Manchester University Press.

Cardwell, D. S. L. (1971) *From Watt to Clausius: the rise of thermodynamics in the early industrial age*. Ithaca: Cornell University Press.

Carnot, S. (1943) *Reflections on the motive power of heat*. New York: The American Society of Mechanical Engineers.

Carslaw, H. S., and Jaeger, J. C. (1959) *Conduction of heat in solids*. 2nd edn. Oxford: Clarendon Press.

Catling, D. C. and Zahnle, K. J. (2009) "The planetary air leak," *Scientific American*, 300(5), pp. 36–43.

Chapman, S. and Lindzen, R. S. (1970) *Atmospheric tides: thermal and gravitational*. Dordrecht: D. Reidel.

Charlier, R. H. and Justus, J. R. (1993) *Ocean energies: environmental, economic and technological aspects of alternative energy sources*. Amsterdam: Elsevier.

Chen, J-P. (1994) "Theory of deliquescence and modified Köhler curves," *Journal of the Atmospheric Sciences*, 51, pp. 3505–16.

Chen, S. and Baker, I. (2010) "Evolution of individual snowflakes during metamorphism," *Journal of Geophysical Research*, 115, D21114.

Churchill, S. W. and Bernstein, M. (1977) "A correlation equation for forced convection from gases and liquids to a circular cylinder in cross flow," *Journal of Heat Transfer*, 99, pp. 300–5.

Churchill, S. W. and Chu, H. H. S. (1975) "Correlating equations for laminar and turbulent free convection from a vertical plate," *International Journal of Heat and Mass Transfer*, 18, pp. 1323–9.

Clausius, R. (1879) *The mechanical theory of heat*. London: MacMillan.

Cohen, I. B. (1964) "Newton, Hooke, and 'Boyle's Law' (Discovered by Power and Towneley)," *Nature*, 204, pp. 618–21.

Colbeck, S. C. (1995) "Pressure melting and ice skating," *American Journal of Physics*, 63, pp. 888–90.

Colbeck, S. C., Najarian, L., and Smith, H. B. (1997) "Sliding temperature of ice skates," *American Journal of Physics*, 65, pp. 488–92.

Cooley, K. R. (1970) *Evaporation from open water surfaces in Arizona*. Agricultural Experiment Station and Cooperative Extension Service Folder 159. Tucson: University of Arizona Press.

Cotton, W. R. and Anthes, R. A. (1989) *Storm and cloud dynamics*. Academic.

Coulman, C. E. (1969) "A quantitative treatment of solar 'seeing'," *Solar Physics*, 7, pp. 122–43.

Crank, J.,1975: *The Mathematics of Diffusion*, 2nd ed. Oxford University Press.

Craven, B. A. and Settles, G. S. (2006) "A computational and experimental investigation of the human thermal plume," *Journal of Fluids Engineering*, 28, pp. 1251–7.

Curzon F. L. and Ahlborn, B. (1975) "Efficiency of a Carnot engine at maximum power output," *American Journal of Physics*, 43, pp 22–4.

Darrow, K. K. (1944) "The concept of entropy," *American Journal of Physics*, 12, pp. 183–96.

Dash, J. G., Rempel, A. W., and Wettlaufer, J. S. (2006) "The physics of pre-melted ice and its geophysical consequences," *Reviews of Modern Physics*, 78, pp. 695–741.

Day, M. A. (1990) "The no-slip boundary of fluid mechanics," *Erkenninis*, 33, pp. 285–96.

de Boer, J. H. (1968) *The dynamical character of adsorption*. Oxford: Clarendon Press.

de Dear, R. J., *et al.* (1997) "Convective and radiative heat transfer coefficients for human body segments," *International Journal of Biometeorology*, 40, pp. 141–56.

Denny, M. (2007) *Ingenium: five machines that changed the world*. Baltimore, MD: Johns Hopkins University Press.

Dorsey, N. E. (1940) *Properties of ordinary water-substance*. New York: Reinhold.

Dorsey, N. E. (1948) "The freezing of supercooled water," *Transactions of the American Philosophical Society*, 38, 247–328.

Durran, D. R. and Frierson, D. M. W. (2013) "Condensation, atmospheric motion, and cold beer," *Physics Today*, 66, pp. 74–5.

Dushman, S. (1962) *Scientific foundations of vacuum technique*. 2nd edn. New York: John Wiley & Sons.

Dutton, John A., 1976: *The ceaseless wind*. New York: McGraw-Hill.

Edward, J. T. (1970) "Molecular volumes and the Stokes–Einstein equation," *Journal of Chemical Education*, 47, pp. 261–70.

Eisenberg, D. and Kauzmann, W. (1969) *The structure and properties of water*. Oxford: Oxford University Press.

Eliasson, A., Blanchard, D. C., and Bergeron, T. (1978) "The life and science of Tor Bergeron," *Bulletin of the American Meteorological Society*, 59, pp. 387–92.

Ellis, F. R. (1979) "Utilization of low-grade thermal energy by using the clear night sky as a heat sink," *American Journal of Physics*, 47, pp. 1010–11.

Emanuel, K. A. (1988) "The maximum intensity of hurricanes," *Journal of the Atmospheric Sciences*, 45, pp. 1143–55.

Emanuel, K. A. (1994) *Atmospheric convection*. Oxford: Oxford University Press.

Emanuel, K. A. (2006) "Hurricanes: tempests in a greenhouse," *Physics Today*, 59, pp. 74–5.

Fast, J. D. (1968) *Entropy*. 2nd edn. Philadelphia: Gordon Breach.

Feinberg, G. (1965) "Fall of bodies near the Earth," *American Journal of Physics*, 33, pp. 501–2.

Feldman, T. S. (1985) "Applied mathematics and the quantification of experimental physics: The example of barometric hypsometry," *Historical Studies in the Physical Sciences*, 15, pp. 127–95

Fermi, E. (1956) *Thermodynamics*. Mineola: Dover.

Feynman, R. P., Leighton, R. B., and Sands, M. (1963) *The Feynman lectures on physics*, vol. 1. Boston: Addison-Wesley.

Fine, R. A. and Millero, F. J. (1973) "Compressibility of water as a function of temperature and pressure," *Journal of Chemical Physics*, 59, pp. 5529–36.

Finney, M. A., *et al.* (2013) "A study of ignition by rifle bullets," Research Paper RMRS-RP-104. Fort Collins, CO: U.S. Department of Agriculture Forest Service, Rocky Mountain Research Station. [A version of this by the same authors was subsequently published as (2016) "A study of wildfire ignition by rifle bullets," *Fire Technology*, 52, pp. 931–54. Both references can be found online.]

Fitts, D. D. (1962) *Nonequilibrium thermodynamics*. New York: McGraw-Hill.

Fourier, J. (1955) *The analytical theory of heat*. Mineola: Dover.

Fox, R. (1971) *The caloric theory of gases*. Oxford: Oxford University Press.

Frank, F. C. (1966) "Reflections on Sadi Carnot," *Physics Education*, 1, pp. 11–18

Franks, F. (1981) *Polywater*. Cambridge, MA: MIT Press.

Fraser, R. G. J. (1938) *Molecular beams*. New York: Chemical Publishing Company.

Fritschen, L. J. and Gay, L. W. (1979) *Environmental instrumentation*. New York: Springer.

Gaggioli, R. A. (1969) "More on generalizing the definitions of 'heat' and 'entropy'," *International Journal of Heat and Mass Transfer*, 12, pp. 656–60.

Galvin, K. P. (2005) "A conceptually simple derivation of the Kelvin equation," *Chemical Engineering Science*, 60, pp. 4659–60.

Garber, E. (1976) "Thermodynamics and meteorology (1850–1900)," *Annals of Science*, 33, pp. 51–65.

Garratt, J. R. (1992) "Extreme maximum land surface temperatures," *Journal of Applied Meteorology*, 31, pp. 1096–1105

Garwood, C. (2007) *Flat Earth: the history of an infamous idea*. London: Macmillan.

Gavroglu, K. (ed.) (2014) *History of artificial cold, scientific, technological, and cultural issues*. Springer.

Gearhart, C. A. (1996) "Specific heats and the equipartition law in introductory textbooks," *American Journal of Physics*, 64, pp. 995–1000.

Giddings, D. C. (1968) "Introduction to calorimetry of non-reacting systems," in McCullough, J. P. and Scott, D. W. (eds.) *Experimental thermodynamics*, vol. 1. London: Plenum, pp. 1–13. [Reprinted in NBS (National Bureau of Standards) Special Publication 300, Vol. 6, *Precision Measurement and Calibration: Heat*, and available online.]

Gilbert, M., *et al.* (2000) "Resuscitation from accidental hypothermia of 13.7 °C with circulatory arrest," *The Lancet*, 355, pp. 375–76.

Gillispie, C. C. (ed.) (1970) *Dictionary of scientific biography*. New York: Scribner.

Goff, J. A. and Gratch, S. (1946) "Low-pressure properties of water from −160 to 212 F," *Transactions of the American Society of Heating and Ventilating Engineering*, 52, pp. 95–121.

Goodman, A. B. and Wolf, A. V. (1969) "Insensible water loss from human skin as a function of ambient vapor concentration," *Journal of Applied Physiology*, 26, pp. 203–7.

Grad, H. (1961) "The many faces of entropy," *Communications on Pure and Applied Mathematics*, 14, pp. 323–54.

Grigul, U., Straub, J., and Scheiber, P. (eds.) (1990) *Steam tables in SI units.* London: Springer.

Gunn, R. and Kinzer, G. D. (1948) "The terminal velocity of fall for water droplets in stagnant air," *Journal of Meteorology*, 6, pp. 243–8.

Hardin, G. (1957) "The threat of clarity," *American Journal of Psychiatry*, 114, pp. 392–6.

Hayward, A. T. J. (1971) "Negative pressure in liquids: can it be harnessed to serve man?" *American Scientist*, 59, pp. 434–43.

Hecht, E. (2003) "An historical-critical account of potential energy: is PE really real?" *The Physics Teacher*, 41, pp. 486–93.

Hecht, E. (2021) "The true story of Newtonian gravity," *American Journal of Physics*, 89, pp. 683–92.

Heilbron, J. L. (ed.) (2003) *The Oxford companion to the history of modern science.* Oxford: Oxford University Press.

Hensel, H. (1982) *Thermal sensations and thermal receptors.* Springfield, IL: Charles C. Thomas

Hildebrand, J. H. and Scott, R. L. (1950) *The solubility of nonelectrolytes.* 3rd edn. New York: Reinhold.

Hinshelwood, C. N. (1951) *The structure of physical chemistry.* Oxford: Oxford University Press.

Hoarau, K., *et al.* (2017) "Did typhoon *Haiyan* have a new record minimum pressure?" *Weather*, 72, pp. 291–5.

Hobbs, P. V. (1974) *Ice physics.* Oxford: Oxford University Press.

Holton, J. R. (2004) *An introduction to dynamic meteorology.* London: Elsevier Academic Press.

Houze, R. A. (1993) *Cloud dynamics.* London: Academic.

Howard, I. K. (2002) "H is for enthalpy, thanks to Heike Kammerlingh Onnes and Alfred W. Porter," *Journal of Chemical Education*, 79, pp. 697–8.

Iribarne, J. V. and Godson, W. L. (1981) *Atmospheric thermodynamics.* 2nd edn. Dordrecht: D. Reidel.

Israelachvili, J. N. (2011) *Intermolecular and surface forces.* 3rd edn. London: Academic.

Jackson, H. L. (1959) "Presentation of the concept of mass to beginning physics students," *American Journal of Physics*, 27, pp. 278–80.

Jacques, A. A., *et al.* (2015) "Central and eastern U.S. surface pressure variations derived from the USArray Network," *Monthly Weather Review*, 143, pp. 1472–93.

Jaffe, B. (1976) *Crucibles: the story of chemistry.* 4th edn. Mineola: Dover.

Jeans, J. (1940) *An introduction to the kinetic theory of gases.* Cambridge: Cambridge University Press.

Jeffrey, C. A. and Austin, P. H. (1998) "A new analytic equation of state for water," *Journal of Chemical Physics*, 110, pp. 484–96.

Johnson, E. (2018) *Anxiety and the equation: understanding Boltzmann's entropy.* Cambridge, MA: MIT Press.

Johnston, D. C. (2014) *Advances in the thermodynamics of the van der Waals fluid.* London: Morgan & Claypool.

Jones, F. E. (1992) *Evaporation of water: with emphasis on applications and measurements.* Boca Raton: CRC Press.

Jones, H. C. (1899) *The modern theory of solutions.* London: Harper.

Joule, J. P. (1884) *The scientific papers of James Prescott Joule.* London: The Physical Society.

Kac, M. and Randolph, J. F. (1942) "Differentials," *American Mathematical Monthly.* 49, pp. 110–12.

Kadoya, K., Matsunaga, N., and Nagashima, A. (1986) "Viscosity and thermal conductivity of dry air in the gaseous phase," *Journal of Physical and Chemical. Reference Data*, 14, pp. 947–70.

Kell, G. S. (1972) "Thermodynamic and transport properties of fluid water," in Franks, F. (ed.) *Water: a comprehensive treatise, vol. I: the physics and physical chemistry of water.* New York: Plenum.

Kell, G. S. (1975) "Density, thermal expansivity, and compressibility of liquid water from to 0 to 150 °C: correlations and tables for atmospheric pressure and saturation reviewed and expressed on 1968 temperature scale," *Journal of Chemical and Engineering Data*, 20, pp. 97–105.

Kennard, E. H. (1938) *Kinetic theory of gases.* New York: McGraw-Hill.

Kennard, R. B. (1941) "Temperature distributions and heat flux in air by interferometry," in *Temperature and its measurement and control in science and industry*, vol. 1. New York: Reinhold.

Kerker, M. (1960) "Sadi Carnot and the steam engine engineers," *Isis*, 51, pp. 257–70.

Kipnis, A. Ya., Yavelov, B. E., and Rowlinson, J. S. (1996) *van der Waals and molecular science.* Oxford: Oxford University Press.

Klein, M. J. (1974) "The historical origins of the van der Waals equation," in Prins, C. (ed.) *Proceedings of the van der Waals centennial conference on statistical mechanics*, North Holland: Elsevier, pp. 28–47.

Knox, J. A., Nevius, D. S., and Knox, P. N. (2017) "Two simple and accurate approximations for wet-bulb temperature in moist conditions, with forecasting applications," *Bulletin of the American Meteorological Society*, 98, pp. 1897–1906.

Knudsen, M. (1950) *The kinetic theory of gases: some modern aspects.* 3rd edn. Methuen.

Koebel, M., Rigacci, A., and Achard, P. (2012) "Aerogel-based superinsulation: an overview," *Journal of Sol-Gel Science and Technology*, 63, pp. 315–39.

Kollias, P., *et al.* (2001) "Radar observations of updrafts, downdrafts, and turbulence in fair-weather cumuli," *Journal of the Atmospheric Sciences*, 58, pp. 1750–66.

Kondo, J. and Yamazawa, H. (1986) "Measurement of snow surface emissivity," *Boundary Layer Meteorology*, 34, pp. 415–16.

Kragh, H. (2008) *Entropic creation: religious contexts of thermodynamics and cosmology.* Farnham: Ashgate.

Kuhn, T. S. (1961) "Measurement in modern physical science," *Isis*, 52, pp. 161–93.

Kuhn, T. S. (1977) "Energy conservation as an example of simultaneous discovery," in Clagett, M. (ed.) *The Essential tension: selected studies in scientific tradition and change.* Chicago: University of Chicago Press, pp. 66–104.

Kurazumi, Y., *et al.* (2008) "Radiative and convective heat transfer coefficients of the human body in natural convection," *Building and Environment*, 43, pp. 2142–53.

Kylili, A., *et al.* (2014). "Infrared thermography (IRT) applications for building diagnostics: A review," *Applied Energy*, 134, pp. 531–49.

Laidler, K. J. (1987) *Chemical kinetics.* 3rd edn. New York: Harper & Row.

Laidler, K. J. (1993) *The world of physical chemistry.* Oxford: Oxford University Press.

Lamb, H. H. (1985) *Climate history and the future.* Princeton: Princeton University Press.

Lambert, F. L. (1999) "Shuffled cards, messy desks, and disorderly dorm rooms—Examples of entropy increase? Nonsense!" *Journal of Chemical Education*, 76, pp. 1385–7.

Lambert, F. L. (2002) "Disorder—A cracked crutch for supporting entropy discussions," *Journal of Chemical Education*, 79, pp. 187–92.

Lau, K. and Wu, H. T. (2003) "Warm rain process over tropical oceans and climate implications," *Geophysical Research Letters*, 30(24) p. 2240.

Lautrup, B. (2011) *Physics of continuous matter: exotic and everyday phenomena in the macroscopic world.* 2nd edn. Boca Raton: CRC Press.

Lawrence, M. A. (2005) "The relationship between relative humidity and the dew-point temperature of moist air," *Bulletin of the American Meteorological Society*, 86, pp. 225–33.

Le Blancq, F. (2011) "Diurnal pressure variation: the atmospheric tide," *Weather*, 66, pp. 306–7.

Leff, H. S. (2007) "Entropy, its language and interpretation," *Foundations of Physics*, 37, pp. 1744–66.

Leff, H. S. (2012) "Removing the mystery of entropy and thermodynamics," *The Physics Teacher*, 50, pp. 28–31, 87–90, 170–72, 215–17, 274–76.

Leff, H. S. (2021) *Energy and entropy: a dynamic duo.* Boca Raton: CRC Press.

Leff, H. S. and Mungan, C. E. (2018) "Isothermal heating: purist and utilitarian views," *European Journal of Physics*, 39, 045 103.

Leibowitz, H. W. (1965) *Visual perception.* New York: Macmillan.

Lennard-Jones, J. E. (1931) "Cohesion," *Proceedings of the Physical Society*, 43, pp. 461–82.

Levenspiel, O., Fitzgerald, T. J., and Pettit, D. (2000) "Learning from the past: Earth's atmosphere before the age of dinosaurs," *Chemical Innovation*, 30(12), pp. 50–5.

Lewis, G. N. and Randall, M. (1923) *Thermodynamics and the free energy of chemical substances.* New York: McGraw-Hill.

Lewis, J. M. (2005) "Roots of ensemble forecasting," *Monthly Weather Review*, 133, pp. 1865–85.

List, R. J. (1949) *Smithsonian meteorological tables.* 6th edn reprint. Washington, DC: Smithsonian.

Loeb, L. B. (1961) *The kinetic theory of gases.* Mineola: Dover.

Lorenz, E. (1993) *The essence of chaos.* Seattle: University of Washington Press.

Lote, C. J. (2012) *Principles of renal physiology.* 5th edn. New York: Springer.

Ludlum, F. (1980) *Clouds and storms.* University Park, PA: Pennsylvania State University Press.

Lugg, G. A. (1968) "Diffusion coefficients of some organic and other vapors in air." *Analytical Chemistry*, 40, pp. 1072–76.

Lyon, W. J. (1938) "Inaccuracies in the textbook discussions of the ordinary gas laws," *American Journal of Physics*, 6, pp. 256–59.

Madey, T. E. and Brown, W. L. (eds.) (1984) *History of vacuum science and technology*, American Institute of Physics.

Magie, W. F. (1935) *A source book in physics.* McGraw-Hill.

Marciano, J. B. (2014) *Whatever happened to the metric system? How America kept its feet.* New York: Bloomsbury USA.

Mass, C. S. and Madaus, L. E. (2014) "Surface pressure observations from smartphones: A potential revolution for high-resolution weather prediction?" *Bulletin of the American Meteorological Society*, 95, pp. 1343–9.

Maurellis, A. N., *et al.* (2000) "Precipitable water column retrieval from GOME data," *Geophysical Research Letters*, 27, pp. 903–06.

Maxwell, J. C. (1866) "The Bakerian Lecture: on the viscosity or internal friction of air and other gases," *Proceeding of the Royal Society of London*, 156, pp. 249–68.

Maxwell, J. C. (1872) *Theory of heat.* Boston: D. Appleton.

Maxwell, J. C. (1952) *The scientific papers of James Clerk Maxwell.* Mineola: Dover.

McAdams, W. H. (1985) *Heat transmission.* 3rd edn. Krieger.

McDonald, J. E. (1953) "Homogeneous nucleation of supercooled water droplets," *Journal of Meteorology*, 10, pp. 416–33.

McDonald, J. E. (1963a) "Early developments in the theory of the saturated adiabatic process," *Bulletin of the American Meteorological Society*, 44, pp. 203–11.

McDonald, J. E. (1963b) "James Espy and the beginnings of cloud thermodynamics," *Bulletin of the American Meteorological Society*, 44, pp. 634–41.

McDonald, J. E. (1963c) "Intermolecular attractions and saturation vapor pressure," *Journal of the Atmospheric Sciences*, 20, pp. 178–80.

McDonald, J. E. (1964) "On a criterion governing the mode of cloud formation in planetary atmospheres," *Journal of the Atmospheric Sciences*, 21, pp. 76–82.

McGlashan, M. L. (1966) "The use and misuse of the laws of thermodynamics," *Journal of Chemical Education*, 43, pp. 226–32.

McKie, D. and Heathcote, N. H. De V. (1935) *The discovery of specific and latent heats*. London: Edward Arnold.

Melikov A. K., *et al.* (1997) "Air temperature fluctuations in rooms," *Building and Environment*, 32, pp. 101–14.

Mendoza, E. (ed.) (1960) *Reflections on the motive power of fire by Sadi Carnot and other papers on the second law of thermodynamics by É. Clapeyron and R. Clausius*. Mineola: Dover.

Mendoza, E. (1990) "The lattice theory of gases: a neglected episode in the history of chemistry," *Journal of Chemical Education*, 67, pp. 1040–2.

Menger, K. (1950) "The mathematics of elementary thermodynamics," *American Journal of Physics*, 18, pp. 89–103.

Menger, K. (1955) *Calculus: a modern approach*. 3rd edn. Oxford: Ginn & Company.

Middleton, W. E. K. (1964) *The history of the barometer*. Baltimore, MD: Johns Hopkins University Press.

Middleton, W. E. K. (1965) *A history of the theories of rain*. London: Franklin Watts.

Middleton, W. E. K. (1966) *A history of the thermometer and its uses in meteorology*. Baltimore, MD: Johns Hopkins University Press.

Middleton, W. E. K. (1969) *Invention of the meteorological instruments*. Baltimore, MD: Johns Hopkins University Press.

Middleton, W. E. K. and Spilhaus, A. F. (1953) *Meteorological instruments*. 3rd rev. edn. Toronto: University of Toronto Press.

Miller, D. P. (2011) "The mysterious case of James Watt's '1785' steam indicator: forgery or folklore in the history of an instrument?" *International Journal for the History of Engineering and Technology*, 81, pp. 129–50.

Mohazzabi, P. and Shea, J. H. (1996) "High altitude free fall," *American Journal of Physics*, 64, pp. 1242–6.

Mohler, N. M. (1939) "The spring and weight of the air," *American Journal of Physics*, 7, pp. 380–9.

Moore, W. M. (1984) "The adiabatic expansion of gases and the determination of heat capacity ratios," *Journal of Chemical Education*, 61, pp. 1119–20.

Moss, S. A. (1943) "American standard letter symbols for heat and thermodynamics," *American Journal of Physics*, 11, pp. 344–9.

Mulligan, J. F. and Hertz, H. G. (1997) "An unpublished lecture by Heinrich Hertz: 'On the energy balance of the Earth'," *American Journal of Physics*, 65, pp. 36–46.

Mungan, C. E. (2015) "Identical distinguishable gas particles in the real world," *Resonance-Journal of Science Education*. 20, pp. 44–6.

Murfitt, J. C., Brown, L. C., and Howell, S. E. L. (2018) "Estimating ice thickness in Central Ontario," *PloS One*, 13(12), 0208519.

Murray, B., *et al.* (2012) "Ice nucleation by particles immersed in supercooled cloud droplets," *Chemical Society Reviews*, 31, pp. 6519–54.

Murray, F. W. (1967) "On the computation of saturation vapor pressure," *Journal of Applied Meteorology*, 6, pp. 203–4.

Nakamura, T., *et al.* (2001) "Spectral reflectance of snow with a known grain-size distribution in successive metamorphism," *Cold Regions Science and Technology*, 32, pp. 13–26.

Neiburger, M. (1957) "Weather, modification and smog," *Science*, 126, pp. 637–45.

Neves, E. B., Salamunes, A. C. C., and de Oliveira, R. M. (2017) "Effect of body fat and gender on body temperature distribution," *Journal of Thermal Biology*, 70, pp. 1–8.

Nicholas, J. V. and White, D. R. (2001) *Traceable temperatures: an introduction to temperature measurement and calibration.* 2nd edn. Oxford: John Wiley & Sons.

Nye, M. J. (1972) *Molecular reality.* New York: American Elsevier.

Nye, M. J. (ed.) (1984) *The question of the atom.* Los Angeles: Tomash.

O'Leary, A. J. (1950) "Enthalpy and thermal transfer," *American Journal of Physics*, 18, pp. 213–21.

Osczevski, R. J. (1995) "The basis of wind chill," *Arctic*, 48, pp. 372–82.

Ostwald, W. (1891) *Solutions.* London: Longmans.

Pagel, W. (1944) ""J. B. Helmont (1579–1644)," *Nature*, 153, pp. 675–6.

Paluch, I. R. (1979) "The entrainment mechanism in Colorado cumuli," *Journal of the Atmospheric Sciences*, 36, pp. 2467–78.

Parthasarathy, S. and Guruswamy, D. (1955) "Sound absorption in liquids in relation to their specific heats," *Annalen der Physik*, 451, pp. 1–42.

Partington, J. R. (1961) *A history of chemistry*, vol. 2. New York: Macmillan.

Partington, J. R. and Shilling, W. B. (1924) *The specific heats of gases.* London: E. Benn.

Patterson, W. S. B. (1981) *The physics of glaciers.* 2nd edn. Oxford: Pergamon.

Pauluis, O. (2010) "Water vapor and mechanical work: a comparison of Carnot and steam cycles," *Journal of the Atmospheric Sciences*, 68, pp. 81–102.

Pearle, P., et al. (2010) "What Brown saw you can too," *American Journal of Physics*, 78, pp. 1278–89.

Pecanha, R. P. (2015) "Fluid particles: a review," *Journal of Chemical Engineering and Process Technology*, 6, p. 1000238.

Perrin, J. (1920) *Atoms.* London: Constable.

Piazza, R. (2019) "The strange case of Dr. Petit and Mr. Dulong," *Quaderni di Storia della Fisica*, 21, pp. 79–110.

Planck, M. (1950) *Scientific autobiography and other papers.* London: Williams and Norgate.

Pollack, H. (2019) "Tip of the iceberg," *Physics Today*, 72(12), pp. 70–1.

Porter, T. M. (1981) "A statistical survey of gases: Maxwell's social physics," *Historical Studies in the Physical Sciences*, 12, pp. 77–116.

Porter, T. M. (1986) *The rise of statistical thinking 1820–1900.* Princeton: Princeton University Press.

Postman, N. (1995) *The end of education.* New York: Knopf.

Poynting, J. H. (1920) *Collected scientific papers.* Cambridge: Cambridge University Press.

Preston, T. (1904) *The theory of heat.* 2nd edn. New York: MacMillan.

Priestley, W. M. (1979) *Calculus: an historical approach.* New York: Springer.

Prospero, J. M. and Mayol-Bracero, O. L. (2013) "Understanding the transport and impact of African dust on the Caribbean Basin," *Bulletin of the American Meteorological Society*, 94, pp. 1329–37.

Pruppacher, H. R. and Klett, J. D. (1980) *Microphysics of clouds and precipitation.* Dordrecht: D. Reidel.

Pryde, J. A. (1966) *The liquid state.* New York: Hutchinson.

Purcell, E. M. (1977) "Life at low Reynolds number," *American Journal of Physics*, 45, pp. 3–11.

Quaranta, E. and Revelli, R. (2018) "Gravity water wheels as a micro hydropower energy source: a review based on historic data, design methods, efficiencies, and modern optimizations," *Renewable and Sustainable Energy Reviews*, 97, pp. 414–27.

Quinn, T. J. (1983) *Temperature.* London: Academic.

Raymond, C., Matthews, T., and Horton, R. M. (2020) "The emergence of heat and humidity too severe for human tolerance," *Science Advances*, 6(19), eaaw1838.

Redhead, P. A. (1999) "The ultimate vacuum," *Vacuum*, 53. pp. 137–49.

Redlich, O. (1970) "The so-called zeroth law of thermodynamics," *Journal of Chemical Education*, 47, pp. 740–41.

Redlich, O. (1970) "Intensive and extensive properties," *Journal of Chemical Education*, 47, pp. 154–6.

Reid, R. C. (1976) "Superheated liquids," *American Scientist*, 64, pp. 146–56.

Reif, F. (1965) *Fundamentals of statistical and thermal physics.* New York: McGraw-Hill.

Reif, F. (1967) *Statistical physics.* New York: McGraw-Hill.

Roche, J. J. (1990) *Physicists look back: studies in the history of physics.* London: Adam Hilger.

Rogers, R. R. and Yau, M. K. (1989) *A short course in cloud physics*, 3rd. edn. Oxford: Pergamon.

Romer, R. H. (2001) "Heat is not a noun," *American Journal of Physics*, 69, pp. 107–9.

Rossby, C. G. (1932) "Thermodynamics applied to air mass analysis," Meteorological Papers, vol. 1(3). Cambridge, MA: MIT.

Rothman, M. A. (1988) *A physicist's guide to skepticism.* Buffalo, NY: Prometheus.

Saunders, P. M. (1957) "The thermodynamics of saturated air: a contribution to the classical theory," *Quarterly Journal of the Royal Meteorological Society*, 83, pp. 342–50.

Schlichting, H. (1968) *Boundary-layer theory.* 6th edn. New York: McGraw-Hill.

Schmidt, Robert F. (ed.) (1978) *Fundamentals of sensory physiology.* New York: Springer.

Scholander, P. F. (1972) "Tensile water," *American Scientist*, 60, pp. 584–90.

Scholander, P. F., *et al.* (1950) "Body insulation of some Arctic and tropical mammals and birds," *Biological Bulletin*, 99, pp. 225–236.

Schriever, W. (1933) "Increased heat emissivity caused by asbestos 'insulation'," *American Journal of Physics*, 1, pp. 48–9.

Shaw, R. A., *et al.* (1998) "Preferential concentration of cloud droplets by turbulence: effects on the early evolution of cumulus cloud droplet spectra," *Journal of the Atmospheric Sciences*, 55, pp. 1965–76.

Sherwood, S. C. and Huber, M. (2010) "An adaptability limit to climate change due to heat stress," *Proceedings National Academy of Sciences of the United States of America*, 107, pp. 9552–55.

Siems, S. T., *et al.* (1990) "Buoyancy reversal and cloud-top entrainment instability," *Quarterly Journal of the Royal Meteorological Society*, 116, pp. 705–39.

Slovis, C. M., Anderson, G. F., and Casolaro, A. (1982) "Survival in a heat stroke victim with a core temperature in excess of 46.5 C," *Annals of Emergency Medicine*, 11, pp. 269–71.

Song, F., *et al.* (2022) "Trends in surface equivalent potential temperature: a more comprehensive metric for global warming and weather extremes," *Proceedings of the National Academy of Sciences of the United States of America*, 119(6), e2117832119.

Squires, P. (1958a) "The microstructure and stability of warm clouds. Part I: The relation between structure and stability," *Tellus*, 10, pp. 256–61.

Squires, P. (1958b) "The microstructure and stability of warm clouds. Part II: The causes of the variations in microstructure," *Tellus*, 10, pp. 256–71.

Squires, P. (1958c) "Penetrative downdraughts in cumuli," *Tellus*, 10, pp. 381–9.

Stern, A. S. (1999) "The lunar atmosphere," *Reviews of Geophysics*, 37, pp. 453–91.

Stone, A. J. (2013) *The theory of intermolecular forces*. 2nd edn. Oxford: Oxford University Press.

Styer, D. (2019) "Entropy as disorder: the history of a misconception," *The Physics Teacher*, 57, 454–8.

Sutton, O. G. (1953) *Micrometeorology: a study of physical processes in the lowest layers of Earth's atmosphere*. New York: McGraw-Hill.

Swendsen, R. H. (2011) "How physicists disagree on the meaning of entropy," *American Journal of Physics*. 79, pp. 342–8.

Swietoslawski, W. (1945) *Ebulliometric measurements*. New York: Reinhold.

Takle, E. S. (1983) "Climatology of superadiabatic conditions for a rural area," *Journal of Climate and Applied Meteorology*, 22, pp. 1129–32.

Taubes, G. (1993) *Bad science: the short life and weird times of cold fusion*. New York: Random House.

Tennekes, H. (1996) *The simple science of flight*. Cambridge, MA: MIT Press.

Touloukian, Y. S., Liley, P. E., and Saxena, S. C. (1970) *Thermophysical properties of matter*, Vol. 3: *Thermal conductivity: nonmetallic liquids and gases*. New York: IFI/Plenum.

Towne, D. (1967) *Wave phenomena*. Boston: Addison-Wesley.

Trenberth, K. E. (1981) "Seasonal variations in the global sea level pressure and the total mass of the atmosphere," *Journal of Geophysical Research*, 86(C6), pp. 5238–46.

Tribus, M. (1968) "Generalizing the meaning of 'heat'," *International Journal of Heat and Mass Transfer*, 11, pp. 9–14.

Truesdell, C. (1969) *Rational thermodynamics*. New York: McGraw-Hill.

Truesdell, C. (1980) *The tragicomical history of thermodynamics 1822–1854.* New York: Springer.

Tsao K. C., Myers, P. S., and Uyehara, O. A. (1962) "Gas temperatures during compression in motored and fired diesel engines," *SAE Transactions*, 70, pp. 136–45.

Tsederberg, N. V. (1965) *Thermal conductivity of gases and liquids.* Cambridge, MA: MIT Press.

Tsilingiris, P. T. (2008) "Thermophysical and transport properties of humid air at temperature range between 0 and 100 °C," *Energy Conversion and Management*, 9, pp. 1098–1100.

Turner, J. S. (1962) "The starting plume in neutral surroundings," *Journal of Fluid Mechanics*, 13, pp. 356–8.

Turner, J. S. (1966) "Jets and plumes with negative or reversing buoyancy," *Journal of Fluid Mechanics*, 26, pp. 779–92.

Twomey, S. (1977) *Atmospheric aerosols.* Elsevier.

U.S. Government Printing Office (1963) *Manual of barometry*, vol. 1. Washington, DC: U.S. Government Printing Office.

U.S. Government Printing Office (1976) *U.S. standard atmosphere.* Washington, DC: U.S. Government Printing Office.

van den Broeke, M., Reijmer, C., and van de Wal, R. (2004) "Surface radiation balance in Antarctica as measured with automatic weather stations," *Journal of Geophysical Research*, 109, D09193.

Van Wazer, J. R. (1958) *Phosphorous and its compounds*, vol. 1: *Chemistry.* New York: Interscience.

Vandam, L. D. and Fox, J. A. (1998) "Adolf Fick (1829–1901), physiologist," *Anesthesiology*, 88, pp. 514–18.

Vandermarlière, J. (2016) "On the inflation of a rubber balloon," *The Physics Teacher*, 54, pp. 566–7

Vigoureux, D. and Vigoureux, J-M. (2018_ "An investigation of the effects of molecular forces in Pascal's barrel experiment," *European Journal of Physics*, 39, p. 025003.

Vollmer, M. "2009) "Newton's law of cooling," *European Journal of Physics*, 30, pp. 1063–84.

Vonnegut, B. (1981) "Misconception about cloud seeding with dry ice," *Journal of Weather Modification*, 13, pp. 9–11.

Vreeman, R. C. and Carroll, A. E. (2008) "Festive medical myths," *British Medical Journal*, 337, a2769.

Wagner, P. E. (1982) "Aerosol growth by condensation," in Marlow, W. H. (ed.) *Aerosol microphysics II: chemical physics of microparticles.* Berlin: Springer, pp. 129–78.

Walcek, C. J. (1994) "Cloud cover and its relationship to relative humidity during a springtime midlatitude cyclone," *Monthly Weather Review*, 122, pp. 1021–35, fig. 1.

Wallace, J. M. and Hobbs, P. V. (2006) *Atmospheric science: an introductory survey*, 2nd edn. London: Academic.

Wang Q., and Albrecht, B. A. (1994) "Observations of cloud-top entrainment in marine stratocumulus," *Journal of the Atmospheric Sciences*, 51, pp. 1530–47.

West, J. B. (ed.) (1981) *High altitude physiology*. New York: Hutchinson Ross.

Williams, E. and Renno, N. (1993) "An analysis of the conditional instability of the tropical atmosphere," *Monthly Weather Review*, 121, pp. 21–36.

Wimbush, M. (1970) "Temperature gradient above the ocean floor," *Nature*, 227, pp. 1042–3.

Wood, R. (2012) "Stratocumulus clouds," *Monthly Weather Review*, 140, pp. 2373–2423.

Wright, P. G. (1970) "Entropy and disorder," *Contemporary Physics*, 11, pp. 581–8.

Wylie, R. G., Davies, D. K., and Caw, W. A. (1965) "The basic process of the dew-point hygrometer," in Ruskin, R. E. (ed.) *Humidity and moisture*, vol. 1. New York: Reinhold.

Yen, Y-C. (1981) *Review of thermal properties of snow, ice, and sea ice. CRREL Report 81–10*. Hanover, NH: US Army Corps of Engineers.

Young, K. C. (1993) *Microphysical processes in clouds*. Oxford University Press.

Zemansky, M. W. (1964) *Temperatures very low and very high*. Mineola: Dover.

Zemansky, M. W. (1970) "The use and misuse of the word 'heat' in teaching physics," *The Physics Teacher*, 8, pp. 295–300.

Zemansky, M. W. and Dittman, R. H. (1997) *Heat and thermodynamics: an intermediate textbook*. 7th edn. London: McGraw-Hill.

Zhang, Y., Held, I., and Fueglistaler, S. (2021) "Projections of tropical heat stress constrained by atmospheric dynamics," *Nature Geoscience*, 14, pp. 133–7.

Index

Absolute humidity, 289–90
Absolute pressure, 73–4
Absolute temperature, 64
  defined, 79
  and mean kinetic energy of ideal
    gas, 66
Absolute zero of temperature, 64, 66
  misconceptions about, 61
  and zero-point motion, 66
Absorptivity, 495
Accommodation coefficient. *See* Mass
    accommodation coefficient
Acid rain, 355
Adiabatic, isobaric mixing of
    parcels, 198–200
Adiabatic liquid water content,
    defined, 425
Adiabatic process
  defined, 25
  and Poisson's relations, 172
Adiabats
  dry, on skew T–log p diagram, 433
  moist, on skew T–log p diagram, 433
Adsorption, 68
Advection, 202, 206
Air
  dissolved in water, Henry's law, 351–5
  dry, composition of, 118
  moist, 1
  still, absolutely, nonexistence of, 28,
    109, 527–8
Altimetry, 122, 133
'Archimedes' principle. *See* Buoyancy
Arrhenius equation, 515
Atmosphere
  Earth's, composition of, 118

  Moon's, composition of, 19
  as unit of pressure, 72
  U. S. Standard, 90–2, 95, 108, 123
Atmospheric dynamics, 1
Atmospheric structure, on
    diagrams, 440
Atmospheric thermodynamics, scope
    of, 1, 2
Avogadro's constant, 117, 124

Balloon, 138, 178
  helium, 222
  hot air, minimum volume to lift one
    kg, 221
  as a parcel, 178
  toy, pressure inside, 124
Barometer, 74
  aneroid, 76
  history of, 76
  mercury, 74–6
Barometric formula, 89. *See also*
    Hydrostatic equation, general form
Bergeron process for rain, 311, 375
Bernoulli, Daniel, 87
Biot number, 503
Boiling
  demystified, 369–72
  and evaporation, 329, 370
  before melting, 330
  nucleation of, 370
  and removal of dissolved air, 372–3
  and superheating, 326
Boiling point
  lapse rate of, 315–16
  measurement of, 326
  scale height for, 315

Boltzmann's constant, 64
Boundary layer
  thermal, 490
  thermal, thickness of, 493, 512
  viscous, 522
Boyle, Robert, 87
Boyle's law, 87
Brownian motion, 31, 52, 266
Brunt-Väisälä frequency, 187
Bubbles
  mechanical equilibrium of, 358
  vapor pressure in, physical
    interpretation, 357
Bulk moduus, 248
Bunsen absorption coefficient. *See*
  Solubility
Buoyancy, 180–2
  effect of freezing on, 299
Buoyancy force, 181
  and stability, 184, 430
Buoyancy waves, 187

Caloric, 37–8, 44
  dietary, 71
Caloric theory, history of, 50–1, 53
CAPE. *See* Convective available
  potential energy
Carbon dioxide, 323
  and acidic precipitation, 355
  concentration in exhaled air, 134
  contraction upon freezing, 377
  critical temperature of, 328
  solid (dry ice) and cloud seeding, 375
  triple point of, 330
Carnot cycle, 271
Carnot efficiency, 273
Carnot refrigerator, 276
Carnot, Sadi, 273, 280
Celsius, Anders, 78, 122
Center of mass, defined, 8
  velocity of, 9
  velocity relative to, 9
Chaos, 60–1
Charles, Jacques-Alexander-César, 86,
  123

Charles's law. *See* Gay-Lussac's law
Chemical energy, defined, 32, 208
Chemical reactions, 32, 37, 102, 138
  temperature decrese in, 42, 208
Chinook, 467
CINE. *See* Convective inhibition energy
Clapeyron, Benoit-Pierre-Émile, 300
Clausius–Clapeyron equation, 298–303
  approximate form, 308
  derived, 298–9
  derived by free energy
    arguments, 342–3
  general form, 299
  history of, 300
  integration of, 300, 308
  molecular interpretation, 300
  and slopes of phase equilibrium
    curves, 220
  for three phase changes, 336
Clausius, Rudolf, 231
Closed system, defined, 29
Cloud condensation nucleus, 367. *See
  also* Condensation nucleus
Cloud droplet
  growth of by diffusion, 531
  and haze, difference between, 368
  nucleation of on soluble
    nucleus, 364–5
  size distributions, width of, 312, 536
  terminal velocity of, 520–1
Clouds
  cumulus, appearance, 451
  cumulus updrafts, 429
  emission by, 499
  entrainment of air by, 450–5
  level of formation, 401
  lifting-condensation level and cloud
    base height, 401–4
  macroscopic properties of, 398
  mixing, 200, 326, 455–7
  radiation from, 499
  stratocumulus, 453
Cloud seeding, 375
Clouds, formation of
  in ascent or descent, 357–60

Cloud-top entrainment, 450–3
Coefficient of performance of a
    refrigerator, 275–7
Coefficient of surface tension. *See*
    Surface tension
Coefficient of thermal conductivity. *See*
    Thermal conductivity
Coefficient of thermal expansion
    of ideal gas, 194
    isobaric, 248
    of water, 195
Coefficient of viscosity. *See* Viscosity
Cold rain process. *See* Bergeron process
Cold trap, 297
Colligative property, 349
Collision and coalescence, 312
Collision cross section, 98
Collisions
    elastic, defined, 15
    mean free path between, 107–9
    as a mechanism for kinetic energy
        transfer, 19
    rate of, in a gas, 110
Compressibility
    adiabatic, 173
    factor, 64
    of ideal gas, 92
    isothermal, 92
    of liquid water, 135
    and speed of sound in liquids, 284
Condensation
    cloud, 364
    in an (idealzed) isothermal
        process, 298
    net, as a warming process, 289
    rate of, 293
Condensational warming. *See*
    Evaporative cooling
Condensation nucleus, 367
Condensation temperature, 327, 379.
    See also Boiling point
Conditional instability, 443–4
Conductivity. *See* Thermal conductivity
Conservation of energy. *See also* Energy
    examples of, 13

local and global, 471
    of point molecule in a gravitational
        field, 6
    of a system of point molecules, 7–13
Conserved parameter diagram, 441–3
    mixing line, 442
    to represent processes, 442
    for a sounding, 442
Constitutive relations, 516, 517
Continuity, equation of
    for enegy, 471
    for mass, 471
Convection, 477–8
    forced, 487, 491
    free, 487, 491
    natural, 492
Convective available potential
        energy, 448
Convective equilibrium temperature
        profile, 260
Convective heat transfer coefficient. *See*
    Film coefficient
Convective inhibition energy, 449
Convective transfer of energy, 477–8
Cooling in air and in water, 510–14
Corresponding states, law of, 321
Critical molar specific volume, 322
Critical pressure
    of carbon dioxide, 339
    of major atmospheric gases, 328
Critical temperature
    of carbon dioxide, 339
    of major atmospheric gases, 328
Cycle
    hydrologic, 32
    on an indicator diagram, 157
Cyclic process, 272, 430
    work done, 274, 430, 438

Dalton, John, 116, 124
Dalton's law of partial pressures, 116,
    124
Defroster, principle of, 296
Dehumidifier, principal of, 297

Density
  of air, decrease with height, 91
  of air, effect of liquid water, 426
  of air, effect of water vapor, 406–7
  of liquid water, variation with
      temperature, 196
Deposition, defined, 295. *See also*
      Sublimation
Derivative of a function, defined, 149
  notation for, 150
Destabilization of a lifted layer, 445
Dew, defined, 294
  difference from guttation, 294
  formation of, 296
Dew point, defined, 294
  depression of, 313
  and human comfort, 314
  lapse rate of, 400–1
  and minimum morning temperature
      forecast, 296
Diabatic process, defined, 25
Differentials
  and infinitesimals, 153–4
  unnecessary in thermodynam-
      ics, 154–5
Diffusion
  Fick's first law of, 523
  Fick's second law of, 524
Diffusion coefficient. *See also* Diffusion
  for self-diffusion, 526
  of water vapor in air, 526
Diffusivity
  for momentum, 522
  thermal, 472
Digit inflation, 70. *See also* Significant
      digits
Disorder, 248–51
Downdrafts, penetrative, 453–4
Drag
  on a small sphere, 520
Drag coefficient, 32, 56, 134, 213
Droplet, vapor pressure
      surrounding, 361
Dry adiabatic lapse rate, 176

and entropy maximization in the
      atmosphere, 266
  and stability, 184
Dry-bulb temperature, 411
Dry static energy, defined, 175
Dynamic viscosity, 516
Dust, Saharan, 25, 53, 268

Eddy conductivity, 491
Eddy viscosity, 522
Effective temperature of
      environment, 498
Emissivity, 495
Energy. *See also* Enthalpy, defined; Free
      energy
  conservation of, global, 471
  conservation of, in mechanics, 12
  conservation of, local, 471
  derived SI unit, 71
  dimensions of, 6
  dry static, 175
  external potential, 11
  flux density of, 40
  internal, 22, 31
  kinetic, 6, 9
  moist static, 419
  motional and positional, 6, 22
  potential, intramolecular, 14
  rotational kinetic, of molecule, 190
  thermal, ambiguity of term, 32
  thermodynamic internal, 55
  total, 6
  transformation of, 7
  translational kinetic, 64, 188, 190
  vibrational kinetic, 66, 190
Energy budget, 13
Energy flux density, 39
Energy flux density, estimates of, 71,
      490
Energy flux density vector, 471
Energy levels
  electronic, 192
  nuclear, 192
  rotational, 192
  vibrational, 192

Energy transfer
  radiative, 493, 496
Energy transformation, 13–14
Enthalpy, defined, 161
  of atmospheric column, 170
  conserved in isobaric, adiabatic
    process, 198, 413–14
  differences, in phase changes, 299,
    303
  and first law, 162
  of fusion, 303
  of solution, 347
  of sublimation, 303
  thermal, 162
  of vaporization, 299
  of vaporization, temperature
    dependence of, 307–8
  of vaporization, vanishing above
    critical point, 307
Enthalpy of vaporization, preference for over latent heat of
  vaporization, 301
Entrainment, 450–4
  at cloud top, 450–3
  and downdrafts, 451
  effect on buoyancy, 450
  parameter, 450
  into stratocumulus, 453, 455
Entropy
  change in an arbitrary, reversible
    process, 241
  change of ideal gas in free
    expansion, 232–5
  change of ideal gas upon heating and
    cooling, 235–7
  change upon mixing gases with
    different temperatur and
    pressure, 245–6
  conservation of in moist adiabatic
    process, 416–21
  definitions and interpretations, 230,
    279
  differences of in phase changes, 306
  and disorder, 248–51
  of ideal gas, 230–2

  and Joule's law, 246–7
  of liquids and isotropic solids, 252–5
  maximization of in the
    atmosphere, 259–70
  maximization of an isolated
    solid, 258–60
  maximum, and stability, 242
  microscopic interpretation of, 251–2
  of mixing, 244
  origin of term, 231
  popular use of term, 229
  and potential temperature, 261
  and stability, 242
  on tephigram, 438
Equations of state
  caloric, 160
  thermal, 160
  van der Waals, 319–23
Equilibrium
  local thermodynamic, 33, 110–12,
    124, 136
  mechanical, of bubbles and
    droplets, 358
  neutral, 180
  stable, 180
  thermodynamic, 32
  unstable, 180
  between vapor and liquid, 287
Equilibrium vapor pressure, 290
  dependence of combined effect of air
    pressure and dissolved air, 355–6
  dependence of on air pressure, 345
  dependence of on dissolved
    substance, 349
  dependence of on temperature, 299
  difference between subcooled water
    and ice, 310–11
  of droplet, 361–3
  expressions for, accuracy of, 308–10
  inside bubble, 361
  of solution droplet, 364–7
  on a thermodynamic diagram, 436
Equipartition theorem, 190–1, 212
Equivalent potential temperature, 421
  conservation of, 427

Equivalent temperature, defined, 413
  approximate expression for, 413
Escape velocity, thermal, 104
  Jeans escape, 106
Espy, James, 401, 431, 461
Evaporation, 288
  net, 293
  rate of, measured, 293, 380
Evaporative cooling, 316–18
Extensive (global) variable, 111

Fick, Adolf, 525, 533
Fick's first law, 523
Fick's second law, 524
Film coefficient, 487
  overall, 498
  physical interpretation of, 489
First law of thermodynamics, 3, 4, 23,
    33, 160
  for adiabatic process, 171
  atmospheric applications of, 201–6
  and enthalpy, 160
  extended to chemical
    reactions, 206–10
  and free energy, 340
  and the possibility of thermodynamic
    engines, 48
Flux density, of any quantity, 39
Forced convection, 491
Force, SI unit of, 93
Fourier, Joseph, 525, 533
Fourier thermal conduction law, 471–2
  as a constitutive relation, 516
  in a fluid at a wall, 487
Free convection, 491. *See also* Cooling
    in air and in water
  buoyancy driven, 511
  contrasted with forced
    convection, 511
Free energy
  Gibbs, 340
  Helmholtz, 343
  many meanings of, 343
  surface, 360

Free expansion, entropy changes of an
    ideal gas, 232–5
Freezing of lakes, 492–3
Freezing point of water. *See* Melting
    point of ice
Frost, defined, 295
  and frost damage, 304, 382
Frost point, 295
Function, 145
Functional, surface, 469

Gas, average separation between
    molecules in, 106
Gas constant
  of a mixture, 117
  specific, 117
  universal, 117
Gas mixtures, heat capacities of, 196–7
Gas, origin of term, 60
Gas pressure
  as momentum flux density, 96
Gay-Lussac, J. L., 86
Gay-Lussac's law, 86–7
Gibbs free energy, 340
Gibbs paradox, 244–5
Gradient operator, 11
Grashof number, defined, 511
  and free convection, 511
  physical interpretation, 522
Gravitational waves, 188
Gravity waves. *See also* Buoyancy
    waves
  external, 188
  internal, 188
Ground Hog Day, physical significance
    of, 538
Guttation, 294

Haze, 363
Heat capacity. *See also* Specific heat
    capacity
  constant pressure, 160
  constant volume, 160
  origin of term, 160

ten possible meanings of, 163
total, 160
Heat capacity, molar, 161, 196
  Petit-Dulong rule for solids, 168
  of water compared with that of
    ice, 168
  of water compared with that of
    organic liquids, 168
Heat content. *See* Enthalpy, defined
Heat flux density, 39
Heating, rate of, as a process, 21, 32
Heat, nonexistence of, 37–44
Heat of solution, 350–1
Heat pump, 57, 283
Heat transfer coefficient. *See* Film
  coefficient
Helmholtz free energy, 343
Henry's law, 351–5
  and Raoult's law, 354
Henry, William, 378
Hertz, Heinrich
  and thermodynamic diagrams, 431,
    432, 461
Hertz-Knudsen relation, 293
Hertz, SI unit of frequency, 71
Heterogeneous nucleation, 364
Homogeneous nucleation, 312, 364, 380
Horsepower, 71
Humidify, function of, 297
Humidity
  absolute, 289
  in heated buildings, 297
  mixing ratio, 290
  relative, 290
  specific, 290
  and vapor pressure, 290
Humidity, measurement of. *See*
  Hygrometry
Hydrostatic equation, general form, 89.
  See also Barometric formula
Hydrostatic equilibrium, 89, 175, 180,
  262, 424, 441, 466
Hygrometry
  based on electrical properties, 415
  chilled mirror, 415

dew point, 294
hair, 415
Lyman-alpha, 416
Hyperthermia
  and core temperature, 514
  survival, record high core
    temperature, 514
Hypothermia
  and core temperature, 514
  survival, record low core
    temperature, 514
Hypsometric equation, 94

Ice
  melting point of, 312, 492
  melting point of, dependence on
    pressure, 338
  nuclei, 310, 492
  polymorphs of, 339
  superheated, 340
  thickness of, on lakes, 492–3
Ice point, 337
Ideal gas, defined, 62
Ideal gas law
  departures from according to the van
    der Waals equation, 332
  derived from molecular point of
    view, 62–4
  history of, 85–8
  invalid at low temperatures, 66
  and temperature dependence of
    specific heats, 193
Ideal solution, defined, 347
Incompressibility, absolute,
  non-existence of, 168
Indicator diagram, 157
  cardiac pressure-volume loop, 158
  history of, 213
Infinitesimals, 152–4
Infrared radiation, 41
  emitted by the Sun, 40
  emitted and absorbed by terrestrial
    objects, 296
  wavelength range, solar, 40
  wavelength range, terrestrial, 41

Insulation, thermal
  of houses, 476, 501
  of porous media, 476, 483
  of water pipes, 500
  of wet suits, 485
Integral of a function, 151
Intensive (local) variable, 111
Intermolecular separation, in a gas, 106
Irradiance
  defined, 71
  solar, 71
Internal energy. *See also* Energy
  of gases, 31
  of solids, 31
Irreversible process, 238
Isentropic process, 237
Isotherm, defined, 470

Joule's law, 164
  entropic derivation of, 246–7

Kelvin equation
  for bubble, 361
  for droplet, 361–3
Kinematics, defined, 5
  contrasted with dynamics, 5
Kinematic viscosity, 522
Kinetic energy
  about center of mass, 10
  of center of mass, 10
  exchanges in molecular collisions, 19
  of point molecule, 6
  rotational, 10, 43, 190
  of system of point molecules, 10
  translational, 64
  vibrational, 190
Kinetic theory, 23, 42, 88, 121, 126, 534
Kirchhoff relation, 307
Kirchhoff's law, 495
Köhler curve, 366–7

Laplace's equation, 528
Lapse rate, defined, 176
  of boiling point, 315, 316
  of dew point, 401
  dry adiabatic, 176

example, 185
  moist adiabatic, 418
  in pure water, 278–9
Latent heat of fusion. *See* Enthalpy,
  defined, of fusion
Latent heat of sublimation. *See*
  Enthalpy, defined, of sublimation
Latent heat of vaporization. *See*
  Enthalpy, defined, of vaporization
Latent heat, origin of term, 301
Latent heat, release of, criticism of
  phrase, 304, 317, 371
LCL. *See* Lifting-condensation level
  (LCL)
Level of free convection, defined, 441
Level of neutral buoyancy, 446
Lifting-condensation level (LCL), 401–4
  and cloud base, 402
  estimate of from dew point
    depression, 401
  expression for temperature at, 404
  simple expression for temperature
    at, 404
  from thermodynamic diagram, 437
Liquid water content, adiabatic, 425
Liquid water loading, 426
Local thermodynamic
  equilibrium, 110–12
Longwave radiation. *See* Infrared
  radiation
Loschmidt's constant, 117
Low-pass filters, soils and thermometers
  as, 505–7

Macroscopic, defined, 84
Macroscopic system, 84
Macrostate, 22, 30
Manometer, 74–5, 128
Mariotte, Edme, 87
Mariotte's law. *See* Boyle's law
Mass accommodation coefficient, 292,
  352, 374
Mass motion, 7
Mass transfer, 523–6

Maxwell-Boltzmann distribution of
  molecular speeds, 98–101
  mean speed according to, 102–3
  most probable speed according
    to, 101
  root-mean-square speed according
    to, 103
Maxwell construction
  and saturation vapor pressure, 333
Mean free path between molecular
    collisions
  defined, 108
  variation with height, U. S. Standard
    Atmosphere, 108
Melting point of ice, 312
  lowering with pressure, 336
Metabolism, 12, 134, 138, 206, 208,
    209, 248, 313, 486, 487
  basal, 313
Metamorphism
  of snowflakes, 379, 535
Metastable states, 249–50, 325–6, 334,
    339–40, 386
Microscopic, scale, 2
Microstate, 22, 30
Millibar, 72
Mixing, and potential temperature
    profile, 268–9
Mixing clouds, 200, 326, 455–7
  and Hutton's theory of rain, 457
Mixing, isobaric, adiabatic of moist
    parcels, 198–200
Mixing ratio, defined, 280. *See also*
    Humidity
Mixture of gases, 115–18
Moist adiabatic lapse rate, 418
Moist air, defined, 1
Mole, defintion, 117
Moist static energy, 419
Molecular collisions, energy exchanges
    in, 19
Molecular flux, of vapor, 291–3
Molecular interactions
  two-body (pairwise, binary), 8
  two-, three-, four body, 8

Molecular solid, 31
Molecular speed. *See also*
    Maxwell-Boltzmann distribution
  mean, 103
  most probable, 101
  root-mean-square, 103
Molecular weight
  mean, of a mixture, 116–17
  mean, versus height, for U.S.
    Standard Atmosphere, 119
Momentum
  angular, 28
  linear, 8
Momentum transfer, 15, 62–4,96, 113,
    538. *See also* Viscosity

Negative pressure, 327, 376
Newtonian fluid, 516
Newtonian viscosity of gases,
    microscopic theory of, 517–19
Newton's law of cooling
  according to many textbooks, 487
  according to Newton, 503
  extended law, 505
Newton's second law of motion, 5
  applied to molecular collisions with a
    wall, 62
Newton's third law of motion, 8
Non-Newtonian fluid, 516
No-slip boundary condition, 132, 478,
    528, 540
Nucleation
  of boiling, 370
  of cloud droplets, 364
  of freezing, 380
  heterogeneous, of ice phase, 380
  homogeneous, of ice phase, 312
  homogeneous, of water phase, 364,
    380
Nuclei
  cloud condensation, 367
  condensation, 367
Nusselt number, 507
  physical interpretation of, 508

Ostwald absorption coefficient. *See* Solubility

Parcel. *See also* Balloon
  buoyancy of, 182
  defined, 28, 178
  and fictitious particles in fluid mechanics, 7
  maximum vertical velocity of, 448
  oscillations of, 187
Partial pressure, 116
Particles, many meanings of, 7
Pascal's barrel, 75, 143
Pascal, unit of pressure, 71
Perspiration
  insensible, 313
  sensible, 313
  sweat, 313, 539
Phase diagrams
  misleading scales on, 334
Phase transition, 318
pH, defined, 355
  of acid rain, 355
  of ocean water, 355
  of pure water, 355
  of rainwater, 355
  of water in equilibrium with atmospheric carbon dioxide, 355
Poisson's relations, 171–4
  supported by measurements, 173–4
Polymorphs of ice, 339, 377
Porosity, defined, 483
Potential (convective) instability, 444
Potential energy
  gravitational, of atmospheric column, 170
  intermolecular, 12
  Lennard-Jones, 98
  of a point mass in a gravitational field, 6
  of a system of molecules, external, 11
  of a system of molecules, internal, 11
Potential temperature, 174–5
  and entropy, 261
  origin of term, 175

profile of, and mixing, 263–9
  and static stability, 185–6, 209
Power, SI unit of, 71
Poynting, John H., 345, 377
Poynting's formula, 345
Prandtl number, 511, 522
Precipitable water, 398
  estimate of, 399
  and specific humidity profile, 399
Precipitation, 94, 355, 375, 412, 425, 442
Pre-melted layer, 28, 50
Pressure
  absolute, 74
  within an interface, 359
  capillary, 358
  critical, 322
  decrease with height, continuum interpretation, 88–90
  decrease with height, molecular interpretation, 95–8
  decrease with height, U.S. Standard Atmosphere, 114
  effect of aircraft on at surface, 114–15
  effect of spherical earth on at surface, 114, 130
  gauge, 74
  intraocular, 358
  kinetic and dynamic, 96
  and kinetic energy density, 64
  kinetic theory interpretation of gas, 96
  mean sea level, 72, 94–5
  measurement of, 73–8
  misconceptions about, 61, 98
  and momentum flux density, 96
  negative, 75, 327
  partial, 116
  reduced, 321
  at surface and weight of the atmosphere, 113–14
  units of, 71–3
Pressure gauge, 74
Pressure gradient, in fluid, 181

Pressure melting, 338
  and ice skating, 377
Processes
  adiabatic, 171
  impossible, 159
  isentropic, 237, 418
  isobaric, 162
  isochoric, 160
  isothermal, 32
  pseudoadiabatic, 426
  quasistatic, 240
  reversible, 172
  on a thermodynamic diagram, 433
Pseudoadiabatic equivalent
    temperature, 461
Pseudoadiabatic process, 424
Psychrometric constant, 411
Psychrometric equation, 411

Quasi-liquid layer on ice, 28. *See also*
    Pre-melted layer
Quasistatic process, defined, 25
  condition for, 26
  irreversible in quasistatic process,
    isolated system, 240

Radiation in porous media, 502–3
Radiative cooling, 35
  and dew formation, 296
Radiative energy transfer, in water and
    air, 512
Radiative heating, 35
Radiative resistance, 500
  external, 502
  internal, 502
Radiative temperature, 497
Radiative thermal conductivity, 503
Rain
  acid, 355
  freezing, 311
  and rapid melting of snow, 375, 382
Rankine, William, 25, 302
  cycle, 277
  temperature, degrees Rankine, 79
Raoult, François Marie, 349

Raoult's law, 349
  and Henry's law, 354
Rawindsonde, 76
Rayleigh number, 511
Reduced variables, temperature,
    pressure, and specific volume, 321
Refrigerator, 275–7
  coefficient of performance of, 276
Relative humidity. *See also* Raoult's
    Law
  above a solution, 286
  definitions of, 291
  as ratio of condensation to
    evaporation, 293
Reservoir
  constant temperaure, 235
  infinite, 236, 238
Resistance. *See* Thermal resistance
Reversible process
  heating and cooling, 240
  phase change, 305
Reynolds number
  and forced convection, 508
  physical interpretation, 545
Rotational kinetic energy. *See* Kinetic
    energy
R-value. *See* Thermal resistance

Saturated adiabatic ascent, history
    of, 431–2
Saturation, criticism of term, 285–6
Saturation equivalent potential
    temperature, 423
Saturation ratio, 366
  dependence of on diameter of solution
    droplet, 367
Saturation vapor pressure. *See*
    Equilibrium vapor pressure
Scale height
  for boiling, 315
  for density, 81
  for pressure, 90
Schaefer point, 312
Sea-level pressure, 72, 96

Second law of thermodynamics
applications of, 255–8
applied to isobaric interaction
between bodies at different
temperatures, 255–7
and stability, 242
Self-diffusion, 526
Sensible heat. *See also* Convection
criticism of term, 418–19, 491
Severe weather, soundings associated
with, 446–8
Shear stress, 515
Shortwave radiation. *See* Solar
irradiance
Significant digits, 70
Size constancy, 6
Skew T–log p diagram, 432–8
basic coordinates of, 432–3
dry adiabats on, 432–3
moist adiabats on, 434
plotting a sounding on, 436
and processes, 433–4
saturation mixing ratio lines
on, 434–5
work and energy from, 437–9
Solar irradiance, 71
Solid atmosphere fallacy, 260
Solubility, 353
of air in water, 373
temperature dependence of, 373
Sorption, 351
Specific gas constant. *See* Gas constant
Specific heat. *See also* Specific heat
capacity
of air, contribution of water vapor
to, 197–8
of gas molecules, 188–94
of mixture, 196–7
molar, 161
temperature dependence of, 193–4
Specific heat capacity, constant
pressure, defined, 162
Specific heat capacity, constant volume,
defined, 160
Specific heat, of gases, 188–94

Specific heat, of solids and
liquids, 254–5
Specific heats, difference between C_p
and C_V
for ideal monatomic gas, 188
for liquids and solids, 247, 252
Specific humidity, 290
Specific thermodynamic variables,
defined, 111
Specific volume
of ice and water at high pressure, 336
molar, 64, 322
Stability
absolute, 443–5
based on density lapse rate, 405
based on dry adiabatic lapse rate, 184
based on potential temperature lapse
rate, 186
based on virtual temperature, 407
conditional instability, 443
of iceberg, 212
Standard Atmosphere, U. S., 90, 92,
95, 123, 133
State
macrostate, 22
microstate, 22
Statistical laws, nature of, 83, 84
Statistics, origin of term, 83
Steam. *See* Mixing clouds
Steam clouds. *See* Mixing clouds
Stefan-Boltzmann
constant, 494
law, 494
Stefan problem, 492, 528
Stokes flow, 532. *See also* Drag
Stokes's law, 520
Subcooling, 310, 312
Sublimation, 294
enthalpy of, 303
Supercooling. *See* Subcooling
Superheated liquid, 326
Supersaturation, 200, 286, 296, 325,
326, 327, 340
Surface conductance. *See* Film coefficent

Surface free energy. *See* Surface tension

Surface functional. *See* Functional, surface

Surface resistance. *See* Film coefficient

Surface tension, 360, 378
  of air-water interface, 363
  dependence on temperature, 329
  and surface free energy, 360

Surroundingsof a system, defined, 29

Swamp cooler, principle of, 297–8

System, defined, 28
  ambiguity of, 29
  closed, 29
  isolated, 29
  large, 30
  open, 29
  surroundings of, 29

Tephigram, 438–40

Temperature
  absolute, 79
  boiling point, 315
  convective equilibrium, 260, 266
  critical, 323
  dew point, 294
  dry-bulb, 413–14
  effective, 498
  equivalent, 413
  equivalent potential, 422
  freezing point, 312
  frost point, 295
  human sensation of, 2
  ice point, 337
  inversion, example of, 305
  mean, of a layer, 204
  measurement of, 79–82
  potential, defined, 174
  radiative, 497
  reduced, 321
  role of in thermodynamics, 1
  scales, 78–9
  Schaefer point, 312, 375
  triple point, 337
  virtual, 405–7
  wet-bulb, 408–14
  wet-bulb potential, 421–30, 432, 434, 438

Temperature scales
  Celsius, 78
  conversion of, 78
  evolution of, 125
  Fahrenheit, 78
  Kelvin, 78
  Rankine, 79

Temperature tendency, defined, 202

Tension in plants, 327, 376

Terminal velocity
  of cloud droplets, 520
  of objects dropped from high altitudes, 134
  of raindrops, 213

Thermal conductivity, 471. *See also* Eddy conductivity
  of air, dependence on temperature, 481
  of air, some myths about, 481–3
  effective, of Arctic mammals, 484, 532
  effective, of porous materials, 483–4
  of a gas, 478–80
  of gases at low pressures, 481
  of ice, 304, 484, 493, 537, 540
  radiative, 503
  of liquid water, 510, 512
  of water vapor, dependence on temperature, 481

Thermal diffusivity, 472

Thermal energy, 32

Thermal expansion. *See* Coefficient of thermal expansion

Thermal resistance, 475
  contribution of radiation to, 502
  of insulated and bare pipe, 500–1
  in parallel, 476
  for radiation, 502–3
  in series, 476
  of skin diver wearing wet suit, 485–6

Thermodynamic diagrams, 430
  area on, significance of, 439
  conserved parameter, 441
  for displaying soundings and
    processes, 441
  for estimating thermodynamic
    variables, 437–8
  history of, 431–2
  skew T–log p diagram, 432
  Tephigram, 438
  vertical structure displayed on, 440
Thermodynamic efficiency
  Carnot, 273
  of hurricanes, 281
  of zero-power (real) thermodynamic
    engines, 277
Thermodynamic laws
  first, 3
  second, 3
  third, 66
  zeroth, 77, 122
Thermodynamic potentials, 343
Thermometer, 45–8
  history of, 122
  infrared, 81
  liquid-in-glass, 80, 82
  as a low-pass filter, 505
  platinum wire, 81
  response time of, 80–1
  thermistor, 81
  thermocouple, 80, 81
  time constant of, 80
Thermoscope, 77, 137
Thickness, 94–5
Tides, atmospheric, 73, 124
Transition region, at liquid-vapor
    interface, 356–7
Translational kinetic energy, 15, 32, 40,
    67, 188
Triple point, 337

Units
  MKS, 69
  SI, 69
Universal gas constant, 117

Vacuum degasification, 373
Väisälä-Brunt frequency, 187
van der Waals equation, 318–34
  and critical point, 322
  and intermolecular forces, 319–20
  isotherms of, 323
  and Maxwell construction, 325
  and molecular volume, 319
  relation to ideal gas law, 331–2
  successes of, 334
van der Waals, Johannes Diderik, 319,
    375–6
Vapor, defined, 327–8
Vapor pressure. *See also* Saturation
    vapor pressure
  for bubbles, 361
  lowering by dissolution, 349
  Poynting's formua, 345
  for solution droplets, 364
Velocity field in a fluid, 9
Vibrational kinetic energy. *See* Kinetic
    energy
Vibrational potential energy. *See*
    Potential energy
Viscosity
  of air, 522
  coefficient of, 515
  dynamic, 516
  eddy, 522
  of a gas, 512
  of a gas, temperature dependence
    of, 519
  kinematic, 512
  of a mixture, 518
  of water, 522
Virtual potential temperature, 407
Virtual temperature
  defined, 405
  dependence on temperature and
    specific humidity, 406
  and stability, 407
Volume fraction, of gases, 110–20
Volume, specific, defined, 65
  molar critical, 322
  reduced, 345

Warm cloud precipitation. *See* Collision and coalescence

Water
  density of versus temperature, 196
  phase transformations of, 306
Water vapor mysticism, 197, 482
Wet-bulb potential temperature, 428, 430, 432
  estimated on skew T–log p diagram, 434
Wet-bulb temperature, 408–14
  adiabatic, 438
  defined, 408
  and human comfort, 430
  killer, 412
  and psychrometric equation, 411
  relative to dew point, 410
  and temperature of a wet bulb, 413–14

Wind
  zonal, meridional, and vertical componenets, 201
Wind chill, 509–10
  history of, 536
Work
  in a cycle, on an indicator diagram, 157
  in a cycle, on thermodynamic diagrams, 439, 446
  to form a surface, 360
Working, as a process, 21, 32
  in compression and expansion of a gas, 23–8

Young-Laplace equation for mechanical equilibrium of droplets and bubbles, 358

Zeroth law of thermodynamics. *See* Thermodynamic laws